Oravas

Lectures on the History of
Technology and Engineering II

Gunhard Æ. Oravas

Lectures on the History of Technology and Engineering

Volume II

2004
Georg Olms Verlag
Hildesheim · Zürich · New York

The publication of this work has been made possible
by a subvention from the Department of Civil Engineering
of McMaster University, Hamilton, Ontario, Canada.

This work and all articles and pictures involved
are protected by copyright. Application outside the strict limits
of copyright law without consent having been obtained
from the publishing firm is inadmissible and punishable.
These regulations are meant especially for copies, translations and
micropublishings as well as for storing and editing in electronic systems.

Das Werk ist urheberrechtlich geschützt. Jede Verwertung
außerhalb der engen Grenzen des Urheberrechtsgesetzes ist
ohne Zustimmung des Verlages unzulässig und strafbar.
Das gilt insbesondere für Vervielfältigungen, Übersetzungen,
Mikroverfilmungen und die Einspeicherung
und Verarbeitung in elektronischen Systemen.

Bibliografische Information Der Deutschen Bibliothek

Die Deutsche Bibliothek verzeichnet diese Publikation
in der Deutschen Nationalbibliografie; detaillierte bibliografische Daten
sind im Internet über *http://dnb.ddb.de* abrufbar.

Bibliographic information published by Die Deutsche Bibliothek

Die Deutsche Bibliothek lists this publication in the
Deutsche Nationalbibliografie; detailed bibliographic data
are available in the Internet at *http://dnb.ddb.de*.

∞ ISO 9706
© Georg Olms Verlag AG, Hildesheim 2004
www.olms.de
Alle Rechte vorbehalten
Printed in Germany
Gedruckt auf säurefreiem und alterungsbeständigem Papier
Herstellung: WS Druck Werner Schaubruch, Bodenheim
ISBN 3-487-11745-2

Table of Contents

Chapter:	Page:	Heading:
	ii	PREFACE AND ACKNOWLEDGEMENTS
	x	PROLEGOMENON
I	1	SUMER: Cradle of Civilisation-Genesis of Writing, Mathematics, Technology and Engineering
II	18	ANCIENT EGYPT: Monumental Engineering
III	41	INDO-EUROPEANS: Men of Action and Engineering
IV	66	PERSIA: Imperial Engineering
V	73	ANCIENT GREECE: Natural Philosophy, Birthplace of Mathematical and Mechanical Sciences
VI	221	ITALIAN ENGINEERING: Rome and First Technological Empire
VII	291	MEDIÆVAL TECHNOLOGY: First Industrial Power Revolution, Mechanical Clocks, and Gothic Cathedrals
VIII	438	RENAISSANCE: Engineering as Practical Art
IX	484	BAROQUE TECHNOLOGY: Ascent of Modern Science, Scientific Technology and Engineering
X	617	AGE OF FRENCH ENGINEERING: Rise of Engineering Sciences
XI	699	ORIGIN OF ELECTRICAL SCIENCES: From Lodestone to Electron
XII	742	BRITISH TECHNOLOGY: Manufacturing and Textile Industries, Industrial Revolution
XIII	760	POWER TECHNOLOGY: From Wind and Hydraulic Power to Steam Power
XIV	803	BRITISH ENGINEERING: Industrial Technology and Engineering
XV	813	TOOLS OF INDUSTRY: From Manual to Machine Tools
XVI	844	SHIP PROPULSION: From Steam-Engine to Diesel-Engine
XVII	862	AGE OF HEROIC ENGINEERING: Advent of Railroads
XVIII	902	AGE OF METAL INDUSTRY: From Steel and Alloy Steels to Light Metals
XIX	945	TELECOMMUNICATIONS: From Telegraph to Telephone
XX	966	INTERNAL COMBUSTION ENGINES: From Industrial Engine to Automobile and Aircraft Engines
XXI	996	TURBINES AS HEAT ENGINES: From Steam-Turbine to Turbojet and Rocket
XXII	1040	ELECTRIC POWER: From Generator to Induction Motor
XXIII	1067	ELECTRIC ILLUMINATION: From Arclighting to Transmission of Electric Power
XXIV	1082	ELECTRONICS IN SPACE: From Wireless to Radio and Radar
XXV	1119	ELECTRIC AND ELECTRONIC TRANSMISSION OF VISION: From Phototelegraphy to Television
	1142	REFERENCES and SELECT BIBLIOGRAPHY
	1152	INDEX

CHAPTER X

AGE OF FRENCH ENGINEERING: Rise of Engineering Sciences
[from 1650 to 1850]

Near the end of the Renaissance period, the Dutch and the Italian engineers were the best engineers in Europe. The Dutch and Flemish engineers had become the outstanding entrepreneurs of the period constructing hydraulic works all over the continental Europe. During the Renaissance the French monarchs imported Italian and Dutch engineers to teach their French counterparts engineering design and construction, in the course of which the Dutch financed and built large hydraulic projects in France. Already by the middle of the 17th century, the French engineers had learnt enough from their Dutch and Italian masters to become the best engineers of Europe. The French began to base engineering designs after the example of Hellenistic engineers on more rationalised procedures. Their aim was to create a science of engineering which would ultimately replace many old uncertainties in the empirical engineering practice, that was essentially a practical art, with sound scientific technology. French contributions to engineering, valuable as they were by their scientific merit, unfortunately often lacked the more practical aspects of engineering. French engineers of the *ancien régime*, like their ancient counterparts, were primarily construction experts and public works specialists, and like Roman engineers their major efforts were with the government undertakings such as building of roads, bridges, aqueducts, and water supply structures. French engineering materials were the traditional materials such as timber, stone and brick, which made French engineers the outstanding construction experts of the premetal age. The mediæval castles and Gothic cathedrals, and the Roman and French bridges were the outstanding achievements of the age of stone. In fact, the French engineers were responsible for perfecting the construction of stone voussoir arches. It appears that up to the time of the French Revolution in 1789, French engineers of the *ancien régime* in their construction practice still represented the last of the ancient engineers.

French engineering during the Ancient Regime, being essentially civil engineering, played a relatively small role in helping and promoting the practical arts of industry and manufacture, which in France were largely small rural operations in a semi-feudal economy. The French autarkic semi-feudal economy and the French guilds retarded, and even prevented, industrial progress in France. There was altogether too much bureaucratic State intervention, and too little room for private enterprise in France to permit industrial development comparable to that which took place in Great Britain. In order to improve industrial conditions, Anne Robert Jacques Turgot (1727-1781), controller general of finance of France, did realise that only paid labour would produce adequate results in industry and, therefore, he abolished *corvée* labour in 1776 and replaced it with money tax.

Turgot, by abolishing *corvée* labour, drew upon himself the wrath of vested interest groups, as well as of Queen Marie Antoinette, and in consequence he was forced to resign his post in 1777. The statutory labour, *corvée*, was immediately reintroduced, and was abolished only in 1798 after the French Revolution. French engineers, particularly after the French Revolution, laid far too great an emphasis on science and theory in their engineering, and as a consequence of it, they tended to neglect practical applications, and all too often failed to carry out solutions-in-practice. In France, engineers were also educated as physicists and mathematicians and, often, their interests were directed more towards abstract theories than towards the more practical matters of engineering.

Early Bridge Building in France

Pont Notre Dame was built from 1500 to 1507 in Paris to connect the island La Cité and the northern bank of the Seine river. It had 6 semicircular arches with spans ranging from 51 to 56.7 feet (15.54 to 17.28 m). Consulting engineer to this bridge construction was the Italian monk Fra Giocondo (1433-1515). He used cofferdams in the building of piers, horse-driven pumps to dewater the foundations and to drive in piles, construction methods which were unknown to French engineers and, thereby, opened a new era of bridge construction in France.

Pont de Pierre in Toulouse, crossing the Garonne river, was constructed by Pierre Souffron and Jacques Lemercier from 1542 to 1632. It had 7 bevelled elliptical arches with spans ranging from 44.19 to 104.00 feet (13.47 to 31.70 m). Piers of the bridge had widths which ranged from 24.05 to 38.39 feet (7.33 to 11.7 m), and the bridge itself was 65.45 feet (19.95 m) wide.

Pont Henry IV in Châtellerault, crossing the Vienne river, was built by Laurent Joguet and Gaschon Belle from 1565 to 1609. It had 9 spans, each 32.32 feet (9.85 m) long. Piers had a width of 15.1 feet (4.60 m) and the bridge was about 67.88 feet (20.69 m) wide.

Pont Neuf at Paris, crossing the Seine river was built by Baptiste Androuet du Cerceau and Guillaume Marchand from 1578 to 1607. It crosses 2 branches of the river over 12 arches, 5 of which cross the narrow branch of the Seine. The spans of the bevelled arches range from 32.0 to 63.7 feet (9.76 to 19.42 m). Pont Neuf is a skew bridge and has a width of 69.6 feet (21.20 m). It is the oldest bridge still extant in Paris.

In 1617, the building of Pont St. Michel in Paris, crossing the Seine river was begun. It has 4 arches with spans ranging from about 32.6 feet (9.94 m) to 46.4 feet (14.14 m). The width of the piers is about 11 feet (3.35 m) and the bridge is 81 feet (27.70 m) wide. It is also a skew bridge.

Pont Royal in Paris, crossing the Seine river, has 5 elliptical arches with spans ranging from 67.5 to 77.2 feet (20.58 to 23.52 m). It was built from 1685 to 1687. The widths of the piers are 15.9 feet (4.86 m) and 16.5 feet (5.04 m), and the bridge is 55.5 feet (16.90 m) wide. It was designed by Jules Hardouin Mansard and Jacques Gabriel I. The Belgian Dominican friar, François Romain

(1646-1735), was hired to solve the difficult foundation problem for this bridge. Romain, who was one of the best foundation engineers of his age, employed a unique foundation construction procedure by using an open caisson and a novel dredging method which used a treadmill type of dipper dredge operated by manpower.

One of the first important French engineers was the mathematician, military engineer, diplomat and architect, François Blondel (1618-1686), who had been trained as a naval engineer. He was entrusted by the French minister Mazarin with the designing of coastal defences of the Iberian peninsula, and the building of forts on the Mediterranean coast of France. Soon after 1651 Blondel was sent on diplomatic missions in Europe, but by 1662 he had been placed in charge of fortification engineering in Dunkirk, in Le Havre, and in Rochefort, and of the design of defences of French islands in the Caribbean. Blondel, who was appointed professor of mathematics of the *Collège Royale* in 1771, and superintendent of all public works in Paris in 1772, designed the major base, Rochefort, at the mouth of the Charente river, planned the city and port, and built the rope factory, forges and the powder magazines of its arsenal. The military defences of the city had been designed by Sebastian Le Prêstre Vauban, the most famous French military engineer. In the course of his work at Rochefort, Blondel had the responsibility of restoring an old Roman bridge at Sainte near Rochefort on the Charente river. Since the clay foundation firm enough to support the bridge lay 66 feet (20 m) below the river bottom which was too deep for construction, Blondel built a false but firm river bottom, thereby improving on the ancient Roman method of construction of bridge foundation. He dammed the Charente river by using a cofferdam construction, dug a trench under the bridge 7 feet (2.13 m) below the bridge from bank to bank, and placed a grillage of oak beams over the width of the river the interstices of which were filled with stones set in mortar. He bolted an oaken platform to the beam grillage on which he constructed 5-foot (1.52 m) high masonry piers. The restored bridge lasted until 1845, when it was dismantled, but this type of bridge foundation was too expensive to be practical.

In 1657 Blondel attempted to correct Galileo's criterion of the strength of beams but his ideas were unsound. The first French architect-engineer to give empirical rules for the dimensioning of arches was the Jesuit monk François Derand who quoted the mediæval rule for arches in his treatise: < Architecture of Vaults or Art > (***Architectura fornicum***,or ***Architecture des voûtes ou l'art***), published in 1643. It gave information on the relation of the thickness of the abutment to the shape of the soffit of the arch but not to the thickness of the arch rib. Moreover, it failed to take into account the weight of the abutment. Blondel published the Derand rules on the arch design in the memoirs of the French Academy of Sciences published for the years from 1666 to 1699, and also a book on his lectures on architecture which contained his rules that set the canons of French architecture until the end of the 18th century.

Early European Canals

The canal between Leyden and Delft in Holland might have been originally dug in part by the Romans. The Dutch and Flemish canal engineers might have been the originators of the first canal locks in Europe, and because of their expertise in such engineering the Dutch and Flemish engineers were in demand for a long time to build hydraulic works in all parts of Europe. In 1602, for instance, the Dutch engineer Jan Lintlaer erected a pump under one of the arches of Pont Neuf bridge, consisting of a waterwheel of 17-foot (5.18 m) diameter and 20-foot (6.10 m) width, with two walking beams operating four pumps of about 9-inch (22.9 cm) bore and 3-foot (10.91 m) stroke and delivering 120 gallons (454.2 liters) of water per minute with a lift of about 77 feet (23.47 m). This pump was known as ***La Samáritaine***, because of the Good Samaritan image on its structure.

In Belgium:

Willebroek canal, designed by the engineer George Rinaldo, was a tidal water connexion between Brussels, Antwerp and the sea, which ran from Brussels to the Rupel river, a tributary of the Scheldt river. Its length was 17 miles (27.4 km), its total fall was from 40 to 50 feet (12.9 to 15.24 m), and it incorporated 5 locks which were large enough for small coastal vessels.

In Italy:

In the 12th and 13th centuries, one of the earliest of the irrigation canals in Italy was the Naviglio Grande, which connected Milan with the Ticiano river, the source of which was the Lake Maggiore. The first definitely known European pound locks, with portcullis-type lock gates, were built by Filippo da Modena and Fioravante da Bologna in the Naviglio Grande canal in 1438. Italian engineers with their knowledge of pound locks, such as Leonardo da Vinci who invented the mitered lock gate, found their way to France, particularly in the 16th century.

In France:

France was the first European country to canalise all its major rivers and to provide a complete system of canalisation by the 18th century. The more than 3,000 miles (4,828 km) of contemporary French canals had their origin in the 16th century with the Briare, the Charolais (the Centre), and the Languedoc (the Midi) navigation canal.

The first canal built in France was built by a French engineer and entrepreneur of Italian extraction, Adam de Craponne (1526-1576), as a private venture and it is known as the Canal of Craponne. This canal was essentially an irrigation canal like the early Italian canals and it irrigated about 125,000 acres (50,600 hectares) of the arid plain of La Crau which lies east of the city Arles. About 1560, this canal reached a small town, Salon, where Craponne had been born. It was later extended to Arles, which made the Craponne Canal 40 miles (64.4 km) long. Craponne invested his entire fortune in the construction of this canal, but owing to some misfortune, he was forced to give up his rights to the canal in 1571. Later, this canal was also used to drive small watermills.

Craponne worked on the drainage of the marshes of Fréjus, the ancient Roman port on the Mediterranean, and also on the defensive fortifications at Nice. He was the first private entrepreneur and canal builder in French engineering, but unfortunately he met a tragic end: he was poisoned in 1576 at a banquet given to honour him as the representative of the king for the corrective engineering work he had done on the fortifications at Nantes, which had been built on sand by Italian engineers.

The Briare Canal was built from 1606 to 1642. It is the first French summit canal with summit water supply and locks, and it connects the Loing river at Montargis (a tributary of the Seine) to the Loire river near Briare. Its length is about 30 miles (48.3 km). The Briare canal surmounted a watershed between these two important river valleys. Where the canal crosses the watershed, 7 small locks were located at the southern end, near Rogny. The Briare canal united 2 of the great rivers of France, the Seine and the Loire.

The Canal du Centre (also known as the Charolais Canal) was built from 1555 to 1792 to connect the Loire river at Digoin, that lies about 120 miles (193.1 km) south of Briare, by way of 2 small lakes with the Saône river at Chalon-sur-Saône, that lies about 80 miles (128.7 km) north of Lyon. The construction of this canal was probably begun by Adam de Craponne. It suffered long delays and interruptions but was finally finished by the famous French engineer Emiland Marie Gauthey (1732-1806) over a span of 9 years with a large workforce of soldiers, thus completing the inland waterways route from the English Channel to the Mediterranean. It was the largest engineering project in France for its time.

The 2 short canals, the Briare and the Charolais, made water communication possible throughout the valleys of the Rhône, of the Loire, and of the Seine, and of their tributaries, and, therefore, over a large area of France.

The Languedoc Canal (also called Canal du Midi, or Canal des Deux Mers) was constructed from 1661 to 1681. It was the first sea-to-sea canal of France, which connects the Atlantic Ocean from the Bay of Biscay to the Mediterranean Sea as it cuts across the Iberian Peninsula a short distance north of the Pyrenees mountains. This canal permitted small vessels of about 1,000-ton displacement to avoid the dangerous fog as well as the pirates, which they would have risked to encounter had they passed through the Strait of Gibraltar. The contractor of the Languedoc Canal was the entrepreneur-engineer Pierre Paul Riquet de Bonrepot (1604-1680), who, before undertaking the actual construction of this canal, built a model of it on his estate to prove the technical feasibility of such a waterway. The length of the Languedoc Canal is 148 miles (238.2 km), the width is 33 feet (10.05 m), and the draught is 6½ feet (1.98 m). The size of its locks is 102 by 19.7 feet (31.08 by 6.00 m). Leaving the river Garonne at Toulouse, the Languedoc Canal rises 206 feet (62.79 m) over 24 miles (38.6 km) in 25 locks to the 3-mile (14.83 km) long summit level, then it descends some

620 feet (189.00 m) over 115 miles (185.1 km) by way of Carcassone and Bézieres in 72 locks to the Gulf of Lyon at Sete (Cette). Its 500-foot (152.40 m) Malpas Tunnel cut through solid rock was the first tunnel in which gunpowder was used for blasting the rock. Malpas Tunnel is also the oldest known canal tunnel. In the 18th century, this canal was extended down the Garonne Valley to Bordeaux, and also lengthened at the other end, which made the Languedoc Canal more than 300 miles (482.8 km) long. This canal was provided with a 250 million cubic feet (7.08 million m^3) storage reservoir, the so-called Bassin de St. Ferréol, at the summit, near the town Revel, which has gravity dam walls and a depth of 105 feet (132.00.m), and numerous feeder canals and aqueducts.

Canal de Bourgogne, which was completed in 1832, connects the Seine to the Saône river at a more northerly point and, thus, provides a more direct waterway route from Paris to Lyons.

The Rhine-Rhône Canal, which was made navigable by 1834, completed a direct north-to-south waterway route across the Western Europe.

Rebirth of Scientific Technology, and Rise of Engineering Sciences

In 1631, a military engineer and mathematician, Pierre Vernier (1580-1637), devised the improved auxiliary scale device, the so-called 'Vernier scale', which was superior to the 'Nonius scale' invented by the Portuguese mathematician Pedro Nunes (in Latin, ***Petrus Nonius***)(1502-1578) in 1542, for accurate graduated scale readings in measurements in which the accuracy was limited by the thickness of its graduation lines. Vernier used it for angular measurements in astronomy, but it was adopted for use in various scientific instruments for angular as well as linear measurements. In 1666, French astronomer and mathematician, Adrien Auzout (1622-1691), proposed the use of the screw-type micrometer combined with movable and fixed wires invented in 1637 by the English astronomer William Cascoigne (c.1612-1644), which led to new operational methods and measuring instruments that substantially improved precision in metrological measurements. From 1669 to 1670 the astronomer, Abbot Jean Picard (1620-1682), used the Vernier scale in his instrument for the triangulation surveying of France from Paris to Amiens, an instrument which had a telescopic sight, cross-hairs and a delicate plumb bob encased in a cover. In 1671 Picard was the first scientist since the Stoic philosopher Poseidonios of Apameia (c.135-c.51 B.C.) in antiquity to measure the size of the Earth by using the method of Eratosthenes of Kyrene (c.276-c.194 B.C.) which he modified by substituting a star for the Sun. Picard calculated from his measurements the radius of the spherical Earth to be 3,950 miles (6357 km), which is close to the presently accepted value.

About 3 decades later, in 1702, the French engineer Alain Manesson Mallet (1603-1706) invented the modern 'dumpy level' by joining the telescopic sight with an air-bubble level. The air-bubble level was probably invented in 1681 by Melchisédech Thévenot (c.1620-1692), physicist and custodian of the Royal Library (***Bibliotéque du Roi***), which consisted of a hermetically sealed tube filled with the spirit of wine. Unfortunately, the bubble in a straight tube was erratic, which delayed its immediate adoption. A long time afterwards the improved curved level tube took its place. The optical surveying instruments were important for accurate layout of large-scale engineering works, such as canals and aqueducts.

The first book on mathematical instruments entitled, < Treatise on the Construction and Principal Uses of Mathematical Instruments > (*Traité de la construction et des principaux usages des instruments de mathématiques*) was published in 1709 by Nicolas Bion (1652-1733), engineer to King Louis XIV for mathematical instruments, in which he described engineering, astronomical, navigational, and military instruments, levelling devices, sundials, clocks, magnets and microscopes, as well as techniques of draughting, of reading of angles, and of measuring distances. Bion's book was translated into English in 1713. The emendated 2nd French editions of this book appeared in 1723, and the 3rd posthumous edition in 1758. The 3rd edition also appeared in its English, as well as in its German translation. It remained a standard reference work on mathematical instruments throughout most of the 18th century.

In 1672, the famous chief military engineer of the King of France (1658), fortress builder, and later Marshal of the armies (1703), Sebastien le Prêstre de Vauban (1633-1707), proposed the establishment of the *Corps Impérial du Génie* (*Corps of Imperial Engineers*) as a professional organisation of military engineers. Vauban was one of the first engineers in Europe to develop standard rules for tenders, contracts and specifications: precise stipulations for execution of particular works, origin and nature of materials, duties and responsibilities of contractors, accountancy, bookkeeping, and other professional requirements. Vauban became the distinguished builder of polygonal and star-shaped fortresses, and the distinguished French successor to the famous Italian military engineer Francesco di Marchi (1504-1577). Vauban built more than 160 fortresses along the border of France from 1678 to 1688, and they represented the very pinnacle of the military engineering craft of the time. Vauban's engineering masterpiece was the fortified port and harbour of Dunkirk, which he constructed and which served as a model to British engineers engaged in fortified harbour design. It is also possible that he may have been the first to work out precise formulae for tunnel engineering. In 1686, he designed two aqueducts for the Languedoc Canal. Vauban was personally responsible for the French army's outstanding skill in military engineering and siege techniques.

In 1678, Arnold de Ville (born 1653), an entrepreneur from Liège, proposed to pump water from the Seine river to supply both Versailles and Marly. De Ville's proposal was first tried out on a small scale in a mill of Palfour near the castle of Saint-Germain. After the successful completion of this experimental project, de Ville hired an assistant, an illiterate Flemish millwright, René Rennequin Sualem (1644-1708) and his brother Paul, to build the huge pumping plant near Marly at the Seine river between 1681 and 1685 for the main purpose of producing an ample water supply for the numerous fountains of the palace of the French King Louis XIV near the village of Versailles, which was located at a distance of 3,900 feet (1,188.7 m) from the plant. Versailles is a low-lying region 10 miles (16.1 km) southwest of Paris. The power plant had 14 undershot waterwheels 39 3/8 feet (12 m) in diameter and 7½ feet (2.30 m) in width each of which operated in its own closely-fitted channel provided with a penstock gate, and arranged in 3 rows driven by the current of the Seine river, which had been improved by a low dam. The 14 waterwheels, producing about 80 effective horsepower at the river, operated a total of 221 lift-and-force pumps which raised the water by stages and were arranged in 3 banks at 3 different levels up a ¾ – mile (1.2 km) long incline of a hill from the river. Two reservoirs were constructed on the hillside, and 64 lift-and-force pumps with a lift of 159 feet (48.46 m) served the first reservoir, whereas a second group of 49 lift-and-force pumps supplemented with 14 pumps taking water from local springs lifted the water 187 feet (57.00 m) to the second reservoir. The 3rd and final lift of 182 feet (55.47 m) to a water tower on the hilltop required 78 lift-and-force pumps supplemented by 16 pumps lifting spring water. The Marly pumping system raised about 11,300 cubic feet (3,200 cubic meters) of water a day through cast-iron pipes to a height of about 531.5 ft (162 m). All these pumps were operated by linked wooden beams driven by 14 waterwheels at the river, which moved alternately back and forth in a 'forest of wood and iron'. In the Marly waterworks the power required to operate the system of power transmission which consisted of linked rigid bars and chains, a system originally proposed by Errard de Bar-le-Ducin in 1584 [but already used for the transmission of power from waterwheels to mine shafts over distances of several miles (3 to 4 km) in German mining in the 16th century], took most of the 80 horsepower produced by the 14 waterwheels at the river. In the Marly plant the transmission of power from the waterwheels was accomplished by a system of suspended chains acting on balance beams connected to the pumps. Such a chain and hinged bar method of power transmission was also used by a Swedish mechanical and mining engineer Christopher Polhem (1661-1751) in 1694 in the ore hoisting mechanism of the Falu mine in Sweden. From the water tower, which was 525 feet (160.00 m) above the Seine river on top of the bridge, the water was carried by an aqueduct to Versailles, Marly and Trianon. The engineers who built the Marly waterworks were paid 9% of the total cost of the project. Sualem is reported to have constructed a model of the proposed mechanism of power transmission to prove its satisfactory functioning before beginning the construction of the Marly waterworks. The power efficiency of Marly waterworks located between Marly and a village was very low, about 6%.

The piping system at Marly was made of cast-iron manufactured in Liège, the first time cast-iron was used for this purpose, for hitherto cast-iron had been used almost exclusively to make cannons. Edmé Mariotte was appointed to advise on the design of the piping system for Marly, in which capacity he carried out his original research on the bending of beams and bursting of pipes giving the earliest technical criteria for assessing the strength of pipes, and on the flow resistance of pipes. Much later, in 1732 the French mathematician and hydraulic engineer, Pierre Couplet de Tartereaux [alias Tartreaux] (? -1743), carried out further experiments on the resistance of pipes to the flow of water at Versailles. He measured the flow resistance of 5 pipes ranging in diameter from about 4 inches (10.2 cm) to 18 inches (45.7 cm), and having lengths from less than ½-mile (0.8 km) to almost 2 miles (3.22 km).

Mariotte carried out experimental research on the force of impact of water jets on blades in an effort to determine the efficiency of the waterwheels used at Marly. The efficiency of the Marly pumping system was very low, for 94% of the power delivered by the waterwheels, each of which supplied about 30 to 40 horsepower at the shaft but only about 10 to 11 effective horsepower to lift the water, was dissipated in moving the rods and chains of the vast mechanism activating the pumps, and, therefore, as a pumping system, it was much inferior to the 'Artificio' of Turriano built in Spain in the 1540's. The cost of machinery in the huge Marly waterworks was enormous, over £ 4,000,000, yet it could not raise more water than one large English 'fire engine' of 1744 which cost only about £10,000. The maintenance of the Marly machine required a team of 60 men working all year round, which made the cost of the operation of the Marly plant very high, about £ 25,000 a year. For reasons of the high operating cost the Marly power plant was entirely rebuilt in 1725, which in turn was replaced by a 60-horsepower steam pump in 1807, and a 90-horsepower steam-engine operated by 3 boilers working at the pressure of 3 atmospheres in 1823; and it was again replaced by a 150-horsepower steam pump from 1853 to 1859.

The Marly waterworks made a contribution to machine tooling, when Nicholas Forq, a French clockmaker, invented in 1751 a planing machine for the planing of the iron pump barrels used in the Marly plant.

Gravity water supply to Versailles had also been under study ever since Riquet de Bonrepos had proposed to build a 30–mile (48.3 km) long canal and aqueduct line to bring an adequate supply of water from the Eure river to Versailles. In 1684, Philippe de La Hire (1640-1718) surveyed levels for such a line, and they were checked by Jean Dominique Cassini (1625-1712) and by other members of the *Académie des Sciences* (*French Academy of Sciences*) in 1685. These efforts stimulated the development of improved levelling instruments, after the invention of a levelling instrument with eyepiece and cross-hairs by Abbot Jean Picard in 1675. The most difficult part of this proposed project was the bringing of the aqueduct over the marshy lowland at Maintenon. Vauban had advocated the use of an inverted syphon pipe to accomplish this, but King Louis XIV wanted a monumental masonry aqueduct carried on arches such as the ancient Romans had built in France. This aqueduct was designed to be a 3¾ – mile (6.0 km) long Roman aqueduct bridge of up to 3 tiers of arches reaching a height of 239½ feet (73.0 m), depending on the topology of the terrain, and to incorporate 242 arches of about 43-foot (13.11 m) span. This monumental project began with a large workforce of troops near the plains of Beuce, but the unhealthy climatic conditions spread a fever epidemic among the troops and, therefore, this ambitious construction project had to be abandoned, and was never finished. Only some huge piers of the first tier of this grandiose aqueduct remain as reminders of the unlimited power of a pompous absolute monarch.

The physician, mathematician, and marine engineer, Hubert Gautier (1660-1737), who served as the engineer of the province of Languedoc for 29 years, created a widespread interest in roads and bridges with his books < Treatise on the Construction of Roads > (*Traité de la construction des chemins*), published in 1693, and < Treatise of Bridges and Roads of the Romans and the Moderns> (*Traité des ponts et chemins des Romains et des modernes*) in 1716. These 2 books became standard works in bridge and road engineering and they ran into many editions. Gautier summarised the earlier and the current construction practices, and discussed floods, foundations, scaffolding, arch centering and machines, and his account revealed that the contemporary procedures of engineering construction had a rather modern flavour. Gautier complained that the rules for proportioning bridges lacked any rational justification and he presented 5 problems for the scientist to solve and explain in a language which is understandable to engineers. His books on bridges and roads aroused wide professional interest in the science of engineering, and they were instrumental in creating a widespread interest in the development of structural mechanics.

In 1717, Gautier published his views on the arch design in a monograph entitled <Dissertation on the Thickness of Abutments of Bridges > (*Dissertation sur l'epaisseur des culies des ponts*) in which he reported on his experiments with a model arch bridge consisting of wooden voussoirs. In this work Gautier pointed out that the frictionless arch theory of La Hire was so far wrong as to be virtually worthless for practical arch design. Gautier suggested the following proportioning for arch bridges: piers $-L/5$; arch thickness $-L/15$ (plus 1 foot for soft stone), where L is the span of the arch. He stressed the importance of cutting the voussoir stones of arches with great accuracy and providing substantial support to the arch spandrels. In his books, Gautier, pointed out the important fact that the existing professional level of road and bridge engineering was rather poor, but particularly poor in road-building, and that progress in the road and the bridge building can be achieved only by applying scientific principles to engineering. In 1720, Gautier invented a system of road building which consisted of ramming clay, gravel and crushed stones between 2 low walls of stone or masonry.

A French engineer of Spanish extraction from Catalan, Bernard Forest de Belidor (1691-1761), who assisted de La Hire and Cassini in tracing the meridian from Paris to the northern coast of France, was an officer in active military service and a teacher at the recently founded military school at La Flère. He later became inspector of artillery in the French army. Belidor published, in 1729, his famous book < Science of Engineers > (*La science des ingénieurs*), which was particularly directed to the attention of structural engineers. It was a scientific handbook of engineering in 5 books with which Belidor tried to meet Gautier's challenge to the profession and treat engineering as a science of mechanics. It contained sections on strength of materials, on beam theory in which he did not go beyond the theories of Galileo and Mariotte but applied them to his experiments with wooden beams, and a somewhat simplified version of La Hire's arch analysis which he applied to arches other than circular. It also contained a section on engineering contracts, tenders, and specifications. Belidor was content to perform a small number of rough-and-ready experiments in engineering, which yielded sufficient data for his immediate engineering requirements.

In 1735 Belidor published another textbook entitled, < New Course of Mathematics for Use in Artillery and Engineering > (*Nouveau cours de mathématique a l'usage de l'artillerie et du génie*), in which he showed how to apply elementary mathematics in mechanics, geodesy, and artillery. In this book Belidor advises students to study infinitesimal calculus.

Belidor's most influential work, a treatise belonging among the dozen outstanding engineering books of all time, was the celebrated < Hydraulic Architecture > (*Architecture hydraulique*), which was published in 1737, and in its 2nd edition in 4 volumes in 1753. It gave illustrations of many engineering works and considered the methods of their construction. Belidor described in this work a class of horizontal waterwheels, first given by Jacques Besson in his book of machines in 1569, known as tub-wheels (*roués à cuve*) which represented a particular form of the Greek mill, also known as the Norse mill. The 'tub-mill' consisted of a horizontal waterwheel with curved blades set at the base of a tall cylindrical chamber. A tapered sluice supplied large quantities of water to the waterwheel tangent to the chamber with considerable rotational velocity which, together with the weight of the water above the wheel made the wheel rotate by the combination of pressure and kinetic energy of the water. This type of 'tub-mill' was built on the Garonne river at Bazacle. Tub-mills were easy to build and convenient to operate, but they were not particularly efficient: at their best, tub-wheels which ran entirely submerged reached an efficiency of only 20%. Belidor expounded the scientific method of engineering in this important engineering treatise which was the first technical book to use integral calculus. This engineering treatise by Belidor served as a textbook to virtually all French and English engineers: for the French in their formal studies; for the English in their self-studies. Belidor, as a teacher of engineering, indirectly exerted a wide-spread influence on the development of the rationally based engineering design through his impressive engineering books, which were not only important engineering treatises but also works of art on account of their artistically magnificent illustrations of engineering works. This classic engineering treatise gives evidence for the existence of a remarkably sophisticated state of mechanical knowledge for its time.

Belidor was not only a theoretical but also a practical engineer. In his wide engineering practice, Belidor was assigned in 1739 the responsibility for the critical examination of the redesign and restoration of the famous waterworks *Samáritaine* built by Jan Lintlaer in 1602 under the 2nd arch of the famous bridge Pont Neuf in Paris. The existing *Samáritaine* waterwheel, which had been improved by René Rannequin Sualem, only made 2 revolutions a minute in the sluggish current of the Seine river and, therefore, it was not a very efficient source of power. Belidor, using Pitot's analysis, advised to increase the number of blades of this waterwheel from 6 to 8 which increased the speed of the wheel to 3 revolutions a minute, thereby making the blades move at ⅓ of the speed of the river and, thus, according to Parent's theory which Belidor accepted as valid, provided the maximum power possible for this undershot waterwheel. Belidor, moreover, also criticised the design of existing pumps for incorporating pipes too small in diameter which made too many sharp turns, and pump chambers and passages which were too obstructive to the passage of water. He preferred unrestricted and steady flow of water in the pumps which was also reflected in their renovated design. His waterwheel which was 20 feet (6.10 m) in diameter and 18 feet (5.50 m) wide supplied the power to drive the pumps that delivered water to the gardens of the *Tuileries* royal palace. When Belidor calculated the impulsive force acting on an overshot waterwheel he discovered that it was about 6 times smaller than the impulsive force acting on the undershot waterwheel, and, for this reason, he used undershot waterwheels in his design. Unfortunately, Belidor completely overlooked the work performed by the water descending in the buckets of an overshot waterwheel which would have made the overshot waterwheel more efficient than the undershot waterwheel.

Engineering as a Learned Profession

In 1716, the famous *Corps des Ponts et Chaussées*, a national highway department instituted by the Duc d'Orleans, was organised under its first chief engineer, architect Jacques Gabriel II (1667-1742), who was also the inspector general of public construction. It was the first professional engineering body in the world organised for the purpose of carrying out the building of a national road network.

Gabriel's father, Jacques Gabriel I (1630-1686) had been one of the prominent French architect-engineers of his time, and, as mentioned earlier, he together with his cousin Jules Hardouin Mansard had built from 1685 to 1687 the bridge, Pont Royal, across the Seine river at the *Tuileries* royal palace in *Paris*. It appears almost certain that Pont Royal, which had 5 elliptical arches with spans ranging from 68 to 77 feet (21 to 23 m), soon developed trouble from the scouring of the foundation of the first pier on the *Tuileries* side, and Gabriel I had to summon the renowned foundation engineer Dominican friar François Romain from Belgium to strengthen the pier foundations with Romain's caisson type of underwater construction. Gabriel I, was probably also the first European bridge engineer after the Roman engineers to use pozzuolana cement made of volcanic ash in mortar which sets under water in the construction of the piers of the Pont Royal.

Almost immediately, Gabriel II had to take charge of the design and construction of the nearly 1,000 feet (305 m) long new bridge of 11 elliptical arches. The central arch of this bridge which crossed the Loire river at Blois spanned 87 feet (26.5 m). The old fortified bridge had been destroyed by a great flood. His resident engineers for the great bridge at Blois were Robert Pitrou (1684-1750) and Jean Baptiste de Regement (?-1725). It took 7 years to build this bridge with slender piers and it still stands. Professionally, the early years of the *Corps* were not entirely satisfactory and, therefore, it was reorganised in 1726, then again in 1750, and in 1754.

In 1743, the superintendent of finances of France, and the director of highways and of the *Corps*, Daniel Charles Trudaine (1703-1769), organised a bureau for the design and coordination of trunk highways of France. In 1747, this bureau was reorganised into the oldest professional engineering school, the *École des Ponts et Chaussées* (*School of Bridges and Highways*), and again reorganised in 1760 under the direction of the outstanding French engineer of Swiss extraction, Jean Rodolphe Perronet (1708-1794). This school was meant to supply competent engineers for the *Corps*. In the beginning, the instruction in this school had been quite informal and practical, and teaching had been done by the best engineer present. In 1775, Jacques Turgot (1726-1781), controller-general of finance of France, made it a formal institution and named it officially the *École des Ponts et Chaussées*. The director of this school, Perronet, was the best stone voussoir bridge engineer of all time, for he brought 2 thousand years of evolution in stone voussoir bridge construction virtually to its practical perfection. There is very little that can be improved in Perronet's stone bridges with modern stress analysis. Perronet was also responsible for the design and construction of the Bourgogne Canal.

By about 1760, Perronet had realised that a quantitative system of organisation can be devised to integrate materials, equipment and men for efficient technological production, and he attempted to use such schemes in his construction and production projects. This makes Perronet also one of the first known production engineers. In his bridge engineering Perronet was inspired by the bridge Ponte Vecchio in Florence, Italy, which had been designed by Taddeo Gaddi, and built in the Late Middle Ages, from 1341 to 1345, by Neri de Fioravante. This old bridge has remarkably flat arches with the rise-to-span ratio of 1– to – 6½, and surprisingly slender piers. The lesson this bridge taught in bridge design had been overlooked by European engineers until Perronet took up bridge building. Perronet realised that the horizontal thrusts of contiguous arches are virtually balanced thus leaving the piers to resist mostly the vertical reactions, a statical condition which considerably reduces the sizes of the piers, a fact that had already been recognised by the Roman engineers who built the palace-fortress of Diocletianus in Spalato (modern Split in Croatia). He was not primarily a theoretician for he designed and constructed his remarkable stone bridges on the basis of his practical experience and shrewd structural intuition. However, he was capable of applying theory to his designs when he found it useful. Perronet compiled a table based on La Hire's method of arch analysis for the proportions of arches and piers of stone bridges in 1750 and 1752, which also drew upon Perronet's extensive practical experience in stone bridge construction, and was later published by P.C. Lesage in 1810 in the book, < Collection of Memoirs Extracted from the Library of Bridges and Highways > (*Recueil des mémoires extraits de la Bibliothèque des Pont et Chaussées*).

In 1771, Perronet designed and constructed the bridge Pont Sainte-Maixence over the Oise river, about 35 miles (56.3 km) north of Paris, which was the most daring stone bridge in the world at the time. Its 3 arches had 72-foot (21.95 m) spans with a rise to span ratio of 1-to-12. In this bridge Perronet used divided piers of 2 cylindrical pillars, each 9 feet (2.74 m) in diameter, both of which were supported on a single foundation. Unfortunately, his remarkable bridge was destroyed in the Franco-German War in 1870. He was assisted in the design and construction of this bridge by Gaspard de Prony and Pierre Antoine Demoustier (1735-1803).

One of Perronet's assistant engineers, François Michel Lecreulx (1729-1812), was even more daring than Perronet, when under Perronet's supervision he designed a bridge at *St. Dié* in 1785. This bridge had 3 flat segmental arches of 38.9-foot (11.86 m) span; each arch had bevelled soffits, the so-called '*cornes-de-vache*', which eased the passage of flood waters, and an amazingly small rise-to-span ratio of 1–to–17 that made this stone arch look more like a haunched beam than an arch. The piers of this bridge had a width of 5.4 feet (1.65 m).

At that time, an open engineering controversy arose concerning the safety of the new structure replacing the Church of Sainte-Geneviève [after the French Revolution this Church was converted under the Revolutionary authority into the *Panthéon Français* to honour the prominent men of the French nation] which was constructed in the university quarter of Paris under the direction of the architect Jacques Germain Soufflot (1713-1780), and his assistant, the engineer Jean Baptiste Rondelet (1734-1814). Soufflot's unconventional design of this structure came under severe criticism of a traditional architect Pierre Patte (1723-1814) for the lack of strength in its unusually slender piers without buttressing which had to support an exceptionally large and lofty dome. Soufflot's learned friend, the inspector general of the *Corps* Emiland Marie Gauthey, a former pupil of Perronet and a docent of mathematics at the *École des Ponts et Chaussées*, was asked for his opinion in this controversy. Gauthey defended Soufflot's design in a paper entitled, < Memoir on the Application of the Principles of Mechanics to the Construction of Arches and Domes > (*Mémoire sur l'application des principles de la méchanique à la construction des voûtes et des domes*) published in 1771 in which he used La Hire's frictionless voussoir arch theory to find the thrust of the dome and its supporting arches, although he sharply criticised La Hire's theory as being valid only for an arch when the mortar of its joints had not yet set. He followed it up with a memoir on the experimental testing of stones, < The Load Carrying Capacity of Stones > (*La charge que peuvant porter les pierres*) published in Abbé Rozier's < Observations on Physics, on Natural History and Arts > (*Observations sur la physique, sur l'histoire naturelle et les artes*), in 1774. In 1772 Gauthey had built a testing machine made of wood, which was similar to the testing machine built by Musschenbroek in 1729, and carried out a series of tests on the strength of stone and brick. It was the first time a testing machine was used in engineering practice. The ratio between the weight used and force applied by his testing machine was 1–to–24. On the basis of his theoretical and experimental results, Soufflot, himself, carried out some further tests on building materials he used in the construction of the dome and piers of the Church of Sainte-Geneviève in a testing machine of his own design which was similar to the testing machine of Gauthey, except that Soufflot's machine was made of iron instead of wood. It was the first time construction materials were tested for the purpose of constructing a particular structure. These strength tests were the first serious experiments on small material specimens since the Musschenbroek's elaborate series of tests in 1729. On the basis of his own tests on materials, Soufflot further increased the size of his dome in his original plans, and in the defence of the soundness of his design he boasted to use a special novel device to strengthen the supporting structure, a device which turned out to be iron-bar reinforcement embedded in the joints of the masonry quite similar to the reinforcement used in modern reinforced concrete structures.

Subsequently, Perronet installed a larger testing machine of about 18-ton capacity at the *École des Ponts et Chaussées*, which was a modification of the testing machine of Gauthey, and capable of testing the tensile as well as the compressive strength of materials. It was the first modern universal testing machine with a satisfactory standard of accuracy. Perronet with the assistance of Rondelet used this machine to test the materials for his next great bridge project, the Pont de Neuilly over the Seine river, which was to be constructed near Neuilly, a suburb of Paris. This bridge had 5 multicentered arches approximating ellipses, each of which had a span of 120 feet (36.58 m), a rise of 30 feet (9.14 m), and piers slightly more than 12 feet (3.66 m) wide. The construction of this bridge was directed by his very capable resident engineer, Pierre Antoine Demoustier, and his theoretical assistants were Antoine de Chézy (1718-1798), Perronet's chief assistant, and Gaspard François Clair Marie Riche de Prony (1755-1839). A horse-driven mortar mixer, one of the first of its kind, was used on this job. Prony helped Perronet to defend the novel structural features of this bridge before the committee of the *Académie des Science*. In testing the materials to be used in the construction of this bridge, Perronet discovered that in ductile metal bars small file cuts on the surface of tensile specimens greatly reduced the ultimate strength of the specimens, which represented the first explicit recognition of the phenomenon of stress concentration in the presence of a sharp dent in the structural material. He also observed a rise of temperature in the test specimen when it was being deformed in tension. The Neuilly bridge was built between 1768 and 1774, and it was dismantled only in 1956 to make room for a modern bridge.

Perronet built from 1785 to 1791 his extant masterpiece of stone arch construction, the renowned bridge Pont de Concorde in **Paris** which has an arch span of 102 feet (31.9 m), and a rise of 13 feet (3.96 m). The piers of this bridge are only 10 feet (3.05 m) thick. It was a daring bridge with narrow piers, for Perronet's original design left 65% of the area below the bridge deck open for flood flows. The arches of this bridge are bevelled and so flat that they appear to be like spandrel beams, despite the fact that Perronet was forced to increase the rise of the arches and enlarge the piers against his own better judgement owing to narrow-minded criticism, perhaps prompted by professional envy. Even then it is not possible to improve upon Perronet's design appreciably with modern structural mechanics. The construction of this bridge continued through the early years of the French Revolution, and it was carried out under the supervision of his assistants, Chézy and de Prony.

As is often the case, the controversy about the structural design of Sainte-Geneviève became intense after the structure itself gave discouraging evidence of its failure to support adequately the loads to which it was subjected. Soufflot had died in 1780 before the construction of the dome had begun, which left the responsibility for the completion of the structure to Soufflot's assistant Rondelet. In 1790 the composite piers supporting the dome began to develop ominous tell-tale cracks which seemed to lend support to Patte's criticism given 28 years earlier. In 1797, Rondelet published a memoir in which he explained the entire history of this building, its structural comportment, the incipient structural failure developed in the structure since its completion, and his proposals for reconstructing the building. In 1798, Pierre Patte revived his past criticism of the building, and recommended the dismantling of the dome and its supporting circular girder, and constructing a new and lighter dome as the only solution to the problem. A number of other proposals were made for the reconstruction of the dome and piers of the structure of the Church of Sainte-Geneviève, called the **Panthéon Français** after the French Revolution.

Since all this criticism and the ominous cracking of the pillars of the **Panthéon Français** called into question Gauthey's professional competence in his early defence of Soufflot's design, Gauthey defended his professional reputation by an acute criticism of the various proposals made for the reconstruction of the dome, and explained the theory of the spherical dome and arches in a paper entitled, < Dissertation on the Deterioration of the **Panthéon Français** > (***Dissertation sur les dégradations du Panthéon Français***) in 1800. In 1799, Gauthey had performed experiments on segmental, semicircular, and 3-centered model voussoir arches which had a clear span of 2.13 feet (0.65 m) and made of 1.1 inch (2.75 cm) voussoirs of wood. The results of his experiments corroborated the concept of Couplet made in 1729, which assumed that voussoir arches fail by relative rotations which open up the joints at certain critical joints, rather than by mutual sliding of the voussoirs as assumed in La Hire's theory of frictionless voussoir arches. Therefore, Gauthey rejected La Hire's frictionless arch theory and used the rotational failure theory of stone arches of Couplet and Coulomb to explain the action of the dome. He attributed the cracks in the piers of the **Panthéon Français** to faulty workmanship. As a result of this renewed controversy, a commission was struck to examine the safety of the structure of the **Panthéon Français**, and after a detailed study of all proposals for the reconstruction, the committee instructed Rondelet to rebuild the cracked outside masonry of the composite piers, to increase their size by a modest amount, and to provide the masonry with effective lateral bracing in the form of lacework of metal bars. In his theoretical calculations, Gauthey had overlooked the eccentric and the wind loading as well as the possible uneven settlement of the building, although the ring tension developed in the dome was adequately resisted by iron chains provided for that purpose. The composite masonry consisting of both hard and soft stones, and poor workmanship had also contributed to the incipient failure in this structure.

Meanwhile, further advances had been made in testing of materials. In 1787, Rondelet redesigned the Perronet testing machine to eliminate the friction at the bolted connexion by introducing a knife edge in its stead, and by installing a screw-jack for the loading mechanism. Rondelet's version of the testing machine maintained the lever in horizontal position all throughout the testing which gave higher accuracy in test results. Rondelet published his improved test results in his work of 5 volumes, < The Art of Construction > (**L'art de bâtir**), from 1802 to 1817.

All these tests on materials were carried out in small-scale like those of Musschenbroek in 1729. From 1738 to 1741, Georges Louis LeClerc, Comte de Buffon (1707-1788), who was a capable mathematician and the French translator of Newton's book on fluxions, and who later became an outstanding natural scientist, performed the first large-scale experiments on wooden beams. Buffon was appointed to assist the versatile Henri Louis Duhamel du Monceau (1700-1782), a prominent botanist, agronomist, technologist and Inspector General of the Marine, in a study of the conservation of timber for naval construction. Buffon, who was critical of Mussenbroek's small-scale tests on timber beams and columns, was particularly concerned about the strength of timber. He was able to perform a long series of experiments on the strength of wood beams with square cross-sections ranging from 4 inches (0.10 by 0.10 m) square to 8 inches (0.20 by 0.20 m) square, and spans up to 28 feet (8.53 m). His experimental beams were simply supported and loaded at the center, the maximum total load was 27,000 pounds (12,150 N). Unfortunately he did not devote meticulous attention to the accuracy of his experimental procedure. Buffon's results confirmed Galileo's assumption that the strength of beams in bending is proportional to the product of the width and second power of the depth of the cross-section of the beam, and directly proportional to the density of timber as material. He measured the deflexion of the beams as a function of time, and discovered that a load which is at least $2/3$ of the load that causes immediate rupture of the beam can cause failure over a certain time interval. He published the details of many of these tests in a memoir of the French Academy of Sciences in 1741. Buffon became the director of the **Jardin du roi** (*Royal Botanical Gardens*) in Paris.

Large-scale experiments of the structural members were resumed by the French engineer Jacques Elie Lamblardie (1747-1797), who was chief engineer of the harbour construction in Normandy. Lamblardie replaced Perronet as the director of the **École des Ponts et Chaussées** in 1792 upon the retirement of Perronet. Lamblardie was made responsible for the reorganisation of this school and other institutions of higher learning into a unified institution called the **École Central des Travaux Publique** (*Central School of Public Works*), but it did not function satisfactorily. Therefore, a group of scientists and engineers, such as Lazare Carnot, Antoine Lavoisier, Lamblardie and others, under the leadership of the outstanding mathematician [later a general], Gaspard Monge (1746-1818), proposed to establish a new type of engineering school, which in 1795 was called the **École Polytechnique** (*Polytechnical School*) admission to which was based solely on the demonstrated ability of the student in difficult competitive entrance examinations. The **École Polytechnique** was patterned after the most advanced type of polytechnical school in Europe, the **École du Corps Royal du Génie** at Mézières, which had been established by Count d'Argenson in 1749 and to which, since 1751, entrance had to be earned through a rigorous entrance examination. This military school had the best-equipped laboratories of physics and chemistry in Europe, and its training staff had included such famous scientists as Gaspard and Louis Monge, Abbé Charles Bossut (1730-1814) and Abbé Jean Antoine Nollet (1700-1770). Among its early graduates were Lazare Carnot, Jean Charles de Borda, Poncelet and Saint-Simon. Tutorials at this school were conducted by senior students called **répétiteurs**, a system of teaching later introduced at the **École Polytechnique**. The syllabus of the new school at first consisted of 2 years of rigorous training in sciences and mathematics, and of condensed engineering courses offered in the 3rd year. The syllabus of the **Bergwerksakademie** (*Mining Academy*) established by Queen Marie Theresa of Austria-Hungary-Bohemia in 1770 had indirectly influenced the founders of **École Polytechnique**. This mining school had been founded in 1733 at Joachimsthal, and then transferred to Prague in 1763. It was reorganised in 1770, and became famous for the high quality of its teaching, which had an early influence on the best French military schools, such as the polytechnical school at Mézières which served as a model for the **École Polytechnique**. **École Polytechnique** was initially designed to take the place of all older advanced technical schools for the military and other kind of engineering schools. Soon, the engineering training was discontinued at the *École Polytechnique* altogether when it became apparent that the school was quite unable to meet this professional responsibility, and it became a school of fundamental sciences and mathematics only, giving the students the necessary preparation to enter one of the restored specialised engineering schools of higher learning, the so-called *écoles d'application* (*schools of application*): *École des Ponts et Chaussees*, *École des Mines* (*School of Mines*), *École d'Arts et Métiers* (*School of Arts and Crafts*), *École d'Artillerie* (*School of Artillery*), *École du Génie Militaire* (*School of Military Engineering*) [moved from Mézières to Métz], *École des Vaissaux* (*School of Marine*), and *École des Ingenieurs Géographes* (*School of Geographical Engineers*). Many famous engineers and scientists such as Monge, Lagrange, Laplace, Prony, Fourier, and others, were teachers at the *École Polytechnique*. Monge, himself, had established the analytic and the projective geometry as engineering subjects, and founded the art of engineering drawing on descriptive geometry, which had been kept a State secret for a long time whilst he was teaching at the *École du Corps Royal du Génie* at Mézières. For a brief period Monge served as Minister of the French Navy under the Revolutionary Convention.

In 1794 the *Conservatoire des Arts et Métiers* (*Conservatory of Arts and Crafts*), an establishment that soon became the greatest industrial museum in Europe in which popular courses in science and trade were given to the general public, was founded. The success of this institution inspired American physicist Count Rumford [Sir Benjamin Thompson] (1753-1814) to found in 1796 *The Royal Institution* in London which was a very successful combined technical school, public lecture forum and learned society. In 1786, the Duc de La Rochefoucauld-Lincourt had established a trade school, later known by the name of the *École Centrale des Arts et Manufactures* (*Central School of Arts and Manufactures*) which rapidly became the best intermediate trade school in Europe after its reorganisation in 1829. The French intended to reduce the industrial lead of Great Britain by educating French engineers in rigorous sciences as a necessary background for the engineering profession. The effect of the *École Polytechnique* on the learnedness of the French engineers was decisive. Through the founding of these schools by the French, a sound foundation for engineering as a learned profession had thus been finally laid in France. By the early 19th century, France was the only country in the world where engineering was firmly established as a learned profession in which an engineer was also trained as a physicist and a mathematician. Unfortunately, too few graduates of the *École Polytechnique* entered private industry mainly, as it appears, on account of the *esprit de corps*. In the period from 1815 to 1850, the *École Polytechnique* came under stern criticism for the excessive emphasis the school laid upon mathematics and large amount of purely theoretical work, which were not found useful in the *écoles d'application*, and for the neglect of humanities. At the time in Britain, engineering was just beginning to emerge as a profession in which an engineer was both a civil and a mechanical engineer; in other countries, engineering was still regarded as only a skilled craft. However, the influence of French engineering was decisive for every modern engineering school is essentially organised according to the model of French *École Polytechnique,* a scientific school of higher learning.

Returning to large-scale testing in 1789, Lamblardie put his assistant Pierre Simon Girard (1765-1836) in charge of the testing project, for he himself was busy with the reorganisation of the French school system. Girard built the large-scale testing machine at Le Havre and carried out numerous large-scale tests, the results of which he published in his book < Analytical Treatise on the Resistance of Solids > (*Traité analytique de la resistance des solides*), in 1798, that contained a long historical preface on mechanics of materials, which proved to be one of the most valuable part of the book. His testing machine was designed to apply loads up to 100 tons, but in actual tests the loads did not exceed 140,000 pounds (627,480 N). Each large-scale test in this testing programme was a laborious procedure. Girard made great efforts in his testing machine design to eliminate directional restraint in column tests, but he was uncertain to what extent the discrepancies between his test results and his calculations were due to experimental error, or to invalid theory. Girard made the same mistake as Mariotte by assuming that the actual location of the neutral axis in the bending of a beam is unimportant to the bending resistance of beams. Euler's writings on the bending of beams inspired Girard to undertake the first experimental determination of the elastic bending stiffness for beams, $B = EI$. Euler, at first, had assumed the flexural rigidity of a beam to be proportional to the cube of its linear cross-sectional dimensions, but later corrected it to be proportional to the fourth power of its linear cross-sectional dimensions. Girard, unfortunately, cleaved to the first, wrong assumption of Euler. The largest beam tested at Le Havre was 17 feet (5.18 m) long, and 11 inches (27.9 cm) square in cross-section.

Girard was later engaged in the hydraulic engineering of the Nile river, in the design and construction of the Canal de L'Ourcq and its branches, in the hydraulic engineering of the water supply of Paris, in the surveying of France, in the gas lighting [which he had examined on his official visit to London in 1819], and many other public works.

By 1850, the *École Polytechnique* and *écoles d'application*, and the *École Centrale des Arts et Manufactures* were intellectually as strong as ever, but soon, the German polytechnical schools, the Swiss Polytechnical School in Zurich, and American engineering schools had risen to rival them in enrolment, and in financial and material resources, which heralded the beginning of the end of the French leadership in all types of technical education.

In German-speaking countries great efforts were made to catch up in engineering education with the French engineers, and in engineering practice with the British engineers. They decided to accomplish this double aim by establishing a system of polytechnical schools which emphasised both engineering sciences and engineering practice. The new polytechnical schools were established in Vienna in 1815, in Berlin in 1821, in Karlsruhe in 1825, in Munich and in Nuremberg in 1827, in Hannover in 1831, in Stuttgart in 1840, and in Zurich in 1857. This effort soon was crowned with a huge professional success, for by the 1870's German engineers were among the best in the world and in many disciplines in engineering, but particularly in chemical engineering, the very best.

Hydraulic Engineering

The superintendent of the *Canal du Midi* in Languedoc, mathematician, astronomer, and a former student of Réaumur, Henri de Pitot (1695-1771), was the designer of the great aqueduct of Montpellier. This aqueduct was constructed sometime between 1752 and 1772. It had an arcade which was 3,280 feet (999.74 m) long, and formed a part of a large hydraulic structure having 11 miles (17.7 km) of canals, and a tunnel. The elevated part of the aqueduct carried on an arcade was one of the finest engineering structures of the 18th century. He was involved in the repair of old bridges, in the drainage of marshes, and in water supply work as his major professional responsibilities. Pitot headed a distinguished line of French hydraulicians, such as Chézy, Bossut, Darcy, Dupuit, Bazin and others, who made important contributions to the advancement of engineering science of hydraulics.

In 1728 Pitot studied the resistance to flowing water in pipes and concluded that it is inversely proportional to the diameter of the pipe. In 1729, he published a memoir in which he attempted to complete Parent's theory of waterwheels by theoretically examining the best arrangement for the blades of undershot waterwheels. He was interested in establishing whether the radial vanes were superior to the inclined vanes, and how many blades should a waterwheel of a prescribed size have for maximum efficiency. Pitot based his analysis on the principle advanced by the French Jesuit savant Ignace Pardies before the latter's premature death in 1673 which related the oblique impulsive force to the normal impulsive force as the angle of incidence is related to the right angle. On this basis Pitot found that the radial blades of the undershot waterwheels are subjected to much larger impulsive forces over a larger arc of rotation than the inclined blades, and that the number of blades in a waterwheel was not arbitrary to obtain maximum power from the stream. As a result of his studies, he presented tabular information on the optimum number of blades for various sizes of waterwheels. Unfortunately, Pitot's analysis was also flawed because he considered only impulsive action on waterwheels which neglected the reaction caused by the weight of water on inclined blades. Despite these shortcomings in the Parent's analysis of the optimum velocity and power of the waterwheels, and in Pitot's analysis of the blade arrangement of the waterwheels which complemented Parent's analysis, both analyses became theoretical standards of design of waterwheels. Neither Parent nor Pitot had performed experiments to corroborate the validity of their theoretical results for waterwheels, and, therefore, since the middle of the 1740's both theories came under question as to their practical value.

In 1732 Pitot invented the 'Pitot tube' for the measurement of water velocity by means of the differential head of two tubes, one straight, the other bent in the direction of the flow. By means of it, the true nature of the velocity gradient in a channel was first discovered, for engineers had thought until Pitot's experiments that the velocity increased with the head of flowing water, an idea based on the Torricelli principle of efflux, $v \propto \sqrt{h}$. The error of this idea was discovered by the use of the Pitot tube, which, however, was not a very accurate instrument, for Pitot had not been as accurate as Daniel Bernoulli in the introduction of the piezometric principle. In the same year, Pierre Couplet de Tartreaux, made experiments on the water flow in pipelines in and about Versailles to determine the effect of friction in long runs of pipes, which showed that the friction losses would be greater the longer and narrower the pipe, and further increased by bends and elbows, and by larger speeds of flow. Belidor also used the Pitot tube to estimate the efficiency of the pumps in the *Samáritaine* waterworks by measuring the speed of the current of the Seine river and, following Couplet, he also attempted to develop a theory of flow of water in pipes based also on Couplet's observations. Pitot, moreover, wrote books on manoeuvring of boats, on pumps, and on the construction of arches.

One of the first serious criticisms of the generally accepted theory of waterwheels as proposed by Parent and by Pitot was advanced in 1746 by a French military engineer Robert Xavier Ansart Du Petit-Vandin (1713-1790). Du Petit-Vandin criticised Pitot's analysis for having neglected the water which passes without any mechanical effect underneath the waterwheel because it never impinges on the blades. He asserted that this lost power could be recovered if additional vanes were installed between the blades the number of which had been specified by Pitot. Du Petit-Vandin recommended the construction practice of Dutch engineers who built waterwheels with 32 up to 48 blades for which Pitot's theory recommended only from 6 to 8 blades. In Du Petit-Vandin's opinion the waterwheel can never have too many blades on the basis of a sound theory. In his own analysis of waterwheels, Du Petit-Vandin tried to take into account the variation of velocities on the blades, or by assuming the speed of flowing water to vary with depth, but his analysis was also inadequate and faulty, and did not improve the theoretical design of waterwheels.

The first serious challenge to the validity of Parent's theory of waterwheels, which was based on experiments, came from a French applied mathematician and technologist, Antoine de Parcieux (1703-1763) in 1754. De Parcieux had been called to assist Count de Buffon in an investigation for the feasibility of a number of proposals which had been made to provide an adequate water supply to the chateau at Crécy of Marquise de Pompadour, the notorious mistress of King Louis XV. The water had to be supplied from a small river, the Blaise, running through the vicinity of the château, and the hydraulic power of the Blaise river had to be used to raise the water from the river to a reservoir at a height of about 163 feet (50 m). De Parcieux found that during the dry season, when the volume of flow of the Blois river was small, only a very small amount of water could be lifted to the reservoir according to Parent's theory of waterwheels, which led de Parcieux to seek for alternative solutions of waterwheels. De Parcieux introduced the crucial 'idea of reversibility' into waterwheel analysis which assumed that if an overshot waterwheel were attached to an identical bucket wheel running in reverse, it should be able to lift back to the original height almost as much water as was used to turn the waterwheel, an idea which his simple experiments with pulleys and weights seemed to support. Considering the water-filled buckets of an overshot waterwheel as an endless series of descending weights, he came to the conclusion that the slower the descent of the weights the closer will the weight to be lifted raise to the descending weights. With this reasoning under ideal, frictionless conditions, the overshot waterwheel by analogy could achieve an efficiency of nearly 100 %. Therefore, de Parcieux concluded that gravity-operated, overshot waterwheels have superior efficiency over the undershot waterwheel at slow rotational speeds.

Since de Parcieux had reached these conclusions through analogy and rational thinking he proceeded to corroborate his theoretical conclusions by model experiments and, particularly, to find the optimum speed of rotation for gravity-operated waterwheels. On account of his discovery that gravity was more efficient than impulse in the action of waterwheels, he became also sceptical about the correctness of Pitot's theory on the arrangement of vanes on the waterwheel. He published his conclusions on the waterwheels in 1754.

De Parcieux constructed out of metal a model waterwheel of 20 inches (0.54 m) in diameter and having 48 buckets with 4 different diameter drums on the axle which permitted him to change the speed of rotation of the waterwheel by a cord wrapped around the drum and supporting a variable weight over a pulley. The waterwheel's output was measured by the product of the weight and the vertical distance of its motion. He made another experimental waterwheel model which had a diameter of 32 inches (0.87 m) and was fitted with 12 hinged blades 7½ inches (0.20 m) wide and 8 inches (0.22 m) high. The hinged blades allowed the inclination of the blades to be varied without changing the diameter of the undershot waterwheel. He carried the blade experiment out on the Biévre river which was 5 feet (1.62 m) wide, 3 feet (0.97 m) deep, and had a speed of flow of 3 feet (0.97 m) a second. De Parcieux was able to demonstrate by this experiment that inclined blades were superior to radial blades, and that 30 degree inclination of blades yielded optimum power. He found that inclined blades always produced superior efficiency in waterwheels in comparison with the radial blades regardless of their number. His experiments indicated that the best angle of inclination depended upon the number of immersed blades and the speed of the stream.

De Parcieux's model experiments also supported Du Petit-Vandin's contention that an increase in the number of blades fitted to the waterwheel increases the output of power. Despite the fact that de Parcieux's experiments only gave rise to general rules for the design of waterwheels, de Parcieux himself recognised that more extensive experiments on model and prototype waterwheels were required to obtain precise quantitative data for design. De Parcieux published his last experimental findings in 1759, and the total effect of his studies was that the earlier confidence in the waterwheel theories of Parent and Pitot had been irredeemably undermined. Unbeknownst to de Parcieux such quantitative information was being obtained by the British engineer John Smeaton (1724-1792), who was at once a craftsman, an engineer and a scientist, in his experiments most of which were performed in 1752 and in 1753. Smeaton's experimental researches were published in 1759 by The Royal Society in London, translated into French in 1810 by Pierre Simon Girard, and soon became the most influential work in hydraulic power engineering of the period.

The practical outcome of the experimental researches of de Parcieux and Smeaton was that they proved the superiority of the overshot gravity waterwheel over the undershot impulsive waterwheel. Smeaton discovered that undershot waterwheels produced about 30% maximum efficiency (about twice the prediction of Parent) with the possibility that the maximum could be raised to 50 % (more than 3 times the prediction of Parent). The overshot gravity waterwheel he found about twice as efficient as the undershot waterwheel (about 4 to 5 times the prediction of Parent). The efficiency of his overshot waterwheel varied from 52% to 76% under various heads and volumes of water. Smeaton's experiments corroborated de Parcieux's conclusion that the slower the speed of the overshot waterwheel, the more its efficiency. For high falls of water the overshot waterwheel was the best, whereas for medium to low falls the breastfed waterwheel was the best. One of the main technical shortcomings of the typical overshot waterwheel was its unsuitability for falls over 40 to 50 feet (12 to 15 m) in height, particularly when the volume of water was rather small. Otherwise the overshot waterwheel was able to run at quarter gate, or at half gate with the same efficiency as at full gate, which is advantageous in small rivers with large fluctuations in the volume of water.

In the period from 1768 to 1775, Perronet's chief assistant Antoine Chézy, a man of extreme modesty, laid the foundation for open channel flow analysis in his planning of the water supply canal from Yvette river to Paris by devising the formula for the mean velocity v of water flowing in a channel, now given as $v = C(DS)^{1/2}$, where D is the mean depth of the channel, S is the mean slope of the channel, and C is a dimensional coefficient which depends on the channel conditions. Chézy, himself, expressed his formula as follows:

$$v = V[(a/A)(h/H)(P/p)]^{1/2}$$

where a, A are cross-sectional areas, h/H is the ratio of slopes, P/p is the ratio of 'wetted parameters', and V,v are the velocities of water of the two channels. Chézy stressed the fact that in order to use this formula, V, A, H, and P for one of the channels must be known. This contribution of Chézy led to a great development of hydraulics in France, for a succession of outstanding French engineers tried to quantify and refine the Chézy Formula. Chézy himself had formulated his equation as a 'similarity parameter' for predicting flow characteristics of one channel from the measurements of another. Since Chézy's study was not included in Perronet's report on the Yvette project, it remained unpublished as an internal report of the *Corps*. This manuscript of Chézy was recovered for posterity in 1803 by Pierre Simon Girard, who called attention to it in his report on the Canal of Ourcq which was built later by Girard. Moreover, Chézy was the first to point out that the motion of water in canals can only then be uniform when the resistance to flow equals the moving force which is due to the force of gravity.

In 1779, the French engineer Pierre Louis Georges, Comte Du Buat [alias Dubuat](1734-1809), introduced the concept of the hydraulic radius R for D:
$$v = C\,(RS)^{½}$$
He tried to quantify the Chézy formula (he was at first unaware of Chézy's contribution), and carried out extensive tests on the resistance of rivers, channels and pipes. He also studied resistances of immersed bodies. He established the so-called 'Dubuat's resistance formula', but, in general, it was far too early to reach his goals, some of which have not been realised till this very day. It was soon recognised that the coefficient C in this formula was not a constant, but rather varied with R and S, and the clarification of this variation became the object of research of many hydraulic engineers in Europe, particularly in Germany.

From 1791 to 1799, the German engineer Reinhard Woltmann (1757-1837) published a series of books entitled, < Contributions to Hydraulic Architecture > (*Beiträge zur hydraulischen Architektur*), which described the practice of different aspects of hydraulic engineering of the period, and contained original measurements on the resistance of pipes and channel flow. He had invented an improved current-meter with which he made his original measurements of pipe and channel resistances. Woltmann proposed an improved and simplified Dubuat's resistance formula:
$$v^n = (133.6)\,g\,(d/4)\,[h-(v^2/478)]\,(1/L) = c_g\,S\;,$$
where n varied from 1.75 for pipes to 2 for open channels, precisely the presently accepted limits.

As early as in 1763, the ingenious French naval engineer who fought in the American Revolution as a volunteer, Jean Charles de Borda (1733-1799) had used the rotating-arm ballistic pendulum, the so-called 'whirling arm' ballistic pendulum of the English mathematician Benjamin Robins (1707-1751) for resistance measurements of bodies moving in fluids, and discovered the great retarding effect of 'drag' on bodies moving through fluids. About 1766 Borda began to experiment on the immersion of perforated vessels because he was a naval engineer. Borda found that the so-called 'live force' (*vis viva*), mv^2, contrary to Daniel Bernoulli's contention, was not always conserved in hydraulic problems. For instance, when water flowed through an abruptly reduced or enlarged tube, or when suddenly brought into contact with other bodies of water, Borda observed that *vis viva* suffered a loss, which he gave by his own method of energy as $Q\,[(V-v)^2/2g]$, where Q is the quantity of flowing water, V and v the speed of water before and after meeting the obstructions mentioned above, respectively.

In the 2nd edition of his < Principles of Hydraulics > (*Principes d'hydraulique*), Du Buat gave the effect of the impulsive waterwheel as $pv = AV(V-v)^n$, where n was a coefficient to be determined experimentally. On the basis of Bossut's experiments he set $n = 5/4$ for confined channels, and $n = 4/3$ for open channels. Borda had demonstrated that resistance of fluids is proportional to v^2, and that in the presence of surface waves the resistance is proportional to even higher powers of velocity v.

In 1766, Borda published a well-known memoir on orifice flow and gave the so-called 'Borda contraction formula' (for a re-entrant tube the cross-section of the jet contracts to half of the cross-sectional area of the orifice which Newton had called **vena contracta**). Borda was one of the first mathematicians after Johann Bernoulli to use explicitly $2g$, where g stood for constant gravitational acceleration, in his hydraulic formulae. He investigated flow through pipe inlets of various shapes, and through submerged openings in the bottom of floating tanks, and formulated useful expressions for the expansion losses in fluid flows. In his theoretical analysis of water conduits Borda introduced stream tubes which ran parallel to the flow of water instead of transverse differential slices of fluid introduced by Johann Bernoulli in 1739, for the analysis of hydrodynamic flow in pipes and passages and through orifices. In 1767, Borda published a memoir, where he applied his theory to waterwheels and demonstrated that the conventional method which used mass and velocity, and his method of kinetic energy which used **vis viva** can lead to identical results. In the conventional part of his analysis, Borda corrected some mistakes in Parent's analysis. Borda asserted that the impulsive force striking the blades of the undershot waterwheel in a narrow channel is proportional to $(V-v)$ and not to $(V-v)^2$ as Parent had assumed. He measured the moving power of water by QH (where H is the head or fall of water) rather than by PV (that was twice the value of QH), which gave the theoretical values for the optimum speed of undershot waterwheels as $v = V/2$, for the optimum load $p = P$, and for the maximum efficiency ½, values which were much closer to experimental findings than those of Parent. Borda's analysis of the overshot waterwheel seemed to follow that of Johann Albrecht Euler (1734-1800) except that he included in his analysis also the impactive effect of water striking the blades. He gave the maximum effect in the form: $Q(H_g - h_w)$ + (½)Qh, where H_g is the height the water falls within the waterwheel, h_w the height which causes the velocity of the overshot waterwheel, and h the height creating the speed of water to strike the blades. In the first part of his analysis, Borda evaluated the dynamic effect of the stream by its linear momentum mv, and postulated that the linear momentum of the water passing through the waterwheel was conserved. However, the linear momentum principle did not reveal why the undershot waterwheel is more inefficient than the overshot, or the breastfed waterwheel. In the second part of his paper which dealt with his method of energy in the waterwheel analysis, Borda used the concept of **vis viva** (i.e., twice the kinetic energy) to explain why energy was lost and not conserved. Borda established the basic formula for the power output of waterwheels:

$$pv = mgH_g + (½)mV^2 - (½)m(V-v)^2 - (½)mv^2,$$

where mgH_g represents the gravitational potential, $(½)m(V-v)^2$ denotes the energy lost in the impact, and $(½) m v^2$ is the energy lost due to the velocity still present in the water leaving the waterwheel.

He suggested that the blades of the undershot waterwheels be curved upstream in order to reduce the energy loss due to the turbulence in the water.

Despite the fact that Borda's original work was a fundamental contribution to hydraulic power, supplying the answers to some basic questions that had been raised in waterwheel analysis, and that it narrowed the gap existing between theory and experiment, it met rather scornful criticisms expressed by such prominent scientists and mathematicians as d'Alembert and Lagrange, and other more practical engineering scientists. The major criticism raised against Borda's theory was that Borda's analysis seemed to violate the principle of equality between cause and effect which had been used in the form of conservation laws over several centuries. In his research Borda had made efforts to put the principle of energy in its application to machines on a more systematic and rational basis, an effort later continued by Lazare Carnot. After the Revolution, Borda was made the leading member of the committees responsible for devising a decimal system of weights and measures. Borda was particularly interested in setting the 'unit of length' as a fraction of the length of the meridional arc, which he named *'meter'*. Borda was for a brief period an instructor of mathematics, but after becoming a military engineer he was at first engaged in harbour construction, then he turned to investigate ballistic problems, and ultimately to perform experiments in hydraulics and hydraulic machinery. After 1767 Borda joined the navy and became a naval engineer where he exhibited his ingenuity in improving the methods for the determination of the longitude and the latitude.

The French military engineer, Lazare Nicolas Marguerite Carnot (1753-1823), was the first scientist to recognise the importance of Borda's pathbreaking contribution, which Borda had used to analyse special cases including the vertical waterwheels. In 1782, Carnot generalised Borda's principle by making it applicable to all hydraulic machines. Borda's analysis had indicated that for maximum efficiency, which approaches 100%, the water should enter the waterwheel without impact and leave it without velocity. Lazare Carnot made it the fundamental principle in the design of hydraulic prime movers: for maximum effect the water must act on the waterwheel without impact and leave it without velocity. He published his work in a long prize-winning essay < Essay on machines in general > (***Essai sur les machines in général***) in 1782, but it was also ignored by the profession until decades later when Carnot had become an important General in the French Revolution, and when for a brief time it came to the attention of the profession upon the publication of the 2nd edition of this work in 1803 under the title, < Fundamental Principles of Equilibrium and of Motion > (***Principles fondementaux de l'equilibre et du mouvement***).

Abbé Charles Bossut (1730-1814), professor of mathematics at the military school of ***Mézières***, was primarily responsible for introducing fluid mechanics as an engineering science into the professional education of engineers. In this field Bossut wrote an important pedagogical textbook, < Elementary Treatise of Hydrodynamics > (***Traité élémentaire d'hydrodynamique***), a work in 2 volumes published in 1771, 2nd edition in 1786 and 3rd in 1795/1796, which gave a new treatment of the subject in the first volume, and new experimental data in the 2nd volume.

In 1775, Bossut, Jean LeRond d'Alembert (1717-1783) and Marquis de Condorcet (1743-1794) carried out scale model experiments on the resistance of bodies in a rectangular towing tank which was 100 feet (30.48 m) long, 50 feet (15.24 m) wide, and varying up to 7 feet (2.13 m) in depth. The experimental method used by the three French scientists was to measure the speeds acquired by the model boats towed through water by given forces. The motive force was provided by a descending weight which was connected to the prow of the model boat by a flexible cord that ran over one pulley on top and another pulley at the foot of a 75-foot (22.8 m) high mast so that the model boat was drawn forwards in horizontal direction. They found in general that the law of resistance varied approximately as the square of the speed, but the resistance increased at a somewhat greater rate than this law implied, which they explained to be an effect of the formation of surface waves by the moving boat. They were able to reduce the resistance by streamlining the stern of the model boat. Attempts were also made to add the effect of air-resistance to the water-resistance of the boat. They also drew the conclusion from their experiments that the resistance to motion increases as the relative area of the channel diminishes, a conclusion which had been reached previously by the American scientist Benjamin Franklin (1706-1790), who had used a miniature towing tank. The results of these experiments were published in a book entitled < New Experiments on the Resistance of Fluids > (*Nouvelles expériences sur la résistance des fluides*) in 1777. Later in 1831, the French engineer Ferdinand Reech (1805-1880), professor of mechanics and director of the *École d'application du Génie Maritime* at Paris, gave the similarity criterion in measuring the resistance of a model vessel, which is now known as the 'Froude criterion of similitude'.

About 1770, Bossut performed extensive model experiments on 3 waterwheels which completed the experiments of de Parcieux, and the results of which helped to discredit Pitot's theory of blade distribution. Bossut tested an undershot waterwheel 3 feet 2 inches (1.03 m) in diameter, and 5 inches (0.14 m) in width, in a closely fitted channel; an undershot waterwheel of 3 feet (0.97 m) in diameter, 5 inches (0.14 m) in width tested in a channel from 12 to 13 feet (3.90 to 4.20 m) in width; and an overshot waterwheel 3-feet (0.97 m) in diameter, and 5 inches (0.14 m) in width. Bossut tested the undershot waterwheels for the effect the number and the inclination of the blades, the varying wheel speed, and the widths of the channel had upon the power output, and the overshot waterwheel for the effect of its speed on the efficiency. He found that for undershot waterwheels in both the wide and narrow channels the optimum speed was about $(4/10)V$, which agreed with Smeaton's experimental results, and that for overshot waterwheels the slower the speed of the wheel the greater the output for a prescribed quantity and fall of water.

In 1797, Bossut determined experimentally that the resistance of water to a moving plane in the confined channel was almost double of that in an open channel. On this basis he gave the power of the undershot waterwheel as,

$$pv = n\,A(V-v)^2 v,$$

where the coefficient $n = \frac{1}{2}$ for open channel and $n=1$ for a confined one. Since 1769, Bossut had been attempting to eliminate the simplifications introduced by Parent by trying to derive a general equation which takes into account many important variables affecting the performance of impulsive waterwheels that had been ignored by Parent such as the radius of the wheel, the size of the blade, the proportions of the immersed radius, different speeds of the impacting stream at different depths, and at different points on the waterwheel's blades, the number and the inclination of blades, and the varying angles of impact. In his analysis Bossut used a revised measure of impulse. Bossut's general formula, which nearly filled a page, was so intricate that in 1819 Navier had to admit that it was virtually impossible to make any practical use of Bossut's solution on account of its enormous complexity. Even Bossut himself was forced to resort to simplifications of his solution by arbitrary assumptions, which contradicted his aim of developing a precise general solution without simplifications, to use it in practice.

The German engineer, Franz Joseph Gerstner (1752-1832), founder and director of the **Technical Institute** in Prague, published, in 1804, a treatise, < Theory of Waves > (***Theorie der Wellen***), in which he developed a rigorous mathematical theory of deep-water waves for irrotational flow and represented the free-surface profiles as lines of constant pressure and as paths of the particles of the surface. His wave profiles were ***trochoidal*** (stretched cycloidal) configurations having cycloidal limits, with circular particle orbits decreasing in diameter with depth below the surface. He gave the velocity of propagation of deep water waves in the form:

$$v = [g\lambda/2\pi]^{1/2}.$$

Gerstner's theory of waves represented a definite advance in hydrodynamics of waves. In 1796, Gerstner had also studied the temperature effect upon the resistance to flow through pipes of small diameter whilst he was consulting engineer to major canalisation projects. Gerstner, furthermore, studied overshot waterwheels, mechanical calculating machines, and introduced the concept of pressure line into arch analysis.

Gerstner's chief work in mechanics was presented in his < Handbook of Mechanics > (***Handbuch der Mechanik***) published in 1831. In the chapter on mechanics of materials he reported on his tensile experiments with piano wires and he gave a nonlinear relation between the stretch of the wire Δ and the tensile force F:

$$F = A\Delta - B\Delta^2$$

where A and B are suitable constants.

He examined the effect of permanent deformation upon the mechanical properties of the wire, and demonstrated that upon reloading the material behaves linearly up to the force which produced permanent deformation on the previous loading. He advised the use of wires in suspension bridges, and pre-stretching of such wires to a certain limiting stress before their use in the suspension bridge.

In 1818, French physicist Alexis Thérèse Petit published a short article in which he used the principle of *vis viva* in the analysis of a number of prime movers including the waterwheel. In the same year Louis Marie Henri Navier published a review of the history of the principle of *vis viva* used in the analysis of machines, in which he stressed the importance of Borda's fundamental contribution. He used Borda's theory of power in his annotated new edition of the first volume of Belidor's < Hydraulic Architecture > (*Architecture hydraulique*) published in 1819. Navier gave the maximum effect of an overshot waterwheel as,

$$pv = n[1,000Q(H - \tfrac{1}{2}h)],$$

where the experimental coefficient $n = 4/5$. About 1820, Navier added a term to Borda's equation to account for losses in the breastfed waterwheel due to the immersion of this wheel in water. The overshot waterwheels made of steel could have an efficiency approaching 90 %. In Great Britain the energy method of analysis of waterwheels became popular only after 1840.

Inspired by the theoretical hydrodynamics of the French engineers, Giorgio Bidone (1781-1839), professor at the University of Turin, Italy, and Giuseppe Venturoli (1768-1846), professor of mathematics at the University of Bologna and later director of the School of Engineering at Rome, undertook the study of the hydraulic jump, a well-known but at that time ill-understood hydrodynamic phenomenon. Bidone in his tracts, < Experiences on the Backwash and Propagation of Waves > (*Expériences sur le remous et la propagation des ondes*) in 1820, and, < Experiences on the Propagation of Backwash > (*Expériences sur la propagation du remous*) 1826, correctly related the velocity v, and incremental height h of the surge to the velocity V and the depth H of the approaching flow as $vh = VH$, but his evaluation of the surge height h in the hydraulic jump was erroneous on account of his faulty use of the linear momentum principle. Therefore, he gave the surge height h in terms of the product of the velocity head $V^2/2g$ and an empirical factor $(2H+h)/h$:

$$h = [(2H + h)/h] (V^2/2g),$$

which made the calculated surge height approximately agree with his experimental results.

His countryman, Venturoli was more successful in deriving an elementary differential equation of the backwash in rectangular channels in his anonymous tract, < Research on the Configuration of Backwash in Uniform Channels > (*Ricerche sulla figura del pelo d'aqua negli alvei di uniforme larghezza*), in 1823. Venturoli was able to trace some branches of the surface profile for gradually varied flows by means of graphical integration of his differential equation of backwash.

The French engineer, Jean Baptiste Belanger (1789-1874), professor at the *École des Ponts et Chaussées* and the *École Polytechnique*, developed the elementary 'backwash equation' in his memoir, < Essay on the Numerical Solution of Some Problems Relative to Permanent Movement of Water > (*Essai sur la solution numérique de quelques problèmes relatifs au mouvement permanent des eux courants*), in 1828, and he applied step-wise integration in its solution. In 1838 he made an attempt in his lectures to improve on Bidone's analysis of the hydraulic jump by using the principle of energy :

$$(V_1^2 h_1)/g = (½) h_2 (h_2+h_1) ,$$

and he gave a solution for the maximum discharge over broad-crested weirs.

In 1825, two German brothers, Ernst Heinrich Weber (1795-1878), professor at University of Leipzig, and Wilhelm Eduard Weber (1804-1891), professor at University of Halle, published in 1825 a treatise giving an account of their experiments on waves, < Theory of Waves Founded on Experiments > (***Wellenlehre auf Experimente gegründet***), which contained a host of various new experimental results on wave motion observed in narrow glass-walled tanks. Interference, reflexion, orbital motion of fluid particles, and profile configurations in water, in brandy and in mercury were thoroughly examined in these experiments which were inventive, adroit and quite unique. These experimental results were qualitative, and no attempt was made by the Weber brothers to correlate their experimental observations with Gerstner's theory.

Gotthilf Heinrich Ludwig Hagen (1797-1884), a German hydraulic engineer, wrote a memoir, < On the Motion of Water in Narrow Cylindrical Pipes > (***Ueber die Bewegung des Wassers im engen cylindrischen Röhren***), in 1839, which was a further development of Gerstner's investigation of flow through small diameter pipes. The diameter of the pipes varied from 0.098 to 0.236 inches (2.5 to 6mm), and the lengths from 18.5 to 43.31 inches (47 to 110 cm), temperature T ranged over 50°F (10° Celsius). It was an experimental study in which Hagen assumed the velocity to vary linearly from zero at the wall to the maximum at the center. He derived on this basis the resistance and the exit loss in the head, and although Hagen ignored the entrance effect, and assumed linear distribution of the velocity, his values are within 1% of modern values for the viscosity variation. He was the first to observe the 2 characteristic regimes of flow: the laminar and the turbulent flow. His enquiry into the 2 different kinds of flow Hagen published in a paper entitled, < On the Influence of Temperature on the Motion of Water in Pipes > (***Ueber den Einfluss der Temperatur auf die Bewegung des Wassers im Röhren***) in 1854, in which he succeeded to correlate his measured resistance to turbulent flow with the loss of the head:

$$h \simeq \rho V^{1.75}/D^{1.25} .$$

Hagen did anticipate several discoveries later made independently by Osborne Reynolds (1842-1912).

Almost at the same time, a French physician, Jean Louis Poiseuille (1799-1869), who was interested in the pumping power of the heart, carried out careful flow-resistance experiments with extremely small diameter tubes ranging from 0.00114 to 0.00559 inches (0.029 to 0.142 mm), and tube lengths as large as 100 times the diameter, under heads as high as 19 ⅔ feet (6m) of mercury. Poiseuille established the empirical quantity of flow:

$$Q = B(1 + \alpha T + \beta T^2)(hD^4/L) ,$$

a result which agrees within ½ % of the modern value.

The theoretical derivation of the so-called 'Poiseuille law' was given by German physicist Franz Ernst Neumann (1798-1895), professor of physics at the University of Königsberg, in 1858, and by a Swiss physicist Eduard Hagenbach (1833-1910) of Basel in 1860. Both physicists arrived at the parabolic distribution of the velocity across the cross-section which took into account the viscosity of the fluid, and both evaluated the correction due to the resulting kinetic energy; however, Hagenbach's correction was wrong.

Giovanni Poleni (1683-1761), professor of astronomy, then physics and mathematics at the University of Padua, was also a linguist, philosopher, theologian, diplomat, and consulting engineer on flood control and water supply. In 1718, Poleni reported experimental measurements on sharp-edged orifices in his work < On Fortresses > (*De Castellis*), and specified the coefficient of contraction to be *0.62*, a great improvement over the Newton coefficient $\sqrt{1/2}$. In his work, < The Motion of Mixed Water > (*De moto aquae mixto*) in 1717, he gave the formula of the rate of efflux per unit width of a rectangular opening, as $Q = (2/3)hbf^{1/2}$ where h and b denote the depth and breadth of the opening, and f now is given as $2gh$. Since this method was later used for sharp-crested weirs, the basic weir equation is usually called the 'Poleni equation':

$$Q = (2/3)Cb\,(2g)^{1/2}\,h^{3/2}.$$

In 1748, Poleni employed Hooke's method of inverted catenary in the study of the equilibrium of the cracked lunes of the dome of St. Peter's Basilica. It had been analysed before by the virtual work method which neglected the deformation of the horizontal circumferential tension ring by 3 Jesuit mathematicians: Thomas Le Seur (1703-1770), François Jacquier (1711-1788) and Ruggiero Giuseppe Boscovich (1711-1787). Poleni, moreover, carried out a large number of strength tests on iron in his investigation of the circumferential reinforcement required for the tension ring of St. Peter's dome. Architectural engineer Luigi Vanvitelli (1700-1773), on the suggestion of Poleni, installed 5 additional circumferential tension rings of iron in the dome to prevent further cracking of the dome.

The German engineer, Julius Albin Weisbach (1806-1871), professor of mechanics and machine design at the **Freiberg Bergakademie** (*Freiberg Academy of Mines*), modernised the engineering theory of fluid mechanics as well as the teaching of engineering mechanics in general to engineers. He followed the method of Jean Victor Poncelet by combining experimental results with engineering design. He wrote his famous, profusely illustrated textbook, < Textbook of Engineering and Machine Mechanics > (***Lehrbuch der Ingenieur- und Maschinenmechanik***), a work in three volumes which was published from 1845 to 1862, and promptly translated into English, Swedish, Polish and Russian languages, but not into French. For this reason, the French engineers for many years remained ignorant of the modern form of the Bernoulli equation as developed by Weisbach in his textbook. Weisbach was interested in the method of teaching engineering mechanics, and pioneered in organising the first laboratory of mechanics in which students themselves had to perform experiments designed by Weisbach. He modernised the Bernoulli theorem for its facile engineering applications, gave numerous worked-out examples and the best available experimental information which included many original results from his own experiments that have never been improved upon. He promoted the use of non-dimensional coefficients in the engineering formulae such as the resistance equation for pipes: $h = \zeta(L/D)(V^2/2g)$, in which the coefficient varies with the velocity, the diameter of the pipe, and the wall material. In 1841, he published the so-called 'Weisbach weir formula' which was later adopted by Henri Emile Bazin. Weisbach was mainly responsible for the popularisation of the Bernoulli theorem in engineering fluid mechanics, and making fluid mechanics an integral part of engineering mechanics.

Hydraulic Power and Hydraulic Turbine

The hydraulic ram principle had been used by an English mechanic John Whitehurst (1713-1788) in a rather crude pumping device supplying domestic water near Derby in which water was led through a pipe of nearly 600-foot length (183 m) and 1½-inch (3.8 cm) diameter down a fall of 16 feet (4.9 m). Whitehurst used a manually operated stop-cock to arrest abruptly the flow of water, and force some water above the load of the source by the resulting pressure. Whitehurst did not use a self-acting valve mechanism necessary for the practical functioning of such a hydraulic ram and, therefore, it was unsuccessful. Moreover his hydraulic ram was noisy and produced severe impacts in its structure.

In an effort to find ways to use water power in a hydraulic device to pump water itself, the French paper manufacturer and inventor, Joseph Michel de Montgolfier (1740-1810), devised his hydraulic ram-pump with a self-acting valve in 1797 which had neither a cylinder nor a piston. The hydraulic ram-pump consisted of a drive-pipe, a valve-box fitted with a pulse-valve and a delivery-valve leading to an air-chamber connected to a delivery-pipe. The flow through the drive-pipe and the pulse-valve was first established, and when the necessary velocity was reached, the pulse-valve was automatically closed and the resulting dynamic pressure-wave overcame the pressure due to the head of water in the delivery-pipe, opened the delivery-valve, compressed the air in the air-chamber, and thus drove the water up the delivery-pipe. When the kinetic energy of the pressure-wave had been dissipated in this hydraulic pulse, the diminishing dynamic pressure caused the water column to surge backwards and open the pulse valve, which again closed when the dynamic pressure exerted by the flow of the supply water was able to overcome the pressure exerted by the column of water over the pulse-valve. This cycle of operation of the hydraulic ram kept repeating itself, often attaining 150 hydraulic pulses a minute, as long as there was a continuous stream of water. Therefore, the hydraulic ram used the momentum of a stream of water falling through a small height to raise some part of this water to a greater height, but the quantity of water raised or delivered, depended on the head and quantity of water flowing into the hydraulic ram, as well as on the total head of delivery and the frictional resistance of the pipes. Montgolfier installed his first hydraulic ram at his paper mill near Voiron which raised water under a 10 foot (3.00 m) head, but soon much larger hydraulic rams were constructed. The hydraulic ram was evidently not a very efficient pumping device, and, therefore, it was mainly used in hilly surroundings to pump small quantities of water to a height above the stream in the valley. It is still being used as a small automatic pumping device for houses in the country owing to its simplicity and reliability. The compound hydraulic ram was patented by Joseph's son, P.F. Montgolfier, in 1816, which had the advantage that when driven by regular river water, it pumped clean well water from a different source. The younger Montgolfier built a hydraulic ram at Mellor, near Clermont-sur-Oisne consisting of a cast iron pipe 4¼ inches (0.11 m) in diameter and 108 feet (33 m) long. It delivered 66 pulses a minute under a fall of 37 ⅓ feet (11.40 m), and lifted water to an elevation of 145 feet (59.5 m) with an efficiency of 65%. The hydraulic ram-pump was much later improved by the French engineer and a pioneer steam-car designer Amédée Bollée (1844-1917).

Joseph Montgolfier became interested in aeronautics after the discovery of hydrogen gas in 1766 which was much lighter than air and made lighter-than-air crafts possible. Since hydrogen was difficult to manufacture, Joseph and his younger brother Jacques Etienne Montgolfier (1745-1799) turned their attention to hot-air balloons first demonstrated by a Jesuit priest, Lourenço de Gusmão in 1709. In 1783 they gave a public demonstration of a large unmanned hot-air balloon measuring 30 feet (9 m) in diameter which rose to 6,000 ft. (1,800 m) altitude and travelled ½ a mile (0.8 km) in 10 minutes. French Academy of Sciences invited Montgolfier brothers to continue their demonstrations in Paris, where they launched another balloon in the same year carrying animals which returned safely to Earth. The first manned flight with the balloon carrying the court historian and a nobleman to an altitude of 3,000 feet (900 m) above Paris and travelling 7½ miles (12 km) in 25 minutes was made in November, 1783. In 1785, the apothecary Jean François Pilâtre de Rozier became the first air fatality when he was killed in a hot-air balloon accidents.

The French physical chemist, Jacques Alexandre César Charles (1746-1823), who was commissioned by the French Academy of Sciences to produce a hydrogen balloon, developed the hydrogen-filled balloon only 10 days after Montgolfier's first manned flight, and together with M.V. Robert (one of the Robert brothers who manufactured the balloon from an air-tight silk fabric varnished with dissolved rubber) he made several ascents with the balloon reaching an altitude of over a mile (1.6 km). In his studies of gases, Charles proved that all gases expand by the same amount under a given rise of temperature. He established that for a degree of rise in temperature in centigrade the volume of gas expanded by (1/273)th part of its volume at zero degree centigrade. Reversing this procedure he determined that when the temperature of the gas is reduced to −273° centigrade (modern value is -273.18° C) the volume of gas reduces to zero according to his law of gases which is now called the 'Charles law of gases' although he did not publish it. This law implied that there cannot be any temperature lower than −273°C, a temperature now called the 'absolute zero'. Charles law of gases is often called 'Gay-Lussac's law', since it was later independently discovered by the French chemist Louis Joseph Gay-Lussac (1778-1850), who also made several ascents in balloons. Although Meusnier was the first to design a balloon with a propeller drive, an enthusiastic French professional balloonist, Jean François Blanchard (1753-1809), was the first to use a propeller-driven large balloon in 1784. Blanchard made many daring flights in United States and Europe, carried air mail from France to England, and invented a parachute. The first successful parachute jump was made in 1797 by a French balloonist André Jacques Garnerin when he descended from an unmanned balloon in a gondola suspended from a parachute. The French military engineer Jean Baptiste Meusnier de La Place (1754-1793), a former pupil of Gaspard Monge in the military school at Mézières who had done original research on the mathematical theory of surfaces, became interested in balloons in 1783. He designed plans for military balloons as aerostations in 1785, and collaborated with the physical chemist Antoine Laurent Lavoisier (1743-1794) in refuting the phlogiston theory, and devising practical means of producing hydrogen for balloons driven by propellers in the course of which de La Place devised a gasometer. Lavoisier and La Place were able to produce enough hydrogen to fill a balloon in 15 hours whereas Charles had taken weeks to do the same with his technique.

Unfortunately, La Place's designs for dirigible balloons propelled by 3 air-screws were far too expensive to build at the time, and soon afterwards Major General La Place was killed in the battle defending Mainz against the Prussian army. La Place had a keen interest in technology, and had designed a still to desalinise sea water, an aerator to improve the taste of distilled water, a machine for engraving bank notes too intricate to be counterfeited, and, together with Lavoisier, an oil lamp which reduced the amount of soot and increased illumination.

The hydraulic-ram type of pumps did not attain immediate popularity for they came into more general use only in the late 1820's. A similar hydraulic ram principle was later used in some hydraulic air-compressors but with mechanically operated valves. High efficiencies of hydraulic rams were possible when the height to which the water was lifted was small in comparison with the fall of water. The efficiency of the hydraulic ram depended upon the number of pulses a minute, the height of the fall of water, and the volume of water to be lifted. Some such hydraulic rams in which all these factors were considered had an efficiency above 90%.

In 1807, another type of hydraulic motor based on Philon's hydraulic reaction wheel, which was called *levier hydraulique*, was developed by Marquis Jean Charles Alexandre François de Mannoury Dectot [alias, d'Ectot] (1777-1822). It operated by the lawn sprinkler principle, an ancient Hellenistic principle of Philon of Byzantion, which Segner had already used in his hydraulic reaction-wheel design. The *levier hydraulique*, consisted of a twin-armed rotor with the 2 arms curved into a spiral, and a central supply pipe. This hydraulic motor became popular as a prime mover for forges and ironworks, but the connexion between the rotor and the supply pipe tended to leak and this problem was never satisfactorily solved. In 1818 Dectot improved his hydraulic motor by incorporating a diffuser into its design.

In France, where there are many rivers with small yields and low falls, and highly fluctuating flow, the undershot waterwheel was still the only feasible solution for hydraulic power. If economy of construction, or high speed of rotation of the hydraulic prime mover was required, such as in industry, then again the vertical undershot waterwheel was indicated.

In 1824, the mathematician and engineer, Jean Victor Poncelet (1788-1876), founder of projective geometry as a mathematical discipline and later commandant of the *École Polytechnique*, as well as a prominent textbook writer in engineering mechanics, followed Borda's advice and designed an undershot waterwheel with blades curved upstream to reduce turbulent energy losses. He delivered a narrow stream of water to the curved vanes through a penstock which made the efficiency of the waterwheel increase from 22% to more than 70%. Since Poncelet's design of the waterwheel was very useful for low falls of water, his design was used near Monserrat in Catalonia, Spain, where a 17-foot (5.18 m) diameter waterwheel was constructed, which was 30 feet (9.14 m) wide and driven by a waterfall which was only 6½ feet (1.98 m) high, yet it was able to develop about 180-horsepower. Poncelet designed his vertical waterwheel theoretically, and corroborated his theoretical findings by model experiments and by tests on the efficiency of large-scale waterwheels of his design with the Prony-brake. The French Academy of Science awarded Poncelet its **Prix de Mécanique** for the theory and design of his waterwheel. The 2nd edition of his famous memoir on his waterwheel was published in 1827, and it was responsible for a rapid spread of his waterwheel which, in a real sense, was a vertical hydraulic turbine, and a radically novel and unique hydraulic prime mover.

Another vertical waterwheel, designed by a French engineer Alphonse Sagebien, became popular in the 1860's. Sagebien designed his breastfed waterwheel theoretically in 1851 basing it on Borda's concept which stipulated that the water should enter the waterwheel without impact. In contrast to Parcieux and Bossut, Sagebien inclined the blades of his waterwheel downstream so that the blades entered the stream on the upstream side almost vertically permitting the water to be forced up the blades by atmospheric pressure, and due to the slow motion of the wheel the exit speed of the water was reduced to a minimum, all of which produced a waterwheel of extremely high efficiency. Sagebien designed his breastfed waterwheel about 1850, and 8 years later he had installed more than 17 of his waterwheels. Tests carried out on Sagebien waterwheels proved that its efficiency reached as high as 93%, the highest efficiency of any waterwheel then in existence. By 1868, Sagebien had constructed more than 60 of his waterwheels only in North-West France, and in 1875 he was awarded the **Fourneyron Prix** by the French Academy of Sciences for this singular engineering achievement. Sagebien's breastfed waterwheels were used in the new Marly pumping plant built in 1850's in preference to small-size, high-speed hydraulic turbines on account of their slow speed of rotation which was more suitable for the driving of the water pumps, their mechanical simplicity, their high efficiency, their ease of installation, their facility for repair, and their reasonable cost. The Marly Plant had 6 Sagebien's waterwheels each about 39 feet 5 inches (12.00 m) in diameter and

14 feet 10 inches (4.50 m) in width which operated under about 10 foot (3 m) fall and a flow of water from 92 to 106 cubic feet (2,600 to 3,000 liters) a second. The vertical slow-turning waterwheel became obsolete when large power, high speeds, and high efficiency of hydraulic prime movers became paramount in industry after 1880. In the undershot waterwheel, water was fed to the blades at one position only, which made the flow enter and leave the wheel at the same point with zero relative velocity, thus creating a reversal of the flow within the wheel that is an undesirable mechanical condition. Therefore, in 1826, Poncelet proposed to obviate this disadvantage by turning the waterwheel horizontal, admitting the water all around the wheel's periphery, or at a single location, and discharging the water without velocity through the interior, which makes this type of hydraulic wheel a radial influx hydraulic turbine.

In 1822, Claude Burdin (1788-1873), professor at the *École des Mines* in Sainte-Etienne who was of Italian extraction, proposed an efflux high-speed hydraulic turbine in a memoir < Hydraulic Turbines or Rotational Machines of Great Speed > (*Les Turbines Hydrauliques ou Machines Rotatoires à Grande Vitesse*) which he submitted to the French Academy of Sciences for evaluation. He called this hydraulic power device a 'turbine' a name taken from the Latin *turbo turbinis*, meaning 'spinning top'. Burdin's turbine consisted of 2 concentric cylinders with an annular space between them occupied by doubly curved blades fixed to the inner, moveable cylinder. Water entered his turbine parallel to the common axis and made the inner cylinder rotate by acting smoothly on the doubly-curved blades. The water left the turbine at right angles to the common axis and in the opposite direction to the rotation of the runner. However, the turbines could not rotate slowly like the breastfed, or overshot waterwheels if they were to be efficient. Burdin's turbine was designed in accordance with the theory of hydraulic machines developed by Lazare Carnot, so that the turbulent impact of water on the blades was minimised and the water left the turbine without any appreciable velocity. Unfortunately, the action of Burdin's turbine was rather inadequate and slow because its moveable vanes lacked sufficient curvature and were, therefore, unable to receive the full impact and reaction effect of the water. In 1827 Burdin submitted his design and discussion of some working machines to the *Société d'Encouragement pour l'Industrie Nationale* (*Society of Encouragement for National Industry*) in competition for the society's 6,000-franc prize for the design of an efficient horizontal waterwheel with curved blades, such as the Belidor Tub-Wheels, for large-scale industrial application. Burdin was awarded only a consolation prize of 2,000 francs because his practical hydraulic machines were really no more than elementary reaction wheels with unexceptional power. This consolation award was granted mainly for Burdin's theoretical designs. Subsequently, Burdin abandoned his work on turbines and instead devoted his efforts to locomotive design.

The modern age of hydraulic turbines began with the reaction turbine of Fourneyron. In 1827, Burdin's former pupil at the *École des Mines*, Benôit Fourneyron (1802-1867), who was a metallurgist, was able to devise a successful arrangement of vanes and to solve the critical bearing problem of a workable turbine by designing a novel lubricated bearing for the annular runner. Although Fourneyron's turbines were of the free-efflux type in which water is expelled from the turbine with considerable speed and energy, he did later consider the submerged efflux type of turbine, in which he employed a radial conical diffuser of his own design, which he patented in 1855, to capture the energy of the efflux and to eliminate the flow separation problem of his turbines. The reaction turbines during their operation are usually filled with water. The entry of water into his turbine was not quite smooth and because of the shape of the passages flow separation tended to occur that led to instability of the turbine which he finally avoided with his diffuser. The efficiency of his efflux-type turbine, which operated under the head of a water column varying from about 1 to 350 feet (0.30 to 109.73 m), was above 80%.

Fourneyon freely admitted the great influence that Poncelet's radically new vertical waterwheel, and the horizontal version of it, had exerted on his invention of the hydraulic turbine, an invention which he made a year after Poncelet's proposal for the horizontal waterwheel that Fourneyron considered fundamentally different from all other hydraulic prime movers.

A model turbine with a wheel of 9½ feet (2.90 m) which produced about 6 horsepower under a fall of water that varied from about 3.9 feet to 4.6 feet (1.20 and 1.40 m) was Fourneyron's first practical success. This turbine had 2 concentric cylinders with curved vanes, but only the outer cylinder was part of the moving wheel. This experimental model turbine was built at Pont-sur-l'Ognon, and it proved to have an efficiency of 80%, which it maintained even when completely flooded by tail water. Fourneyron then moved to Besançon where his development of the hydraulic turbine was financially supported by a forge master F. Caron. Fourneyron's 2nd turbine erected at Dampierre had an exterior diameter of about 3 feet (0.90 m), and operated under heads which varied from 9.8 feet to 19.7 feet (3.00 to 6.00 m) whilst producing 7 to 8 horsepower at about 80 % efficiency. His 3rd turbine at Fraisans built in 1832 had an exterior diameter of 9½ feet (2.90 m) and a height of 1.2 feet (0.36 m). It operated under the head which varied from 9 inches to 4.6 feet (0.23 to 1.40m), developed 50-horsepower with an efficiency of more than 80 %, and discharged from 53 to 117 cubic feet (1.5 to 5.0 cubic meters) of water a second. After all these successes, Fourneyron patented his turbine in 1832, and submitted a memoir on his turbine design to the *Sociéte d'Encouragement pour l'Industrie Nationale*, and was subsequently awarded the 6,000 francs first prize by this Society.

In 1834 Fourneyon constructed a turbine for the spinning mill at Inval which had an exterior diameter of 9½ feet (2.90 m) and a height of 1 foot (0.30 m), it operated under the fall ranging from 1 foot (0.30 m) to 3.8 feet (1.15 m), and produced 30-horsepower with an efficiency about 80% even when completely submerged. The Inval turbine received wide public attention, and was subjected to a great number of experimental tests carried out by a number of French engineers.

Already by 1837, Fourneyron had 2 water-driven turbines in operation in a spinning mill at Saint Blaisien in the Black Forest, which were capable of delivering 60-horsepower under the heads of 354⅓ feet (108 m) and 374.0 feet (114 m) with the water delivered to the turbine under pressure in a 1,640 foot (500 m) long pipe, and running at the speed of 2,300 revolutions a minute with an efficiency of more than 80 %. The movable cylinder of the turbine had a 1½ foot (0.46 m) diameter, and the turbine itself weighed only 40 pounds (18.2 kg). One such turbine could operate about 180 textile machines. Fourneyron tested the efficiency of his turbine by means of the so-called 'Prony-brake'. This dynamic friction brake had been originally invented by 2 men, Hardy and Piobert, in 1821, but it soon was called 'Prony-brake' after an article entitled, < Note on a Means to Measure the Dynamic Effect of Machines of Rotation > (*Note sur un moyen de mesurer l'effect dynamique des machines de rotation*) published in 1822 by Riche de Prony, who was the director of the *École des Ponts et Chaussées*. To obtain reliable results for his turbine, Fourneyron had to develop the Prony-brake into a rather precise testing instrument. Fourneyron published his method of using the Prony-brake in 1829, and the first detailed account of his turbine with a theoretical analysis in 1834. Fourneyron built more than 100 turbines all of which were amazingly small machines for the great power they delivered. The most powerful turbines built by Fourneyron were the 2 very low-head turbines, which developed 220-horsepower and drove 30,000 spindles and 800 looms in a textile mill in Augsburg, Germany. The shafts of these turbines were so heavily loaded that the bottom bearings had to be replaced about every fortnight. Fourneyron turbine which provided the air-blast for the Fraisians ironworks had a wheel of 7.9 feet (2.4 m) in diameter and delivered 50-horsepower under 4¼ foot (1.30 m) head.

American civil engineer Uriah Atherton Boyden (1814-1879) built his first turbine in 1844 in a mill in Lowell, near Boston, and beginning in 1846 he supplied additional improved versions of his turbines to the same mill. Boyden's efflux turbines were similar to Fourneyron's turbines except that Boyden had supplied his turbine with a submerged conical diffuser which reduced the efflux turbine's instability, and increased its efficiency. Boyden tested his first turbines with a Prony-brake and found their efficiencies to be between 75% and 88% at optimum gate. He made a small fortune in the commercial exploitation of his turbine.

Several French engineers attempted to improve Fourneyron's efflux turbine concept, but failed. It turned out that Fourneyron's efflux-turbines performed successfully under limited conditions: only when running at full flow under submerged or above the surface of the tailrace, for which condition they had been designed, were his turbines highly efficient. When the water flow was diminished to reduce the power output, Fourneyron's turbines developed turbulence in the water leaving the guide vanes which reduced their efficiency dramatically. Moreover, if the load were suddenly removed, the reaction-turbine could pick up speed and race to its destruction. Fourneyron's turbines soon went out of favour, mainly because the efflux-type of turbine lost its popularity on account of its fundamental shortcomings. Also for small streams with fluctuating flow, the all-steel overshot waterwheels as well as breastfed waterwheels (such as the Sagebien waterwheel) could have efficiencies nearing 90 % under varying head and volume of water, whereas small turbines seldom had efficiencies higher than 70%. Moreover, Fourneyron turbines were expensive to build because of their complex construction.

In the 1840's, European engineers cleaved to the concept of axial-flow hydraulic turbines, which ultimately led to the propeller-type turbine of Kaplan. An eminent hydraulic engineer, James Bichens Francis (1815-1892), a British engineer who had emigrated to the United States and had been called to measure the power of Boyden's turbines in collaboration with Boyden, turned his attention in 1849 to the improvement of the axial-flow hydraulic turbine, a type of turbine which had been invented in 1837 by the German engineer Carl Anton Henschel (1780-1861) and installed in 1841 in a stone-polishing works in Brunswick, Germany. The Henschel turbine, which did not run submerged like the Fourneyron Turbine, was patented in France in 1843 by Nicolas J. Jonval, who had seen this turbine in Brunswick and had copied it in his workshop Koechlin in Mülhausen, Germany. A somewhat similar axial-flow turbine was made independently by a French mechanic Pierre Lucien Fontaine-Baron [alias Fontaine] (1809-1895) in 1845. In this axial-flow turbine, which were popularly known as the 'Jonval turbine', the guide vanes were placed above the runner and the downwards-flowing water remained at about the same distance from the axis of the turbine and was, therefore, little affected by the centrifugal forces. The so-called 'Jonval turbine' became popular for a while because the efflux-turbines were inherently unstable and difficult to control. Poncelet's idea to have the rotor inside the annulus of fixed guide vanes was patented as an inwards-flow, or influx turbine in the United States, and built in a crude form by an American mechanic, Samuel B. Howd, in 1838. It was much improved in 1840 by Francis, under licence from Howd. The Francis turbine was an influx-type of turbine in which the water flowed downwards and inwards. Later Francis also developed a mixed-flow turbine. Both types of turbines, which were complex and expensive to build, are still called Francis turbines. James Thomson (1822-1892) [elder brother of the physicist William Thomson, later known as Lord Kelvin], professor of engineering at the University of Glasgow, patented in 1850 a kind of double-discharge Francis turbine, called the 'vortex-wheel influx turbine', which had a rotor with curved blades, and pivoted guide vanes that could be coupled together and controlled by a governor. It was a very efficient influx flow radial-turbine.

Louis Dominique Girard (1815-1871), an outstanding French hydraulic engineer, demonstrated in 1850 how the action principle in mechanics can be applied to axial-flow turbines in a partial-admission impulse design to achieve an advantageous regulation of his turbine, called the 'syphon turbine'. Impulse turbines under operating conditions are always partially filled with air, and their practical success came with the device invented by Girard for ventilating their vanes. This axial flow partial-admission impulse turbine is easily adaptable to both variable flows and variable head, and in this versatility and good efficiency at partial-admission lay its great practical value.

The vertical wheel driven by a jet of water applied at one point of the wheel periphery in which the water-flow is inwards and radial, as Poncelet had developed, led to another category of hydraulic turbines – the hydraulic impulse wheel. A British engineer, Lester Pelton (1829-1908), who worked in the California gold fields about 1870 discovered as a result of an accident to a simple waterwheel that fast and powerful impulse wheels can be worked by direct high pressure water jets impacting on the edge of two hemispherical cups with central partition located around the circumference of the wheel. By 1890, a 800-horsepower 'Pelton wheel' worked by a jet fed by a 400-foot (121.92 m) head of water was in use in Alaska. A 220-pound (99.79 kg) 'Pelton wheel' which was portable by 2 men could deliver 125-horsepower. Efficiencies of 85% or more have been obtained with 'Pelton wheels' after the American engineer William Abner Doble, who ran a rival Doble Company which in 1912 merged with the Pelton Company, fitted it with an improved nozzle equipped with a needle and ellipsoidal buckets. The 'Pelton impulse wheel' was a definitive solution to the high head problem in hydraulic turbines.

In 1916 an American engineer, Forrest Nagler, had realised that the limiting speed of a turbine was mostly a result of frictional resistance, and that for higher speeds the wetted surface-area of the turbine had to be reduced. He removed the stiffening ring around the outer circumference of the runners which produced no thrust but was subjected to significant skin friction, and made the runner vanes radial which made the turbine into an axial-flow turbine and the runner into a propeller- runner. It represented a new type of turbine which achieved almost double the speed of the comparable Francis-type of runner under the same conditions of head and power.

An even more advanced and completely novel type of turbine was designed by an Austrian engineer of Czech stock, Victor Kaplan (1876-1934), in his attempts to improve the Francis turbine operating under very low heads. The 'Kaplan turbine', which is a complement to the Pelton impulse wheel, had a vertical axle and a propeller-form rotor with variable-pitch blades so that the angle of the guide vanes and the wicket gate could be automatically adjusted whilst the turbine is running to make the turbine as efficient under variable load as the Nagler turbine was under full load. Kaplan's turbines work with best efficiency under relatively low heads of water, up to about 100 feet (30.50 m). Kaplan patented his turbine in 1912, and in 1913. The first commercial Kaplan turbine which had a 25.8-horsepower output under 7½-foot (2.3 m) head was installed in a textile mill at Velm, Austria, in 1919. Next year several other Kaplan turbines were installed, but soon thereafter a serious shortcoming of Kaplan turbine became apparent: the blades of the Kaplan turbine in Velm had become heavily 'pitted'. This unexpected setback forced Kaplan to embark on a careful study of the 'pitting' problem, and after some years of intensive research he discovered that the reason for the 'pitting' of rotor blades in his turbines was 'cavitation'. Kaplan found that when the water under high pressure and velocity passed through the guide vanes and the runner of his turbine into the draught tube which led to the tailrace, it lost its energy as well as pressure so drastically that the pressure of water fell beneath the vapour pressure and, as a result, vapour-filled bubbles, or 'cavities' formed in the water which, when they came into contact with the metal surfaces of the runner or of the passages, collapsed and, thereby, delivered powerful blows to the surface of the metal over areas of pin-point size. After millions of such blows had been delivered to the metal surfaces by the collapsing vapour bubbles during the operation of the Kaplan turbine, the structure of the metal finally broke down and resulted in 'pitting' which made the metal spongy and ultimately led to its failure. Turbines working under low heads and with high speeds, such as Kaplan turbines, are peculiarly prone to 'cavitation'. The chief engineer of a Swedish company charged with the manufacture of Kaplan turbines for the power station Lilla Edet at the Göta river in Sweden, Olov Englesson (1884-1953), was able to solve the cavitation problem of the Kaplan turbine in close collaboration with Kaplan by developing a new guiding system for the turbine in the form of a hydraulic servomechanism which operated the guide vanes and turned the blades of the turbine in unison. The rotor of the Kaplan turbine has two or more blades which can be turned around their own axis under the control of a governor, whilst the angles of the inlet and outlet guide vanes are also adjusted automatically during the running of the turbine. The efficiency of the Kaplan turbine can be maintained by means of this automatic guidance system nearly constant within the range of 85% to 90% over the range of load from 30% to full load. Kaplan's turbine is most efficient under low heads of water supply, whereas the Pelton impulse wheel is most efficient under high heads of water supply. In 1926 the first completely successful Kaplan turbine weighing 62½ tons and having a 19-foot (5.80 m) diameter rotor which supplied 11,000-horsepower under the modest head of 21.3 feet (6.49 m) was put into service in Sweden. Soon after this successful installation, Kaplan turbines became universally popular since his turbine was the definitive solution to the extremely low head problem in hydraulic turbine design, although the Kaplan turbines are quite expensive to build. In the 1940's a bulb-type Kaplan turbine was developed in Germany and Switzerland for tidal power plants. In the bulb-type Kaplan turbine, the electric generator of a waterproof type is installed horizontally in a streamlined bulb of the waterway and coupled directly to the horizontal Kaplan turbine. The turbine itself is located under the building of the power plant which stands directly above the tubular waterway.

The British company English Electric developed the 'Deriaz turbine' which incorporated the design features of both the Francis and the Kaplan turbine. These inventions for all practical purposes concluded the essential development of waterpower technology.

The real future of the water-driven turbines lay in driving the electric generators in electric power stations. These hydraulic turbines became much more complex in their technical details but involved no essential change in their fundamental mechanical principle. The work done on the hydraulic turbine contributed directly to the development of the steam turbine at the close of the 19th century by the Swedish engineer of French stock, Carl Gustaf Patrik de Laval (1854-1931), by the American engineer Charles Gordon Curtis (1860-1953), by the French engineer Auguste Camille

Edmond Rateau (1863-1930), by the Swiss engineer Hans Zoelly (1880-1950), and by the Swedish engineers Birger (1872-1948) and Fredrik (1875-1964) Ljungström.

Gaspard Gustave de Coriolis (1792-1843), who was a talented mathematician already in his teens, became professor of hydraulics at the *École des Ponts et Chaussées*, an examiner at the *École Politechnique*, and professor of analytic geometry and general mechanics at the *École Centrale des Arts et Manufacture*. His major work was his book < Treatise of Mechanics of Solid Bodies and of the Calculation of the Effects of Machines > (*Traité de la mécanique des corps solides et du calcul de l'effect des machines*) published in 1844, of which the mechanical part had already appeared in print in 1829. In 1844, Coriolis found in his study of the hydraulic turbine the correct relative acceleration expressed in terms of the rotating frame of reference, by correcting the calculation error of Euler and restoring the factor 2 in the so-called 'Coriolis acceleration'. Coriolis was the first engineering scientist to introduce the expression $mv^2/2$ for kinetic energy as the replacement for 'live force' (*vis viva*), mv^2, which was used in mechanics before his researches on the energy principle, and the expression $\int \mathbf{F} \cdot \mathbf{dr}$ for work done by force $\mathbf{F}$ over its displacement $\mathbf{dr}$. Coriolis used the principle of energy, which he renamed the principle of transmission of work, in most of his researches on mechanics. In 1830, he analysed waterwheels in which the speed of the exiting water was larger than the speed of the blades of the waterwheel.

In 1835 Coriolis published his book, < Mathematical Theory of the Effect of the Game of Billiards > (*Théorie mathématique des effets du jeu de billard*) in 1835, in which he developed the general theory of billiards. In 1836 Coriolis published his researches on the open channel flow containing his well-known correction coefficient of the mean velocity, and on the backwater equation which was based on the principle of energy. All researches of Coriolis were distinguished by their mathematical elegance.

Soil Mechanics

Soil mechanics as an engineering science had its beginnings in the design of retaining walls. The retaining walls are vertical walls designed to retain earth, which is dumped on one side of such walls, from sliding down. Retaining walls had been built already by ancient Sumerians, and their dimensions and structural details were based on previous practical experience with such walls.

The mechanics of soils as an engineering science was first developed in France. The famous military engineer of France, Sebastien le Prêstre de Vauban still gave empirical rules in 1658 for the safe dimensions of retaining walls, which ranged in height from 6 feet (1.83 m) to 80 feet (24.40 m), in tabular form. He specified the thickness of the retaining wall at the top to be 5 feet (1.52 m) and the batter in the wall to be 1–to–5. These rules gave unnecessarily heavy sections for low walls, and unstable walls which were higher than 35 feet (10.70 m) unless special care was taken in backfilling. It was the architect-engineer Pierre Bullet (1639-1716), a contemporary of Vauban, who made the first rational attempt to establish a mechanical model for the retained soil in his < Practical

Architecture > (*L'Architecture pratique*) published in 1691, by comparing a heap of sand and gravel to a pile of round artillery shots which could be piled up at a stable slope of 60 degrees to the horizontal. Bullet proposed to use a 45 degree slope as stable for earth fill because of the irregular size of sand and gravel. He regarded the wedge of soil above the stable slope to produce the force of pressure on the retaining wall because he thought the wedge of earth would slide as soon as the retaining wall began to move or tilt. Therefore, Bullet is the originator of the 'sliding wedge' theory of earth pressure.

No further progress was made in the theory of earth pressure until 1726. The French mathematician and mechanical engineer, Pierre Couplet de Tartreaux [alias Tartereaux], elaborated Bullet's wedge theory of earth pressure further by calculating from statics the pressure force as $F = wH^2/6$, and the overturning moment exerted by the soil wedge against the retaining wall as $M = wH^3/6$, where w denoted the specific weight of the earth and H the height of the wall. Couplet demonstrated that tetrahedral and square pyramids of soil were denser than Bullet's triangular soil prism and gave different values of pressure against the retaining wall. Couplet considered the resultant pressure F acting at a distance $\frac{1}{3}H$ from the top of the wall, that is, at the center of gravity of the soil wedge, which is incorrect. In this work Couplet did express an important idea: the failure plane in soil need not coincide with its plane of natural slope. However, he did not take advantage of this idea in his work.

The next advance in soil mechanics came in 1729 in the book < Engineering Science > (*La Science des ingénieurs*) of Belidor. In the chapter on retaining walls Belidor noted that a shot in a state of equilibrium on a 45-degree slope in the absence of friction was subjected to a horizontal force equal to its weight. Dividing the soil wedge into strips parallel to the stable 45-degree slope of the unretained earth, he showed that in the absence of friction the horizontal reaction exerted by the wall at its vertical face is equal to the weight of the prism of the soil, which in the absence of the retaining wall would slide down an approximately 45-degree slope. Therefore, in Belidor's soil pressure theory, the earth pressure increases linearly from the top of the wall like a liquid pressure. He showed that the resultant earth pressure was located at $\frac{2}{3}H$ from the top of the wall, the force of pressure was $F = wH^2/2$, approximately equal to the weight of the wedge, and the overturning moment which determines the thickness of the retaining wall was $M = wH^3/6$. Belidor, then, proposed to reduce the lateral pressure of the retained earth to half on account of the internal friction present in the soil which he had neglected in his analysis. Then the overturning moment of the soil pressure reduced to $M = wH^3/12$, and the resultant soil pressure to $F = wH^2/4$. In this way, Belidor was able to determine theoretically the proportions of retaining walls which were, generally, in agreement with the prevailing practice.

The first experiments on the earth pressure exerted against retaining walls were carried out in 1745 by a French military engineer Col. Gadroy. His small-scale experiments were carried out in a wood box filled with dry sand which was 10 inches (0.25 m) long and 3 inches (7.6 cm) square in cross-section. The box was supplied with a movable shutter representing the retaining wall at one of its ends. Gadroy observed that when the shutter of his box failed, it tended to overturn about its base which gave Gadroy the erroneous notion that the pressure of the earth was greatest at the top of the wall, a contradiction to Belidor's theory of earth pressure. Gadroy also noted that whenever the end-wall failed a crack parallel to the end wall appeared on the surface of the sand at a distance from the end-wall which measured about half the height of the end-wall. This crack always ran down to the bottom of the end-wall along a plane. This represented the first experimental verification of the existence of the so-called 'plane of rupture' which did not necessarily coincide with the plane of natural slope as had been assumed by Couplet in the failure of earth embankments.

About 40 years later, in 1784/1785, Gauthey undertook experimental testing of soil pressure in small scale. His experimental apparatus consisted of a box filled with sand which was 30 inches (0.76 m) long, 30 inches (0.76 m) high and 12 inches (0.30 m) wide, and closed at its end by a door pivoted in the bottom. At both sides of the box a cord was attached to the door at ⅓ of its height, and passed over a pulley to suspend a weight. Gauthey regarded the amount of sand which flowed out of the box when the door failed by pivoting about its base as the weight of the wedge of the sliding earth. However, since sand, which has a low angle of repose, leaked past the door, and thus, gave unreliable results, Gauthey repeated his experiments with shot in a smaller box. In this experiment the end-door was replaced by a stack of horizontal slabs which allowed Gauthey to determine the thrust exerted by shot on each horizontal slab and, thereby, prove that the pressure exerted by shot on the end-wall was indeed proportional to the depth below the surface like in a liquid, and that the resultant of the earth pressure was located at ⅔ depth from the top surface as Belidor had inferred from his theory, and not at ⅓ depth as Col. Gadroy had assumed.

Next advance in the theory of soil mechanics and retaining walls was made by Charles Augustin Coulomb (1736-1806), a French military engineer who became a famous physicist. Coulomb originated the theory of the critical equilibrium state in soil mechanics by formulating its basic postulates, and he applied it to the design of retaining walls. He gave the first satisfactory wedge theory of earth pressure in soil mechanics by means of an extremum principle in 1773. He introduced cohesion and friction at the plane of rupture, and found the horizontal reaction exerted by the wedge of soil on the retaining wall to be:

$$F = (wH^2/2) \left[(1 - \mu \tan \alpha)/(1 + \mu \cot \alpha) \right],$$

where $\mu = \tan \psi$ denotes the coefficient of internal friction in soil. He found the maximum F for an angle of repose α by using its necessary condition, the stationary value, $dF/d\alpha = 0$, which resulted in the basic relation between α and μ:

$$(1 - \mu \tan \alpha)/(1 + \mu \tan \alpha) = \tan^2 \alpha,$$

and gave the solution for α:

$$\tan \alpha = -\mu + (1 + \mu^2)^{1/2}.$$

He obtained the same solution when cohesion was neglected. For cohesive soil Coulomb gave its shear resistance:

$$\tau = c + \sigma \tan \psi ,$$

where τ is the shear stress, c the internal cohesive stress, σ the compressive stress in the soil, and ψ the 'angle of internal friction' of the soil. He also considered the case of surcharge, when an extra load P acted on the top of the surface of the soil wedge. Moreover, Coulomb studied the case when the failure surface was not a plane, but a curved surface, and he examined the shape of such a failure surface.

De Prony simplified Coulomb's earth pressure theory by simplifying the relation between α and μ:

$$\cot (\alpha + \psi) = \tan \alpha ,$$

which resulted in the relation,

$$\alpha = \tfrac{1}{2} (90° - \psi) ,$$

and made the plane of rupture bisect the angle between the vertical wall and the 'angle of repose' of the soil α.

The most drastic simplification in the Coulomb's theory was introduced by the German engineer, Reinhard Woltmann in 1794. Besides neglecting cohesion, $c = 0$, and setting $\mu = \tan \psi$, Woltmann equated the internal friction of all soils with their angle of repose: $\psi = \alpha$. By means of this simplification, Woltmann obtained the resultant earth pressure:

$$F = (wH^2/2) [(1 - \sin \alpha)/(1 + \sin \alpha)] ,$$

which is often mistakenly called the 'Rankine formula' after the Scottish engineer and physicist, William John Macquorn Rankine (1820-1872), who derived a similar result in 1856 by a less satisfactory method. Woltmann's hypothesis considerably simplified the earth pressure and slope stability analysis because it obviated any testing of soils apart from determining the angle of repose of the soil, which was relatively easy. However, already Coulomb had realised by 1773 that clay can stand vertically, at least for a short time, to a certain critical height H for which he gave the correct algebraic expression:

$$H = (c/w) \{4 \cos \psi / (1 - \sin \psi)\} ,$$

where w is the 'density of clay', and ψ is the 'internal angle of friction'.

In 1806-1807, a French military engineer K. Mayniel, who was in charge of the construction of fortifications at Juliers, carried out larger scale experiments on earth pressure against retaining walls. He built a box about 10 feet (3 m) long and 5 feet (1.5 m) square with a door hinged at the bottom closing the end of the big box. The box was filled with earth. The door was supported by a horizontal wooden strut which at its other end pushed against a tank filled with water and loaded with additional weights. This tank rested on a wooden platform and was fitted with a tap. The testing procedure was to drain water through the tap until the thrust exerted by the earth against the door, and transmitted to the tank by means of the strut, made the tank slide. The force in the strut was then indirectly measured by experimentally finding the force exerted by a rope passing over a pulley to a loading platform which made the tank slide. The point of application of the force was varied by changing the location of the strut on the hinged door. Altogether a series of 33 experiments were performed, but only when the strut was applied to ⅓ of the height of the door were consistent results obtained. For all other cases the unknown reaction exerted by the hinges made definite conclusions impossible. Mayniel was able to verify from his experiments that Coulomb's theoretical, and Gauthey's experimental conclusions in earth pressure theory were correct. He also concluded that only Coulomb's theory predicted his experimental results correctly among all the existing theories of soil pressure, and that the plane of rupture in the soil was virtually unaffected by any surcharge.

Coulomb had knowingly neglected the friction which exists between the soil and the retaining wall which made the resultant soil pressure to act normal to the wall, because he was reluctant to complicate his solution particularly because this neglect of friction gave a larger margin of safety to the design. This deficiency in Coulomb's theory was first removed by Mayniel in his report entitled < Earth Pressure > (*Poussée des terre*), published in 1808, but his solution was complicated. The German engineer, Reinhard Woltmann, as stated above, went even further by neglecting cohesion, and equating internal friction of all soils with their angle of repose which made this theory even less applicable to clayey soils.

Another French military engineer, Français, published a memoir in 1820 in which he recognised the importance of cohesion in clayey soils, and generalised the Coulomb formula for a clayey bank having a slope β:

$$H = (c/w) \{4 \cos \psi \sin \beta / [1 - \cos (\beta - \psi)]\},$$

which for $\beta = 90°$ becomes the Coulomb formula. This solution of Français gives the correct relation

$$(c/wH) = f(\psi, \beta),$$

which implied that for cohesionless soil for which $c = 0$, the condition of limiting equilibrium is obtained when $\beta = \psi$, that is, when the Woltmann hypothesis is satisfied, and that a cohesive soil for which $c = 0$, there exists no unique 'angle of repose' α. Français conclusions were correct, but at the time no experimental means existed for the determination of c and α. Moreover, as already Coulomb had noticed, clays tended to soften in cutting or by seepage in earthen dams, which meant reduction in the value of c.

All these observations were ignored by Jean Victor Poncelet in his monograph < Memoir on the Stability of Retaining Walls and their Foundations > (*Memoire sur la stabilité des revêtements et de leur foundations*) published in 1840. In it Poncelet gave an ingenious graphical construction for finding the resultant earth pressure, and its point of application when the friction between the wall and the earth is taken into account, but this solution was adequate only for sand fills, for it gave quite misleading results for clayey soils. Rankine's solution helped to propagate this incorrect theory, since he believed that cohesion was important only in earthen banks of low height and, therefore, he dismissed the effect of cohesion from all his works on earth pressure.

The French engineer, Alexandre Collin (1808-1890) who had recognised the serious shortcomings in the soil mechanics of Poncelet in its application to clayey soils, carried out a series of field studies on the behaviour of unstable clayey soils and surveyed about 15 slips in the slopes of clay embankments, railway cuttings, and earthen dams. He discovered, perhaps by excavation for the counterforts, that the failure surface in slipped earth embankments was approximately cycloidal, and not a plane as assumed in Poncelet's theory of earth pressure. Collin developed the Coulomb theory of cohesive soils in its application to clays, and devised methods to evaluate the cohesion and internal friction through field observations and laboratory tests. He invented a shear box apparatus for the testing of shear resistance of clay soil samples in their undisturbed state. Collin also developed an approximate method for the stability analysis of soils in which account was taken of the curved failure surface, and published his results in a remarkable memoir entitled, < Experimental Research on Spontaneous Landslips in Clay Soil, Accompanied by Consideration on Some Principles of Soil Mechanics > (*Recherches expérimentales sur les glissements spontanés des terrains argileux, accompagneés de considération sur quelques principes de la mécanique terrestre*), in 1846.

In this memoir Collin not only gave the scientific foundation for modern soil mechanics and the experimental methods necessary for its practical implementation, but also showed the fallacy of the angle of repose theory in its application to clayey soils. Unfortunately, this significant contribution to the scientific study of soil mechanics was completely ignored by the profession, which continued to use the incorrect earth pressure theories of Poncelet and Rankine. The modern soil mechanics as sketched by Collins reappeared in the work of the Swedish engineer Wolmar Knut Axel Fellenius (1876-1957) in 1922, and in the work of the Austrian engineer Karl Terzaghi (1883-1963) in 1925, more than 70 years later.

The French engineer, and member of the *Corps*, Henri Philibert Gaspard Darcy (1803-1858), municipal engineer of his native city Dijon, built several aqueducts and water supply systems. He made several important contributions to hydraulic engineering, but his great fame rests on his celebrated equation for the flow through such porous media as permeable soils :

$$v_r = k_r (\partial h_r / \partial r),$$

where $r = x, y, z, h_r = p/\gamma$ is the hydraulic head, k_r is the coefficient of permeability, γ is the specific weight of water, and p is the pressure. In Darcy's equation the loss in the hydraulic head is postulated to be proportional to the speed of flow v rather than to the square root of the speed, $\sqrt{v}$, as was commonly assumed before Darcy published his pathbreaking research in 1856. Darcy, moreover, made further improvements upon the 'Pitot tube' to increase the accuracy of his hydraulic measurements.

French Aqueducts

Aside from the famous aqueduct of Montepellier as part of the 11 miles (17.7 km) long canal designed by Pitot and constructed between 1752 and 1772, there were a few other aqueducts built by French engineers which are worthy of mention.

The water supplied to the city of Marseille was brought over 57 miles (91.7 km) from the Durance river, a tributary to the Rhône river. The aqueduct was largely an open channel, called the Marseille Canal, which passed through 40 tunnels, and near *Aix* was carried by a huge aqueduct bridge that dwarfs any similar Roman structure.

The aqueduct of Rocquefavour was built from 1839 to 1847 under the supervision of the engineer Jean François Mayor de Montricher (1810-1858). The aqueduct bridge which crossed the valley of the Arc river has 3 tiers of arches, and still stands as the highest stone bridge in the world that has the length of 1,300 feet (396.24 m) and the height of 270 feet (82.30 m). Montricher also directed the modern draining of Lake Fucino in Italy.

French Highway Engineering

Extensive highway construction in France started under Trudaine, who is the creator of the modern French highway system since his founding of the *bureau* in 1744, which ultimately became the *Corps*. In 1775, an inspector general of the *Corps*, Pierre Marie Jerôme Tresaguet (1716-1794), described a practical method of 'hard-surface' road construction in a memoir to the *Corps*, a method of construction which used flat stones set as voussoirs in a shallow arch across the width of the road as its foundation on which rested 2 layers of stones, the top layer providing an impervious surface for traffic. Tresaguet's improved construction method produced roads that were much shallower (9 to 10 inches: 23 to 25 cm) yet had better drainage and lower camber, which made them cheaper to build and to maintain. He claimed that his roads properly built and maintained should last 10 years with minimal maintenance. Tresaguet's method, in a slightly modified form in which foundation stones were laid horizontally and not in an arch, was used later by the famous Scottish engineer Thomas Telford (1757-1834), and it is generally known as Telford's roadbuilding method.

Tresaguet's method of road construction came too late for France, for soon after, the French Revolution broke out and put a stop to road construction but by 1788, about 12,000 miles (48279 km) of constructed roads had already been built in France. Nevertheless, the highway transport was still too expensive because water transport through the canals was several times less in cost than highway transport. However, by that time France already had the best road system in Europe. Tresaguet was dismissed from his position in the *Corps* during the French Revolution, and he died in poverty and obscurity.

French Harbour Construction

Since the Late Middle Ages, the Dutch engineers were the leading hydraulic engineers in Europe, but they directed their major efforts towards building sluices, barriers, canals and pumping mills as a defence against the sea encroaching upon their low-lying land. The French rather than the Dutch became the first modern harbour builders in Europe after the Romans. The French harbours at Boulogne, Calais, Dieppe and Le Havre at the mouth of the Seine river had basins supplied by locks to store water for the scouring of the entrance channels. At Le Havre extensive fortifications were constructed beginning in 1509 as a defence against the English fleet of Henry VIII. The Le Havre harbour had to shelter 135 French warships which were meant to defend Boulogne against the possible attack of the English fleet. The best natural harbour in Europe, which served as a French naval base, was built at Brest under the tenure of Cardinal Richelieu. Its naturally sheltered deep waters only required the construction of tidal locks to make the harbour fully effective.

Another well-known port and its fortifications was planned at Dunkirk in 1681 by the famous French military engineer Vauban, who employed floating reservoirs for the scouring of the access channels. The well-known French military engineer, Belidor, improved the Dunkirk harbour by designing new wet-lock facilities for it. Belidor was also one of the earliest authors to treat harbour engineering in his popular treatise in 4 volumes < Hydraulic Architecture > (*Architecture hydraulique*) in 1737. Dunkirk, after its destruction in the Franco-British wars, lost its importance as a strategic French harbour.

The strategic location of a harbour often justified great expense in harbour construction as was the case with Cherbourg in France and with Plymouth in England. The planning of the Cherbourg harbour commenced in 1692 when French engineers were called upon to report on this project. The 2 harbours lying on the opposite sides of the English Channel assumed such great strategic importance during the later Franco-British wars, but particularly during the Napoleonic wars, that great efforts were made at enormous expense to bring the construction of these harbours to an early conclusion, particularly at Cherbourg. The Cherbourg harbour not only had to be protected against savage storms coming from the North Sea, but also against British naval attacks. Since the coastal indentation at Cherbourg was relatively shallow the greatest engineering problem facing the engineers was to construct a large breakwater to protect the harbour installations and the large French fleet from both the storms and the enemy attacks. This huge breakwater of 2½-mile (4.0 km) length had to be built more than a mile (1.6 km) off the shore in waters that were from 36 to 42 feet (11.00 to 12.80 m) deep. On account of the savage storms that swept this coast, the breakwater had to be rugged.

In 1781 the French engineer Louis Alexandre de Cessart (1719-1806), who had earned fame in harbour engineering by having rebuilt the locks and made important improvements in the port at Le Havre, reconstructed the quay at the port of Rouen and other ports, was given the task to report on the plan of the Cherbourg harbour and devise a method of constructing the gigantic breakwater.

De Cessart rejected the earlier proposal to construct a breakwater of loose rubble as too weak to resist the storms, and, instead, suggested the construction of the breakwater from prefabricated units to be floated into position and sunk to the sea-bed, a method of construction which had been already used by Roman engineers in the breakwater of the *Portus Claudius*, the outport of Rome at Ostia. He proposed to build 90 huge timber frameworks in the shape of huge truncated cones measuring about 150 feet (45.72 m) in diameter at the base, 60 feet (18.29 m) in diameter at the top, and 70 feet (21.34 m) in height. These conical frameworks were made of squared oak and beech baulks, 13 inches (0.33 m) square at the bottom reduced to 8 inches (0.20 m) at the top, and was provided with horizontal walings at certain intervals, and two platforms at the top for the ballasting operation. De Cessart had planned to sink 90 such conical structures adjacent to one another to make up a chain of stone-filled conical islands constituting the huge breakwater. Owing to economic and military exigencies only 21 such cone frameworks were built from which only 18 were used, despite de Cessart's warning, when actual construction began in 1784. The spacing between the cones was increased varying from 159 feet (48.46 m) to 1,845 feet (562.36 m) over a total breakwater length of 12,470 feet (3.8 km). The next stage of the construction began in 1791 when the spaces between these timber-stone islands were to be filled out with rubble tipped from barges, a work which was interrupted by the Revolution. Work on the breakwater was reassumed in 1803 when huge stones of 20 to 50 cubic feet (0.57 to 1.42 cubic meters) were sunk in the middle of the breakwater to create a barrier on which gun platforms were built. A great storm in 1808 swept over this barrier inflicting severe damage to the breakwater. In 1813, and 1817 work on the breakwater was again interrupted when storms flattened the slopes of its 2 rubble arms. Since the conical timber frameworks had broken up in the rough weather and the damage done by marine borers, in 1830 a new, fourth plan of the breakwater was proposed in which this structure consisted of a solid vertical masonry wall with 36 feet (10.97 m) base. This wall was supported on a concrete foundation poured in boxes on the rubble base, and built 6 feet (1.83 m) above the high water level. Finally, this huge breakwater was completed in 1858, after it had taken more than ¾ of a century to bring this imposing French maritime structure to its completion.

Rise of Structural Mechanics

In 1695, the French engineer and mathematician, Phillipe de la Hire [alias Lahire](1640-1718) in his treatise < Treatise of Mechanics > (*Traité de mecanique*), and his pupil Antoine Parent in the latter's impressive memoir of 1713, were the first scientific engineers after Huygens to study the equilibrium of an arch as a mathematical problem of statics, by assuming the arch to consist of frictionless *voussoirs*, which led to certain theoretical anomalies. It represented the rise of structural mechanics as a rational science.

Following Hubert Gautier's model analysis of voussoir arches in 1717, a definite improvement in the conception of the stability of arches was proposed in 1729 by Pierre Couplet de

Tartreaux [alias Tartereaux], who after examining the static stability of voussoir arches without the neglect of friction between the contiguous voussoirs came to the conclusion that the voussoir arch fails by relative rotation of certain adjacent voussoirs which opens up the joints of the arch at certain critical joints rather than by relative sliding of adjacent voussoirs. Couplet, himself, made an unsuccessful attempt to determine the minimum thickness of a voussoir arch so that the arch would not fail through relative joint rotations in certain critical joints at postulated locations in the arch. Couplet's idea for the relative rotational mode of the arch failure rather than sliding mode of arch failure motivated another French engineer, Augustine Danisy (1691-1777), to check the validity of these modes of arch failure in model experiments on voussoir arches. In his model experiments, Danisy also considered failure of arches under applied loads which had been ignored by earlier experimenters. Each of Danisy's 13 model arches, which were either circular or elliptical in shape, had 16 plaster voussoirs. Danisy found from his model experiments that Couplet had been correct: all his model arches failed through relative rotation of joints which opened up certain critical joints and converted the arch into a failure mechanism before the ultimate collapse of the arch occurred. Danisy reported his experiments to the Academy of Montpellier, but his report was published only in 1769 by the French engineer and mathematician Amadée François Frézier (1982-1773), in his <Treatise of Stereometry> (*Traité de stéréométrie*). Frézier, being still dissatisfied with Couplet's theory of arches, consulted the famous Swiss mathematician Johann Bernoulli (1667-1748) who advised Frézier to use the principle of virtual work for the study of the failing arch, a method which was later used by Perronet and Coulomb to establish the first sound general theory of arches.

In 1773, a French military engineer, who later became a famous physicist, Charles Augustin Coulomb published his researches which he had done on the French island of Martinique in the Caribbean Sea whilst discharging his duties as military engineer. His memoir dealt with the earth pressure theory, the limiting strength of voussoir arches, and the tests and statical analysis of beams, a work in which all topics were considered in relation to practical design considerations. Coulomb's beam theory was complete and included Parent's theory of shear stresses, and the balance of the longitudinal forces. Coulomb realised that the shear stresses depend on the orientation of the plane at a point, an original idea, and he proved that shear stresses become maximum on 45 degree planes. Coulomb employed simple extremal methods in this memoir and his entire work on structural mechanics was based on scientific theory. Like Belidor, Coulomb was satisfied to carry out a few rough-and-ready experiments in support of his theoretical results. Coulomb, with this research, became the founder of 'structural mechanics as a scientific discipline' which is regulated by structural engineering practice.

As stated above, the prominent French engineer of the period, Charles Augustine Coulomb, took up the theoretical problem of the failure of arches, in which he examined both the relative sliding and the relative rotation between the 4 segments into which the collapsing voussoir arch breaks as was known from all previous experiments with arches. Coulomb used an extremal principle in order to find the maximum and the minimum horizontal thrust for both cases of failure, and he demonstrated that the mode of failure of arches which determines the critical horizontal thrust for stable equilibrium is the relative joint rotation. He did not give any definite rules for the design of arches, but only information for the upper and lower limits of the thrust necessary for the stability of the arch. Despite the scientific merit of Coulomb's theory, his omission of any definite information on the proportioning of voussoir arches made French structural engineers to prefer the use of Perronet's tabulated empirical information on the proportioning of voussoir arches in their design practice. Only when in the middle of the 19th century graphical methods were developed for the evaluation of Coulomb's limiting arch thrust did his method of arch analysis become popular in structural engineering practice.

The first engineer to carry out tests in large scale on an actual arch bridge was Jean Rodolphe Perronet who was in charge of all public works in France. When Perronet directed the construction of the bridge of Sainte-Edmé at Nogent-sur-Seine, which consisted of an elliptical arch with a span of 96 feet (29.28 m) and a rise of 29½ feet (8.99 m), he ordered the spandrel walls near the support of this arch bridge built up before the arch itself was closed at the center. This spandrel load near the support deformed the centering of the arch so much that the arch ring and its centering separated after which the arch was immediately keyed, the spandrel walls cut free from the abutment, and the arch decentered. The resulting settlement and the deformation of the initially level lines on the sides of the bridge were carefully measured. From this information it was easy to trace the line of thrust and find the locations where the critical joint rotations would have taken place had the arch failed. It represented the first full-scale demonstration of the failure mechanism of the voussoir arches which had been previously observed only in small-scale model experiments. Perronet reported on this full scale experiment in the memoirs of the French Academy of Sciences in 1773. One of Perronet's assistant engineers, Lecreulx carried out a similar test in 1784 on a bridge at Frouart which also confirmed the existence of joints of rotational failure for voussoir arches.

In 1799 as described earlier, Gauthey, who was much chastened after his adverse experience with the dome of the **Panthéon Français**, performed model experiments on voussoir arches himself and his results corroborated the validity of the idea of Couplet and Coulomb that the relative rotation between voussoirs as a failure mechanism represents the actual collapse mechanism of such arches. Therefore, in his later work Gauthey used only Coulomb's arch analysis.

Soon after Gauthey's experiment, in about 1800, a French engineer, Boistard, performed large scale experiments on semicircular, elliptical and segmental arches consisting of polished brick voussoirs 4 inches (10.2 cm) thick and 8 inches (20.3 cm) wide, and having a span of 8 feet (2.44 m). Boistard carried out 22 experiments on arches, each repeated 3 times, in which he produced failure in the arch by lowering the centering, by increasing the loading, and by reducing the thickness of the abutment. Boistard learnt from his experiments that the voussoirs of an arch never slide relative to one another in the failure of the arch, and, therefore, he advised discarding La

Hire's theory of arches and replacing it by the empirical rules based on Perronet's tabulated information on arch proportions, or on Boistard's own experiments. He demonstrated in his experiments that the voussoir arch always fails through relative joint rotation by breaking into 4 segments with the intrados opening at the springing and at the crown, and the extrados at the haunches of the arch. Such was the situation in structural mechanics before the important contributions to structural theory were made by Gauthey's nephew, Henri Louis Navier.

In 1803, a French engineer, Louis Poinsot (1777-1859), developed the general theory of couples, which was an important basic concept hitherto missing in the classical mechanics.

Jean Baptiste Joseph Fourier (1768-1830), a prominent French mathematician and physicist, was mostly self-educated in higher mathematics, and had initially studied to be a priest, but later the Bishop of Auxerre had helped him to enter the military academy of *Auxerre*. Fourier was so successful in his mathematical studies that already in 1789 he was appointed teacher of mathematics at the military academy in *Auxerre*, and in 1796 he was appointed professor at the military school, and soon after that at the *École Normale* and at the *École Polytechnique*. He accompanied Napoleon Bonaparte to Egypt, and became the Secretary of the *Institut d'Égypte*. After returning to France he occupied the post of prefect for department of Isère and then of Rhône. After 1817 he became the secretary of the French Academy of Sciences. Fourier was a mathematical physicist who published in 1822 his celebrated treatise, < Analytic Theory of Heat > (*Théorie analytique de la chaleur*), a landmark in mathematical physics which was important to the solution of the boundary-value problems, that is, fitting solutions of differential equations to prescribed initial boundary conditions which was the central problem of mathematical physics. Fourier used infinite periodic trigonometric series in his solutions by assuming that any function can be represented in terms of it. This approach was not new for Daniel Bernoulli had used it in vibration problems in 1753, and Euler had used it in 1777 in a mathematical problem consisting of expressing a prescribed trigonometric series in terms of a trigonometric series with unprescribed coefficients in the solution of which Euler established the so-called 'Fourier integral formula' to evaluate the unprescribed coefficients of the trigonometric series. At the time, a controversy had raged between d'Alembert, Euler and Daniel Bernoulli concerning the ability of trigonometric series to represent any arbitrary function. Although Fourier could not give a rigourous mathematical justification for his method, which was supplied in 1828 subject to certain restrictions by the German mathematician Johann Peter Gustav Lejeune Dirichlet (1805-1859), his method influenced the conception of mathematics, created interest among mathematicians in mathematical rigour, and led to a much more general definition of functions. Fourier also showed in his work a number of piecewise continuous functions the graphs of which obviously enclosed 'areas'. Fourier then interpreted such an 'area' to be a definite integral of a function, now considered as a sum in which the function represented by trigonometric series was not everywhere differentiable. These examples of Fourier revealed the shortcomings of the current 'antiderivative definition of definite integrals' and suggested the versatility of Leibniz's definition of integrals as sums. Fourier believed that the tendency to explain everything in terms of forces and dynamics, such as was the case with Laplace, Ampére and Poisson, was inadequate and doomed to failure, because dynamics in his opinion had its limitations in some fields of physics such as heat. Therefore, he developed the analytical theory of heat. Fourier's theorem had a wide field of application, and could be used in any wave phenomenon. The mathematics resting on the Fourier theorem is now called 'harmonic analysis'. Fourier, moreover, was the first to point out the requirement that scientific equations must be 'homogeneous in units' which led to 'dimensional analysis' as a new discipline of mathematical physics.

Sophie Germain (1776-1831) was a self-educated French mathematician and philosopher who was able to overcome the prejudice of the contemporary scientific world against women through a remarkable personal triumph. She attended the lectures of Lagrange at the *École Polytechnique* under the name of Auguste-Antoine LeBlanc(1775-1797), a young man and first year student of the Polytechnical School whom Miss Germain helped in mathematics, and who had died at 22 years of age before he could attend the *École des Ponts et Chausées*, because women were not admitted to higher studies. She corresponded for many years with the great German mathematician Carl Friedrich Gauss (1777-1855), a former infant prodigy, about the latter's number theory before Gauss discovered to his great amazement that his correspondent was a young lady.

In 1808 the German experimental physicist, Ernst Florens Friedrich Chladni (1756-1827), gave a demonstration in the French Academy of Sciences on the vibration of elastic plates by covering the vibrating plate with sand and demonstrating the formation of nodal lines on the plate. Napoleon, who attended this demonstration was fascinated by this experiment and suggested to the French Academy to set a prize competition for the mathematical theory of the vibration of plates which could theoretically predict the nodal lines. The French Academy set the competition in 1809, in which the winning essay was to earn the prize of a gold medal worth 3,000 French francs. By the closing date of the competition, in late 1811, only Sophie Germain had submitted a solution in which she had derived the differential equation for the vibrating plate by minimising by means of Lagrange's variational calculus the middle surface integral representing the strain energy stored in the bent plate, an idea she had obtained from Euler's work on the elastic curve:

$$A \int [(1/\varrho_1) + (1/\varrho_2)]^2 \, da = minimum,$$

where ϱ_1 and ϱ_2 denote the principal radii of curvature of the bent plate, and the differential area $da = dx\, dy$ of the plate surface. Unfortunately, she had made a formal mistake in calculating a variation, and, therefore, she was not awarded the prize. Lagrange, who was one of the judges of the prize committee, corrected her mistake and obtained the correct differential equation of transverse vibration of a thin elastic plate :

$$B \nabla^4 w + (\partial^2 w / \partial t^2) = 0,$$

where $w(x, y, t)$ denotes the transverse displacement field of the vibrating plate, B represents an appropriate constant, and the potential and the bipotential differential operators are:

$$\nabla^2(\cdot) \equiv \frac{\partial^2(\cdot)}{\partial x^2} + \frac{\partial^2(\cdot)}{\partial y^2} \quad \text{and} \quad \nabla^4(\cdot) \equiv \nabla^2\nabla^2(\cdot) \equiv \frac{\partial^4(\cdot)}{\partial x^4} + 2\frac{\partial^4(\cdot)}{\partial x^2 \partial y^2} + \frac{\partial^4(\cdot)}{\partial y^4}.$$

The French Academy set the late 1813 as the new closing date for this competition, and again only Sophie Germain submitted a solution. This time, she had been able to derive the correct differential equation for the plate, but now the judges required a physical justification for her strain energy integral, and again her essay was not accepted.

The French Academy set once again 1816 as a new closing date for this competition, and again only Sophie Germain submitted a solution. This time she was awarded the prize, although the judges were not entirely satisfied with her solution and the derived boundary conditions of the plate. The French mathematician, Denis Siméon Poisson (1781-1840), who had been one of the judges in the second competition, criticised Germain's hypothesis and demonstrated in his memoir published in 1814, that Sophie Germain's plate equation can also be obtained by minimising the integral:

$$A \int \{ [(1/\varrho_1) + (1/\varrho_2)]^2 + m[(1/\varrho_1)^2 + (1/\varrho_2)^2] \} da = \text{minimum}.$$

This integral can be expressed in the form which is in current use:

$$\int \{ C_1 [(1/\varrho_1)^2 + (1/\varrho_2)^2] + C_2 (1/\varrho_1)(1/\varrho_2) \} da = \text{minimum}.$$

Of course, with proper choices of A and m Poisson's integral can represent the strain energy stored in the bent elastic plate, but Poisson's solution was conceptually not entirely correct either. Another French mathematician and engineer, Henri Louis Navier (1785-1836), who defended Sophie Germain's hypothesis against Poisson's criticism, gave a more satisfactory theory of plates in 1820. In Navier's molecular theory of elastic solids the molecules of the plate were distributed throughout the thickness of the plate which gave an improved differential equation:

$$D\nabla^4 w = p,$$

where p is the transverse load intensity on the plate.

The mean of the 2 curvatures at a point of a surface, $(½)[(1/\varrho_1) + (1/\varrho_2)]$, is the 'Sophie Germain or mean curvature' and the product of the 2 curvatures, $(1/\varrho_1)(1/\varrho_2)$, the 'Gaussian or total curvature' of the surface.

Sophie Germain, as well as Poisson and Navier, never did obtain the entirely correct general boundary conditions for vibrating plates, which were first correctly established only in 1850 by the German physicist and mathematician Gustav Robert Kirchhoff (1824-1887) with a variational method. It was a very courageous venture on the part of Sophie Germain to undertake this difficult theoretical problem the mathematical solution of which had not been previously attempted by any of the great French mathematicians of her time. Subsequently she took a great interest in number theory and in <Arithmetical Enquiries > (***Disquisitiones arithmeticae***), Leipzig, 1801, by the great German mathematician Carl Friedrich Gauss, with whom she had a long series of correspondence about higher arithmetic.

The actual founder of modern structural mechanics and mechanics of materials as an engineering science of mechanics is, however, Louis Marie Henri Navier, a graduate of the *École Polytechnique* in 1804, who received his early education and practical engineering knowledge from his uncle, the famous French engineer Gauthey. He entered the *École des Ponts et Chaussées* in 1804 and graduated in 1808 with an excellent theoretical background tempered by considerable practical engineering knowledge. Navier was a professor of structural mechanics at the *École des Ponts et Chaussées* until 1830, and professor of calculus and mechanics at the *École Polytechnique* from 1830 until his death in 1836.

Navier rejected the ultimate strength approach to the study of structures, and considered instead the elastic behavior of structures. In 1821, Navier published his famous work on elastic theory of solids and fluids by considering the elastic continuum to be a static, or moving system of mass-points. This molecular type of conceptual model as a basis for explaining all terrestrial, as opposed to celestial, phenomena was promoted by the French physicist and mathematician Pierre Simon Laplace (1749-1827). In 1809 Laplace had made a claim that the behaviour of light, the flow of heat, electricity and magnetism, fluid statics, capillary action, chemical affinities, the physics of gases, as well as the deformation of elastic bodies will all admit molecular analysis. Laplace himself had established his theory of capillarity in 1819 by means of a molecular model in which he imposed the requirement that molecular forces diminish rapidly with increasing distance between molecules. During the first decade of the 19th century the molecular method of modeling appeared to possess the theoretical power to describe all phenomena. Owing to Laplace's influential position and great authority within the French scientific community at the time his scientific opinions influenced younger French scientists, such as Poisson and Navier, who came to believe in Laplace's claim that molecular modeling of various physical phenomena can explain these phenomena in mathematical terms. Later when Laplace's scientific importance began to wane, so did the molecular mentality of French scientists. Although Navier's mathematical model of the elastic continuum was questionable and his results were subject to unnecessary restrictions, he did succeed in establishing the basic equations for solids and fluids known by his name, and it triggered the subsequent brilliant work of Cauchy in continuum mechanics. Navier was the first to solve the problem of bending of elastic plates, now known as the 'Navier solution', by using a double trigonometric series for w, which is valid for simply supported rectangular plates. He, moreover, established the correct differential equation for lateral buckling of thin elastic plates subject to compressive boundary forces N applied in the plane of the plate:

$$D\nabla^4 w + N \nabla^2 w = 0.$$

Therefore, it was Navier who gave the first satisfactory solutions to the problem of bending of plates.

In 1826, Navier published his celebrated lectures on structural mechanics, < Resumé of Lectures given at the *École des Ponts et Chaussées* on the Application of Mechanics to the

Establishment of Construction and Machines > (*Résumé des Leçons données à l'École des Ponts et Chaussées, sur l'application de la méchanique a l'établissement des constructions et des machines*) which is an epoch-making work, for modern engineering still essentially follows the theoretical treatment of structural mechanics and elastic deformation given in these lectures in which Navier had rejected the attempts of the 18th century engineers to find ultimate strength solutions for such problems after a few futile attempts of his own. Everything in Navier's book looks quite familiar to a modern engineer. In his lectures Navier gave an important analysis of the stresses in arches which he assumed to be linearly distributed over the cross-section of the arch and to be zero at the intrados of the crown and at the extrados of the haunch of the arch, at the cross-sections where Coulomb's theory predicted failure by relative joint rotation. Navier found by this method, which neglected the deformation of the arch, a somewhat larger maximum horizontal thrust H than that provided by Coulomb's theory, and on this basis he calculated the width of the abutment of the arch. His method of arch analysis was generally approved by the French engineering profession.

In his lectures Navier solved the deflexion problems of simply supported and statically indeterminate elastic beams, as well as statically indeterminate curved elastic bars, and thin shells. Surprisingly, he never applied his theory of elastic curved bars to stone arch analysis, which was first suggested in 1852 by Jean Victor Poncelet in his memoir on the critical history of arch analysis long after the death of Navier. Navier did not use bending moment and shear force diagrams and consequently in some cases the maximum bending moment was not correctly found. Navier also neglected the study of shear stress distribution over the cross-section of the beam, which was first given in 1856 by a young Russian railroad engineer Dimitri Zhuravskii (1821-1891), perhaps because Coulomb had emphasised that shear stresses become important only in short beams. For this reason, Navier only considered shear stresses in the 2nd edition of his book in 1833, when he analysed bending of short beams in which he incorrectly used an average shear stress distribution over the cross-section of the short cantilever beam. Neither did Navier consider the effect of shear stresses on the deflexion of the short elastic cantilever beam, which was first given by Poncelet in his unpublished lectures at the Sorbonne in Paris sometime before 1848.

In 1821 Navier presented his famous memoir on the equilibrium and vibration of elastic solids in which he followed Laplace's idea by treating an elastic solid as an isotropic molecular medium obeying Hooke's law of linear elasticity over the distances between molecules in which any change in the molecular distance brings forth a proportional change in the central forces interacting between molecules with the proportionality rapidly diminishing over the increasing molecular distance. Navier's theoretical method consisted of applying Lagrange's analytical mechanics to molecular medium as a discrete model of a solid, and it served as an inspiration for similar investigations of solids to Poisson and Cauchy.

For the purpose of studying the engineering of a suspension bridge in Great Britain, Navier was sent to Britain by the French government, where he inspected the suspension bridges of Capt.

Samuel Brown (1776-1852) and the famous suspension bridge of Thomas Telford (1757-1834) over the Menai Strait in Wales. Upon his return from Britain, Navier published in 1823 his classic treatise on suspension bridges, < Report and Memoir on the Suspension Bridges > (*Rapport et Memoir sur les Ponts Suspendus*) in 1823, containing a historical review of such bridges, a description of most important suspension bridges, and a development of the first structural theory of suspension bridges which included an investigation of vibration of such bridges under moving loads. Unfortunately, the suspension bridge with a span of 508 ½ feet (155 m) over the Seine river at Paris, called the *Pont des Invalides*, which Navier himself had designed and described in this treatise proved inadequate as a structure. The bridge showed early signs of incipient failure of the suspension cables in 1826, before it was opened to the public, owing to an inadequate anchorage of the cables, an engineering disaster which grieved him for the rest of his life, but which fortunately did not destroy his considerable professional reputation. In the 2nd edition of this treatise Navier discussed all the particular features of his suspension bridge which ultimately justified the dismantling of this bridge before its completion. Marc Isambard Brunel (1769-1849) had anticipated the collapse of this suspension bridge after a glance at the drawings of Navier's first suspension bridge, *Pont des Invalides* : "You would not venture, I think, on that bridge unless you would wish to take a dive." Marc I. Brunel himself had designed 2 suspension bridges in 1822 to span the du Mât river in the Ile de Bourbon (now called Réunion) which were prefabricated in parts by the Milton Iron Works at Sheffield and reassembled at the construction site. His was a prophetic voice of experience. However, Navier's book on suspension bridges remained for half a century the most important treatise on the design of suspension bridges.

Besides editing Belidor's *L'Architecture hydraulique* in 1819 and *La Science des ingénieurs* in 1830, Navier also edited Gauthey's < Treatise of the Construction of Bridges > (*Traité de la construction des ponts*) which his uncle had left unfinished, in 3 volumes in 1809, in 1813 and in 1816 with many editorial notes.

Mathematics and Mathematical Theory of Elastic Solids:

One of the most prominent French mathematicians of the period, Augustin Louis Cauchy (1789-1857), was educated as an engineer, after having been a distinguished student of classical studies at the *École Central du Panthéon*. He attended the *École Polytechnique* from 1805 to 1807, and the *École des Ponts et Chaussées* from 1807 to 1809, and after his graduation he was employed as hydraulic engineer at the works of the Ourcq Canal under the direction of Simon Girard, then on the construction of the bridge of Sainte-Cloud, and from 1810 to 1813 at the building of the harbour of Cherbourg. After 1813, Cauchy devoted most of his intellectual efforts to research in mathematics and theoretical sciences.

His interest in the foundations of mathematics had been aroused by his former teacher, the Italian mathematician of French stock, Joseph Louis Lagrange (1736-1813). Lagrange was the first major mathematician who felt that the foundation of calculus was unsatisfactory as it then existed, and that a new, exact foundation of calculus should be created which is not geometric but algebraic in nature. The basic axiom of statics in Lagrange's < Analytical Mechanics > (*Mécanique analytique*) (1788) is the principle of virtual work, from which dynamic follows if assigned forces are supplemented by inertial forces arising from accelerations subject to the assumption that the constraints do no work. His presentation of mechanics restricted to systems of mass-points is strictly algebraic without any explanation of concepts. Lagrange had reduced much of mechanics to differential equations with his introduction of generalised coordinates in 1782 and by achieving invariant formulation of analytical dynamics now called Lagrange's equations. Lagrange, whose mathematics was strictly manipulative, had deep insight into algebraic arrangements and transformations, and his analysis is purely formal and unrigorous. One of his most lasting contributions was to the algebra of infinitesimal calculus, and he felt that if calculus could be made rigorous then his analytical mechanics would also become rigorous. He believed that all particular results in mathematics, like in mechanics, ought to be particular cases of a general principle. For this reason, Lagrange himself developed a theory of calculus which he based on the algebra of power series, particularly, the 'Taylor series'. Lagrange took this basic idea from Euler's 'theory of approximations' which Euler had based on infinite Taylor series. Lagrange, instead, used infinite Taylor series with his 'remainder term', and computed explicitly error bounds by the use of inequalities in his studies of the approximation techniques in algebra, since Euler had already used inequalities to find the necessary and sufficient conditions for relative extrema. Lagrange developed his algorithmic rule for calculus by basing it on Taylor series expansion of arbitrary functions, in which he identified all derivatives as functions with the coefficients of the Taylor power series, thereby avoiding the limit concept. Thus, Lagrange was the first mathematician to apply the delta-epsilon method of Eudoxos of Knidos, a powerful intellectual tool, to calculus. The basic mistake committed by Lagrange was, as Cauchy soon demonstrated, that different functions can have the same Taylor series expansion about a point. Lagrange published his first paper on the foundations of calculus in 1772, and more completely in his treatises < Theory of Analytical Functions > (*Théorie des fonctions analytiques*) in 1797, and < Lectures on the Calculus of Functions > (*Leçons sur le calcul des fonctions*) in 1806. Lagrange was so concerned about the lack of sound foundation for calculus, that whilst he was at the Berlin Academy of Sciences he was mainly responsible for having the Academy set a prize competition in 1784 for an essay to solve the logical difficulties in the foundation of calculus as the major unsolved mathematical problem so that mathematics could continue to be respected for its precision and rigour. Among the essays submitted was one by the French mathematician Simon Antoine Jean L'Huilier (1750-1840), and another by the French military engineer Lazare Carnot. Both essays were later expanded and published in two books <Elementary Exposition of Calculus> (*Exposition élémentaire du calcul*) in 1787 by L'Huilier and < Reflexions on the Metaphysics of Infinitesimal Calculus > (*Réflexions sur la métaphysique du calcul infinitésimal*) in 1797 by Carnot. The first prize was given to L'Huilier for his essay, but only as the best among the many unsatisfactory submissions. The books by Lagrange, L'Huilier, and Carnot were the only treatises on the foundations of mathematics published in the 18th century. The single significant contribution of L'Huilier was the introduction of the 'two-sided limit' for alternating series which permitted the variable to oscillate about its limit. Before L'Huilier's definition, only 'one-sided limit' in which the variable approached its limit from one side had been used. Carnot claimed to be able to demonstrate that errors introduced by treating differentials as if they were zeros always mutually cancel one another in calculus. Bishop George Berkeley (1685-1753), an Irish philosopher and clergyman, had already shown in 1734 in his famous, and very effective criticism of Newton's fluxions, < The Analyst, or a Discourse Addressed to an Infidel Mathematician >, which he wrote as a defence of religion against attacks by scientists such as astronomer Edmund Halley (1656-1742), how two errors resulting from using increments cancelled each other in the problem of the tangent to a curve. However, Lagrange, had again stressed in 1760, that such compensation of errors in calculus cannot be proven for all cases. In the face of this, Carnot proposed to change this criticism into the foundation of calculus, which was, of course, unacceptable in the absence of a general proof.

Lagrange in proving a lemma that a function with a positive derivative on an interval is increasing at every point of the interval, used a method of proof which is very close to the modern delta-epsilon proof. Lagrange used finite Taylor series in calculus, and by employing his remainder term he was able to estimate its value in his error estimates by a sequence of inequalities.

French theoretical physicist André Marie Ampére (1775-1836), the founder of electrodynamics, was also a first-rate mathematician, and it was he who made the 'Lagrangian inequality' in calculus a defining property in a memoir published in 1806, in which he attempted to free calculus not only from infinitesimals but also from Lagrange's Taylor series foundation. In this paper Ampére gave his inequality proofs of the basic properties of the derivative of a function, and the first inequality definition of the derivative, which, unfortunately was not entirely satisfactory. Ampére believed, mistakenly, that he had demonstrated both the uniqueness and the existence of the derivative. However, his method of proof was valuable because his former pupil, Cauchy, used it to establish a rigorous theory of derivatives. Cauchy had recognised that the delta-epsilon method which had been considered useful by Lagrange and Ampére, was actually essential as the defining property of the derivative. However, he had been anticipated in some fundamental contributions to calculus by Bernardus Placidus Johann Bolzano (1781-1848), a Bohemian cleric in Prague. Bolzano, already in 1817, had defined a derivative of a function $f(x)$ as $f'(x)$ which the ratio $[f(x+\Delta x) - f(x)]/\Delta x$ approaches indefinitely close as Δx approaches zero through its positive and negative values. Bolzano had pointed out that $f'(x)$ was a 'number' and not a 'quotient of zeros', or a 'ratio of evanescent quantities', an idea which had been effectively criticised by Berkeley. Cauchy gave a definition similar to that given by Bolzano in 1823. Cauchy then defined the differential dx, as one of his teachers, Silvestre François Lacroix (1765-1843), had done in 1797, to be a finite quantity and $dy = f'(x)\,dx$, where $y(x) = f(x)$. In this viewpoint differentials as finite quantities had a meaning only in terms of the derivative $f'(x)$ and were mere auxiliary notions which could be logically dispensed with. Cauchy gave the relation between $(\Delta y/\Delta x)$ and $f'(x)$ by the mean-value theorem:

$$\Delta y = f'(x + \theta \Delta x)\Delta x,$$

where $0 < \theta < 1$. Cauchy's proof of the mean-value theorem (already known to Lagrange) rested on the continuity of $f'(x)$ over the interval Δx.

Both Bolzano and Cauchy had given a rigorous definition of 'continuity' and 'derivative', but only Bolzano understood the fine distinction between the two. Bolzano gave an example for a continuous function which had no finite derivative at any point of the interval. Besides the definition of continuity and derivative, Bolzano also anticipated Cauchy in giving the 'intermediate-value theorem', and the so-called 'Cauchy criterion' requiring every Cauchy sequence to converge. Cauchy defined convergence in terms of inequalities in 1821. Newton had shown how areas could be calculated by reversing differentiation, which led to the concept of integrals as antiderivatives. Leibniz defined areas and volumes as infinite sums of infinitesimal areas and volumes. Poisson had shown in 1820 that the value of a definite integral:

$$\int_{-1}^{1} \frac{dx}{x}$$

can be different for different paths of integration if one path contains an infinite value of the function. Therefore, the value of the integral over an interval is no longer the difference between the values of the antiderivative at the endpoints of the interval $[a,b]$:

$$\int_a^b f(x)\,dx \neq F(b) - F(a), \quad \text{where } F(x) = \int f(x)\,dx.$$

Poisson showed that the definite integral as a sum is still the difference between the antiderivatives at the endpoints of the interval if the function is finite over the interval. Fourier regarded integrals as sums in handling discontinuous functions in his analytical theory of heat, although the analytical meaning of the integral when the function $f(x)$ is discontinuous had to be considered.

In 1823 Cauchy abandoned the definition of the definite integral as antiderivative, and, instead, defined the integral as the limit of a sum. Cauchy proved that the integral of a continuous function as a limit of a sum exists, and is itself continuous. He proved by mean-value theorem that the derivative of the integral is equal to the function in the integrand, which is the fundamental theorem of calculus. However, his proof was not entirely rigorous because he lacked the 'concept of uniform continuity' required for a completely rigorous proof.

He also defined improper integrals, the so-called 'Cauchy principal value' of an integral with a singular integrand, and integrals of infinitely large functions over infinitely small paths of integration [the so-called ' impulse' and 'delta functions']. He extended his theory of integration to complex domain, and established his powerful calculus of limits (*calculus des limites*), and calculus of residues (*calculus des résidus*) or polar singularities for the solution of differential equations with constant coefficients. He gave the first systematic theory of complex numbers, and he founded the complex function theory in which he established the so-called 'Cauchy-Riemann differential equation' which had first appeared in the hydrodynamic theory of d'Alembert. He made 3 studies of error theory and proved the 'central limit theorem' by 'Fourier transforms' and gave the first correct formulation of its 'inversion theorem'. He had published, in 1812, the first comprehensive treatise of determinants which contained the product theorem of determinants which also had been given by Jacques Binet (1786-1856) at the same time, inversion of a matrix, and basic theorems on determinants consisting of subdeterminants that are presently important in numerical discretisation of many engineering problems.

Cauchy developed the method of permutation groups in his calculus of substitutions (*calcul des substitutions*) which contained the notions of element order, subgroup and conjugateness, and his theorem for finite groups. He invented the so-called 'Jacobian' for 2 and 3 dimensions before the German mathematician Carl Gustav Jacobi (1804-1851) applied it to any number of dimensions, and gave the orthogonal transformation of a quadratic form into principal axes simultaneously with Jacobi in 1829.

In his work on differential equations, Cauchy insisted that the existence of the solution be proven, and uniqueness be enforced on it by specifying the initial and boundary conditions, something which he probably discovered in his study of waves in liquids in 1815. Cauchy was aware of the so-called Cauchy-Lipshitz method of approximation by difference equations when he solved ordinary differential equations, and of the iteration principle. In 1819 Cauchy discovered the 'method of characteristics' for the solution of first-order partial differential equations, almost at the same time as the German mathematician Johann Friedrich Pfaff (1765-1825). He obtained analytical solutions of ordinary differential equations with his calculus of limits (*calcul des limites*), and treated simple boundary value problems with the so-called 'Green functions'.

In 1801, English physician, linguist, scientist and a former child prodigy, Thomas Young (1773-1829) had suggested, after having performed his famous experiments on the interference of light, that light be considered an undulation of a subtle gaslike medium called æther, rather than a flow of corpuscles. This longitudinal wave theory of light had been originally proposed by Huygens, whilst the corpuscular theory of light had been postulated by Newton, although Newton also had been compelled to introduce æther waves into his theory in order to explain the periodicity in the flow of corpuscles in the so-called 'Newton's rings', where such waves were not regarded as light but merely served as the means to induce periodicity into the flow of corpuscles of light. Later Euler had been the only supporter of Huygens' 'wave theory of light', in which theory he had assumed æther to behave like an elastic fluid that can only transmit longitudinal light waves. In 1817, a young French civil engineer, Augustin Jean Fresnel (1788-1827), who was building roads in Brittany at the time, won the prize competition for explaining theoretically the light diffraction phenomenon set by the French Academy of Science in the same year with a memoir that was a masterpiece in mathematical physics. Instead of a corpuscular explanation expected by the Academy, Fresnel explained diffraction of light by a combination of the principle of interference of Young and the Huygens wave theory of light. His mathematical wave theory of light gave certain new predictions which were afterwards experimentally corroborated. Unfortunately, the elastic fluid theory of æther used by Fresnel could not explain polarisation of light discovered by the physicist Étienne Louis Malus (1775-1812) in 1808. Both Young and Fresnel had hit upon the idea of transverse light waves acting perpendicular to the direction of propagation of the light waves, an idea first suggested in 1671 by Robert Hooke (1635-1703). Since an ideal fluid cannot resist distortion and, therefore, cannot resist transverse action, Fresnel, in 1821, hypothesised the subtle æther to possess the properties of an isotropic elastic solid. Using his elastic solid theory of the luminiferous æther, Fresnel founded transmission of light by transverse waves, although longitudinal waves still subsisted in his elastic wave theory of light. His theory of transverse waves in an elastic æther, which behaved like an elastic solid, supplied satisfactory explanation to all known contemporary optical phenomena. Thus Fresnel's single principle connected all known light phenomena, such as polarisation, diffraction, interference, refraction and other optical behaviour, which made Fresnel one of the most ingenious men of science of his time in France for Young's wave theory of light had been unable to represent polarisation of light.

Fresnel showed his versatility by making valuable contributions to industrial chemistry and to lighthouse technology. In 1811, he developed an entirely new method of making sodium carbonate, an alkali used in the manufacture of such commodities as soap and glass and many other alkaline products. Fresnel's ammonia-soda process was based upon the chemical reaction of carbon dioxide, salt and ammonia. Unlike the method of Nicholas Leblanc (1742-1806), then in current industrial use, Fresnel's method was fuel efficient and allowed industrial utilisation of its byproducts. Unfortunately, contemporary practical difficulties prevented its industrial application until the 1860's, when the Belgian salt refiner Ernest Solvay (1838-1922) succeeded.

In 1819, after being appointed to the lighthouse commission of the Public Works, Fresnel developed a powerful beacon light which was produced by a special refractive lens of his own invention, rather than by the inefficient metal reflectors then in general use. Lenses had been tried before, but nobody had been able to make an ellipsoidal lens thin enough to have the required refractive properties and sufficient curvature. Fresnel, using his profound theoretical knowledge of optics, designed a thin lens now called the 'Fresnel lens' that consisted of a series of refracting prisms, which concentrated light and produced an intensely bright beacon from the light source. The beacon light produced by Fresnel's lens rapidly replaced the old reflector produced beacon light. Unfortunately, the life of this versatile and talented young man was cut short by consumption.

Thus, Fresnel had regarded the luminiferous æther not as an elastic fluid but rather as an elastic solid continuum, and gone far towards constructing a continuum theory of small elastic deformations. Cauchy, having been influenced by Fresnel's brilliant theoretical work, recast Coulomb's and Navier's work on the elastic continuum, solid or fluid, by introducing his 'stress tensor' which is independent of the constitution of the continuum, into the phenomenological theory of continuous bodies, a concept which had already been foreshadowed by Fresnel. Cauchy introduced the general concept of strain, the concept of local rotation, and, most importantly, the stress tensor [anticipated by Fresnel], which unifies all continuum theories and takes the stress vector to act obliquely on any arbitrary section in the continuum. Cauchy's introduction of the concept of stress tensor is comparable in importance to Newton's introduction of force into mechanics as an *priori* mathematical concept. Cauchy applied Euler's principles of the balance of the linear and the rotational momentum to a continuous body, and expressed the resulting axioms of motion in terms of the stress tensor. He also introduced the constitutive equations for the isotropic linearly elastic continua which are still used today:

$$[\sigma] = 2k[\epsilon] + K(\Sigma \epsilon_{ss})[1],$$

where $[\sigma]$ and $[\epsilon]$ are the stress and the strain tensor respectively, $[1]$ is the identity tensor, $\Sigma \epsilon_{ss} = \epsilon_{xx} + \epsilon_{yy} + \epsilon_{zz}$ the trace of the strain tensor, and k and K are suitable elastic constants. This constitutive equation can also be expressed by means of Gibbs' direct tensor (polyatic) algebra:

$$\sigma = 2k\epsilon + K[\epsilon : 1]1,$$

where $\sigma = \sigma_{ij} e_i e_j$ is the stress tensor, $\epsilon = \epsilon_{ij} e_i e_j$ is the strain tensor, $\epsilon : 1$ is the trace of the strain tensor, $1 = e_k e_k$ is the identity tensor, and i, j, k go over x, y and z. The trace of strain tensor ϵ is given in terms of Cartesian unit base vectors e_x, e_y and e_z:

$$\epsilon : 1 = (\epsilon_{ij} \ e_i e_j) : (e_k e_k) = \epsilon_{ij}(e_i \cdot e_k)(e_j \cdot e_k) = \epsilon_{kk} = \epsilon_{xx} + \epsilon_{yy} + \epsilon_{zz},$$

$$\epsilon \cdot 1 = (\epsilon_{ij} e_i e_j) \cdot (e_k e_k) = \epsilon_{ij} e_i (e_j \cdot e_k) e_k = \epsilon_{ij} e_i e_j = \epsilon,$$

where $e_i \cdot e_k = 0$ if $i \neq k$, or $=1$ if $i = k$, and repeated indices such as kk mean summation over x, y, z. Cauchy reaffirmed the distinction, which had been made earlier by Jacob Bernoulli and particularly by Euler, between general principles and constitutive equations usually known as stress-strain relations. Therefore, by 1822, Cauchy had created the necessary mathematical means of the theory of elasticity, a theory which was axiomatic and independent of Navier's molecular theory. Cauchy arrived in his continuum theory at the same equations obtained by Navier for isotropic elasticity both of which had one elastic constant, but without any reference to the molecular substructure of the continuum. Cauchy soon added another elastic constant to his theory by introducing a relation between the volumetric stress and the volumetric strain. For anisotropic continuum, Cauchy introduced a general linear dependence between stress and strain which required 36 elastic constants. Since he never proposed the concept of elastic potential for the deformed elastic medium, as was done by the English mathematician George Green (1793-1841) in 1837, Cauchy could not demonstrate that the number of elastic constants reduces from 36 to 21 for isotropic linearly elastic solids. At the same time, Cauchy did not neglect Navier's molecular concept of the solid, but established a molecular theory of anisotropic elastic solids which had 15 elastic constants. The controversy soon arose about which one of the 2 theories of elasticity is correct: the axiomatic multiconstant theory, or the molecular rariconstant theory. This controversy lasted until 1900 when it was finally settled by the careful experiments of the German physicist Woldemar Voigt (1850-1919) in favour of the multiconstant theory of Cauchy.

Cauchy did not neglect Fresnel's wave theory of light, since he was particularly interested to explain dispersion, reflexion and refraction of light in which he attempted to eliminate the spurious longitudinal waves. He developed 3 different theories in which one theory had a negative compressibility of the elastic æther. Cauchy devoted a great deal of his energies to the light theory in elastic medium, and it remained one of the last such efforts before physicists became convinced that any light theory in elastic medium is inadequate for the description of the general phenomena of light.

Cauchy applied his elasticity theory to plane diaphragms, to rectangular beams, and to the bending of plates. In his study of the torsion of rectangular bars, he succeeded in finding a satisfactory solution only for a bar of narrow cross-section. He also studied large deformation problems in elasticity. Cauchy's contribution to rigid body mechanics such as the first rigorous proof that any infinitesimal motion is a screw motion, the momental ellipsoid and its principal axes, and the surfaces of momentaneous axes of rigid motion were rather minor. He made numerous, lengthy contributions to celestial mechanics such as the development of the perturbation function and the solution of the Kepler equation.

Cauchy was a prolific author second only to Euler. However, most of his works were hastily written, because he was an impatient man like Descartes, and lavishly published in *Exercices de mathematiques*, a private journal he himself had founded, the 12 yearly issues of which he himself filled with his own mathematical researches.

French mathematician and scientist, Denis Siméon Poisson, was a few years older than Navier and Cauchy, and had graduated from the *École Polytechnique* in 1800. He was at first made an instructor at this school, then in 1806 professor of analysis and mechanics, and later professor of rational mechanics at the Sorbonne. Poisson, who was a rival to Navier in solid mechanics and to Cauchy in mathematics and solid mechanics, was a protégé of Laplace, and the most prominent promoter of Laplace's molecular theory of physical phenomena. Although Poisson did important work in mathematics, professionally he worked in a transition period when mathematics underwent fundamental conceptual changes owing to the researches of the brilliant German mathematician, astronomer and physicist Carl Friedrich Gauss and, particularly of the ingenious Cauchy, which reduced the lasting value of Poisson's contributions to mathematics.

Poisson developed exclusively the so-called 'molecular theory of physics' which when applied to mechanics he called 'physical mechanics'. In 1824 and 1831, Poisson published his two most important memoirs on elastic bodies and fluids in which he considered such bodies to consist of molecules which are subjected to the action of centrally interacting molecular forces much like Navier had done in his famous memoir in 1821. Poisson's researches were quite influential in laying the foundation for molecular theory of elasticity, although he did not contribute such fundamental ideas to his subject as Navier and Cauchy. Poisson obtained the equation of equilibrium and the associated boundary conditions in elasticity which were similar to those previously established by Navier and Cauchy. Poisson proved that these conditions are not only necessary but also sufficient to guarantee equilibrium of any arbitrary part of the body. Poisson was able to integrate the equations of motion of the body and demonstrate that a local disturbance in the body produces two kinds of elastic waves: the faster pressure wave, and the slower distortional wave. He proved theoretically the presence of the lateral strain '$\nu\epsilon$' accompanying the impressed strain ϵ, the so-called 'Poisson's effect', specified by the so-called 'Poisson's ratio ν', which had first been observed in practice by the Hellenistic military engineer Philon of Byzantion about 240 B.C.. He solved the free radial vibration problem of a sphere, and the deformation of a hollow sphere. He also obtained the equation of flexure for a thin elastic plate:

$$D\nabla^4 w = p.$$

Poisson's kinematic boundary conditions for simply supported and for fixed edges of the plate were correct, but he had 3 boundary conditions for forces instead of 2, which defied solution. Poisson solved correctly the rotationally symmetric plate under radially varying transverse load, and a symmetrically vibrating circular plate. Unlike Cauchy, Poisson loathed to give credit where credit was due to his predecessors, and this reluctance led him into a particularly long and bitter controversy with Navier.

Poisson wrote an influential book on rigid body mechanics in 2 volumes entitled, <Treatise of Mechanics> (*Traité de mecanique*), which was first published in 1811, and in a 2nd enlarged edition in 1833 which contained many practical problems from physics, astronomy and ballistics. Poisson's book gave a simpler presentation of mechanics than Lagrange's <Analytical Mechanics> (*Méchanique analytique*), but he made no use of Poinsot's important theory of couples probably because of his personal dislike of Poinsot. In this treatise Poisson solved a simply supported beam by the application of trigonometric series which represented the first such solution in structural mechanics. He published his treatise on the mathematical theory of heat entitled <Mathematical Theory of Heat > (*Théorie mathématique de la chaleur*) in 1835, which was the first such treatise to rival that of Fourier. Poisson published altogether more than 300 long memoirs on mathematics, physics, and astronomy.

The actual founders of the modern structural, machine, and industrial mechanics were Navier, Coriolis, and Poncelet. From the three men, Jean Victor Poncelet, the most talented student of the geometer Monge, was the most ingenious. Poncelet was a first-rate mathematician who founded the new projective geometry. His book, < Treatise of Projective Properties of Figures > (*Traité des propriétes projectives des figures*), published in 1822, laid the foundation for modern geometry. Poncelet's geometric researches were immediately subjected to quite unfair and openly malevolent attacks by Cauchy concerning Poncelet's principle of continuity, and by the mathematician and astronomer Joseph Diez Gergonne (1771-1859), who wanted to replace Poncelet's theory of reciprocal polars with his own incompletely founded principle of duality. On account of these persistent academic intrigues motivated by envy, Poncelet virtually abandoned his research in geometry and directed his considerable talent to technical fields such as hydrodynamics, machine dynamics, and industrial mechanics to which he supplied the necessary foundation of applied mathematics.

In 1825 Poncelet was appointed professor of engineering mechanics at the *École d'Application de Metz*, where he also gave popular evening lectures on geometry and mechanics in their application to arts and crafts. In 1838, he was appointed professor of mechanics at the Sorbonne in Paris. From 1848 to 1850 General Poncelet was commandant of the *École Polytechnique*, and also chief commander of the National Guard in the Department of Seine.

In 1829, Poncelet published his lectures entitled < Introduction to Industrial, Physical and Experimental Mechanics > (*Introduction à la mécanique industrielle, physique ou expérimentale*), which brought Poncelet quick fame since it contained much of his original work on mechanics of materials and which went through several editions. In technical mechanics Poncelet was the first to introduce the concept of work, and use it effectively in all fields of engineering mechanics. Poncelet's valuable contributions to machine design and dynamics are given in his book, <Mechanics Applied to Machines > (*Mécanique appliqué aux machines*), published in 1848. Many important engineering theories of hydraulics are treated with considerable originality in his book. He created the theory of unsteady flow of water in canals which was based on his own careful experiments in collaboration with his colleague, Capt. Lesbros, and the Prony Formula. Poncelet's important work on hydraulic turbines and soil mechanics has been already discussed. Poncelet was an excellent teacher like Euler, and he is generally regarded as the true founder of technical mechanics.

Poncelet, himself, considered his friend Baron Pierre Charles François Dupin (1784-1873) the founder of industrial education in France. Dupin, like Poncelet, had been an excellent mathematician in his youth. After graduation from the *École Polytechnique* and the *École de Marine*, Dupin became a naval engineer, but for a while he was active in philology and classical studies. Having been a student of Monge, the great geometer, Dupin did original research on differential geometry of surfaces, and gave the well-known 'indicatrix of Dupin' in the mathematical theory of surfaces.

The National Convention had decreed in 1794 to establish a conservatory for the collection of instructive models for mechanical and chemical industries. In 1819, a royal order converted this conservatory into an institute for free public lectures, similar to the Royal Institution in London which had been established in 1801 by Benjamin Thompson, alias Count Rumford, (1753-1814), and named the *Conservatoire des arts et métiers* (*Conservatory of Arts and Crafts*). Since 1816 Dupin had been travelling through Great Britain and Ireland by the order of French government as an official observer, and upon his return to France in 1819, he called the attention of the French government to the importance of improving the efficiencies of craftsmen and workers by offering them popular science lectures, a practice he had observed in Great Britain. Dupin published his reports on his journeys through Great Britain in 6 volumes. Dupin himself was appointed in the same year professor at the *Conservatoire des arts et métiers*, where free public lectures were given on geometry, mechanics, chemistry and national economy. Dupin's own celebrated lectures on the geometry and mechanics of arts and crafts which he published in 3 volumes entitled < Geometry and Mechanics of Arts and Crafts > (*Géometrie et mécanique des arts et métiers*) from 1825 to 1827, were distinguished by a clear, simple manner of presentation, and by a great number of practical examples to which were applied geometrical and mechanical principles. Dupin, who was an excellent teacher, lectured to groups of managers of factories and workshops, artists, craftsmen, and workers of all ages and trades usually numbering more than 600 persons. Dupin's influence on the industrial education in France was quite decisive, and Poncelet judged him to be the founder and promoter of industrial education in France. In addition to his educational duties, Dupin also served as Inspector General of the maritime engineering.

Both Gabriel Lamé (1795-1870) and Emile Benôit Clapeyron (1799-1864) had graduated from the *École Polytechnique* in 1818 and from the *École de Mines* in 1820, when they were recommended to the Russian government as talented young French engineers to teach applied mathematics and physics in the new Russian engineering school, the Institute of Ways of Communication in St. Petersburg, which had been founded by the French engineer Augustin de Bétancourt (1758-1824) in 1809, and to help with the design of various important structures planned by the Russian government. During their stay in Russia, Lamé and Clapeyron wrote a long and famous memoir < On the Internal Equilibrium of Homogeneous Solid Bodies > (*Sur l'equilibre interieur des corps solides homogènes*) in 1828, published in France in 1833, in which they derived the equations of equilibrium for solids by Navier's molecular theory, and demonstrated that the same equations are obtained by means of Cauchy's continuum theory. This paper contained many theoretical solutions which have practical value, such as the well-known Lamé stress ellipsoid by means of which the stress vector acting on any plane passing through a point in a body can be found. The two authors used the so-called 'rariconstant theory of elasticity' of Navier and showed how this constant can be found from the tensile test, or from the compression test. Then they examined the following problems: hollow circular cylinder subjected to uniform internal and external pressure, simple torsion of circular shafts, a sphere stressed by gravitational forces directed towards its center, semi-infinite body bounded by a plane on which normal tractions are acting, a cylinder of infinite length with the first use of cylindrical coordinates, and torsion of the cylinder by the tangential surface forces normal to the central axis. This long memoir of Lamé and Clapeyron contained all the theoretical solutions for isotropic, linearly elastic materials known at the time.

Lamé was assigned the task to test the mechanical properties of iron produced in Russia, and used in the design of iron structures in that country. The Swedish physicist Per Lagerhjelm (1787-1856) had designed the first hydraulically operated tensile testing machine in 1824 which had been so successful that Lamé had a similar tensile testing machine built in 1825 for Lamé's own tests. In Lamé's machine the magnitude of the load was measured by means of weights placed at the end of a horizontal lever.

During their stay in Russia, Clapeyron and Lamé were called to examine the structural stability of the Cathedral of St. Isaac in St. Petersburg. Their study of the cylindrical arches and the dome of this cathedral, which was under construction at the time, was published in their memoir in 1823. They modified Coulomb's theory of arches by taking the cross-sections of the arches vertical, then found the location of the fractured sections by a simple graphical method, and then determined the horizontal thrust.

After their return to France in 1831, Clapeyron and Lamé were engaged in 1835 in the planning and construction of the railroad between Paris and St. Germain. Lamé soon relinquished engineering, and became professor of physics at the *École Polytechnique* where he published his course of physics in 1837, and some important memoirs on the theory of light. Lamé used his memoirs written in collaboration with Clapeyron in the first book on theory of elasticity,< Lectures on the Mathematical Theory of Elasticity of Solid Bodies > (*Leçons sur la théorie mathématique de l'elasticité des corps solides*) in 1852. In the meantime Lamé had rejected the rariconstant theory of elasticity and introduced Cauchy's multiconstant theory which for isotropic linearly elastic bodies has 2 elastic constants now called 'Lamé elastic constants':

$$\mu = E/2(1+v), \text{ and } \lambda = [2v/(1-2v)]\mu,$$

where E denotes the modulus of elasticity first defined by Euler. In this book Lamé gave an example of the Clapeyron's principle of conservation of work in calculating the displacement of a joint in a simple truss. In 1829 Clapeyron had established his energetical principle for elastic members: $U = W$, where U denoted the strain energy stored in the deformed member in its equilibrium state, and $W = Pu/2$ represented the work done by the applied force P over its displacement u. Clapeyron had developed his principle of conservation of work in his design of military equipment in Russia. Unfortunately, only one scalar unknown could be found from this single scalar equation. In 1860, the brilliant Scottish physicist, James Clerk Maxwell (1831-1879) invented his 'method of auxiliary free-bodies' by means of which the Clapeyron's principle of conservation of work could be used to solve structural problems having any number of scalar unknowns.

Lamé introduced in 1837 curvilinear coordinates for the solution of differential equations of mechanics, heat and elasticity. In 1859 he collected all these results into a book entitled, < Lectures on the Curvilinear Coordinates and Their Diverse Applications > (*Leçons sur les cordonnées curvilignes et leurs diverses applications*). In his study of the 3 orthogonal coordinate surfaces Lamé had established 3 partial differential equations, which, surprisingly, contained the integrability conditions for surfaces, although it had not been the intention of Lamé to establish such equations. The fundamental equations of surfaces as such were first established by the Livonian mathematician Karl Peterson (1828–1881) in his doctoral dissertation in 1853 at the University of Dorpat, in Estonia, which basically constituted the integrability conditions for the surface expressed in terms of intrinsic surface coordinates, which in modern direct vector notation represent the equations:

$$\partial^3 r / \partial u^2 \partial v = \partial^3 r / \partial u \partial v \partial u,$$
$$\partial^3 r / \partial v \partial u \partial v = \partial^3 r / \partial v^2 \partial u,$$
$$\partial^2 n / \partial u \partial v = \partial^2 n / \partial v \partial u,$$

where r denoted the position vector of the surface, n designated the unit normal vector to the surface, and u and v are the intrinsic surface coordinates. Lamé's partial differential equations were equivalent to the first 2 equations of Peterson. The 3rd Peterson equation had already been given by Gauss in 1827. Lamé closed his book with a discussion of the principles upon which the basic equations of elasticity are founded. He had by now exclusively adopted Cauchy's multiconstant linear elasticity theory, which for isotropic bodies reduced to 2 elastic constants, the so-called Lamé elastic constants μ and λ still in use today.

In 1849, Clapeyron found the first general solution for an elastic beam continuous over many unyielding supports, also known as the 'Clapeyron equation', which he developed in his design of the railroad bridge over the Seine river at Asniers, and published in 1857. Clapeyron's equation was transformed into a form containing 3 moments over the successive supports by a French engineer Jean Henri Bertot (born 1824) in 1855, and is presently known as the 'three moment equation'.

Jacques Antoine Bresse (1822-1883), who taught engineering mechanics at the *École des Ponts et Chaussées*, made many contributions to applied mechanics, but his major fame stems from his theory of curved elastic bars and its application to the design of arches.

In 1853, a former student of Navier and a deputy of Coriolis at the *École des Ponts et Chaussées*, Jean Claude Barré de Saint-Venant (1797-1886), perhaps the foremost elastician in the history of the subject, published his epoch-making memoir on torsion, a problem for which all great elasticians, including Cauchy, had failed to find the general solution. Saint-Venant introduced his novel 'semi-inverse method of solution' in solving this general and difficult problem of torsion. In 1857 he published an accurate and complete treatment of elastic beam theory. Saint-Venant studied the dynamic action of loads on elastic beams and bars and made many other valuable and original contributions to various fields of theoretical mechanics, such as to fluid and gas dynamics where he was the first to give in 1834 and in 1842 the fundamental equation for viscous flow of fluids, and to plasticity, in the course of his long and illustrious professional career.

Rise of Thermodynamics and Theory of Heat Engines

Few men in history have done as much to found an entirely new science in natural philosophy as did Sadi Nicolas Léonard Carnot (1796-1832) in his founding of the science of thermodynamics. Sadi Carnot was the precocious son of Lazare Nicolas Marguerite Carnot, a famous General and the 'organiser of victory' in the war between the French revolutionary Jacobin army and the armed forces of the European coalition. Lazare Carnot had also distinguished himself as an outstanding engineering scientist and mathematician, and among his other researches had published in 1803 an important treatise on the theory of hydraulic engines entitled < General Principles of Equilibrium and Movement > (*Principes généraux de l'equilibre et du mouvement*) that was based on the principle of work which Lazare Carnot promoted in mechanics.

Sadi Carnot examined the British steam-engines with the intention of establishing a general theory of the motive power of heat. At this time there existed no theory at all for heat engines, and the physicists with their concept of molecular motion were in no frame of mind to be able to found a science of thermodynamics, which had to be established as a phenomenological theory based on the measurable concept of work. Sadi Carnot had been educated at the *École Polytechnique* as a scientifically trained military engineer, and his intention was to produce a general theory of heat engines which any ordinary educated person, who is not necessarily a scientist, can understand. After some years of research and study, he published his epoch-making monograph, < Reflexions on the Motive Power of Fire and on Proper Machines to Develop this Power > (*Réflexions sur la puissance matrice du feu et sur les machines propres à developper cette puissance*) in 1824, which contained the results of his researches on heat engines. It represents one of the most original scientific works ever published for it created a new science called 'thermodynamics'. In this study, Sadi Carnot's intention was to create a general theory of heat engines applicable to any heat engine regardless of its working substance, and to identify the theoretical quantities which determine, in general, the production of mechanical work from heat. He was particularly interested to find out whether or not there is an upper limit to the work obtained from heat, and under what conditions will such work be available. In carrying out his intention, Sadi Carnot founded and illustrated a programme of thermodynamics, and thereby became the outstanding creator of its concepts and principles. His approach to the general heat engine theory was entirely new, for there existed no such theory when the younger Carnot launched his research on heat engines. He was the first to introduce the concepts of process, of heat absorbed, and of heat emitted. In fact, his is one of the most remarkable accomplishments of abstraction in the history of natural philosophy. Carnot had the genius to abstract from the performance of the steam-engine by analogy the conceptual essentials necessary for the general theory of the heat engine. Sadi Carnot formulated his novel theoretical approach to the heat engine in analogy with hydraulic engines, such as the waterwheel, working in cycles and with their theory based on the concept of work, a theory which had been systematically developed by his father, Lazare Carnot. Lazare Carnot's conditions for the maximum efficiency of the hydraulic engine was stipulated on the condition first established by Borda that water enter the machine without impact, and leave it without relative velocity. Sadi Carnot in his theory of heat engines was indebted to his father's hydraulic engine theory for the concepts of efficiency, reversibility, as well as their mutual relationship. The younger Carnot approached the heat engine theory in analogy to his father's hydraulic engine theory in which the head (or, equivalently, the velocity) through which the water must fall to drive the hydraulic machine was in some way analogous to the difference of temperatures applied to the working medium to drive the heat engine. Sadi Carnot expressed this analogy in the following way: "We may justly compare the motive power of heat with that of a fall of water: both have a maximum that cannot be exceeded, whatever be, on the one hand, the machine employed to receive the action of the water, on the other, the substance used to receive the action of the heat." Lazare Carnot had demonstrated for hydraulic engines, and subsequently his son did the same for heat engines, that the greatest mechanical effects will be produced regardless of the details of the design provided certain fundamentals are observed.

Sadi Carnot was also indebted to his friend Nicolas Clément (1778/1779-1841), who, since 1819, had given a series of lectures on the theory of heat and the operation of the steam engine in the ***Conservatoire des Arts et Manufacture*** (*Conservatory of Arts and Manufacture*) in Paris, which incorporated lectures on the principle of the expansive operation of the steam and on the problem of boiler feed-water. In these lectures Clément had stressed the point that, ideally, the temperature of the steam must fall during its expansion to that of the condenser, or the 'cold body'; implying thereby that this process must be adiabatic (a terminology invented much later, in 1859, by the Scottish engineering scientist Rankine). Carnot advanced a theory in which heating and cooling occurs according to whether a hypothetical medium he called ***calorique*** (caloric), first introduced by the French physicist Guillaume Amontons (1663-1705), increases or decreases. The caloric theory assumed the materiality of heat. This caloric could be defined thermally for certain substances such as perfect gases.

Carnot assumed that in a heat engine work is produced by passing heat from a hot body [the boiler] to a cold body [the condenser], in analogy to a waterwheel in which work is done by water passing through a difference of gravitational potential. The caloric theory represented heat as a subtle substance called 'caloric' in which heat is regarded to be an imponderable, massless fluid which is conserved in any process. Carnot said: "Thus the motive power of heat is due not to any real consumption of caloric, but its transport from a hot body to a cold one." He further stated that the motive power "is fixed solely by the temperatures of the bodies between which is effected, finally, the transfer of caloric." He calculated this motive power for air, steam and alcohol and found them to be almost the same for different fluids. He found that a good Cornish pumping engine's duty was only one twentieth of the motive power of coal as the combustible. In analogy with the hydraulic engine, Carnot introduced the hot body as a 'heat source', and the cold body as a 'heat sink', between which the caloric changes take place. He postulated that in the operation of the heat engine there is no useless flow of caloric in the process, and that for a change of caloric to take place an operating difference of temperatures between the heat source and the heat sink is essential.

Carnot introduced an ideal, 4-stage, thermodynamic process for heat engines which he described by a simple model consisting of a cylinder, a piston, a working substance that expands or contracts upon the application of heat or cold, a heat source and a heat sink. He assumed uniform density and temperature in the working substance throughout his ideal thermodynamic process, called the 'Carnot cycle', which consisted of an isothermal expansion of the working substance when it absorbs heat from the heat source at the higher temperatures, an adiabatic expansion when the working substance cools and heat is converted into work, followed by an isothermal compression when the working substance emits heat to the heat sink at the lower temperature, and by an adiabatic compression when the working substance returns to the higher temperature of the heat source. The temperatures of the heat source and the heat sink are the operating temperatures of the Carnot cycle. Carnot assumed that his cycles exist for any pair of operating temperatures. Since each stage of Carnot's ideal cycle is reversible, the entire cycle is reversible. In answering the question: "What is the meaning of maximum motive power of a heat engine?," Carnot formulated the concept of reversibility of the heat engine based on the infinitesimal temperature difference between the operating temperatures of the heat engine. He showed that the Carnot cycle produces a definite positive amount of work.

Carnot demonstrated that among all thermodynamic cycles which absorb heat, Carnot cycles alone with their reversible changes in the heat engine are the most efficient in the performance of work for a given amount of heat, that is, for a given operating temperatures. Carnot derived the motive power of his thermodynamic cycle without the use of caloric theory, by only relying on the doctrine of latent and specific heats:

$$Q(t) = L_v(v,\theta)(dv/dt) + K_v(v,\theta)(d\theta/dt),$$

where $L_v > 0$ describes the latent heat with respect to volume, and $K_v > 0$ designates the specific heat with respect to volume, v and θ denote the volume and the temperature of the working substance respectively, t denotes time, and Q stands for heat. Latent heat is heat absorbed by the working substance at constant temperature: $\theta = constant$. Hence, Carnot demonstrated that maximum effect is obtained when the heat engine operates reversibly.

Carnot based his proof of maximum efficiency of the heat engine on the impossibility of perpetual motion, an argument used by his father in the theory of hydraulic engines. When a Carnot cycle is run in reverse, it consumes as much power as it generates when run forwards. If it is assumed that a cycle other than the ideal, reversible Carnot cycle is more efficient than the Carnot cycle, and if the Carnot cycle is connected to such a cycle which is more efficient, then a perpetual motion machine would result, which Carnot considered physically unsound. This realisation, in effect, indirectly expressed the second law of thermodynamics.

Carnot's analysis is a crucial concept underlying the operation of all theoretical heat engines: the motive power of heat is theoretically independent of the working substances that develops it, and its quantity is determined solely by the temperature difference between the heat source and heat sink, which according to his caloric theory is responsible for the transfer of the heat (caloric) from the heat source to the heat sink. Of course, in a real heat engine the working substance from an engineering point of view is important because its physical properties are essential for the mechanical design of the engine. In Carnot's theory of heat engines, the evaluation of motive power of heat which makes no use of caloric theory is correct because the origin of power is not considered, and, therefore, it allows an experimental check on caloric theory, the first law of thermodynamics, and 'specific heats of the working substances'. In Carnot's heat engine theory, two Carnot cycles with the same operating temperatures do the same amount of work, which expresses the universal efficiency of the Carnot cycle. For a given quantity of heat the work produced is greater for greater difference in operating temperatures. Moreover, for the same difference of operating temperatures, the heat engine operates more efficiently the lower the temperature scale. Carnot's proof for the use of lower temperature required a proof that the isothermal expansion of a gas through a given volume ratio is greater at higher temperatures. Carnot's proof is erroneous on account that he used the erroneous caloric theory and some faulty experimental values, but fortunately for him the 2 errors in his theory cancelled each other, thus leaving his conclusions correct despite his partially faulty reasoning.

When Carnot enquired about the source of power, he had to consider particular working substances. Carnot concluded from his theory, which did not impose any limitations upon the efficiency of the heat engine, that working substances which expand the most for a given change of temperature are the best, which implied that gases are better working substances than liquids or solids. For this reason, he considered air a more effective working substance if practical difficulties with these engines could be overcome. Sadi Carnot developed his general thermodynamic theory of heat engines in terms of the air engine, or the ideal gas engine, but all his fundamental reasonings were based on ideal gas as the working substance. In the course of his reasoning, Carnot made the following statement: "... to give rise to motive power it is not enough to produce heat; cold must also be produced; without it the heat would be useless." This statement of Sadi Carnot foreshadows the second law of thermodynamics. In this connexion Carnot also made a statement which is another one of the various different forms of statement of the second law of thermodynamics: "It is impossible for a heat engine to have done positive work, yet have restored to the hot body [heat source] all the heat it previously absorbed from it and have withdrawn from the cold body [heat sink] all the heat it previously emitted to it."

Carnot regarded his ideal cycles only as a limiting condition on real thermodynamic cycles, for he fully recognised that thermal and thermochemical changes in nature are in general irreversible, and irreversibility reduces efficiency. Turning his attention to the source of work which compelled him to use the current caloric theory of heat, he realised that only if the working substance in the heat engine is brought back exactly to its original state after each cycle, can it be assumed according to the caloric theory that all the heat which passed through the heat engine has performed work.

Considering heat as an indestructible caloric, he derived from his thermodynamic theory of ideal heat engines definite conditions relating the specific heat and the motive power of a working substance. It was his intention to obtain the value of the motive power of a working substance by comparing its specific heats with the experimental value. Since the working substances examined by Carnot were ideal gases, his theory involving the source of power based on caloric theory led him to conclude that at least one of the specific heats of any ideal gas has to be a decreasing function of the volume, a theoretical conclusion that contradicts later experimental results. Therefore, Carnot's theory gave wrong results when the origin of the work from heat was considered. Carnot's theory gave the correct result when the conditions under which maximum work could be obtained from the transfer of a given quantity of heat from the heat source to the heat sink was considered, because in this case the origin of work was not involved.

Carnot was not the first to use the work concept in calculating the power delivered by an engine. The first such calculations had been done by an Englishman Henry Bayton for the Newcomen atmospheric engine in 1721, that were published in 'The Ladies Diary' of which Bayton was the editor. In 1791, an English mathematician, David Giddy [alias Gilbert], gave the mathematical analysis of the work done by the expansive operation of a steam engine in support of Hornblower's patent renewal in the latter's litigation with Watt. A French physicist, Alexis Thérése Petit (1791-1820) published a memoir, < On the Employment of the Principle of Live Forces in the Calculation of the Effects of Machines > (*Sur l'emploi du principle des forces vives dans le calcul de l'effet des machines*), in 1818, in which Petit claimed to calculate the effect of an engine by equating the power of the working body to the rate of increase of the kinetic energy of the piston. Carnot calculated the motive power of heat by using an expression for the work done by the fluid body (in his case an ideal gas) in transversing an isothermal path at constant temperature. Carnot cites, too generously it seems, the paper by Petit. Since Carnot used an infinitesimal difference of temperatures, the adiabatic paths are infinitesimally short and contribute nothing to the work done, which is why he used infinitesimal temperature differences for his ideal cycle. He was unable to analyse a cycle having a finite difference in temperature. Since Carnot at the time was not in possession of the concept of the mechanical equivalent of heat, he excluded from his theory engines which were not of the heat-transfer type such as engines worked by man or beast, or by waterfalls, or by air currents. Under the influence of the caloric theory of heat, Carnot drew the erroneous conclusion that 'all' the heat is transferred from the heat source to the heat sink in each cycle of operation of his ideal heat engine. It is at this point that Carnot's analogy between the heat engine and the hydraulic engine ceased to be valid, because not all the heat which leaves the heat source in the heat engine arrives in the heat sink. In the hydraulic engine the same amount of water entering at the upper elevation leaves at the lower elevation with work appearing as a result of the fall of water. By analogy, Sadi Carnot postulated that work was done in the heat engine as a result of the fall of heat (caloric) through the difference of the temperature between the heat source and the heat sink. In the caloric theory there was no 'conversion' of heat – only 'conservation' of heat! In other words, Carnot did not observe the first law of thermodynamics in heat engines. Considering the source of work, the working substances have certain internal energies when heated, and it is this 'internal energy' rather than the 'heat' which is conserved in Carnot cycle.

Sometime between 1824 and 1826, after the publication of his celebrated memoir, Carnot came to realisation, after having acquired a clear idea of the first law (the law of conservation of energy) in thermodynamics and calculated with surprising accuracy the mechanical equivalent of heat as 370 mkg/cal., that only part of the heat supplied by the heat source is delivered to the heat sink with the difference having been converted into work done by the heat engine. He even suggested some experiments that would show the equivalence of heat and work as different forms of energy. Similar experiments were actually carried out by the Scot James Prescott Joule (1818-1889) in 1847, although Joule was unaware of Carnot's suggestions. It appears that Carnot had rejected by that time the caloric theory of heat in favour of the kinetic theory of heat, for he wrote:

"Heat is simply motive power, or rather motion that has changed form. It is a movement along particles of the body. Wherever there is destruction of motive power there is, at the same time, production of heat in quantity exactly proportional to the quantity of motive power destroyed. Reciprocally, wherever there is destruction of heat there is production of motive power."

In analogy with the hydraulic engine, the running of which can be reversed in operation to make it into a pump, Carnot clearly recognised that reversal of the operation of the heat engine converts it into a heat pump, which pumps heat from the heat sink into the heat source by adding to the heat delivered to the heat source the heat-equivalent of work furnished to the operation of the reversed heat engine. This idea of Sadi Carnot supplied the working principle of the compression-type refrigerator. According to Carnot's theory a heat engine of efficiency close to one becomes, if run backwards, a refrigerator of efficiency near to one.

In 1699, French physicist Guillaume Amontons had suggested air as a working substance in engines. Carnot's idea of air as a working substance anticipated the development of the internal combustion engine. He opined that if practical engineering difficulties with air can be overcome, air as the working substance would offer notable advantages over the vapour of water. Carnot suggested hydrocarbons, or vapours of all bodies, such as alcohol, mercury, sulphur, and so on, which are capable of evaporation as effective working substances in heat engines involving internal combustion. He foresaw the possibility of igniting the fuel in an internal combustion engine by compression alone, an idea first successfully used in practice by Herbert Akroyd Stuart (1864-1927) in 1890, and by Rudolf Diesel (1858-1913) in 1897.

Carnot also proposed a compound engine in which heat, for instance, acts on air and vapour of water: "We might conceive even the possibility of making the same heat act successively upon air and vapour of water. It would only be necessary that the air has, after its use, an elevated temperature, and instead of expelling it immediately into the atmosphere to make it envelope a steam boiler, as if it issued directly from the furnace." This concept of a compound engine by Carnot was first exploited in the so-called 'Still-engine', a Diesel-engine combined with a steam boiler and a reciprocating steam-engine, 100 years later. Recent compound engines combine a gas turbine with another turbine run by steam, or by another vapour.

Carnot's concern with the air or gas engine was not at all capricious, but rather motivated by the miserable industrial conditions existing in France at the time, a country where economy was centrally controlled by the heavy hand of the parasitical State bureaucracy. As a result of this economic stifling, private enterprise in France was in a straightjacket and, consequently, venture capital was unavailable for large private undertakings because the state bureaucracy had arrogated the privilege for such capital and confiscated it for their pet State monopolies. This circumstance kept the French industry and French factories small by British standards. For this reason, French manufacturers were small businessmen who required small prime movers, a requirement quite adequately met by air and gas engines.

Carnot was a scientifically educated engineer, but he self-educated himself in every other area of culture including physical culture, because he was a firm believer in the Renaissance ideal of the *uomo universale*. He, singlehandedly, laid down the programme, the framework, and physical concepts for thermodynamics as a new discipline of physical science. His contemporary physicists were far removed from conceiving such a science because they were ensnared by the conception of molecular motion in physics, and their theories lacked one important idea, the concept of mechanical work, a quantity which can be measured experimentally. It was Sadi Carnot who invented a way to introduce it into the physical theory of heat, and with this theory began the era of really modern physics. It was a stupendous achievement by an engineer, an achievement which has never been quite equalled. In the viewpoint of modern continuum mechanics, Carnot's approach to thermodynamics is quite modern because Carnot realised that the theory of work produced by heat necessarily restricts the class of allowable constitutive relations. He treated the principles of thermodynamics as restrictions placed upon materials, and not upon processes in a body of given materials. Carnot's own restriction for all ideal gases was:

$$v\, L_v / R = \textit{the same function of temperature}.$$

Like any important creator of a new science, Carnot accomplished much that was correct, but he also failed in much. His heat engine theory was not quite complete nor quite correct, and his own notes indicate that he himself became quite aware of the fact, and yet it represents an extraordinary effort of an original individual thinker whose ideas were expressed in general terms which were also relevant to physics, chemistry and meteorology.

Carnot decided to write his monograph in common language so that every educated person, including engineers, could read it with understanding. He even went as far as having his younger brother Hippolyte, who was neither a scientist nor an engineer, read it to check whether it can be understood by an educated layman. It is quite probable that Carnot composed the initial draught of his monograph by using mathematical analysis, and then replaced it, *expressis verbis*, by verbal analysis. This unfortunate decision of Carnot has brought about many sundry interpretations of the meaning of Carnot's theory till this very day.

In 1832, Sadi Carnot fell ill with scarlet fever which developed into a 'brain fever'; then he contracted cholera and died before he was able to publish his revised ideas on thermodynamics. Sadi Carnot was buried with his personal belongings including his papers, which was the customary public health measure in cholera epidemics. A few fragments of his notebook, which had been overlooked, were published posthumously in the 2nd edition of his monograph in 1872.

Sadi Carnot's contribution was saved from oblivion after Carnot's death by the French engineer, Émile Clapeyron. In 1834, Clapeyron put Carnot's arguments into semi-mathematical language and demonstrated their great significance for the physics of gases and vapours. In his study of vapours, Clapeyron established the well-known Carnot-Clapeyron theorem. Clapeyron introduced the graphical form of Carnot's cycle by means of the James Watt's indicator diagram, the P-V diagram, which is now a standard intellectual tool in engineering thermodynamics. However, Clapeyron's formulae for specific heats at constant pressure and constant volume were invalid, for he relied on some misleading experimental results of Francis Delaroche (1775-1813) and Jacques Étienne Bérard (1789-1869), which implied that specific heats vary with the volume or with the pressure of the gas. Clapeyron was interested in Sadi Carnot's work on the thermodynamics of heat engines because Clapeyron was engaged in designing locomotives, and teaching a course on steam-engines at the *École des Ponts et Chaussées*.

Sadi Carnot's work was later further clarified, and extended by the original researches of the Scottish physicist William Thomson (1824-1907)[later Lord Kelvin], of the German physicist Rudolf Clausius (1822-1888) in physics, of American physicist and mathematician Josiah Willard Gibbs (1839-1903) in physical chemistry, and of English physicist William Napier Shaw (1854-1945) in meteorology. The theory of thermostatics of heterogeneous substances by Gibbs made important technological advances in metallurgy, high explosives, refrigeration, and aviation possible.

Thermodynamics and Giffard's Injector :

One of the great achievements of scientific engineering of this period which depended upon thermodynamic considerations, and helped to discredit the caloric theory in favour of the mechanical theory of heat was Giffard's steam injector, a device that intrigued many engineers and scientists. Henri Jacques Giffard (1825-1882) was a French inventor interested in aviation and steam power. He was a self-educated scientific engineer, having learnt mathematics, sciences and engineering on his own from the notebooks of his friends who were students at the *École Centrale* in Paris. Giffard after a formal high school education had become a locomotive engineer which inspired him to become a professional engineer. Giffard's first inventive effort was his design of a high-speed steam-engine which was built in 1849 by H. Flaud in the latter's machine shop in Paris. This engine operating at the speed of 3,000 rpm, had an output of 3 horsepower and weighed only 99 pounds (45 kg). Giffard's engine provided power to a number of machine tools in Flaud's shop.

However, Giffard's major interest was aeronautics. He is responsible for developing the first manned steam-powered dirigible, called the ***aérostat***, which he flew successfully around Paris in 1852. Giffard developed his flying machine rationally by basing his engineering of all the components of the *aérostat* on statics, dynamics, hydraulics, deformable body mechanics, steam-engine theory and scientific engineering design. He began with a careful analysis of the technological problem involved in the design of the powered dirigible. Giffard examined the aerial steam navigation in its broadest context and established what he regarded as the fundamental scientific principles required for its engineering design: a steam engine with high power-to-weight ratio to drive a screw propeller, and an *aérostat* with minimal air-resistance and facile dirigibility. For the first requirement he designed a simplified, lightweight, high-speed, and very efficient steam engine. For the second requirement he designed a cigar-shaped balloon which gradually tapered to a point at both of its ends. He then began to design a 40-foot (12.19 m) *aérostat*, as far as possible by means of scientific theories. In this effort, he developed from the basic principles of mechanics the theory of the propeller by means of which he determined the proper size and angle of inclination of the propeller blades, and then calculated theoretically the air resistance produced by the aerostat in motion. Giffard determined the theoretical deflexion of the propeller blades, which he calculated to be 0.10 feet (0.03 m) during the flight, and pre-inclined the blades in the opposite direction by the same amount of deflexion to counteract this deformation. This deflexion of the propeller blades was discovered independently by the Wright Brothers almost 50 years later, and they solved the deflexion problem of the propeller blades the same way as Giffard had done. Giffard published his analysis of his dirigible in 1851.

In connexion with his steam engine Giffard saw the necessity of designing a new boiler-feed apparatus which had no moving parts. Hitherto mechanical pumps had been used for that purpose but such apparatus had great limitations and were usually a source of considerable trouble. Such a force pump in a dirigible would have been highly undesirable because it would have used up an appreciable amount of power of the steam engine and produced additional friction. After a careful study of the methods of boiler-feed, Giffard came to the conclusion that the principle of induced current would provide the solution to this technological problem.

During his years as a locomotive engineer, Giffard had become familiar with the operation of the blast pipe first introduced into locomotives by the English engineer Trevithick in 1804, which had become an indispensable component of locomotive design by 1840's. The blast pipe consisted of a steam exhaust pipe fitted with a convergent nozzle into the base of the smoke stack of the locomotive. The hot steam expelled with high velocity into the smoke stack collided with the hot waste gases from the furnace and induced them to accelerate upwards. This lowered the pressure in the smoke stack and created an upward draught which increased the rate of combustion in the furnace. This practical experience together with his scientific background afforded Giffard the technological insight that the power of steam-jet lay in its momentum rather than in its pressure, which led him to regard the induced current principle as a momentum transfer principle, and as a mechanism of heat exchange. Giffard's scientific understanding of the principle of induced currents was vital for his conception of the injector. The injector he devised consisted of an arrangement of two nozzles, and a diffuser carrying out several different functions. The steam nozzle converted a portion of the heat and pressure of the steam into kinetic energy by imparting to the steam an extremely high velocity. In the combining tube, the principle of induced currents was used to collide the steam and feedwater and to produce a heat exchange and momentum transfer between them. The delivery tube with the diffuser nozzle transformed the kinetic energy obtained by the water into static pressure which was able to overcome the pressure within the boiler and thus feed water into the boiler. The injector was a very sophisticated device, and from its theory, Giffard determined the limits within which the injector would operate. He published his injector theory in 1860.

The momentum-transfer equation used by Giffard was based on the inelastic collision concept between the steam and the water particles, a collision in which the momentum is conserved, and, therefore, he was able to evaluate the rebound momentum of the mixture of steam and water. He used the heat-exchange equations to obtain the heat transfer between the steam particles and the water particles and the Regnault Formula experimentally established by the French physical chemist Henri Victor Regnault (1810-1878) to find the total heat developed by expanding steam. Since a theory of the motion of steam did not yet exist, Giffard used the theory of incompressible fluid, that is, hydrodynamics and, therefore, he calculated the velocity of the steam through the nozzles by means of the Torricelli efflux principle, which resulted in the most prominent errors in his theoretical work. He also used the assumption of steady flow, and the uniform velocity distribution over the cross-section of the flow entering the delivery tube, conditions which are not satisfied with real fluids. He was familiar with the use of diffusers from hydraulic turbines and recognised that turbulence can be avoided in such tubes by designing a small angle of divergence for the diffuser nozzle. Although all these theoretical limitations affected a successful design of the injector, it did afford Giffard the insight to overcome the ensuing practical problems with the operation of the diffuser. It took Giffard a year to refine his design of the injector to make it function in a satisfactory manner in practice during which time he carried out many experiments to validate, extend, and improve the accuracy of his theory.

In the meantime, in 1857, Giffard had also produced another design to feed boilers which consisted of a turbine running a centrifugal pump mounted on the same axle. This device which could be used on a 100 horsepower steam engine, weighed only about 6½ pounds (3 kg) and had a diameter of almost 4 inches (10 cm). Giffard's priority in the invention of this turbine-operated feeder was contested by the hydraulic engineer Louis Dominique Girard, and to avoid delays in producing a boiler feeder for his aerostat he dropped the turbine-operated boiler-feed system. Giffard injector was an almost immediate commercial success, and it created a technological sensation because according to the caloric theory of thermodynamics it seemed to act like a *perpetuum mobile*, a concept which had been rejected by Sadi Carnot in his caloric theory of heat. This paradox in the caloric theory of thermodynamics in its application to Giffard's injector led to the gradual rejection of the caloric theory of heat by the scientific community, and the injector came to be regarded as a proof for the validity of the mechanical theory of heat. Giffard's own theory had used the notion of convertibility of heat and mechanical work, but his method had been indirect for he did not convert the quantity of work done in his injector into heat units by using the concept of equivalence between heat and work, an equivalence which had been first established by Sadi Carnot in France, and later by James Prescott Joule in England. Giffard's injector revealed the urgent need for the availability of the laws of steam flow because this was the first mechanical device which required precise knowledge of such laws. Therefore, the injector inspired extensive research in thermodynamics, which in turn made thermodynamics an important discipline of engineering science. In 1860's German engineer Gustav Zeuner (1828-1907), French engineer Charles Combes (1802-1872), French engineer Ferdinand Reech, and Franco-German manufacturer Gustav Adolf Hirn (1815-1890), wrote papers on the thermodynamics of the injector. In 1908, the great French mathematician Henri Jules Poincaré (1854-1912), analysed the injector on the basis of the first 2 laws of thermodynamics in his book on physics, because he still believed that the existing theoretical treatments of the injector were thermodynamically not entirely complete.

CHAPTER XI

ORIGIN OF ELECTRICAL SCIENCES: From Lodestone to Electron

Unlike other fields of engineering, electrical engineering almost from its very beginning was rooted in science and, therefore, it is useful to study the early emergence of basic knowledge of electric and magnetic phenomena to understand the development of electrical technology and engineering.

Man has probably always been aware of the effects of lightning and the Aurora Borealis, the electric eel, the electric fireball called 'St Elmo's fire' at the tip of ship's mast, and the attractive power of natural resins when rubbed. The early practical use of the lodestone was in navigation, and the electric eel was used in medical treatments. In general, the ancient natural philosophers considered electricity and magnetism to be quite separate phenomena.

The Ionian natural philosopher and mathematician Thales of Miletos (c.625-c.547 B.C.) knew that rubbing amber (in ancient Greek, *elektron*) creates electrostatic attraction that he appears to have confused with the magnetic attraction of the lodestone. Almost 2 millennia later the infamous Italian Renaissance mathematician, Girolamo Cardano (1501-1576), did distinguish between these 2 kinds of attraction.

The famous ancient Greek dramaturge, Euripides of Salamis (c.480-406 B.C.), referred to the natural magnetic ore as magnesium stone, named perhaps after Magnesia, a city in Thessaly where he resided for a period and where lodestones (in ancient Greek, *magnes*) were usually found.

The magnetic effect was also recognised in ancient China. Apparently, by the 3rd century A.D., a magnetic needle floating on water was used as a compass by Chinese mariners. The Chinese appear to have been aware during the Sung dynasty, about A.D.1000, that iron can be magnetised by rubbing, and they seem to have also discovered the polarity of magnetism and the magnetic deviation.

The year A.D.1600 was an important milestone in electrical science since in that year William Gilbert [he himself wrote his name Gilberd] (c.1544-1603) of Colchester, a graduate at St. John's College, in Cambridge, with B.A., M.A. and M.D. degrees and the Chief Physician to the Court of Queen Elizabeth I of England, published his pathbreaking book < On the Magnet, Magnetic Bodies, and the Great Magnet the Earth; New Physiology Demonstrated with Many Arguments and Experiments> (*De Magnete, magneticisque corporibus, et de magno magnete tellure; Physiologia nova plurimis et argumentis et experimentis demonstrata*), where physiology meant natural philosophy, which besides magnetism dealt in one chapter with a full range of electric phenomena known at the time. Gilbert stressed the difference between the attraction of amber and that of lodestone by pointing out that lodestone attracts only iron, whereas amber attracts only small light bodies excluding iron. Gilbert was the first person to make a clear distinction between magnetism and electricity, but the basis for his distinction was wrong because the mediæval metaphysics and the Neoplatonic philosophy still influenced his thinking according to which electricity binds the particles of a body together whereas magnetism gives it its shape. He believed that all action by means of matter must be by contact, and that if there appears to be action at a distance there must exist a material *effluvium* responsible for it. Since electric attraction could pass through solid matter, it could not be due to material *effluvium*. He applied his theory of effluvium to the Earth's attraction for falling bodies by assuming the atmosphere to be the *effluvium* in this case. In his studies of the lodestone, he determined the poles and its lines of force. After having found the poles of a long magnet, he broke it into pieces and found the poles of the fragments. He boldly postulated the existence of the magnetic field of force (*orbis virtutis*) surrounding the magnet, and demonstrated that the strength and range of the magnetic attraction of the lodestone is proportional to its size. He increased the power of attraction of the lodestones by 'arming' them with steel caps which he called *magnes armatus*. Gilbert employed the Latin term *electrum* for the Greek term *electron* when he referred to 'amber'. He showed that the attractive power of rubbed amber had nothing to do with amber as a substance but was shared by a very large number of other substances, and he called all other substances which behaved after rubbing like amber, *electrica*. The term 'electricity' was first used by Thomas Browne (1605-1682) in his book < Epidemic of Pseudo-Opinion > (*Pseudodoxia Epidemica*) published in London in 1646.

In the 2nd book of < On the Magnet > which contained a detailed study of amber, Gilbert distinguished substances as '*electrica*' and '*non-electrica*', and referred to the electric force for the first time, and to the absence of electric poles in contrast to magnetism. But this classification was not quite correct because he did not know the properties of electrical conductivity and insulation. The substances, mostly metals, he had classified '*nonelectrica*', but after being properly insulated such substances could no longer instantly lose their charge by conduction to the ground and, therefore, could be electrified by rubbing.

Moreover, Gilbert carried out, for the first time, a series of systematic experiments to enquire about a limited part of physical phenomena by means of a 'scientific method' based on the method of induction. Gilbert, through his magnificent research lasting 17 years, made 2 major and 31 minor discoveries about electricity, and changed what had previously been an accumulation of fancy, superstition, mystery and folly about 'magnetics' and 'electrics' into a body of sound knowledge supported by experimental evidence. Gilbert was able to duplicate the phenomenon of terrestrial magnetism by making a small spherical iron model of the Earth magnetised by a lodestone called *terrella*. He attached to *terrella* a series of pivoted iron needles called *versorii*, the first electroscopes which functioned like magnetic compasses, and he concluded from his experiments with *terrella* that the Earth itself is a magnet since Gilbert was able to observe the magnetic 'dip' on his *terrella*. The magnetic 'dip' of compass needles on the surface of the Earth had already been discovered in 1544 by Georg Hartmann (1489-1564) of Nuremberg (*Nürnberg*), Germany, but he never published his manuscript about this observation. Gilbert predicted from his experiments with the *terella* that the magnetic dip would increase as the pole is approached, a contention later corroborated by Henry Hudson (died 1611) who found in 1608 that the dipping needle became almost vertical at the latitude of 75 degrees. Gilbert also remarked that the Earth exerts a couple, and not a single force, upon a needle, an observation already made in 1581 by the compass-maker Robert Norman, who had described the magnetic dipping-needles, the *versorii*, in his booklet < Newe Attractive > (1581), but it was not accepted by his contemporaries. The concept and mechanical theory of the couple was first introduced into mechanics by the French mathematician Louis Poinsot (1777-1859) in 1804. Gilbert erroneously assumed that the magnetic poles of the Earth coincide with the geographic poles which vitiated his arguments concerning the changes in the inclination and declination in different geographic locations. He was the first to surmise that the Earth itself is a giant magnet which is surrounded by a magnetic force field that produces tangible effects in space. By postulating the atmosphere to be the *effluvium*, he explained Earth's attraction of falling bodies with his *effluvium* theory, and attributed the orderly movements of the Sun and the planets in the Copernican theory to the interaction of their *effluvia*. He adopted many ideas of Nicholas of Cusa (1401-1464) in cosmology. The diurnal rotation of the Earth he attributed to magnetic energy. Therefore, Gilbert explained the material world in magnetic terms.

Gilbert had a great influence on Galileo, who in his < Dialogue ...> (*Dialogo ...*) thought Gilbert "worthy of extraordinary applause." Gilbert's book, which served as the foundation of the science of magnetism, was not only the first book on experimental magnetism and electricity but also the first printed book in English on experimental physics. In 1629, Niccolo Cabeo (1585-1650), an Italian Jesuit cleric, reported in his book < Philosophy of Magnetism > (*Philosophia Magnetica*) his discovery of electric repulsion by observing that electrified bodies repelled after being attracted, which had been denied by Gilbert.

Gilbert was scientifically indebted to 4 men: he had carefully studied < Letter about the Magnet > (*Epistola de Magnete*) written in 1269 during the siege of a Muslim city Lucera in Apulia, in southern Italy by the French military engineer, crusader and a nobleman Pierre de Maricourt [his Latin name, *Petrus Peregrinus* (meaning Peter the Pilgrim)] to his friend Siger de Foucaucourt in which the globular lodestone, the *terrella*, is first described, and from whose work Gilbert had largely derived his systematic experimental method. Gilbert demonstrated that the magnetic needle is attracted by the magnetic pole rather than by the celestial pole as Pierre de Maricourt had thought. Pierre de Maricourt, who was keenly interested in all branches of technology and mechanical arts of his time, knew that unlike poles attract and like poles repel each another, that pieces of a magnet become new magnets with new poles. He floated lodestones on water in wooden vessels.

Gilbert knew the studies and observations in the books: < The Newe Attractive > (1581) by the hydrographer, Robert Norman, a retired expert mariner who was an artisan and compass-maker; annexed to Norman's booklet was < A Discourse of the Variation of the Compas, or Magneticall Needle > by William Borough (born 1536),Comptroller of the Queen's Navy in 1583, and commander of a British ship in the battle with the Spanish Armada; and <The Navigator's Supply> published in 1597 by the Archdeacon of Salisbury, William Barlow (died 1625), with whom Gilbert had corresponded, and who was later praised as a forerunner of Gilbert. Besides being very much interested in the practical problems of ocean navigation, Gilbert was also competent in mining engineering, and quite familiar with the techniques of metallurgy and forging. Gilbert's studies opened up the new scientific field of terrestrial magnetism and some aspects of empirical geophysics.

His cosmological work < New philosophy about Our World beneath the Moon, Posthumous Work > (*De mundo nostro sublunari philosophia nova, opus posthumum*) > composed after 1591was edited by his younger brother, and published posthumously in 1651.

In 1660, Otto von Guericke [Otto Gericke before he was ennobled], an engineer-physicist and burgomaster of Magdeburg in Germany, built the first triboelectric generator, which consisted of a sulphur ball revolving about a shaft and made electric when rotated against a dry hand or cloth. It was the first machine able to generate a continuous supply of man-made electricity. He demonstrated the attractive and the repulsive property of electric charges, and discovered electric sparks and their crackling sound. He observed electric conduction and induction, but apparently did not emphasise them. Guericke believed that gravitational attraction was electric rather than magnetic as Gilbert had assumed.

The English scientist and instrument-maker, Francis Hauksbee (died c.1713), who was an assistant to Robert Hooke (1635-1703) at The Royal Society in London, built a combined air-pump and electric machine incorporating a spinning glass globe to perform a series of experiments, the results of which he published in 1709 in a book entitled < Physico-Mechanical Experiments on Various Subjects >, in London. He was able to create a luminous shower of mercury globules in a partially evacuated glass vessel, and a purple glow in an evacuated inner glass vessel of his new electric machine of revolutionary design. In it was the first experimental demonstration of fluorescent lighting. About 1709, Hauksbee built an electroscope consisting of a number of woollen threads suspended from a wire hoop which was concentric with a glass globe and when the globe was rotated and electrified with friction, the threads pointed with their free ends towards the center of the globe. Moreover, Hauksbee was probably the first to investigate experimentally the transmission of sound through water, and the intensity of sound transmitted through the air of different densities.

In 1731 and in 1732, English physicist Stephen Gray (1670-1736), a former assistant to mathematician Roger Cotes (1682-1716) at the Cambridge observatory and to physicist John Théophile Desaguliers (1683-1744) in London, published 2 memoirs on his electrical experiments, in which he demonstrated the process of electrical induction and proved that some substances were conductors of electricity and some were nonconductors, or insulators. He demonstrated electric conductivity in 1729 by proving that electricity travels along a conductor from one place to another. He built an electroscope which consisted of 2 pieces of thread suspended side by side from a silk line. The 2 threads diverged when an electrically charged body was brought to close proximity of the threads.

The French scientist, Charles François de Cisternay du Fay [alias Dufay] (1689-1739), a former soldier turned diplomat, chemist, anatomist, mechanic and botanist, became the superintendent of the King's Royal Botanical Garden. He wrote memoirs on all recognised fields of science. Du Fay devoted his attention to electricity after learning of Gray's experiments, and he succeeded in creating the first theory of electric phenomena in his reports published in 1733, in 1734 and in 1737. He carried out careful experiments and proposed the idea that electricity has 2 distinct forms: the 'vitreous' electricity and the 'resinous' electricity. He established the rule that similarly electrified bodies were mutually repellent, whereas dissimilarly electrified bodies were mutually attractive. Du Fay's concepts replaced the ideas of earlier investigators who maintained that electric phenomena were natural emanations, a kind of material effluvia issuing from the electric bodies. The vitreous electricity was created by rubbing glass, rock crystals, precious stones, animal hair and wool. The resinous electricity was created by rubbing amber, copal, gum-lac, silk, thread and paper. Neutral substances, according to du Fay, contained equal amounts of both electricities. Du Fay's theory explained many common electrical phenomena, but it did not take into account the fact that the electric charge also depends upon the substance used in rubbing. It was he who introduced the term *isolée* (insulated) into the language of electrical science.

The principle of an electrical reservoir, or condenser, called the 'Leyden Jar', was discovered simultaneously, and independently, in 1745 by 2 scientists, Dutch physicist Pieter van Musschenbroek (1692-1761), professor of mathematics and physics at the University of Leyden, and a few months earlier by a German amateur scientist, Ewald Georg von Kleist (1700-1748), Dean of the Cathedral of Kammin in Pomerania. Musschenbroek coated the outside of a glass jar with tin foil, representing an insulator, which contained water, representing a conductor, and a metallic rod which passed through the lid dipped into the water inside the jar. He was able to store considerable electric charge within the jar by applying an electric machine to the rod which made the jar a 'condenser' (now called a 'capacitor') for static electricity. It was an important method of storing electric charge.

Later on, Leyden Jars were arranged into groups, called 'batteries', put on metallic bases and arranged with multiple connexions whereby the electric discharge was considerably multiplied. The Leyden Jar batteries, with their higher potential for discharge permitted experimenters to try longer distances of discharge.

In 1747, the American political philosopher, inventor and the first international scientist, Benjamin Franklin (1706-1790), used the Leyden Jar to demonstrate experimentally his new theory of electricity. Franklin theorised that electricity consisted of 2 possible states of electrified matter, and his theory soon replaced du Fay's two-fluid theory of which he was unaware at the time. Franklin's theory was successful because it did explain many electrical phenomena without difficulty. Franklin believed that all matter contains electricity which he called 'electric fire', and that charging a body positively, or negatively, merely adds to, or subtracts from the quantity of electricity normally present in the body. Franklin, moreover, advanced an amazingly modern idea about the nature of electricity: "The electrical matter consists of particles extremely subtle, since it can permeate common matter, even the densest, with freedom and ease as not to receive any appreciable resistance."At present a body which is negatively charged is considered to have a surplus of electrons since electrons are regarded as negatively charged. Franklin's particles of 'electric matter' have in modern times been identified and experimentally measured as 'electrons'. If in Franklin's terminology 'positive' and 'negative' are interchanged and electrons are substituted for his 'electric fire', then his theory becomes in certain regards similar to the modern theory of electricity. For instance, if static electricity is regarded as an accumulation of electrons or a deficiency of electrons, then the modern understanding of static electricity is similar to what Franklin proposed. For this reason, it is justified to consider Franklin the founder of the theory of electricity. In 1749, Franklin hypothesised that lightning is an electric phenomenon, a supposition which he corroborated in 1752 by his famous kite experiment. He also made the first practical application of electricity when he invented the 'lightning rod' as the lightning conductor. Franklin's simple theory of electricity, which was useful in electric experimentation, was largely ignored for more than a century.

Since the availability of Leyden Jar batteries with their high potential made brief discharges over longer distances possible, the English physician Dr. William Watson (1715-1787), attempted to determine the velocity of electric conduction by installing an 8,000-foot (2,438.4 m) circuit in which the Earth was the 'return circuit'. In 1748, he found the velocity of the electric conduction to be nearly instantaneous over a 4-mile (6.44 km) circuit. He was the first person to demonstrate that electricity can pass through vacuum. Watson also improved the Leyden Jar by dispensing with the water content of the jar and lining the inside of the jar also with tin foil. Watson referred to the 'plus' and the 'minus' electricity in his introduction of Franklin's theory of electricity to European scientists.

Despite such efforts as that of Watson, the 2 fluid theory based on du Fay's ideas became popular among the continental physicists in the 18th century. This theory presumed the existence of 2 'weightless' fluids which acted as positive and negative electricity. Material bodies in an unelectrified state were assumed to contain equal amounts of these imponderable fluids but became charged positively or negatively when the positive fluid or the negative fluid respectively was in excess. Although this theory could be applied to the explanation of various observed phenomena of electrified bodies, it had the effect of divorcing electricity from the matter itself which was conceptually a backward step in comparison with the theory of Franklin.

In 1754, John Canton (1718-1772), a private schoolmaster and a Fellow of the Royal Society who gave the first experimental demonstration that water is incompressible, devised an improved electroscope consisting of 2 suspended, equally charged cork or elder pith balls, the repulsion of which was used for the relative measurement of electric charge. Abraham Bennett (1756-1799), an English clergyman, introduced the most sensitive 'electroscope' still in use which consisted of 2 much lighter gold leaves suspended at the lower end of a metal bar.

In 1767, the English clergyman, philosopher and autodidactic chemist, Joseph Priestly (1733-1804), speculated in his book, < The History of Present State of Electricity >, about the quantitative law of the electric force of attraction or repulsion acting between 2 spheres charged with electricity, and conjectured it to be inversely proportional to the square of the distance between the 2 spheres, as in Newton's theory of gravitation. Priestley, moreover, advanced an explanation for the oscillatory nature of the discharge from the Leyden Jar which had been first noticed by Benjamin Franklin. Further progress in the study of electric force resulted from the invention of the 'torsion balance' by Reverend John Mitchell (1727-1793) in 1770.

In 1771, English physicist and recluse Henry Cavendish (1731-1810) by using Mitchell's torsion balance discovered experimentally that 2 electrostatic charges produce a pair of equal and opposite electric forces which are proportional to the inverse square of the distance between them, but true to his reclusive habit he did not publish his findings. In 1772, Cavendish determined the capacity of a conductor: the ratio existing between the quantity of electricity stored, called the 'charge' and the degree of electrification of that conductor, called the 'electric potential'. He established that 2 bodies of different shapes connected by a conducting wire have the same electric potential without carrying the same charge, which made the capacity of a conductor dependent upon the geometry of that conductor. Cavendish, who must be counted among the greatest experimental scientists of all time, kept a copious set of laboratory notes which his relatives published only about a 100 years after his death. These notes revealed that Cavendish had discovered all the laws of electric and magnetic interactions at the same time as Coulomb, and his basic contributions in physical chemistry were comparable to those of the French physical chemist Antoine Laurent de Lavoisier (1743-1794). Cavendish also applied a balance to the study of extremely weak gravitational forces acting between small bodies and, on the basis of these experiments, he was able to determine a rather exact value of the mass of the Earth.

In 1785, the French engineer and physicist Charles Augustin de Coulomb (1736-1806), invented a more accurate torsion balance in 1780, and by means of it he was able to make more accurate experimental measurements of the electric force than Cavendish. Coulomb established the inverse square law of electric force between 2 electrically charged bodies known as the 'Coulomb law': 'The force of attraction between two electrically charged bodies is directly proportional to the two electric charges and inversely proportional to the square of the distance between them'. It represents the first quantitative law in electricity which constitutes the quantitative basis of electrostatics since the electrostatic unit of charge can be defined by the Coulomb law. Electricity became a genuine science with the establishment of the Coulomb law. Moreover, Coulomb proved by the use of his torsion balance that his inverse square law is also valid for magnetic interaction: 'That the force of attraction or repulsion between two magnetic poles is directly proportional to the strength of the poles and inversely proportional to the square of the distance between the poles'.

The most impressive and the largest electrostatic generator was designed by the Dutch physician and botanist Martin van Marum (1750-1837) for the Teyler Museum in Haarlem, in The Netherlands. It was constructed with a magnificent workmanship by the English instrument-maker John Cuthbertson in 1784. In 1789 Cuthbertson added to the machine the auxiliary battery of 25 giant Leyden Jars each 20 inches (50 cm) high. This gigantic electrostatic machine consisted of 2 glass disks, each 65 inches (1.65 m) in diameter and spaced 7½ inches (19.1cm) apart, and it could strike a spark 2 feet (0.61 m) long between its large spherical brass electrodes. The disks had 8 rubbing pads of waxed taffeta, and the machine had the charge-collecting metal combs which came within 1 ½ inches (3.81 cm) distance from the inner surface of the glass disk. The receiving standard terminated in 2 spheres which were 1 foot (0.30 m) in diameter. The 2 glass disks were rotated by 2 or 4 operators using compound cranks. Employing this gigantic electrostatic generator the relative melting of metals was determined, the branched nature of long electric sparks was observed, earthquakes were simulated, and the effects of electrical discharge on magnetism and on gases were investigated. Cuthbertson was able to melt a 6 inches (15.2 cm) long iron wire 1/40th of an inch (0.06 cm) in diameter with this giant electrostatic generator which is still extant and on display in Haarlem.

In 1786, an Italian scientist and professor of medicine at the University of Bologna, Luigi Galvani (1737-1798), discovered that a frog's leg contracted when he simultaneously touched the muscle and its connecting nerve with the ends of a metallic conductor made up of a rod of zinc and a rod of copper. When the rod was made of one metal, no contraction of the leg occurred. Galvani considered 2 possible explanations for this experience: either the metals were creating the electric charge and the frog's leg acted as an electrometer, or the frog's leg supplied the electricity and the rods acted as conductors. Galvani decided for the second possibility and called it 'animal electricity' in his memoirs of 1791. Of course, he was not entirely in error, for the existence of 'animal electricity' in living tissue has been proven in recent times. Actually, the frog's leg muscle acted both as an electrolyte and a current indicator in Galvani's experiment.

Already in 1678, a noted Dutch scientist, Jan Swammerdam (1637-1680), had experimented with the effect of electricity on animal nerves and muscles by bringing the nerve which protruded from a glass tube into contact with a copper ring and silver wire, upon which the muscle contracted. In 1762, a Swiss philosopher, Johann Georg Sulzer (1720-1779), had observed the electric effect 2 joined metals have on living tissue when he placed his tongue between 2 pieces of different metals joined together, such as silver and zinc or copper and tin, and experienced an itching sensation in his tongue. However, Sulzer did not draw any definite physical conclusions from this experience.

Italian physicist Conte (Count) Alessandro Giuseppe Antonio Anastasio Volta (1745-1827), professor of physics at the University of Pavia, was the first scientist to prove that 2 different metals in contact produce electricity. He constructed a sensitive electroscope in 1782 for the detection of the presence of minute amounts of static electricity, and by means of it he discovered the presence of electric charge when two different metals were in contact. Volta called the generation of electric current when the ends of a composite wire, formed by 2 wires of different metals joined together at their juncture, are in a water solution of salt, 'galvanism', in honour of Galvani. Volta produced his important invention, known as the 'Volta Pile', a primary electric battery which provided a continuous flow of electric current in contrast to the Leyden Jar which did not. It consisted of pairs of zinc and silver disks separated by brine-soaked cloth or paper. To provide even a stronger continuous flow of current, Volta devised his 'crown of cups' consisting of a ring of cups filled with brine into which were placed alternating strips of silver and zinc. The Volta Pile evolved in time into various forms of wet electric batteries. Volta's battery proved to be the most important instrument for the development of electrical engineering. He introduced the term 'condenser' for devices like the Leyden Jar because electricity at that time was considered an 'imponderable fluid' which could be stored or condensed. The modern name for a condenser is 'capacitor'. In 1800, a German pharmacist, chemist and amateur physicist, Johann Wilhelm Ritter (1776-1810), who made several fundamental discoveries in physics such as the ultraviolet rays in 1801 and the principle of electric accumulator in 1802, was the first to attribute the generation of electric current in the Volta Pile and other electric batteries to chemical reaction.

The electrolytic decomposition was first discovered in 1797 by the German naturalist Alexander von Humboldt (1769-1859) in an experiment with a vessel or cell containing a silver and a zinc electrode placed in a layer of water. Ritter used this electrolytic method in 1799 to separate copper from a solution of cupric sulphate.

In 1801, an English physician Dr. H. Moyes first observed an electric spark generated between charged carbon rods in his experiments with the Volta Pile. In 1802 Etienne Gaspard Robert [alias Robertson] also observed an electric arc springing between 2 carbon rods. In 1808 the English chemist Humphry Davy (1778-1829) built a huge Volta battery made of 2,000 zinc and copper plates, and with it he extracted by means of electrolysis a new metal from caustic potash which he named 'potassium'. He used electrolysis to isolate sodium from caustic soda, and to reduce alkaline earth to such metals as calcium, magnesium, barium and strontium. Davy's many experiments in electrolysis helped to found 'electrochemistry'. In 1809 and 1810, Davy broke new ground in electrical engineering when he created a powerful electric arc 3 inches (7.6 cm) long between 2 charcoal electrodes, which had a light intensity comparable to that of the Sun. However, his wood charcoal electrodes burnt up rapidly. This experiment of Davy showed a way to the first practical electric illumination by arclight, and to the electric furnace. Davy could not develop the electric arc any further because of the intrinsic limitations of his source of electric power, the Volta battery.

In 1797, an English physician, George Pearson, had used the short discharge from Leyden Jars to decompose water into hydrogen and oxygen and then, by passing a spark through the vessel containing the mixture of these gases, to reunite them into water again. In 1799 Johann Wilhelm Ritter produced oxygen and hydrogen by electrolysis of water. The electric current in his electrolysis was provided by the discharge of a Leyden Jar.

In 1800, Dr. Anthony Carlisle (1768-1840), a well-known surgeon and professor of anatomy at the Royal Academy, privately learnt about Volta's long report on his Volta Pile experiments which Volta had sent to the president of The Royal Society for publication, because Volta was a Fellow of The Royal Society. Since Carlisle was chiefly interested in physiological effects of electricity, he consulted a schoolmaster, chemist and civil engineer, William Nicholson (1753-1815) who published a scientific periodical < Nicholson's Journal > devoted to general science. Carlisle and Nicholson repeated Volta's experiments and built a Volta Pile which they then used to supply the electric current in Pearson's experiment on the electrolytic decomposition of water instead of the Leyden Jars used by Pearson. Carlisle and Nicholson immediately published the results under their own names in Nicholson's periodical before Volta's memoir was read at The Royal Society, but this deception did not succeed for Volta's investigations of the Volta Pile became known through other sources. Carlisle and Nicholson were accused of scientific plagiarism.

The next important, and the greatest advance which triggered the precipitous development of electrical engineering, was made in 1819 by the Danish scientist Hans Christian Ørsted (1777-1851), professor of physics at the University of Copenhagen, who wanted to find out if electric current from the recently invented chemical batteries possessed the same attractive power as static electric charge, but instead he discovered accidentally the mutual influence that exists between electricity and magnetism. He discovered that the electric current in a conducting wire does not attract objects but instead causes a magnetic needle under the wire to rotate into perpendicular position relative to the wire. Before Ørsted's revolutionary discovery that electric current does interact with magnetism, Coulomb had argued that electricity and magnetism do not interact, and his views were accepted by all physicists. In 1812, Ørsted had published a memoir, <View of Chemical Laws>, in which he had discovered the electrochemical forces. Ørsted correctly postulated that both the magnet and the electric current build up a 'field of force', and that the 2 force fields can act upon each other. He called this phenomenon 'electromagnetism'. This idea introduced a new physical force into science for hitherto it was believed that most if not all physical phenomena could be explained by gravitational forces alone. Ørsted published his discovery of electromagnetism in the memoir <Experiments about the Interacting Effect of Electricity on Magnetic Needle> (*Experimenta circa effectum conflictus electrici in acum magneticam*) in 1820, which was promptly translated into French, Italian, German, English and Danish languages.

In 1825, Ørsted produced for the first time metallic aluminium by means of an electrochemical method. Ørsted was versatile man of learning: his doctoral dissertation was about the philosophy of Immanuel Kant (1724-1804).

In 1820, the precocious French mathematician, André Marie Ampère (1775-1836), professor of mathematics at the **College de France** and the **École Polytechnique** and a man of genius, after learning of Ørsted's fundamental discovery on the mutual influence between electric current and magnetism, carried out a number of remarkable experiments of his own within a few short weeks on the basis of which he published an epoch-making series of memoirs, as well as his renowned treatise < Collection of Electrodynamic Observations > (*Recueil d'observations électro-dynamique*) in 1822, in which he gave a complete and original theory of electrodynamics that relied heavily on theoretical concepts used in classical mechanics, a subject in which Ampère was a recognised expert, and the rotational effects of 2 circuits. He established the general law of electrodynamics which is analogous to the Newton's law of gravitation and consisted of the law of forces exerted by 2 infinitesimal elements of electric current-carrying wires upon each other from which the quantitative dynamics of all electric and magnetic phenomena had to be obtained by integration. He showed that the connexion between electricity and magnetism is not an isolated effect between electricity and magnetism but rather constitutes a fundamental aspect of it which besides electrodynamics also manifests itself in electrostatics and electrochemistry. He demonstrated that an electric current not only acts on a magnetic needle but that 2 electric currents also act upon each other; that there is an attraction between 2 parallel wires carrying electric currents if the currents run in the same direction and a repulsion if the 2 currents run in opposite directions; that a coil of copper wire which is free to rotate around a vertical axis assumes a north-south direction like the compass needle if an electric current is run through the coil; and that 2 copper coils carrying electric currents interact with each other like 2 bar-magnets. Ampère introduced the concept that natural magnetism is created by electric currents present inside the magnetised body. He postulated that each molecule of a magnetic material which internally carries a circular electric current represents a tiny electromagnet. When the material body is magnetised the molecular electromagnets are oriented, at least partially, in one direction thereby causing their magnetic attraction or repulsion. If the material body is not magnetised, the individual molecular electromagnets are randomly oriented in all directions thereby producing no magnetism. These profound concepts of Ampère were completely corroborated hundred years later by modern physics which considers the magnetic properties of atoms and molecules to be created by electrons spinning round the nuclei of atoms or spinning rapidly round their own axes. He determined the attraction and the repulsion for straight, and spiral conductors, explored the electromagnetic behaviour of 'solenoids', a name he coined for helical wires carrying electric current, and proposed an astatic galvanometer. He showed that electric current influences a magnet but was itself incapable of producing permanent magnetism. Ampère demonstrated that 2 solenoidal wires carrying current behaved like true magnets in polar attraction or repulsion. He also showed that a current-carrying wire can set a magnet into motion and *vice versa*. Ampère illustrated his rule for relating the direction of the flow of electric current in a wire to the direction in which the adjoining compass needle would point by a little 'manikin', called by his colleagues '*le bon homme d'Ampère*'. If his manikin swims in the current direction to face the needle, then the north pole is always in the direction of the manikin's left arm. Ampère's theory laid the foundation for 'electrodynamics', a name which he himself coined, a theory which still stands today as one of the perfect creations in science. Ampère was the first scientist to apply higher mathematics to electric and magnetic phenomena, and his theories were initially received by his contemporaries with great skepticism. Maxwell, another man of genius, considered this series of memoirs by Ampère to be one of the most brilliant achievements in science. Since Ampère had accomplished with his theory of electrodynamics what Newton had done for gravity and Coulomb for electrostatics, Maxwell regarded Ampère worthy to be called the 'Newton of electrical science'. Ampère's electrodynamic theory is one of the most brilliant single achievements in science, but it was mathematically so complex that many physicists could not understand it. In 1820, Ampère also promoted the idea of the electromagnetic telegraph, first suggested to him by the experiments of the famous French mathematical physicist Pierre Simon Laplace (1749-1827), in which the response of the magnetic needles to tiny magnets produced by electric current-carrying wires would identify letters. Moreover, Ampère's studies of the magnetic properties of solenoids made it possible to produce strong electromagnets important in electrotechnology. The magnetic effects of electric currents afforded the means of measuring the electric current strength, and electrodynamics made possible a 2nd system of electrical units which was independent of the Coulomb law.

Ampère's intellectual talents and interests were very profound and broad. He wrote on the foundation of mathematics, physics, chemistry, psychology and natural history, and developed a new and extensive classification of all sciences. He was also a talented portrait artist.

The French physicist of Spanish extraction, Jean Dominique François Arago (1786-1853), professor of mathematics and physics at the *École Polytechnique*, and the French chemist Joseph Louis Gay-Lussac (1778-1850) discovered that an electric current passing through a solenoidal wire coil induces magnetism in an iron needle placed within the coil. Arago also noted that the electrically induced magnetism in steel was permanent, whereas wrought iron rapidly lost its induced magnetism. Thus, Arago discovered 'magnetic induction' by electric current, and a new technological means of making electromagnets with current-carrying solenoids. He also performed the famous 'Arago disk experiment', in which a copper disk is rotated rapidly beneath a magnetic needle which begins to follow the rotation, an action later explained by Faraday to be due to the 'electric eddy currents' induced in the disk. The puzzle provided by the action of Arago's disk was important in electrotechnology because it led to the development of electric generators and electric motors.

In 1820, one of the first practical applications of Ørsted's discovery was made by Johann Salomo Christoph Schweigger (1779-1857), professor of physics and chemistry at the University of Halle in Germany, with his invention of the 'galvanic multiplier'. It consisted of an insulated coil of 100 turns which produced a magnetic effect on the magnetic needle that was 100 times as large as that produced by Ørsted. It was the first true galvanometer.

A physician, Dr. Thomas Johann Seebeck (1770-1831), a Baltic German native to Estonia and a friend of poet and amateur scientist Johann Wolfgang Goethe(1749-1832), experimented with this galvanic multiplier and came to the conclusion that the effective power of a current flowing in a coil is not directly dependent upon the number of turns because the increased length of the wire also increases the resistance of the wire. Besides his studies of phosphorescence Seebeck discovered in 1821 the so-called 'Seebeck thermoelectric effect', that is, the conversion of heat into electricity, which is important in modern semiconductor devices. Seebeck discovered that when 2 conductors made of different metals [in his apparatus copper and bismuth wires] are joined at their ends to form a continuous circuit, and the 2 junctions are maintained at different temperatures, a small steady electric current flows continuously round the circuit. The magnitude of the electric potential and the resulting thermoelectric current appeared to depend upon the nature of the metals used in the conducting wire loop, and to be proportional to the difference in temperatures maintained between the 2 junctures. The conversion of heat into electricity was not quite thoroughly interpreted by Seebeck himself and, therefore, the 'Seebeck thermoelectric effect' was not followed up for more than a century.

In 1834, a Paris watchmaker, Jean Charles Athanase Peltier (1785-1845) discovered the so-called 'Peltier thermoelectric effect' which is the reverse of the 'Seebeck thermoelectric effect'. Peltier discovered that if a weak current generated by a battery is sent round a wire loop made of 2 wires of different metals when both junctions of the wires in the loop have the same temperature, heat is created at one junction and absorbed at the other. The electric current brings about cooling instead of heating if it runs through a junction in a loop of 2 different metal wires in the same direction as would a thermoelectric current generated by heating the same junction.

Later the Italian physicist Leopoldo Nobili (1784-1835) built the so-called 'thermopile', or 'thermomultiplier', by linking together a large number of thermo-elements, which became of great importance in the investigation of infrared rays since thermocurrent was preferred to battery current because the batteries producing the current were subject to frequent changes of voltage.

In 1824, William Sturgeon (1783-1850), an English inventor and educator, produced his first electromagnet, both the straight and the horseshoe type, by applying Arago's observation on the magnetisation of a soft iron bar by passing an electric current through a solenoidal conductor wrapped around the iron bar. Sturgeon's electromagnet was able to carry a weight of 9 pounds (4.08 kg). Sturgeon also published the first journal on electrical science in English, < Annals of Electricity >.

In 1826, Georg Simon Ohm (1787-1854), an ingenious German gymnasium teacher, formulated the fundamental law known as the Ohm's law, which established that the current I in a wire is directly proportional to the electric potential V, and inversely proportional to the resistance R of the wire: $I \propto V/R$, which was as important a discovery in electricity as that of Ampère. Ohm used Seebeck's thermoelectric method for the production of the precise quantities of current necessary in his experiment. It appears that Henry Cavendish had also discovered this law from his rough-and-ready experiments but he never published it. In 1827, Ohm published the theoretical deduction of his law in his memoir < The Galvanic Chain, Mathematically Worked Up > (***Die galvanische Kette, mathematisch bearbeitet***) in which he taught the correct use of terms such as electromotive force, current intensity and resistance. He showed clearly how the resistance of the wire depends upon its length, cross-section and material.

Ohm made just as fundamental and important contribution in acoustics. The musical quality and transmission of sound are understood today through the analysis of the distribution of energy among the components of sound. Ohm published a pathbreaking paper on sound in 1843 in which he gave what is known today as 'Ohm's law of acoustics': 'Musical sounds depend only upon the distribution of energies among the harmonics, and have no dependence upon the differences of phase'. Ohm's law of acoustics solved the riddle of musical tone and, thereby, saved acoustics from a fundamental confusion.

Despite his brilliant scientific work, Ohm was completely ignored in Germany, until The Royal Society in London awarded Ohm in 1841 their Copley medal, the coveted Blue Ribbon of science at the time, and elected him to membership in 1842. His work had been so unfairly criticised in Germany that Ohm felt forced to resign his gymnasium teaching position, and live for 6 years in penury and bitter disappointment. After the honour accorded to Ohm by The Royal Society in London, Ohm was finally granted the professional recognition and respect he so richly deserved also in his own country by being appointed to the chair of physics at the University of Munich in 1849.

In 1828, American scientist Joseph Henry (1797-1878) improved dramatically the electromagnet of Sturgeon by adopting, for the first time, insulated wire which was wound over itself in a manner that compensated for the effects of obliquity of the turns, so that the final effect of the current was at the required right angle. In 1829, when Henry carried out experiments on the capacity of electromagnets he observed the occurrence of a strong spark on the breaking of the current in a long wire wound into a coil around an iron core. It was the first observation of 'self-induction' in history. Henry constructed the first electric motor also in 1829, and gave a demonstration of a telegraphic instrument over a mile (1.61 km) length of wire. In 1830, Henry built an electromagnet with a holding power of 650 pounds (2,89 N), and in 1832 he made an electromagnet weighing 100 pounds (45.36 kg), which had a holding power of 3,500 pounds (15,569 N). He attempted to use still greater numbers of turns, but he discovered as Seebeck had found before him that turns added beyond a certain number became less effective in increasing the lifting power of the magnet, since he did not yet quite understand the limitations of current in his coils as stipulated by the Ohm's law. Henry discovered in 1831 the principle of electrical induction which he applied to the transformation of magnetism into electricity. In the same year he also discovered in his research on 'intensity electromagnets' the principle of the 'step-up' and the 'step-down' transformer. Despite the fact that Henry published in 1832 all his electrical discoveries in the < Silliman's Journal >, they did not have the professional influence they deserved.

In 1832, William Jenkins informed the great English self-educated chemist and physicist, Michael Faraday (1791-1867) of his experience of having suffered a great electric shock when breaking the current in a circuit containing a wire coil. Faraday immediately began to investigate this phenomenon of the so-called 'extra current' generated by making and breaking the electric current in a circuit. Faraday discovered that the electric spark at the switch is much greater when there is a coil in the circuit. Experiments with this circuit led Faraday to his independent discovery of self-induction. Faraday published the results of his investigation on self-induction and the 'extra current' in 1835 which was very influential, and led to the technological development of induction coils for electric discharge.

In 1832, Faraday made electrolysis which had been established by Humphry Davy an exact science by announcing what are now called the 'Faraday laws of electrolysis'. These laws of Faraday reduced electrolysis to quantitative rules which can be expressed in modern terms as follows:

1. The mass of substance liberated at an electrode during electrolysis is proportional to the quantity of electricity driven through the solution.
2. The mass liberated by a given quantity of electricity is proportional to the atomic weight of the element liberated and inversely proportional to its 'valence', that is, to the combining power of the element liberated.

These laws of Faraday established the close connexion that exists between chemistry and electricity, and they were of great technological importance to the contemporary electroplating industry. These laws also implied that electric current consists of particles, a point of view which had been favoured by Benjamin Franklin. In 1843 Faraday was able to demonstrate the law of conservation of electricity but it did not receive the recognition it deserved by the profession.

In 1842, Henry had observed that a one-inch spark could magnetise needles over a distance of 30 feet (9 m) or more, which led him to hypothesise the distance of an electric plenum which is distributed by an electric spark in the form of a diffusing motion in the electricity of space that is comparable to that of a spark in the case of light – an astounding analogy made more than twenty years before light was demonstrated to be an electromagnetic phenomenon by Maxwell. He built the first model electromagnetic telegraph and relay, as well as the first model induction motor.

Principle of Electric Motor and Electric Generator

Michael Faraday, was absolutely convinced that the process in which magnetism is created by electricity can be reversed, that is, magnetism can create electricity. In 1821, Faraday succeeded in producing electromagnetic rotation in an experiment in which a suspended wire with its free lower end dipped in a bowl of mercury was revolving continuously round a permanent magnet supported in mercury, when the wire and mercury were connected to the opposite poles of an electric battery. This fundamental experiment which proved that electricity can produce motion, served as a discovery which led to the invention of the electric motor.

In late 1831, Faraday, being inspired by Arago's magnetic phenomena, made fundamental discoveries of the following basic principles: mutual induction, self-induction, transformer, and generator. The principle of mutual induction and the principle of self-induction had been independently discovered by Joseph Henry before Faraday, but Henry published his discoveries only in 1832. However, it was Henry who constructed the first functioning model of a reciprocating electric motor in 1829. The very first electrically operated toy motor had been built by a Scottish cleric and teacher, Andrew Gordon (1712-1751), in 1745. It consisted of a metal star pivoted about its center, and having all its tips inclined in the same direction relative to their radii. An electric charge to the tip of the star sets it in motion like a pinwheel as the charge leapt from one tip of the star to the next.

In 1845, Faraday succeeded in demonstrating that a powerful electromagnet had an effect on light rays: a strong magnetic field deflected the plane of polarisation of a suitably 'polarised' ray of light. When he passed a polarised light through a piece of the so-called 'heavy glass' [a special kind of glass consisting mainly of lead silicate he had made years ago] between the poles of an electromagnet, the plane of polarisation rotated, and this rotation became a maximum when the light-ray was directed along the magnetic lines of force. This phenomenon is known as the 'Faraday effect' which he described as magnetisation of light and illumination of magnetic lines of force. Faraday interpreted this to mean that in some sense, 'light was electromagnetic in nature'. His attempts to find the effect of the magnetic field on the colour of light were not successful. In addition to the Faraday effect, later two other phenomena were discovered which also indicated a close relationship between electricity and magnetism on the one hand, and between electricity and light on the other hand, or the magneto-optic and electro-optic effects of the Scottish physicist John Kerr (1824-1907).The magneto-optical effect is the rotation of the plane of polarisation of polarised light on reflexion from the reflecting end-piece of a magnet. The electro-optical effect is the double refraction produced in certain dielectrics in an electric field which results in their behaviour as uniaxial crystals with their optical axis being parallel to the electric lines of force. Faraday's great discoveries such as electromagnetic induction, laws of electrochemistry, dielectric phenomena, magnetic conductivity or permeability, diamagnetism, and magneto-optical rotation emanated from his search for a general correlation of forces in physical nature. In his studies on dielectric and electrolytic processes he became convinced that such processes are transmitted by contiguous action of a medium in the space between charged bodies. By not believing in 'action-at-distance' which is still generally accepted as the explanation of electrostatic forces, Faraday assigned the medium between 2 separated bodies, which are subjected to electric or magnetic attraction or repulsion, physical properties analogous to the properties of an elastic material that can transmit forces through its deformation. Since electric and magnetic forces can act through vacuum it was necessary for him to introduce an intangible universal medium called the 'æther' which fills the entire space. From that time forwards the effect of the medium became an important part in the theory of electricity. His greatest theoretical contribution came from his ideas about the lines of electric and magnetic force which he thought to traverse the entire space. The image of the lines of force, Faraday obtained from the well-known figures of iron filings in the vicinity of the magnet. He regarded the lines of force to fill the entire space and the tangents to these lines to give the direction of the intensity of the magnetic field at every point. He discovered that the electromagnetic force of induction induced in a wire which cuts magnetic lines of force is proportional to the number of lines of force that were cut in a unit of time. In his remarkable conceptual memoir of 1852, Faraday extended the principle of contiguous action in a general qualitative analysis of the magnetic and the electromagnetic phenomenon which was based upon the notion that lines of magnetic force have the physical property of shortening themselves and repelling one another laterally, an idea which was quantitatively formulated by Maxwell in 1861. Faraday assumed that the 'lines of force' had such properties as tension, attraction, repulsion, motion and some others. The idea of lines of force was not quite new, for in 1629 an Italian Jesuit priest and scientist Niccolo Cabeo (1585-1650), had introduced *lineae virtutis* (*lines of force*) in his treatise < Philosophy of Magnetism > (***Philosophia Magnetica***). Since every line of force is a closed curve in space, then the lines of force which intersect any small closed curve in space form a tubular surface returning into itself, and such a surface is called a 'tube of force'. Information on the direction and magnitude of the magnetic intensity can be derived from such a 'tube of force'.

Faraday's last discovery in electricity was diamagnetism, which he also made by using 'heavy glass'. After comparing diamagnetism with paramagnetism he came to the conclusion that magnetism was a general force of nature which could be evoked in all bodies and not only in iron. The phenomenon of diamagnetism had been already observed by the French physicist Antoine César Becquerel (1788-1878) and Thomas Johann Seebeck, but the basic principle of diamagnetism had never been understood before Faraday. Faraday's researches on the magnetic and electric phenomena were responsible for opening the new era of modern physics.

Following Ørsted's discovery of the interaction between electric current and magnetism Henry and, particularly, Faraday established the fundamental principles of the electric generator, the electric motor, and the electric transformer – the 3 pillars of electrical power technology.

Based on his ingenious experiments on electricity and magnetism, Faraday, a supreme experimenter, finally proved that relative motion between the magnetic field and the conductor induces electric current to flow in a conductor, which Faraday expressed in his law: 'The voltage induced in a conductor is directly proportional to the rate at which the conductor cuts the lines of magnetic flux'. To maximise the rate, the conductor in an efficient generator should pass through the magnetic field at right angles to the magnetic lines of force. Faraday also showed that coils wound on an iron ring gave an induced current much greater than coils wound on a wooden core. The current was much stronger because the iron ring provided a superior magnetic circuit by compacting the magnetic flux so that more magnetic lines of force passed through the iron, but neither Faraday nor any of his successors realised this important aspect of magnetic circuits, particularly in the early development of dynamos when the character of magnetic circuit was more or less ignored. In the late 1831, Faraday had discovered electromagnetic induction and all its governing laws. He also made a working model of an electric generator by rotating a copper disk between the poles of a strong magnet and discovered that a steady electric current was induced across the copper disk, before moving on to other research. These scientific discoveries laid the foundation necessary for electrical power technology. It took almost half a century before Faraday's theories and all his technical hints were incorporated into the design of really efficient electric generators of the 1880's.

In 1834, Heinrich Friedrich Emil Lenz (1804-1865), a Baltic German native to Estonia who investigated electric induction at the same time as Faraday and Henry, proposed the law overlooked by Henry and Faraday, now called the 'Lenz law', which established a connexion between the phenomenon of self-induction and the rest of the electrodynamic theory: 'The self-induced voltage always opposes any change in the impressed voltage'. The rising and falling of current is accompanied by the rising and falling of magnetic force lines which not only cut the secondary coil but also the primary coil itself thereby creating the self-inductance and inducing a voltage in it which according to the Lenz law opposes the change in the impressed voltage and, therefore, reduces the peaks in the impressed voltage. This smoothing-out effect of the self-induced voltage called 'counter-voltage' is used in filter circuits of radio and television sets. The Lenz law, which indicates that a current induced by electromagnetic forces always produces effects that oppose those forces, must be taken into account in the design of all electric equipment. In 1833, Lenz discovered that the resistance of a metallic conductor changes with the temperature: it increases with the rise and decreases with the fall of temperature.

Electromagnetic Field Theory

Electromagnetic field theory is of such fundamental importance to modern scientific engineering because it is one of the great monuments of human ingenuity, ranking with Newton's gravitational theory, which had a decisive influence on modern scientific engineering. An account of its historical background is valuable for the knowledge of how and why it was created.

Faraday, who had not been feeling well from 1840 to 1845, probably because of mercury poisoning, was back at work in 1845 investigating his idea that the forces of electricity, magnetism, light and gravity are intimately related, and was able to show that polarised light is affected by a magnetic field. He was unable to achieve similar results with an electric field, which, however, was successfully demonstrated by the Scottish mathematical physicist John Kerr (1824-1907) in 1875, who discovered the so-called 'Kerr electro-optical effect' in which the electric field makes substances such as glass and other insulators capable of double refraction when placed in an intense electric field, since molecules of the substances have directional properties and tend to align themselves with the electric field thereby causing optical changes in the characteristics of the material. The Kerr electro-optical phenomenon is used in very-high speed shutters, called the 'Kerr cells', which can open or close under an electric impulse in 10^{-8} seconds and transmit an accurately timed short burst of light.

Faraday, inspired by Ørsted's ideas of fields and of the unity of natural forces consisting of a correlation of natural forces of all kind, began to study electric and magnetic forces in their mechanical aspect in which such forces were created by a network of 'lines of force', a network which he assumed to exist in an all-pervading medium called the 'æther'. He had already arrived at both the theory of polarisation of dielectrics and the treatment of electrostatic phenomena by means of electric lines of force. Faraday, like the founder of the mathematical field theory Leonhard Euler (1707-1783) before him, insisted that only a single medium is required for all phenomena, in this case for both the electromagnetic and the light phenomena. Faraday's method of conceiving the electromagnetic phenomena was also an original mathematical method in which global statements of physical laws provide the local effects of the phenomenon by a purely formal mathematical analysis. Maxwell emphasised that Faraday as a theorist was an original mathematical thinker who preferred global statements of physical laws to local statements. Since the analytical methods for interconverting global and local statements [such as Green's transformations] of physical laws did not exist in Faraday's time, as they did later in Maxwell's time, the great power of Faraday's preference of global principles in science was first fully appreciated only in the 20th century because global principles wherever found are more fundamental.

Faraday was the first scientist to believe that the principle of action-at-distance was inadequate to account for electric and magnetic forces, and that a global field theory was necessary for this purpose since the field theory took into account the surrounding medium which the action-at-distance theory did not. The most complete theory of electricity and magnetism which was based on the action-at-distance postulate was the theory of Wilhelm Weber. It required the force between 2 electric charges to depend upon their relative velocities as well as upon the distance separating them. Both Faraday and Maxwell refused to accept this theory as the final one for the electromagnetic phenomenon. The continental theoretical physicists rejected Faraday's global field conception of electromagnetic theory, and instead adopted the synthetic approach of Ampère and Wilhelm Eduard Weber (1804-1891), professor of physics at the University of Göttingen in Germany, which presumed the existence of centers of force in an analogy with Newton's gravitational theory, with the power to act instantaneously on electrified particles over distances in space without the intervention of any transmitting medium. Therefore, the continental physicists attempted to construct by synthesis a general global theory of electromagnetic phenomena from particular local effects. The great power of Faraday's global viewpoint is fully appreciated by modern theorists because it is fundamental.

The Newtonian viewpoint had been adopted by Ampère in 1826 in formulating his inverse square law for the electric force dF acting between two elemental electric currents $i_1 \, dl_1$ and $i_2 \, dl_2$:

$$dF = G\left(\frac{i_1 i_2}{r^2}\right) dl_1 \, dl_2 ,$$

where r denotes the distance between the two elemental currents, and G represents a geometric factor related to the angles between r, dl_1, and dl_2.

In 1845, the German physicist, Franz Ernst Neumann (1789-1895), extended Ampère's theory to electromagnetic induction by deriving a potential function related to Ampère's electric force, and found that forces generated by electric currents could be given as functions of their relative velocities and accelerations. Neumann regarded the electric current as the flow of 2 equal-and-opposite groups of charged particles subjected to an equal-and-opposite pair of central forces F_{12} acting on each pair of electric particles e_1 and e_2, the magnitude of which depends upon their relative velocity (dr/dt) and relative acceleration (d^2r/dt^2) along the line joining the 2 particles:

$$F_{12} = \left(\frac{e_1 e_2}{r^2}\right)\left\{1 - \left[\left(\frac{dr}{dt}\right)^2 - 2r\frac{d^2r}{dt^2}\right]\frac{1}{c^2}\right\} ,$$

where c denotes a constant speed. Neumann determined that the ratio of the 'electrostatic' to the 'electrodynamic' force gives a constant speed.

In 1856, the German physicists Rudolf H.A. Kohlrausch (1809-1858) and Wilhelm E. Weber measured experimentally the ratio of electrostatic force to electrodynamic force, and found that this constant speed was essentially the speed of light c expressed in the special units of Weber's theory, by discharging a condenser through a ballistic galvanometer. This ratio, called the 'Weber constant', turned out to be $3.1(10^8)$ meters a second, a value which is quite close to the best experimental measurement of the speed of light, $3.15(10^8)$ meters a second.

Fizeau, Foucault and Speed of Light:

The speed of light was found experimentally in 1849 by the French physicist Armand Hyppolyte Louis Fizeau (1819-1896) by means of a rotating toothed wheel, which seemed to indicate that a relationship exists between light and electricity. However, Fizeau's measurement of the speed of light was about 5% too high by the modern standard.

A friend of Fizeau, Jean Bernard Léon Foucault (1819-1868), a physician and an ingenious and versatile self-educated physicist who had assisted Fizeau to measure experimentally the speed of light, decided in 1862 to measure the speed of light more accurately than Fizeau. Since the speed of light is so immense it was more accurate to measure time over a very long distance of travel of light. He achieved the long-distance travel of light by a method suggested by Arago a decade earlier consisting of 2 mirrors of which one was stationary and the other revolving thereby bouncing a light-ray back and forth between the 2 mirrors in a simulation of long-distance travel. This ingenious method allowed Foucault to determine the speed of light to within 1% of the most accurate value known today, $2.9973(10^8)$ meters a second. Foucault presented his ingenious experiment as his doctoral thesis.

Foucault, Inertial Pendulum, Gyroscope, and Foucault Currents :

Foucault was also the first to demonstrate the existence of electric eddy currents induced by a continuously changing magnetic field in a massive conductor. The inducted currents in turn create a magnetic induction the flow of which opposes the flow of the magnetic field that produced it. For this reason the magnetic force produced by these eddy currents, called the 'Foucault currents' is always a resisting force. Although the 'Foucault currents' cause losses of energy, they are profitably used in modern electric induction furnaces, in electromagnetic braking systems, and in many other electrical applications.

In 1851 Foucault also performed a series of ingenious experiments, the so-called 'Foucault inertial pendulum' experiment, consisting of a large iron ball about 2-feet in diameter which had a spike towards the floor, and was suspended by a steel wire of more than 200 feet (61 m)in length from the dome of the ***Panthéon Français*** in Paris. Since the plane of oscillation of the inertial pendulum is independent of the motion of the Earth, the spike of the swinging pendulum scored a mark in a layer of sand spread on the floor of ***Panthéon*** the rotation of which demonstrated that the Earth rotated underneath the Foucault inertial pendulum making one rotation in 31 hours and 47 minutes as expected for the latitude of Paris. In the 1840's Foucault built a tracking telescope incorporating a mechanical clockwork regulator that turned the telescope in the opposite direction to the rotation of the Earth thereby making photographing of stars possible. In 1845, Foucault produced the first photograph of the Sun, and in 1851 the first photograph of the eclipse of the Sun. In 1852 Foucault also invented the so-called 'gyroscope' by demonstrating experimentally that a wheel with a heavy felloe set into rapid rotation maintains the direction of its axis of spin in the inertial space like the axis of spin of the Earth. When the spinning wheel was tipped the effect of gravity produced a motion at right angles that was equivalent to the precession of the equinoxes. The spinning wheel of Foucault could also be used to demonstrate the rotation of the Earth.

Faraday, Thomson, Maxwell and Electromagnetism :

The equations of Ampère and Neumann, and Neumann's potential theory, all based on the concepts of centers of force and instantaneous action, served as the source of virtually all research done on electromagnetism in continental Europe until the 1870's.

The first significant progress towards a mathematical theory of electromagnetism was made by a precocious undergraduate at Cambridge University in 1841, when a young Scottish physicist, William Thomson (1824-1907) [who later became Lord Kelvin] used mathematical analogy as a source of his analytical technique. Thomson established a formal analogy between the basic equations of electrostatics and equations of heat flow in an infinite solid, in which the source of heat or fluid is the analogue to the electric charge, magnetic pole or source of electric current, and equipotential surfaces are analogues to isothermal surfaces.

In 1846, Thomson demonstrated that Faraday's representation of the electric phenomena in terms of lines of electric force were mathematically consistent with the inverse-square law and that the lines of force are mathematically defined, without requiring any hypothesis about their physical character when the distribution of the centers of force is given. In this paper Thomson examined the analogies existing between electric phenomena and elasticity by comparing forces in an incompressible elastic body under a state of rotational strain. He compared the forces in this deformed elastic body with the electric, magnetic and galvanic forces exerted by an electrified body, and showed that the distribution of the elastic displacement vector is analogous to the distribution of the electric force. Moreover, Thomson demonstrated that the elastic displacement can be made to correspond to a vector **'a'** defined in terms of magnetic induction vector **'B'** expressed in modern vector calculus by the cross-product:

$$\nabla \times \mathbf{a} = \mathbf{B},$$

where **a** is equivalent to the vector potential in the works of Neumann and Weber on the induction current, and $\nabla(-) \equiv \partial(-)/\partial \mathbf{r} \equiv [\partial(-)/\partial x]\mathbf{i} + [\partial(-)/\partial y]\mathbf{j} + [\partial(-)/\partial z]\mathbf{k}$ denotes the directed derivative operator in orthogonal Cartesian space $\{x, y, z\}$.

In 1851, Thomson published a paper in which he obtained the results of the theory of magnetism of the French mathematician Siméon Denis Poisson (1781-1840) from experimental data without resorting to the hypothesis of magnetic fluids. In this work, he introduced for the first time the vectors later called 'magnetic induction' and 'magnetic force' by Maxwell. In another fundamental memoir, Thomson established the energy of a system of permanent and temporary magnets as volume integrals throughout the space. Similar results were given by him for a system of circuits carrying steady currents. These remarkable results of Thomson's investigations suggested that in both cases, energy could be considered as stored throughout the field.

Philipp Johann Reis (1834-1874), a German schoolmaster and the inventor of the first crude telephone, had observed that when a Leyden Jar discharges through a wire wrapped around a knitting-needle of his telephone, sometimes one end of the needle became north-seeking, and sometimes the other end. Similarly, British scientist William Hyde Wollaston (1766-1828), who used the discharge of the Leyden Jar to decompose water by electrolysis, instead of finding oxygen at one electrode and hydrogen at the other electrode, discovered a mixture of both gases at each electrode. The German physicist Hermann Ludwig Ferdinand von Helmholtz (1821-1894) speculated in 1847, that these anomalies could be explained if the discharge of the Leyden Jar is oscillatory as Benjamin Franklin had already noted. The oscillatory nature of a condenser discharge in association with self-induction had been first discovered by Joseph Henry in 1842. Thomson took up this question in 1853, and not only proved the oscillatory nature of the discharge from the Leyden Jar, but also determined the frequency of such oscillations by using an analogy to mathematical pendulum. Subsequently the oscillatory nature of the discharge of the Leyden Jar was studied experimentally by a German physicist of Danish stock, Berend Wilhelm Feddersen (1832-1918) of Leipzig beginning in 1857. He was able to demonstrate successfully the oscillatory character of the discharge in 1861, and in 1862, by using revolving mirrors, and he was the first to prove it completely and correctly. The wavelength of the oscillations obtained by Feddersen from his Leyden Jars was about 30 meters, which was too long in comparison with the dimensions of the laboratories of the period, and that was one of the reasons why the oscillatory nature of the propagation of the electric discharge from Leyden Jars escaped the notice of experimental physicists.

In 1851, the imaginative Scottish civil and mechanical engineer, physicist and poet, William John Marquorn Rankine (1820-1872), introduced 'molecular vortices' into his new theory of matter, an idea which he applied to thermodynamics and to properties of gases. In Rankine's theory the molecular vortices were responsible for centrifugal forces that produced longitudinal contraction and radial pressure. Thomson, inspired by Rankine's idea of molecular vortices, devised an alternative theoretical interpretation of magnetism in which magnetism possessed rotatory character. In this theory of Thomson, the molecular rotations were to account for the Faraday effect, that is, the magneto-optical effect of the slight rotation in the plane of polarisation of light passing through a block of glass located between the poles of a magnet. Employing an analogy with a pendulum suspended from a spinning arm, Thomson concluded that this slight rotation in the plane of polarisation of light could be caused by the coupling between the vibrations of the æther and the spinning of molecules of glass about the lines of force.

It can be learnt from this account, that Thomson had considerable talent of devising disconnected yet important ideas in the subject of electromagnetism, but he lacked the ability to develop these ideas into a comprehensive theory. However, such universal abilities were possessed by another brilliant Scot who was responsible for the establishment of electromagnetic field theory. His name was James Clerk Maxwell (1831-1879), the precocious son of a Scottish laird. Maxwell, a man of short but robust stature, was an intellectual giant, and deserved the nickname, the 'Little Giant', which had been given to the famous railroad, bridge, and naval engineer, Isambard Kingdom Brunel (1806-1859). Not only did Maxwell have remarkable mathematical talent, exceptional creative imagination, profound intellectual grasp of anything physical, and a wide scope of professional interests, but he was also a talented amateur artist and an incisive writer of prose.

In electromagnetic phenomena Maxwell preferred the global field viewpoint of Faraday to the instantaneous action-at-distance without an intervening medium, a concept favoured by the continental mathematical physicists. He began his investigations of the electromagnetic phenomena in 1855 which led him in 1865 to the so-called 'Maxwell electromagnetic field equations'. Maxwell began his research with an intention to find out all the mathematical consequences of Faraday's global field conception of the electromagnetic phenomena and to amalgamate the ideas of Faraday with Thomson's theories of electricity. Maxwell went far beyond the conceptions of Thomson whose results were inadequate owing to the absence of an assumption for the source of strain in the intervening medium. Following Thomson's example, Maxwell considered analogies to be the important means of invention and development of physical ideas which obviate an early commitment to definite abstract scientific hypotheses, a commitment which he thought had led Rankine astray. He recognised that the method of analogy cross-fertilises the scientific techniques between different fields of science and, therefore, affords an expedient means between analytic abstraction and the method of hypotheses by providing partial resemblances between distinct fields.

In 1855 and 1856, Maxwell wrote on Faraday's lines of force and established partial analogies between electric and magnetic lines of force and the lines of flow of an incompressible fluid in elastic tubes of flow. By applying Euler's hydrodynamic theory to the motion of the perfect fluid and theory of elasticity to the tubes of flow, he was able to derive 6 laws which accounted for many observed facts in electricity. Faraday who was deeply impressed by Maxwell's results, had considered the longitudinal and transverse forces of the magnetic field as physical manifestations of the tubes of force which suggested to Maxwell that energy was stored in the space surrounding the magnet and the circuits carrying electric currents, and that such energy storage needed mechanical models such as the elastic tubes of force.

In 1861 and 1862, Maxwell invented another mechanical model for the electromagnetic field. Following Thomson's ideas, Maxwell identified magnetism with the rotational kinetic energy and electrification with the linear elastic displacement to which he grafted Faraday's concept of the tendency of tubes of force to contract longitudinally and expand laterally, a behaviour associated with centrifugal forces of a rotating fluid in an elastic tube. The angular velocity of each molecular vortex he assumed to be proportional to local magnetic intensity, and this rotation created centrifugal forces which made each vortex contract longitudinally and expand radially thereby exerting radial pressure in agreement with the stress distribution proposed by Faraday for physical lines of force. Thus, Maxwell visualised that in a magnetic field the medium is in a state of rotation about the lines of magnetic force in which each unit tube of force may be represented by an isolated elastic vortex. By means of this mechanical model Maxwell was able to establish formulae identical to the existing theories of forces acting between magnets, diamagnetic bodies and steady currents. In this work, the so-called 'electric displacement current' which occupies the central place in Maxwell's electromagnetic theory is first introduced.

In order to deal with the electromagnetic induction he had to take into account the action of electric current on the molecular vortices representing magnetism. In his mechanical model, Maxwell represented the molecular vortices, the spins of which measured the magnetic field intensity, by means of spinning octagonal cylinders separated by a layer of counter-rotating minute particles representing electric particles capable of displacement which kept the cylinders in gear by behaving like the idling-wheels of a gear train such as were used by William Siemens (1823-1883) in his design of 'chronometric' governor of steam-engines. These idling-wheel type electric particles could be displaced by the different rotations of the magnetic cylinders. The elastically restrained displacement of the electric particles can be interpreted as bringing about the electric current, since when the electric displacement increases or decreases, the effect is equivalent to an electric current moving in two opposing directions. Maxwell was able to show by means of this mechanical model the interaction between the electric current and magnetic field: the production of magnetic field by an electric current because displacing the idling-wheel type electric particles produces spins of the cylinders, and the production of the electric current by the change in the magnetic field because the different spins of circumjacent magnetic cylinders produce displacements of the electric particles.

Maxwell's mechanical model for the electromagnetic field resembled somewhat the model of æther proposed in 1736 by Johann Bernoulli Jr. (1710-1790), which consisted of an enormous number of extremely small whirlpools that lent elasticity to the æther by continually striving to press against the neighbouring whirlpools. Maxwell's mechanical model for the medium was so constructed that it produced all known experimental results, including the effect of the magnetic field on polarised light discovered in 1845 by Faraday. Maxwell was able to show with the aid of his mechanical model of the electromagnetic medium that the comportment of the magnetic medium in space is similar in some respects to that of an elastic medium and the same as that of the luminiferous medium, if the two media were coexistent, coextensive, and equally elastic, and that the speed of transverse undulations in Maxwell's electromagnetic medium were exactly the same as the speed of transverse undulations of light. Therefore, Maxwell concluded that light consists of transverse undulations of the same medium which is the seat of the electric and the magnetic phenomenon. His mechanical model made it possible for Maxwell to derive the mathematical relations existing between the magnetic and the electric phenomenon, and he demonstrated that the necessary properties of the electromagnetic medium were the same as those of the luminiferous medium. According to Maxwell's theory there is no empty space, that is, there is no space without a field.

Already in 1832 Faraday had predicted the electromagnetic wave nature of light: "I am inclined to compare the diffusion of magnetic forces from a magnetic pole to the vibrations upon a surface of disturbed water, or those of air in the phenomenon of sound, i.e., I am inclined to think the vibratory theory will apply to these phenomena as it does to sound, and most probably to light."

Maxwell was thus able to represent electromagnetic waves by means of his mechanical model of the electromagnetic field. Already Faraday had demonstrated that the type of insulating material made a great difference in the condenser's capacity to keep a charge. Maxwell proposed a bold hypothesis for such a local electric phenomenon taking place in dielectrics (insulators) by regarding such a phenomenon as displaced electric particles restrained in their motion by the elastic forces exerted by the cylinders as magnetic vortices. Removal of the electric force which created this electric displacement releases the electric particles which then oscillate for a brief interval of time about their undisplaced position whilst this local oscillation is transmitted through the dielectric like a wave representing a current. Since the æther as the medium pervades everything, it no longer matters whether it is the well-known conduction currents or the so-called 'displacement current' ($\partial \mathbf{E}/\partial t$), where $\mathbf{E}(\mathbf{r}, t)$ denotes the electric field strength, because in both cases the æther is set in motion, and this motion is transmitted mechanically in the medium, and its effects appear as mechanical force, heat, light, or any other phenomena of electricity and magnetism. Maxwell subjected these complex phenomena to the Principle of Least Action: 'Every action in the electromagnetic field takes place with the least expenditure of energy, in order to correlate quantitatively all these diverse effects of different force fields'. Without this universal action principle governing all actions in the electromagnetic field, the mechanical representation of the electromagnetic phenomena would have been impossible. Therefore, according to Maxwell's field theory, the 'total current' $\mathbf{S}$ in the medium is given by:

$$\mathbf{S} = \mathbf{s} + \frac{\partial \mathbf{E}}{\partial t},$$

where $\mathbf{s}$ is the 'regular electric current', and $\partial \mathbf{E}(\mathbf{r}, t)/\partial t$ is the 'displacement current' of Maxwell.

Maxwell used vector analysis, which he extended and generalised for the purposes of his electromagnetic field theory. The basic ideas of Faraday's theories such as the theory of lines of force and the concept that the energy of the electromagnetic field is stored in the dielectrics as well as in the conductors he was able to express in his vectorial language. In Maxwell's electromagnetic theory such Faraday's terms as 'polarisation of dielectric' became 'dielectric displacement', and 'change of polarisation' became 'displacement current'. By means of his mechanical model, vector analysis and Faraday's ideas on electricity and magnetism, Maxwell was able to formulate a complete theory of electromagnetism resulting in a few fundamental equations called the 'Maxwell electromagnetic field equations'. From these equations, Maxwell deduced mathematically that in a medium in which dielectric displacements can occur, waves of periodic displacements can be produced the speed of propagation of which is equal to the speed of light-waves. This led Maxwell to assume that the two media, one carrying the light-waves, and the other the electromagnetic waves were in reality one and the same medium. It expressed the fundamental principle of electromagnetic theory of light. Hence, according to Maxwell, light is a self-propagating periodic disturbance in the equilibrium of the electric and magnetic intensity in which both vectors were perpendicular to the direction of propagation, thus making the wave movement automatically transverse. The transverse nature of the electromagnetic wave removed the main objection against the undulatory theory of light.

One objection to Maxwell's theory of electromagnetism was that it did not give a really convincing explanation for such fundamental phenomena of light as reflexion and refraction of light. Helmholtz indicated a way to accomplish this, and a mathematical treatment of it was given by the Dutch physicist Hendrik Antoon Lorentz (1853-1928) in his thesis of 1875 entitled < On the Theory of Reflexion and Refraction of Light > which refined Maxwell's electromagnetic theory concerning light. In 1865, Maxwell published his famous memoir on the dynamic theory of the electromagnetic field. In this memoir he had removed all traces of his mechanical model by means of which he had obtained his electromagnetic field theory and, thereby, reduced his electromagnetic field equations to their abstract mathematical form.

Maxwell derived the following two scalar-valued electromagnetic field equations for an electromagnetic field:

$$\nabla \cdot \mathbf{E} = 0,$$

where the directed derivative operator $\nabla(-) \equiv \partial(-)/\partial \mathbf{r} \equiv [\partial(-)/\partial x]\mathbf{i} + [\partial(-)/\partial y]\mathbf{j} + [\partial(-)/\partial z]\mathbf{k}$, and $\mathbf{E}(\mathbf{r}, t)$ denotes the electric field strength, implying that the electric field strength is conserved in space in the absence of electric charges, and:

$$\nabla \cdot \mathbf{H} = 0,$$

where $\mathbf{H}(\mathbf{r}, t)$ denotes the magnetic field strength, implying that the magnetic field strength is conserved in space. These 2 equations express the spatial continuity of the electric and the magnetic field.

The vector-valued equation:

$$\nabla \times \mathbf{E} = -\frac{1}{c}\frac{\partial \mathbf{H}}{\partial t},$$

represents Maxwell's form of the Faraday's law of electromagnetic induction, where c denotes the speed of light which enters the Maxwell equations as the conversion factor between electrostatic and electromagnetic units, which states that a changing magnetic field creates an electric field at right angles to the change in magnetic field.

The vector-valued equation:

$$\nabla \times \mathbf{H} = \frac{1}{c} \frac{\partial \mathbf{E}}{\partial t},$$

represents Maxwell's form of the production of the magnetic field strength $\mathbf{H}$ by the time-rate of change in the electric field strength ($\partial \mathbf{E}/\partial t$) that represents the Maxwell's 'displacement current'.

The last 2 equations of Maxwell show how a change in one field produces a change in the other field. These equations can be reduced to the vector-valued equation of the propagation of $\mathbf{H}$:

$$c^2 \nabla \cdot \nabla \mathbf{H} = \frac{\partial^2 \mathbf{H}}{\partial t^2}.$$

These equations can also be reduced to a similar vector-valued equation of the propagation of $\mathbf{E}$:

$$c^2 \nabla \cdot \nabla \mathbf{E} = \frac{\partial^2 \mathbf{E}}{\partial t^2}.$$

These 2 equations are the so-called 'Maxwell electromagnetic wave equations'.

It is important to recognise that the electromagnetic waves of Maxwell are not material waves. At every point in the field the vectors $\mathbf{E}(\mathbf{r}, t)$ and $\mathbf{H}(\mathbf{r}, t)$ are orthogonal to each other as well as to the direction of propagation of the electromagnetic wave. Maxwell indicated that in the neighbourhood of an oscillating electric current the electric field of intensity $\mathbf{E}(\mathbf{r}, t)$ and the magnetic field of intensity $\mathbf{H}(\mathbf{r}, t)$ are subjected to rhythmical fluctuations in synchronism with the oscillating current itself, but that these fluctuations in the intensities of $\mathbf{E}$ and $\mathbf{H}$ spread out like waves in a direction perpendicular to both vectors $\mathbf{E}$ and $\mathbf{H}$. This spreading disturbance in the intensities of the electric and of the magnetic fields Maxwell called an 'electromagnetic wave' which propagates in the direction transverse to both vectors $\mathbf{E}$ and $\mathbf{H}$ and, therefore, represents a transverse wave, a kind of wave which had been anticipated for the propagation of light by the English experimental scientist and engineer Robert Hooke (1635-1703). Maxwell, himself, urged that the mechanical analogy of his abstract electromagnetic field equations should be regarded as merely illustrative and not explanatory of the electromagnetic phenomena.

In 1873 Maxwell also published his electromagnetic field theory in a book entitles < Treatise on Electricity and Magnetism >, a work which was difficult to understand because he himself had no clear idea of the nature of electricity itself because the subatomic particle called the 'electron' which is the carrier of electric current was only discovered near the close of the 19th century.

The electromagnetic field theory of Maxwell, but particularly the displacement current ($\partial \mathbf{E}/\partial t$), was not accepted by the majority of his most capable contemporaries such as William Thomson, and most of them remained unconvinced long after the death of Maxwell in 1879 until the theory was corroborated by the experiments of a young ingenious German physicist Heinrich Hertz, although the simple experiments of David Edward Hughes (1830-1900) in 1878 and 1879 could have provided some evidence for the existence of electromagnetic waves if properly interpreted. The brilliant French theoretical physicist, science philosopher and historian of science, Pierre-Marie-Maurice Duhem (1861-1916) admired the genius of Maxwell, but as is characteristic of the French scientists in general, he preferred a completely abstract, axiomatic theory built by the method of hypotheses and logic. Duhem loathed the mechanical foundation of Maxwell's theory even after Maxwell had made it completely abstract, a foundation which Duhem said he could still 'smell' in the abstract set of Maxwell's equations. Duhem expressed his views on this matter by saying that when he began his studies of Maxwell's theory of electromagnetism, he expected to enter a scientific mansion as a peaceful, carefully ordered seat of abstract hypotheses, precise reason and rigorous logic but, instead, he found himself in a 'factory'. However, Duhem's own theory of electromagnetism based on a strictly axiomatic approach had some awkward aspects to it and it did not succeed. Although Duhem, one of the founders of physical chemistry, was not only one of the most capable and ingenious scientists but also the leading philosopher and historian of science of his time, he did not appreciate the inventive power of analogy as well as Maxwell. The famous German physicist and mathematician Hermann von Helmholtz, after many years, came to accept Maxwell's electromagnetic theory, but the Scottish physicist William Thomson who later became Lord Kelvin never accepted it completely.

Maxwell predicted on the basis of his theory that light as electromagnetic radiation exerts a pressure in full Sunlight equal to $5(10)^{-5}$ dynes per square centimeter, a claim which was ultimately proven experimentally by a Russian physicist Pyotr V.N. Lebedev (1866-1912) in 1907, which Albert Einstein (1879-1955) was able to use in the corroboration of his famous formula: $E = mc^2$.

Maxwell was a versatile talent and the leading physicist in a century of genius. He had a remarkable scientific imagination, and according to his renowned university coach, the English geophysicist William Hopkins (1793-1866), he was virtually incapable of wrong thinking in science. Before he was 19 years of age he had developed a complete method of photoelasticity for the experimental study of strain. In 1860, he proved the stability of the Saturn rings in an essay which won the coveted Adams Prize, and made him world famous. He created quantitative colourimetry, and was the first to produce a trichromatic colour photograph. From 1860 to 1863, he developed the kinetic theory of gases based on statistics and probability, thereby eclipsing the molecular vortex theory of gases of William John Macquorn Rankine. In 1864, Maxwell gave the general theory of reciprocal figures and reciprocal force diagrams for the graphical static analysis of trusses, a theory which had been initiated by Rankine. In the same year, Maxwell published a complete solution for the deformation and equilibrium of statically determinate and indeterminate linearly elastic trusses on the basis of the principle of conservation of energy and by means of his ingenious 'auxiliary free-body method'. In 1868, Maxwell published a memoir on the governors as a feedback systems and determined the conditions of their stability which laid the theoretical foundation for servomechanics and cybernetics. In 1870, he extended the method of reciprocal functions to continuous media, and gave the 3 stress functions, now called the 'Maxwell stress functions', as a method to solve the general equations of 3-dimensional elasticity. Maxwell made outstanding contributions also to rheology, thermodynamics, and geometrical optics. He was a man of broad culture and immense scope of learning.

Contemporary leading physicists were particularly incredulous of the existence of Maxwell's displacement current ($\partial \mathbf{E}/\partial t$), so much so that in 1879, the year Maxwell died, the Berlin Academy of Sciences offered a prize for the experimental detection of the displacement current in dielectrics. Helmholtz urged his assistant Heinrich Rudolph Hertz (1857-1894), a former student of engineering who became a brilliant physicist, to undertake this challenge, but Hertz at the time saw no possibility of bringing this phenomenon within the reach of experimental testing. At that time, Hertz was unaware that in 1870 a German physicist and meteorologist, Friedrich Wilhelm von Bezold (1837-1907) of Munich, had been successful in detecting experimentally the transmission of electric waves along wires with one free end, and in wire circuits with a spark gap, indicating the finite speed of propagation of the electric potential along the wire.

In 1886, Hertz noticed that an open circuit with a small air gap when connected to a circuit with a spark discharging induction coil passed a spark through the air gap, and that a spark could be passed across the air gap even if the induction coil circuit as a primary circuit is not connected to the open circuit as the secondary circuit. Unbeknownst to Hertz, this kind of observation had been made by David Edward Hughes in 1879, who detected signals produced by a spark from an induction coil transmitted over a distance of 1,500 feet (457 m) to a microphone, which was later called a 'coherer', consisting of loose carbon granules, that was connected to a telephone to make the signals audible. Hughes claimed that these signals were transmitted by electric conduction through the air. In the course of his experiments Hughes discovered the phenomenon of standing electromagnetic waves brought about by the interference between the incident and the reflected waves, although at the time he did not realise it. Hughes demonstrated his experiments to physicists William Spottiswoode (1825-1833), George Gabriel Stokes (1819-1903), and Thomas Henry Huxley (1825-1895). Stokes, who failed to connect the experiments of Hughes with Maxwell's electromagnetic theory, mistakenly thought that all these effects could be explained by the known theory of electromagnetic induction.

Since already in 1842, the American physicist Joseph Henry had observed that the inductive effect of the discharge of the Leyden Jar could be detected at considerable distance from the Leyden Jar. Actually, Hughes' simple apparatus represented a sender and a receiver of radio signals. Hughes became discouraged because of Stokes' wrong opinion and discontinued his experiments.

A conclusive experimental proof of the validity of the electromagnetic theory was given by Hertz in 1887. Hertz regarded the Maxwell electromagnetic equations as a description of the electromagnetic field in space and time, and he thought that the real crux of this theory is to find the experimental means to prove that electromagnetic waves really exist in space. He discovered a method of generating electromagnetic waves with the 'Ruhmkorff induction coil', an electric discharge coil perfected in 1851 by the German instrument-maker Heinrich Daniel Ruhmkorff (1803-1877). Hertz used as a theoretical basis for his experimental equipment a formula deduced by the Scottish physicist William Thomson [later Lord Kelvin] in theoretical investigation of the discharge of condensers, which gave the period of oscillation T of an oscillating circuit in terms of its electrostatic capacity C and self-inductance L:

$$T = 2\pi (LC)^{1/2}.$$

This formula of Thomson served as a guide to the experimental arrangement designed by Hertz, so that a wavelength could be obtained which was short enough for the measurement in the laboratory.

Hertz built an electric oscillator consisting of 2 metal sheets set in the same plane with each connected to a stiff rod ending with a knob, the 2 knobs being separated by a short air gap. The 2 metal sheets, which can be considered to represent the 2 coatings of the Leyden Jar insulated by the air gap between them, were excited by being connected to the 'Ruhmkorff induction coil' Hertz produced sparking discharges across the air gap between the knobs with the aid of the Ruhmkorff induction coil and the two metal plates. When this kind of Leyden Jar discharges, the electricity surges from one plate to the other with a period proportional to $(CL)^{1/2}$, where C is the electrostatic capacity of the two metal sheets and L denotes the self-induction of the connexion. In order to reduce the period, the capacity C and the induction L have to be made as small as possible. The detector, or 'tuned receiver' of Hertz, was a wire bent into an incomplete circle with a small air gap which was not connected to the primary or the transmitting circuit. It represented a resonant circuit or 'resonator', built in such proportions that its free period of oscillation was the same as that of the 'primary', or 'oscillatory' circuit. His primary circuit oscillated between the two metal balls separated by a small air gap, and each time the potential reached a peak in one direction or the other it sent an electric spark across the air-gap. Hertz also noted that when ultraviolet light shone on the negative terminal of the spark gap, the electric spark generated was stronger and more readily emitted. Although he did not follow up this line of research, it was the first observation of the photoelectric effect which was explained by Albert Einstein a generation later by means of quantum theory. The oscillating electric spark generated electromagnetic radiation which produced small electric sparks jumping across the gap in his detector coil as a result of the current created by the electromagnetic current in space. He was able to generate standing electromagnetic waves with his apparatus by means of reflexion and interference of electromagnetic waves, and the nodes and anti-nodes of these waves he was able to determine with his 'resonator', or 'tuned-receiver', particularly built for that purpose. He was also able to demonstrate reflexion and refraction of electromagnetic waves by his apparatus, and prove that they were subject to the same laws as the light-waves. Hertz found the shape of the electromagnetic waves by moving his detector to various locations in the laboratory, and noting the intensity of electric spark formation in the gap of the detector. He was able to calculate from this information the wavelength of the electromagnetic wave transmitted from his spark-gap sender to be 2.2 feet (0.66 m) which is a million times the wavelength of the visible light. This ingenious experiment of Hertz proved the soundness of Maxwell's electromagnetic theory.

In 1888, Hertz proved that electromagnetic waves in air are propagated with a finite speed by transmitting the electric disturbance from the primary circuit through the air, and also along the wire to the 'detector', and then observed the interference between the two. From these observations, he could determine the ratio of electromagnetic wave speed in the air to the electric wave speed in the wire, which turned out to be in the ratio of 45-to-28. He could calculate the speed of the electric wave in the wire from his measurements which turned out to be $200(10^6)$ meters a second. Hertz was able to demonstrate that the speed of electromagnetic waves in the air is finite, and has the same order of magnitude as the speed of light. It was later shown that the numerical values given by Hertz were not quite correct. The famous French mathematician Henri Poincaré (1854-1912) showed in 1890 that Hertz's period was actually $\sqrt{2}$ times the actual period, which made the propagation of the electromagnetic wave through the air equal to $\sqrt{2}$ times the speed of light. In the same year, the Austrian physicist Ernst Lecher (1856-1926) found from his experiments that the speed of propagation of electric vibrations along the wire was within 2% of the speed of light. Finally, the experiments of the Swiss physicists Edouard Sarasin (1843-1917) and Lucien de La Rive (1834-1924) in 1893 proved that the speeds of propagation of electric waves in the air and along the wire are actually the same.

In 1888, Hertz demonstrated experimentally that electromagnetic waves obey most of the fundamental laws of light-waves such as reflexion, interference and stationary waves, as well as confirm the fundamental laws of optics. According to Maxwell's theory electromagnetic radiation is produced by the oscillation of electric charges, and Hertz had proven this to be the true for radio-waves. Moreover, Hertz's experiments proved that electromagnetic radiation in space did not attenuate according to the law of inverse square of the distance, and that it was not instantaneous, thereby refuting the action-at-distance theory of spatial electromagnetic phenomenon. Hertz succeeded in establishing experimentally the fundamental principle of radar, when he demonstrated with the help of a big concave mirror and a zinc-plate reflector that radio-waves could be focussed on, and reflected from objects like a beam of light. In 1887, Hertz also did experiments on the photoelectric effect, and discovered that electric charges were emitted from the negatively charged alkali metals such as sodium and potassium under the action of light. This discovery led to the invention of the photoemissive cell by German physicists Hans Friedrich Geitel (1855-1923) and Julius Elster (1854-1920) in 1887, which they perfected by 1913. Subsequently the photoemissive cell became fundamentally important in the development of electronic technology, particularly in television technology.

The discovery of the so-called 'Hertzian waves' was one of the most important discoveries in electrical sciences at the close of the 19th century since it led to the technological developments of wireless, radio, radar, and telecommunications in general.

English physicist Oliver Joseph Lodge (1851-1940) repeated Hertz's experiments from 1889 to 1894. In 1892, Lodge replaced Hertz's spark-gap detector by a more sensitive instrument, named 'coherer' by Lodge, which was connected to a 'decohering device', which he called the 'trembler', consisting of an electric bell mechanism incorporated into the battery circuit. Lodge demonstrated that two metallic spheres in light contact with each other obey the same cohering phenomenon as metal filings, and that the trembler automatically restored the high resistance of the coherer after it detected an electric wave. Lodge realised, as the American engineer of Croatian stock, Nikola Tesla (1856-1943) had recognised before him, that Hertzian waves can be used for telegraphic transmission without wires. Lodge also showed in 1889 that a variation of a magnetic field produces an electric field.

In 1894, Lodge constructed an apparatus which incorporated a receiver resonantly tuned to the sender. Lodge's receiver consisted of a relay for amplifying the signals, a coherer, a trembler and a printer to record the dots and dashes of the Morse code. His coherer, however, proved to be a rather erratic instrument.

Since 1889 Augusto Righi (1850-1920), professor of physics at the University of Bologna in Italy, was able to improve Hertz's experiments by setting up equipment which generated much shorter electromagnetic waves than those of Hertz in his effort to demonstrate the relationship of the Hertzian waves to light-waves. Hertz had produced electromagnetic waves with a wavelength of about 2.2 feet (0.66 m), whereas Righi was able to generate ultrashort electromagnetic waves with a wavelength of 1-inch (2.5 cm). Righi's sender had 4 metal balls in its spark gap, instead of the 2 metal balls used by Hertz, which were enveloped in parchment bags containing a mixture of vaseline and its oil, thereby reducing the tarnishing of the balls due to continuous sparking. Righi used mirrors for the reception of these ultrashort electromagnetic waves.

If light was electromagnetic radiation similar in nature to Hertz's radio-waves, then where were the electric charges which caused the oscillations? Since by 1890 it seemed quite likely that the electric current consisted of charged particles, Hendrik Antoon Lorentz thought that material atoms also could consist of charged particles. He suggested that the oscillation of charged particles within the atom produced the visible light. If this were true, then placing a source of light in a strong magnetic field should affect the oscillations within the atoms and, therefore, change the wavelength of the light emitted. This was experimentally demonstrated in 1896 by the Dutch physicist Pieter Zeeman (1865-1943) who was a pupil of Lorentz. He repeated Faraday's investigation with a modern spectroscopic aid, the so-called 'Rowland grating' (about 15,000 lines per inch) devised by the American physicist Henry August Rowland (1848-1901) in late 1870's, with which he could observe a widening of the yellow line in the spectrum of sodium, now called the Zeeman effect. This effect had been observed a year before by Charles Fieviez, but it attracted no attention in the absence of theoretical formulation. In 1896, however, it was a discovery of enormous importance and formed the beginning of very extensive investigations which made an important contribution to the origin of modern atomic theory. Lorentz formulated hypotheses on the structure of the atom in the so-called 'electron theory' by essentially following Maxwell's line of thought. He found the specific charge of electrons to be approximately 1,000 times as large as the charge of the hydrogen ions, and from this result he predicted the existence of particles smaller than the atom, a claim which was experimentally substantiated in 1897 by Joseph John Thomson in his experiments with cathode rays.

Lorentz also studied the motion of charged particles and postulated like the Irish physicist George Francis FitzGerald (1851-1901) did in 1895, that contractions of length accompany motions, which are now called the 'Lorentz-FitzGerald contractions'. Therefore, the volume occupied by the mass of a charged particle in motion such as an electron must increase with speed since its volume is reducing with its speed. According to Lorentz's theory at 161,000 miles (259,097 km) a second, the mass of an electron is twice its mass at rest, and at the speed of light, 186,282 miles (299,784 km) a second, the mass of electron must become infinite because its volume becomes zero. This was another indication that the speed of light in vacuum is the greatest speed at which any material object can move. By 1900, measurements of mass of speeding subatomic particles showed that the Lorentz Equation specifying the variation of mass with its speed was accurate. In 1905, Einstein gave his special theory of relativity from which the Lorentz-FitzGerald contraction could be deduced and from which it could be shown that the Lorentz mass increase with speed is not only valid for charged particles as Lorentz had postulated, but for all objects regardless whether they are charged or uncharged. In 1902, Lorentz and Zeeman shared the Nobel Prize in physics for their contributions to the theory of electron, magnetism and light.

Science of Electronics and Atomic Physics

Another line of research of the electrical phenomena, which became as important to the 20th century engineering as the electromagnetic theory of Maxwell, emerged from the study of cathode rays which are emitted from the electrically charged cathode of a vacuum tube.

Already in 1838, Faraday had noticed the phosphorescent glow accompanying the discharge of electric current through a partial vacuum. About 1850, Heinrich Geissler (1815-1879), a German glassblower who became a manufacturer of laboratory equipment, showed this effect in glass tubes of different shapes by building a partially evacuated glass container fitted with two sealed electrodes which produced attractive lighting effects depending on the shape of the container when electric current was supplied to the electrodes from a battery. Geissler's low-pressure discharge tubes were long, small bore glass tubes usually shortened by many coils and bends, which were filled with various gases at low pressure, and originally excited by high-voltage alternating current. Since beautiful effects could be produced by Geissler's tubes filled with different gases, they were often used for decorations. In science, Geissler's tubes were used in spectral analysis and lecture demonstrations.

In 1854, the German mathematician and physicist, Julius Plücker (1801-1868), professor of physics at the University of Bonn, studied the Geissler's tube phenomenon by experimentally investigating the luminous effect of electric discharge through gases at low pressure. He discovered that the glow in the tube changed its position in a strong electromagnetic field, and when the poles of the electromagnets were shifted, the glow also shifted with the poles. In 1859 Plücker reasoned that some kind of electrical 'ray' from the cathode was involved in the production of the greenish fluorescence from the glass itself in a glass tube under very high vacuum because he was able to produce a shadow in the fluorescent glass wall by interposing a solid object between the heated negative electrode, the cathode, and the fluorescent glass of the tube. In 1869, Plücker's student, Johann Wilhelm Hittorf (1824-1914), demonstrated that the stream of electric current between 2 electrodes sealed in an evacuated glass vessel flowed in a straight line and produced phosphorescence where it met the glass walls of the vessel. If a solid object was placed in the path of the stream of electric current flowing from the negatively charged electrode (cathode), a shadow was projected on the glowing wall. If the solid object was placed in the path of the positively charged electrode (anode) no shadow was cast. Therefore, the electric current forming a ray flowed from the cathode to the anode, but not in the opposite direction. He demonstrated that the rays could be deflected by a magnetic field. Hittorf thought that the cathode rays were electromagnetic waves like light-waves but of a different wavelength. In 1871, the English physicist, Cromwell Fleetwood Varley (1828-1883), discovered that the discharged electric ray in a cathode ray tube was negatively charged.

In 1876, the German physicist, Eugen Goldstein (1850-1931), followed the example of Plücker and Hittorf and investigated the luminescence produced by the cathode in an evacuated tube. He was able to deflect the electrically charged ray in an electric field, and it was Goldstein who named the stream of electric current emitted by the cathode the 'cathode rays'. In 1886, he used a cathode perforated by holes called 'canals' and discovered that there were streams of particles he called the 'canal rays' passing through the canals in the opposite direction to the cathode rays. In 1895 the French physicist Jean Baptiste Perrin (1870-1942) who had shown that a conductor becomes negatively charged when placed in a cathode ray, also proved that the particles of the 'canal rays' originating in the gas discharge between the anode and cathode and travelling from the anode towards the cathode were positively charged ions. The study of these positively charged rays ultimately led the New-Zealand physicist Ernest Rutherford (1871-1937) to the recognition of the existence of 'protons'. Nevertheless, the physicists in general remained unconvinced of the physical nature of the cathode rays, despite the experimental evidence presented by Varley and Goldstein.

In 1879, the English physicist and chemist, William Crookes (1832-1919), built the so-called 'Crookes tube', consisting of a partially evacuated glass tube which contained 2 electrodes: a platinum cathode at one end of the tube and an anode at the other. When the Crookes tube was connected to a high-voltage source, the tube began to glow. Crookes proved that the rays which create the glow have a definite mass and are emitted by the cathode as the negatively charged electrode in the glass tube since a small sensitive pinwheel when placed in the path of the cathode ray began to revolve in the direction of the anode as the positively charged electrode. Crookes also demonstrated that when his tube was placed between the poles of a powerful magnet, the cathode rays were deflected from their original path. Since electromagnetic waves did not behave this way, he postulated that the cathode rays were made up of charged particles, which in his opinion represented the fourth state of matter. These sensational experiments and the articles of Crookes helped to gain acceptance of the idea that the cathode rays are corpuscular in nature among the contemporary physicists. He demonstrated that cathode rays have momentum as well as definite energy.

In 1889 Hertz was appointed professor of physics at the University of Bonn, where he studied cathode rays with his assistant Philipp Eduard Anton Lenard (1862-1947), another outstanding pupil of Helmholtz. In 1892 Hertz discovered the remarkable ability of cathode rays to penetrate through the metal foils which led him to believe that a cathode ray was a longitudinal electric wave radiation rather than being a stream of particles, since it seemed to him that particles even as small as atoms would not be able to penetrate metallic solids. In 1893, Lenard devised a cathode ray discharge tube with a window made of thin aluminium foil through which cathode rays, called the 'Lenard rays' could penetrate into the open air thereby ionising the air and making it electrically conductive. For his studies of the 'Lenard rays' and their absorption and diffusion in different matter, Lenard was awarded in 1905 the Nobel Prize in physics.

The first convincing theoretical and experimental explanation of the cathode rays was given by the English physicist, Joseph John Thomson (1856-1940), who was professor of physics at the Cambridge University. Although Crookes, and others had offered some evidence that cathode rays may consist of negatively charged particles, their demonstrations still remained inconclusive since no one had been able to show, apart from Goldstein, that the cathode ray was also affected by an electric field, which it had to be if it consisted of electrically charged particles. The idea to subject the cathode ray simultaneously to the influence of both a magnetic and an electric field occurred almost at the same time to two men: Thomson and the German physicist Emil Johann Wiechert (1861-1928). In 1897 several investigators, including Joseph J. Thomson, the German physicist Wilhelm Wien (1864-1928), the Irish physicist George FitzGerald, and Emil J. Wiechert showed the ratio of the mass of the cathode ray particle to its charge to be about 1,840 times smaller than the same ratio for the hydrogen atom, the smallest atom known. In 1898, Wilhelm Wien concluded from the deflexion measurements concerning the ratio of mass to charge in the case of the 'Goldstein canal rays' just as he had calculated the magnitude of electrolytic ions from the Faraday equivalence law in electrolysis, that the 'canal rays' were either electrically charged atoms, or molecules.

Thomson's interest in Maxwell's theory of electromagnetic radiation led him to the study of cathode rays as a new form of radiation that was not electromagnetic in character. Thomson used a highly evacuated cathode ray tube in which he subjected the cathode ray both to the action of the magnetic and the electric field. From the measured deflexion of the cathode ray in the magnetic field he was able to find (mv/e), where m denotes the mass, v the speed, and e the electric charge of the electron. From the measured deflexion of the cathode ray in the electric field he was able to find (mv^2/e). From these two results he obtained the ratio of mass to the electric charge of the electron: (m/e). Thomson used a glass cylinder fitted with a piston and a metal disk connected to an electroscope which he filled with moist air exposed to X-rays in order to find the electric charge e of the electron. When the piston was suddenly pulled up causing a less than 30% expansion of the moist air, a cloud of fog formed only on negative ions (atoms with a negative electric charge) by water condensation which slowly settled on the disk below thereby allowing the measurement of the total electric charge attached to the disk with the electroscope. If the initial amount of water vapour in the cylinder and the average size of fog droplets are known, it is possible to find the total number of fog droplets produced by water condensation on negative ions. The total number of fog droplets produced also gives the total number of ions. The fog droplets, however, were too small to be visible and, therefore, Thomson found their size from the settling speed of the fog. The smaller the spherical fog droplet, the slower it settles down according to a formula in fluid dynamics first established by the Irish physicist Georg Gabriel Stokes which gave a relationship between the speed of fall, the radius of the average droplet, and the viscosity of air. By dividing the total electric charge measured on the disk with the number of fog droplets, Thomson found the electric charge of the electron e to be $(4.77/10^{10})$ electrostatic units, the same value as in the electrolysis of liquids, and the mass of the electron m to be $(0.9/10^{29})$ grams, which is 1,840 [present value 1,837] times smaller than the mass of the hydrogen atom. Therefore, the electron is a particle that is 1,840 times lighter than the lightest atom known, the hydrogen atom. In 1906, Thomson was awarded the Nobel Prize in physics for his outstanding research of electrons and cathode rays. After 1906, Thomson studied the Goldstein canal rays consisting of streams of positively charged ions by using the same principle. He subjected the canal ray simultaneously to the action of a magnetic and an electric field which should have produced a trace of a parabola on a fluorescent screen. Instead of producing one parabola, there were two or more parabolas, which indicated the presence of atoms with different masses for the same element. Atoms of the same element having different atomic weights were called 'isotopes' by the English chemist Frederick Soddy (1877-1956), meaning that they occupy the 'same place' in the periodic table of elements established in 1869 by the Russian chemist Dimitrii Ivanovich Mendeleev (1834-1907). It was another important contribution of Thomson to the foundation of modern physics. Thomson concluded from his experimental results that the cathode rays are streams of negatively charged particles which are much lighter than atoms. He believed that these charged particles which the Dutch physicist Hendrik Antoon Lorentz had called 'electrons', a name suggested in 1891 by the Irish physicist George Johnstone Stoney (1826-1911) for the minimum electric charge thought to constitute electricity, over Thomson's objections, were present in all matter. Electron was the first, and the smallest known subatomic particle discovered at the time. Thomson was able to explain the 'Edison effect' by means of his electron concept, and in 1903, he experimentally corroborated his claim of 1897 that electrons were the carriers of electric current. Thus Thomson demonstrated that the particles forming the cathode rays were free electric charges, whereas the Faraday's ions were atoms carrying an electric charge. The notion of the existence of the smallest quantity of electricity for which the term 'electron' was introduced by Stoney followed quite logically from the ideas that Faraday had about electrolysis.

In 1899, Emil Wiechert confirmed the values found from the deflexion experiments for the ratio of the mass of the cathode ray particle to its charge by means of electrical oscillations which enabled him to take direct measurements of the speed of cathode rays. This confirmation definitely established the existence of electrons. Since 1896, a number of physicists were attempting to determine the charge of the electron in absolute terms which is independent of the mass of the electron. The best direct method for the measurement of the electric charge of the single electron was essentially devised in 1907 by the Austrian physicist Felix Ehrenhaft (1879-1952), a method which was skilfully used by the American physicist Robert Andrews Millikan (1868-1953) in 1913, and with further improvements in 1940. Millikan determined the electric charge of an electron from the falling of an oil droplet ionised by X-rays in the gravitational and the electromagnetic field. The electromagnetic attraction was imposed by an electrically charged plate positioned above the falling oil droplet. He was able to determine directly the electric charge 'e' of a single electron by balancing the downwards gravitational attraction and the upwards electromagnetic attraction acting on the ionised oil droplet which influenced its motion. Furthermore, he proved that any electric charge always existed as a 'whole number' of the electric charge of the single electron. It was the final proof of the particle nature of electricity. He also verified by very careful experiments the validity of the equations of the photoelectric effect Albert Einstein derived in 1905 by means of the concept of photons of light and quantum theory. In 1923 Millikan was awarded the Nobel Prize in physics for his remarkable experimental work on the electric charge and the photoelectric effect.

The mass of an electrified particle such as an electron was not a fixed quantity, since already in 1881 Thomson had demonstrated theoretically that such a mass increases with its speed. After 1905, many physicists investigated the change of mass of a moving electron, and found that the mass of the electron was entirely electrical and dependent upon its speed as Lorentz had predicted.

In 1896, the German engineer, Wilhelm Conrad Röntgen (1845-1923), who became a self-educated physicist and professor of physics at the University of Würzburg in Germany, investigated the behaviour of cathode rays in a discharge tube the 2 electrodes of which were charged with high voltage. Röntgen, almost accidentally, discovered that besides the cathode rays his discharge tube emitted also another kind of radiation in the form of mysterious rays having a great power of penetration. Röntgen's momentous discovery of these rays, which he called 'X-rays', was not only important in medicine, but also in the theory of matter since it led to intensive investigation of the source of X-rays. He had discovered X-rays by the fluorescence they created in a chemical which also fluoresced when exposed to cathode rays. It is important to point out that already in 1895 Nikola Tesla had taken an X-ray picture of a hand at a 40-foot (12.19 m) distance using his high-frequency 'Tesla coil'.

The French physicist, Henri Antoine Becquerel (1852-1908), professor of physics at the *Museum d'Histoire Naturelle* and the *École Polytechnique*, continued the investigations of his father Alexandre Edmond Becquerel (1820-1891) on the fluorescence phenomenon. Henri used in his experiments a compound of uranium (*potassium uranyl sulfate*) and being intrigued by Röntgen's experience, he was curious to learn if any other fluorescent material such as his uranium compound also emitted X-rays besides the visible light rays. In 1896 Becquerel discovered to his great amazement that the radiation his uranium compound emitted was independent of its fluorescence. In 1899 he proved that the radiation emitted by the uranium compound could be deflected by a magnetic field which implied that the radiation contained charged particles. Becquerel found that only substances containing uranium emitted this mysterious radiation, but soon other elements were discovered which also emitted natural radiation, the most important of such elements was 'radium' which gave 2 million times more natural radiation than uranium, discovered by Marie Curie (1867-1934) and Pierre Curie (1859-1906) who shared in 1903 the Nobel Prize in physics with Becquerel. In 1898 Marie Curie named Becquerel's radiation 'radioactivity'. Since Becquerel's radiation did not consist of X-rays, Becquerel and other physicists attempted to determine its precise physical nature and source. In 1898 Rutherford found that this radiation contained 2 kinds of high-speed charged particles: the positively charged alpha particles and the negatively charged beta particles. The beta radiation was about 100 times as penetrating as the alpha radiation. In 1900, the French physicist Paul Ulrich Villard (1860-1934) demonstrated that the Becquerel's radiation also contained electromagnetic radiation which was unaffected by either a magnetic or an electric field. Villard called it 'gamma radiation' which was about 100 times as penetrating as the beta radiation. In 1900, Becquerel, who had determined the radioactive part of his compound to be the uranium, identified the negatively charged beta particles with electrons. This seemed to indicate that the only place these electrons could come from was the inside of the uranium atom, and, therefore, it implied that the atom has an internal structure. It heralded the very beginning of the modern atomic physics. In early 1899, the German physicists Julius Elster and Hans Friedrich Geitel rejected the hypotheses for the cause of radioactivity of Crookes, and of the Curies. Elster and Geitel proposed that the cause for radioactive decay was similar to how a molecule with unstable bond passes into a stable state by giving up energy. This conception, in their opinion, requires the assumption of a gradual transformation of an active substance into an inactive one which is accompanied by a consistent change in its elementary properties. Elster and Geitel also discovered the law of decay of natural radioactivity which states that the number of particles emitted from the element per second decreases exponentially with time. This time-rate, however, varies with the element. The so-called 'half life', representing the time in which the number of particles emitted by the element reduces to one-half, characterises each element and it varies between enormously wide limits, from a minute fraction of a second to almost 'countless' number of centuries. In 1902 the fundamental theory of 'atomic disintegration' was established in a detailed exposition by Rutherford and Soddy.

In 1900, Lenard developed the 'dynamid' theory of bodies, which assumed the material part of an atom to consist of 'dynamids', a form of neutral doublets of a positive and a negative charge in order to account for the penetrability of cathode rays, a theory which had many features in common with the model of atom announced in 1911 by Ernest Rutherford. In his explanations, Lenard was the first physicist to assume that an atom was mostly empty space, and that it contained evenly distributed electrons and positively charged particles later called 'protons' by Rutherford. It was Lenard's work which by and large convinced most physicists that atoms contained electrons as part of their inner structure. In 1903 Rutherford asserted that every radioactive process of atomic disintegration results from a spontaneous transmutation of one chemical element into another – something that every mediæval alchemist had hoped in vain to produce by means of their experimental methods. Rutherford discovered from his experiments on the scattering of extremely high-energy alpha particles [positively charged helium ions] when passing through metal foils that an atom was mostly empty space as Lenard had assumed but it had a tiny, yet massive and heavily charged central core which Rutherford named the 'atomic nucleus'. The positive charge of the nucleus and, therefore, the number of electrons swarming in circles around the nucleus under the action of Coulomb attraction is equal to the atomic number of the element in Mendeleev's periodic system of elements. It was demonstrated in 1926 that the emission of gamma rays is produced after the transmutation of the elements when the disturbed nucleus of the atom, brought about by the emission of positively or negatively charged particles, passes into a normal state. The age old dream of the alchemist – the transmutation of metals – was rendered more than a dream by the discovery of 'natural radioactivity', which indicated that such transformations do indeed occur in nature. In 1919, Rutherford's experiments showed a possible method to achieve 'artificial transmutation' by bombarding nitrogen gas with alpha particles which produced entirely new materials: hydrogen and an isotope of oxygen. Rutherford used alpha particles from a natural radioactive element. In 1932, John Douglas Cockcroft (1897-1967) and his assistant Ernest Thomas Sinton Walton (1903- ?) produced artificially accelerated protons with which they bombarded lithium to produce helium. In 1934 Irene Curie (1897-1956) and Frédéric Joliot (1900-1958) bombarded aluminium with alpha particles and discovered that the reaction continued after bombardment. An isotope of phosphorus was created, which subsequently broke up to form silicon, and in the process produced another type of emission, the 'positron'. This kind of transmutation after bombardment is known as 'artificial radioactivity'. It showed that any element could be radioactive if the proper isotope was used.

In 1902, Lenard followed up Hertz's study of the photoemissive phenomenon by demonstrating that the electrical effects produced by light falling upon certain metals was the result of emission of electrons from those metals. He demonstrated that only certain wavelengths of light could bring about emission of electrons, and that for any particular wavelength electrons of fixed energy were emitted, and that the increased intensity of light only increased the number of electrons emitted but not the energy of individual electrons.

In his work on the incandescent light bulb in 1883, Thomas Alva Edison (1847-1931) made his only scientific discovery, the so-called 'Edison effect', which became the very source of the science of electronics. Edison noticed that after prolonged use his carbon filament lamps became blackened with carbon deposits on the inside surface of the glass bulb. He also observed that under certain conditions of the vacuum and the voltages within the bulb there existed a peculiar bluish glow in the light bulb. Upon closer examination Edison discovered that this strange glow was caused by an electric current passing through the vacuum between the two leads to the filament which flowed in the opposite direction to the main current passing through the filament. In an experimental bulb made to test this strange current, he found that a small electric current passed through vacuum from the carbon filament in his incandescent bulb to an electrode sealed into the bulb when the electrode was positively charged. Yet, no electric current flowed when the electrode was negatively charged. Although Edison was unable to explain this puzzling electric phenomenon of thermionic emission, he took out a patent in 1886 for an electric indicator based upon the Edison effect, and used it to regulate the supply voltage in his electrical power station. It was the first patent in electronics. Two Englishmen, the electrical engineers William Henry Preece (1834-1913) in 1885 and the physicist John Ambrose Fleming (1849-1945), a former student of Maxwell, in 1890 and in 1896, studied the Edison effect extensively, but they also failed to give a sound theoretical explanation of it.

In the Edison effect the electrons were emitted by the incandescent filament in a physical process known as thermionic emission. The theory of thermionic emission was established by the English physicist and a former student of Thomson, Owen Willans Richardson (1879-1959), in 1903. Richardson's theory showed that electrons are emitted from hot metals such as hot cathodes by a process analogous to evaporation, and that the character of this emission was determined by a property of the substance analogous to the latent heat of evaporation. He elaborated a theory of electron and ion (electrically charged atom) emission by showing that the electrons emitted from hot bodies have the same kinetic energy as molecules of a gas at the same temperature, that the process of emitting electrons involves cooling as in the process of evaporation, that the process of absorbing electrons involves heating of the absorbing body, and by investigating the emission of positively charged ions from hot filaments – a detailed research which made possible the rapid technological development of the vacuum and cathode ray tubes. In 1928 Richardson was awarded the Nobel Prize in physics for his pioneering work on thermionic emission.

The basic discovery of the electron in the electrical science opened up a new and very important discipline of electrical engineering, called 'electronics', a specialty in electricity which dominates contemporary engineering science.

The revolutionary advances made in the understanding of the physical nature of the cathode rays, the discovery of the electron, and the natural radioactivity inaugurated the second scientific revolution in physics as well as in electrical sciences which launched the remarkable development of atomic physics in the 20th century.

CHAPTER XII

BRITISH TECHNOLOGY: Manufacturing and Textile Industries, Industrial Revolution
[from 1730 to 1860]

When the King of England, Henry VIII (1491-1547), abolished the catholic monasteries and banished the monks, it at first retarded the economic development in the British Isle, since monks had been the principal leaders of many public works. A revival of British engineering took place during the reign of Queen Elizabeth I (1558-1603), who left the people to their own devices and, thereby, promoted free enterprise. A new group, the English surveyors and mathematicians, emerged. Hugh Myddleton built his celebrated water supply system for London, which was inaugurated in 1613, as a private venture. Fen drainage began, and foreign craftsmen and engines were imported from the continental Europe to launch new industrial ventures. The Cromwellian period in the early 17th century, the Restoration and the 'bloodless Revolution' of 1688, finally achieved a workable constitutional monarchy, which led to a remarkable age of engineering and industry. Englishmen rapidly learnt about silk twisting from Italy, about canal building from France, about land drainage and agriculture from the Low Countries, and about mining from Germany.

Despite the fact that ironmaking had not been expanded but rather severely curtailed to save timber for shipbuilding since the old small iron furnaces had virtually denuded the British Isle of its once rich forests, a new mode of progress emerged that was prompted by the expanding foreign trade which, in turn, promoted specialisation in trades and division in labour. Although the ancient Greeks had used specialisation quite effectively, the British made it an effective means of industrial progress through free enterprise which changed the old agricultural and static way of life into a progressive manufacturing economy on an international scale promoted by British reliance on free trade. Efficient British shipping provided the handy means for the exchange of merchandise and manufactured goods. First canals, then roads, and finally railroads opened up the British Isle to new industries that could be located anywhere in the country where ready means of transportation were available.

Iron industry using coal as fuel, and the steam-engine as a source of power gave a tremendous boost to the British industrial progress. Private enterprise, and gifted individual inventors and craftsmen brought on a revolutionary change in the methods of manufacture and the way of life of the common man in Britain, whilst France was stagnating under the central heavyhanded control of a parasitical bureaucracy. Consequently, a Statist and sluggish territorial country like France never went through an 'industrial revolution' manifested by a precipitously accelerated but continuous evolution of industry, and was surprisingly slow to adopt the steam-engine as the primary source of industrial power. Therefore, France remained a large, unprogressive country relying primarily on manual industry.

The rise of British technology and engineering in the 18th century created a new era in the history of man by joining technology and industry in the creation of a new industrialised production system. The so-called 'industrial revolution' was a direct result of this union in which science played a subordinate role. Various progressive developments that dominated the 18th century of the West emerged from the British industrialisation.

Serfdom had vanished in Britain much earlier than in the continental Europe, and consequently many farm workers became wage earning labourers. The freedom of movement of the British worker made the recruitment of modern labour force possible. The iron-clad stranglehold the guilds had on their craftsmen in the continent, a holdover from the Late Middle Ages, suffocated industrial development, because the craftsmen were not free to move, but rather like children, they required the permission of their guild to do so. In England, in contrast, there were no rigid barriers between the town and the countryside: workmen could move freely from one part of the country to another. British landowners were prepared to exploit their own mineral resources, and they took an active part in commercial and manufacturing enterprises. Moreover, successful manufacturers from the towns could buy country estates, and in this way enter the ranks of the landed gentry. As a result of this freedom, a sizeable middle class developed which became the source for many energetic entrepreneurs and managers of the newly emerging factories. Since an energetic middle class represents the future of any nation, the industrial progress was rapid in Britain, particularly as loans at moderate interest rates were readily available to the entrepreneurs.

After the Cromwellian period, the Clarendon Codes forbade any member of non-Anglican religion to hold a position in local government, civil service, or university, mainly because the religious Nonconformists, such as Quakers, Presbyterians, and Unitarians refused to swear the required loyalty oath. These Dissenters were allowed to engage in trade, in which they soon prospered because through their diligence and application they throve in industry and finances. The Quakers and the Unitarians promoted a system of education which put heavy emphasis on personal excellence and success. Therefore, the Dissenters founded their own academies in which the Nonconformists were permitted to teach, and they were the ones who developed the first modern curriculum which provided the best education as a preparation for industry. In these academies, the first general scientific education was taught. The outcome of all this was that the most successful entrepreneurs and engineers in Britain were the Dissenters, constituting about half the successful entrepreneurs in Britain at the close of the 18th century.

In consequence of all these developments, England became the place of the greatest and most rapid engineering progress since the Middle Ages. The joint-stock companies, in which anybody could invest any amount of their savings, became the powerful instruments of industrialisation. Fecund voluntary collaboration, without the crippling intervention of the parasitic State bureaucrats in engineering, technological industry and trade, under the aegis of an energetic and enterprising middle class, brought about a dramatic expansion of industry. In England a general belief in the free enterprise entrusted everything to the ingenuity of the individual, who sought his rewards in the public demand alone. Englishmen refused to divert too large a portion of their technological resources to the sterile State projects. A large class of artisan-inventors interested in mechanics and science, and in constant contact with it, provided the initiative to technological inventions, which made the British Industrial Revolution possible.

A succession of great inventions from 1730 to 1800 revolutionised the textile, the metal and the transport industry. Ingenious spinning, weaving and carding machines appeared. New ways of smelting and purifying iron ore, and methods of making crucible cast-steel were devised. The cylinder-boring machine, the hydraulic press, the steam-engine, and the steam-hammer transformed the technological industry. Invention in one industry stimulated invention in another. By 1800, the great technical inventions produced by the British intelligent craftsmen, including the most important of them all, the steam-engine, had become practical realities, and by this outstanding accomplishment, Britain had become the 'workshop of the world'.

The Industrial Revolution in Britain developed the all-important efficient production system: a fusion of technological inventions with the factory system, and the free enterprise outlook created the new production system which improved the economy in a fundamental manner. The triple alliance of engineering, industry, and individual free enterprise inaugurated a new era in the life of man. During the period of Industrial Revolution, engineering science began to enter technological development in a more fundamental way: it was a period of growing cooperation between science and technology. The basic achievements in the Industrial Revolution were:

1. Replacement of manual tools by machine-tools.
2. Introduction of new stationary and mobile prime movers.
3. Factory as a new form of organised production.

The situation in France was quite different. Despite the fact that the old relic of the Holy Roman Empire, the *corvée* labour, had been temporarily eliminated in France by the Finance Minister Turgot in 1766, but reintroduced when Turgot resigned a year later, and finally abolished in 1798, the French peasant proprietors and tenant farmers were doggedly attached to their land, and stubbornly influenced by their family ties. The spirit of the late Feudalism still prevailed over the French population, and, therefore, it was very difficult to attract them to factories. The thrifty French people preferred to invest heavily in land, and in moribund government bonds that suffocated creativity and production, because they were suspicious of banks and reluctant to risk their savings in industrial shares or in other joint-stock enterprises. Consequently, France remained essentially an agricultural, and largely autarkic country dominated mainly by small unprogressive industries whose production depended heavily on manual labour until the 19th century. In France, everything was centrally controlled : the State bureaucrats, a parasitic and autocratic coterie of employees in the government service, dictated the economy, promoted government favouritism that placed certain manufacturers to arbitrary and undesired preeminence by granting a government imposed monopoly to a manufacturer, which secured for the latter the involuntary patronage of the people, and, thereby, promoted inefficiency, waste and industrial stagnation. For instance, construction of railroads was seriously retarded in France by the State politics, and industrial development in general was suffocated by the arbitrary and incompetent central bureaucratic control from Paris. Therefore, in France, nothing important could ever take place in other centers of the country since anything of real significance could only take place in Paris. For a French engineer or scientist to be sent away from Paris meant almost the same as to be exiled. In Britain, conditions were the exact opposite: important industrial developments took place all over the country.

Early Water Supply Systems

In 1581, a Dutch mining engineer, Peter Maurice (alias Morris), publicly demonstrated in an experiment how to throw a jet of water from the Thames river over the steeple of St. Magnus Church. As a direct result of this mechanical feat, Maurice was granted a 500-year lease of two arches in the Old London Bridge for the installation of a battery of waterwheel-driven pumps to supply water to London. In 1582 Maurice had successfully erected his great pumping engine in the archway of the Old London Bridge which was driven by an adjustable waterwheel, 20 feet (6.10m) in diameter and 14 feet (4.27m) wide, turned by the rising and falling of the tide. Cogwheels and cranks converted the rotary motion of the waterwheel into reciprocal motion for the pumps with pump-barrels of 5 3/8-inches (13.7cm) bore and about 4-foot (1.22m) length. The descendants of Maurice sold their pumping privilege in 1701 for 3,800 pound sterling.

Whilst the city of London grew, the need for more water than was supplied by Maurice's water pumps became so acute that the British Parliament finally authorised to supply more water for Hertfordshire in a leat. Among the more notable early British civil engineering works, this leat, as part of the so-called 'New River' water supply system for the city of London, was built and financed by a capable entrepreneur Hugh Myddleton (1555-1631), a London goldsmith and a friend of Sir Raleigh, with the later financial assistance of King James I. Myddleton's leat, known as the 'New River', was an open channel, 10 feet (3.05m) wide and about 4 feet (1.22m) deep, which tapped the fresh water springs of Chadwell and Amwell in Hertfordshire and brought fresh water to a circular pond at Islington, known as the 'New River Head', by following the contour line of the terrain the channel had a length of 38¾ miles (62.4km), which was later shortened to 24 miles (38.6km). The average gradient was 2 inches (5.1cm) to a mile (1.6 km), which made it necessary to carry the water over low ground and roads by timber troughs or flushes lined with lead. The distribution system consisted of elmwood pipes with spigot and socket joints which were later replaced by lead pipes. This work was finished in 1613 under the direction of the engineer Edward Wright (1559-1615). Myddleton later leased some flooded mines, drained them, and in mining them, made a large profit. He also hired Dutch engineers in 1621 to drain land on the Isle of Wight.

One of the early successful British engineers was George Sorocold of Derby. He was a versatile engineer and most famous for building of public water supplies. He used waterwheels to work sets of pumps that forced water into elevated cisterns from which water was distributed through pipes by gravity. In 1702 he replaced Peter Maurice's water pumping machinery, which had been badly damaged in the 'Great Fire' of London in 1666, and by 1701 was in a bad state of disrepair.

Sorocold, who tended to be a perfectionist in his engineering works, was responsible for the waterworks at Derby (1692), Leeds (1694), Norwich (1694), Exeter (1695), Bristol (1996), Sheffield (1697), and Bridgenorth (1706). He built the great waterwheel driving the Derby Silk Mill, one of the birthplaces of the silk-spinning industry, and built a waterwheel with a crank and beam to drain the Clackmannanshire colliery of the Earl of Mar. The versatile Sorocold was not only engaged in draining mines, but he was also the consulting engineer for the first Liverpool Docks in 1708, and probably the first wet-docks at Rotherhithe.

Canal Era: 1760 -1830

The early roads in Britain, like the early roads in France, were terrible. Even in the 18th century, it was said in jest that it was easier to make the roads navigable than to give them a hard surface. For this reason, all great cities in Britain were located next to coastal waters where shipping was easily available. The development of British canals for water transport had lagged far behind a similar development in France. The first important step taken towards the industrialisation of Britain was the building of canals which became important supply-and-demand arteries for markets.

The founder of the British canal construction era was an unlettered, self-taught, yet skilled millwright and builder James Brindley (1716-1772), who relied on his native abilities in his engineering works. In 1761, Francis Egerton, the Duke of Bridgewater, hired Brindley to construct the 10½-mile (16.9 km) long Worseley Canal from his colliery at the Mersey river to the city of Manchester. This canal incorporated a 600 feet (182.88 m) long and 36 feet (10.97 m) wide Barton aqueduct, which carried the canal 39 feet (11.89 m) across the Irwell river on 3 semicircular stone arches. The Duke extended his canal to Runcorn which connected southeastern Lancashire with Liverpool, and as a result of the availability of the canal to transport coal from the Duke's colliery to Manchester and Liverpool, the price of coal was reduced by 50 %, and that was the main reason why Manchester became the first industrial center of Britain.

Brindley preferred to drive his canals along the contour lines of the terrain rather than cut them through the hills since longer canals were not uneconomical as long as the wages of boatmen were low. In his time boring a tunnel was more economical than cutting through a hill. The longest tunnel engineered in this period was on the Huddersfield canal at Standedge which had a length of 16,245 feet (4,951 m).

From 1776 to 1777, Brindley was building under the sponsorship of the Duke of Bridgewater and Josiah Wedgewood (1730-1795), a scientific manufacturer and a tireless experimenter on the industrial production of pottery at his Etruria works, his most extensive project, the Grand Trunk Canal (also called the Trent and Mersey Canal), which was 139½ miles (224.5 km) long and connected the Trent river with the Mersey river. The Grand Trunk Canal had 5 tunnels, the longest of which was the Harecastle tunnel that was 210 feet (64.00 m) below the surface of the ground, had a 9 by 12 feet (2.74 by 3.66 m) cross-section, and was 2,888 yards (2,640.79 m) long. The Grand Trunk Canal incorporated into its system 160 minor aqueducts and 109 road-bridges. In this project Brindley collaborated with a young associate engineer, John Smeaton (1724-1792), who was the first engineer to call himself a 'civil engineer', and with whom began the professional development of British engineering. In his entire engineering career, Brindley built about 500 miles (804.7 km) of canals, 298 locks, 847 bridges and 12 canal tunnels. In his engineering practice, Brindley made a beginning of separating the engineer from the builder. Both Brindley and his collaborator Smeaton learnt their engineering indirectly from the engineering done by the Dutch and the French engineers. Smeaton, who had a first-rate education in law, was an avid student of the excellent books on French engineering published by the French military engineer Bernard Forest de Belidor (1691-1761).

Smeaton built the Forth and Clyde Canal from 1768 to 1775, and finished it in 1777. This canal, which was 35 miles (56.3 km) long, 7 feet (2.13 m) deep, and had 39 locks, remained the largest canal in Britain until Telford's Caledonia Ship Canal constructed along the line first surveyed by James Watt (1736-1819) [who later became famous for the invention of his steam-engine] was opened in 1822.

The 'Canal Era' in Britain lasted until 1830, a year which marked the beginning of the 'Railroad Era'. By that time, Britain had a network of nearly 2,000 miles (3,218 km) of canals.

Whilst the canal building, which was promoted by manufacturers, was progressing, a great advance was overtaking the textile and iron industries, an advance that played an important role in the phenomenal industrial progress in Britain.

Industrialisation of Textile Manufacture

In the textile industry the spinning had been done manually for countless centuries. The very first improvement in spinning was the introduction of the spinning-wheel from China, but weaving was still done by the same crude loom which had been in use since primitive times.

The remarkable industrial development in Britain may be considered to have begun in 1733, when an inventor and a clockmaker, a son of an owner of a small Lancashire textile works, John Kay (1704-1774), invented the 'flying shuttle' [a shuttle on wheels struck through the warp threads by a hammer] that dramatically improved the speed of weaving on the loom also invented in China. This invention made it possible to produce a double-width weave and to increase the speed of weaving so much that the weaver almost doubled his daily production. Kay also developed a machine to make cards, which were used to entangle the fibres before spinning. In 1747, the local weavers opposing Kay's invention smashed his machines, and Kay himself had to flee to France to save his life, where he ultimately died in poverty. Kay's invention, however, which made spinning and weaving separate operations, was appropriated and used with profit in weaving by the weavers who had driven him into exile. It now took 4 or 5 spinners to supply yarn to a single weaver using a loom with Kay's 'flying shuttle'.

In 1764, a versatile weaver, carpenter, blacksmith and toolmaker, James Hargreaves (1720-1778), from Stanhill near Blackburn, invented a manually driven spinning-machine, that he named 'Spinning Jenny' after his daughter Jenny, in which one spinner could operate 8 spindles at the same time instead of only one as was the custom before his invention. The 'Spinning Jenny' intermittently produced a fine, but not very strong yarn, and the yarn so spun was suitable only for the weft but not for the warp. Therefore, the yarn for the warp still had to be spun manually by the hand-wheel. Hargreaves' machine operated intermittently in a similar sequence to manual spinning, and it was soon improved, particularly by Haley of Houghton Tower. In 1768, Hargreaves was attacked by a mob of hostile spinners, and his 'Spinning Jennies' and looms were burnt. In order to save his life, Hargreaves had to flee to Nottingham, where he went into partnership with Mr. James in a new mill, which became a profitable venture. Hargreaves patented his 'Spinning Jenny' in 1770, a few weeks after Arkwright, but his patent was declared invalid because he had sold a few 'Spinning Jennies' before that date. Six years after the death of Hargreaves in 1778, there existed about 20,000 'Spinning Jennies' in England.

The inventions of both Kay and Hargreaves were still manually operated machines. However, the 'Spinning Jenny' of Hargreaves had its difficulties : It required great power from the spinner to operate it, and it was difficult to drive all the spindles at the same speed. Moreover, only softly twisted yarns could be spun on it, because under greater pull, the fibres in the roving became locked and thus were prevented from being drawn out. The yarn produced by the 'Spinning Jenny', therefore, was only suitable for the weft.

During the Mediæval Age, about 1300, silk-thread machines driven by waterwheels had been invented in Italy. The problems of spinning silk are relatively simple since the silk fibres were very long (sometimes up to several miles long), and by being sticky they held together without much difficulty when being spun. In the case of common textiles, such as cotton and wool, the fibres are much shorter than silk fibres, so that giving them the right amount of twist to form a uniform, strong thread is a much more critical operation. The mediæval invention of the spinning wheel had increased productivity but had not reduced the need for the skills of the spinner, called the 'spinster'. It seemed impossible to try to mechanise spinning with the rather clumsy wooden machinery of the 18th century because it would have snapped the fragile thread.

The first attempt to mechanise spinning and catch up with the demand of the weavers for more thread was made in 1738 by John Wyatt (1700-1766) and Lewis Paul (?-1756), a son of a

French Hugenot, when Paul patented their spinning machine. These machines showed a strong resemblance to the silk-throwing machines in Lombe's mill in Derby, a mill opened in 1721, the machines of which were not greatly different from those invented in mediæval Italy about 1300. The one distinctive original contribution made by Wyatt and Paul in their machine was the use of rollers to draw out the carded sliver of fibres before twisting it into thread. The cotton or wool was fed into their machine in rovings, a form of loose rope of extended and twisted fibres, through a series of pairs of rollers, in which each successive pair of rollers revolved faster than the previous pair to draw out the sliver, and then passed to a 'flyer' that twisted it and reeled it onto a bobbin on a spindle. The roving was supposed to be drawn out into a thread between the pairs of rollers and the 'flyer'. Wyatt and Paul used two donkeys to drive their machine, but unfortunately, their machine never worked quite satisfactorily. The spinning machine of Paul and Wyatt failed most likely because the drawing stage and the twisting, or spinning stage of the operation were not mechanically separated. It appears that this machine was essentially the invention of Paul. In 1748, Paul invented an incomplete carding machine. In 1758, Paul patented a partly successful new spinning machine with only one pair of rollers, in which the attempt to draw out the sliver between the rollers had been abandoned.

Successful spinning of cotton and wool threads with mechanical means was perfected by a former barber and wigmaker, Richard Arkwright (1732-1792), who invented in 1769 his spinning frame, popularly called the 'water frame' which gave the thread a twist by drawing it through 4 pairs of rollers running at different speeds. The yarn produced in a continuous process by the water-frame was rough but strong enough to be used for both the weft and the warp in woollen textiles. A shortcoming in the 'water frame', which to some extent was later inherited by the 'ring frame', was the friction of the bobbin created by the pull on the yarn that made it impossible to spin fine or soft, twisted yarns by this method. For this reason, the 'Spinning Jenny' and the 'water frame' remained complementary spinning machines. From 1775 to 1785, Arkwright devised an engine which carded the textile fibre material by a crank, a comb and two cylinders provided with carding teeth. Unfortunately, his patents were declared unoriginal and invalid in 1785, and as a result, Arkwright lost the inventor's right for a limited monopoly, but by that time, he had become a wealthy industrialist who lent money even to Prince of Wales.

Arkwright's machines were no longer manually operated machines, since they had to be driven by power. At first Arkwright used six horses to operate a 27-foot (8.23m) horsewhim which drove his textile machines by means of wooden shafts and pulleys. In 1771, his factory at Comford used water power to operate his textile machines. In 1777, he established a new factory with Jedediah Strutt (1726-1797) of Derby, in which his expensive machines were powered by waterwheels. In 1781, Arkwright used a Newcomen engine to pump water into an elevated reservoir for the driving of the waterwheel, which produced the rotative motion necessary to turn the textile machines in his factory. For this reason, Arkwright's invention effectively ended the domestic system of manufacture of textiles, and created the factory system of manufacture. By restricting licences to 1,000 spindle units, the spinning machines could only be operated economically in water-powered factories. After 1785, when Arkwright lost his monopoly because his patent was cancelled, any large-scale enterprise which could use Watt's steam-engine became economically advantageous. Watt's steam-engine was first used to drive the spinning machines in the Robinson factory at Nottinghamshire, but after 1790, the use of steam power in textile factories spread with the rapid expansion of the cotton industry.

These labour-saving devices in the textile industry were regarded by textile workers to be a threat to their livelihood, and as a result of this fear, they resorted to violence and destroyed a number of the textile machines in the riots of 1779. During these violent disturbances, Arkwright had to flee to save his life.

In 1779, Samuel Crompton (1753-1827), a spinner, created a synthesis of the textile machines of Arkwright and Hargreaves, for which reason he called his spinning machine a 'mule', by adding a moveable carriage to the spinning machine that housed the spindles. Crompton's 'mule' was able to spin very fine and strong thread which was suitable for both the weft and the warp. Unfortunately, owing to the hybrid nature of his spinning machine, Crompton was unable to patent it. The 'mule' was a versatile machine because the relationship between the motion of the carriage and the speed of the rollers could be changed so that every kind of yarn could be spun on it. The intermittent spinning process with Crompton's 'mule', which could handle up to 900 spindles, had now become so efficient that the bottleneck of the textile industry under the new circumstance was once again weaving, which desperately needed a power-loom. The problem with Crompton's 'mule' was not the spinning but rather the winding-on operation. When the yarn was wound too slowly, a length of it became wasted; when it was wound too quickly, there was no length of it available to form the spirals up the spindles for spinning the recommence. Many modifications of the 'mule' were patented over the years, but only in 1835 was the winding-on problem ultimately solved by Richard Roberts (1789-1864).

Johann Georg Bodmer (1786-1864), a Swiss engineer and a former child prodigy who for a time was a resident of Lancashire, invented in 1824 a device which made spinning continuous with carding. The industrial value of this invention was ignored in Britain, but not in United Sates and continental Europe.

The first less expensive spinning machine to replace the rather expensive 'water frame' appeared in 1786 and was called 'Billy'. It looked rather like a 'Jenny', but the spindles on 'Billy' were on the carriage and the twisting and winding-on took place like in the 'mule'. The 'Billy' was also a manually operated machine, but its inventor is not known.

In 1800, John Kennedy (*floruit* 1790-1800), a Manchester spinner and machine builder, rendered the mule almost self-acting by leaving only the winding-on operation manual. The complete automation of the 'mule' came as a result of the spinners strike in 1824, when Richard

Roberts, an ingenious Welsh machine toolmaker and inventor, produced the first completely self-acting 'mule' upon the urging of the textile manufacturers.

In 1828, John Thorp (1784-1848), in the United States, invented continuous spinning with his ring-spinning frame, in which the flyer was replaced by a small 'traveller' that worked on a ring set in a frame round the bobbin. The yarn passed through the feed rolls, then through a guide and finally, through a 'traveller' working round the flange ring. The 'traveller', consisting of a thin wire sprung from the ring, held the roving at the winding point as it passed from the drawing rollers to the spool on the spindle. This enabled the spinning to proceed continuously and at a great speed, 11,000 revolutions of the spindle a minute. Although no other fundamental device in the textile industry has been the subject of so many patents and improvements, yet it became influential in textile industry only after 1850. A new form of 'water frame' to make strong yarn for use in the power-loom, called the 'throstle', came into manufacture after 1815. An American version of the throstle was invented by Charles Danforth (1797-1876). Danforth's throstle had a conical cap over the spindle to conduct the thread to the bobbin, but it produced a softer yarn.

In 1785, a minister and an Oxford don, Reverend Edmund Cartwright (1743-1823), who knew virtually nothing of the textile industry but was a remarkably versatile, enterprising and inventive man, solved this difficult weaving problem by devising a mechanical loom, but his machine was rather crude and needed considerable improvements before it could become fully effective in manufacture. Since it required the strength of two powerful men to work the machine at a slow rate, and even then only for a short time, his loom had to be driven by animal or water power. Cartwright improved his loom in 1786, in 1787 and in 1792, and he set up his own spinning and weaving mill in Doncaster and later, a much larger mill in Fishergate. He installed a steam-engine in the Doncaster mill in 1788, but only after 1790 was the steam-engine generally used as a power source for Cartwright's looms. New devices of sizing and drawing the warp, devices which were unsatisfactory in the mechanical loom of Cartwright, were urgently needed for the general adoption of the mechanical loom. On account of these shortcomings of his loom, Cartwright's mills were commercial failures, but despite the commercial failure of his mills, Cartwright made several improvements to spinning and wool-combing machines.

In 1747, Cartwright improved the performance of the steam-engine by inventing the first completely metallic packing for it, which ensured tight joints and prevented serious leaks. Cartwright's metal packing was later replaced with a metallic piston ring by the English inventor John Burton, who filed a patent for it in 1816. Cartwright also invented rope-making, bread-baking, and brick-making machines, and various agricultural implements, among them a 3-furrow plough. For his original essays and inventions in the field of agriculture, Cartwright was awarded 2 silver medals and one gold medal. He collaborated with Robert Fulton (1765-1815), the renowned American painter, civil and marine engineer, in the latter's experiments on the application of steam-engines to navigation. Moreover, Cartwright was also a prominent literary critic and a poet.

In 1803 William Radcliffe of Stockpart, with the help of Thomas Johnson, improved the mechanical power-loom further by inventing a dressing-and-drying machine. Radcliffe took up the woven cloth in a manual loom by a ratchet-wheel connected with batten, thereby making the cloth beam move automatically. This invention which contributed to the development of the power-loom, was patented by Thomas Johnson, an assistant of Radcliffe, in 1803 and in 1805. This type of loom was called the 'dandy loom'. William H. Horrock (1776-1849) improved the 'dandy loom' in 1813 by a method of carrying the speed of battens so as to increase the period over which the shed remained open for the shuttle to pass through.

The first completely successful automatic power loom based on the 'dandy loom' of Horrock was invented in 1830 by the resourceful Welsh machine toolmaker Richard Roberts. The result of all this mechanisation was that the textile industry, but particularly the cotton industry, increased its production 100 times between 1760 and 1830.

Iron Industry

The availability of ample supplies of iron for machine building and other manufacturing needs was an important essential for industrial progress. Whilst the industrial development was taking place in the textile industry, the development in iron manufacture was paving the way for steam power, and locomotion by steam power.

The iron smelting technology, and the production of cast-iron was brought to England from Germany, where expert iron mining and metallurgy flourished as early as in 1543 and, as a result, the number of blast-furnaces grew so rapidly that by the 17th century, the forests had been laid to waste in order to provide charcoal, the essential fuel of ironmaking. A simple iron smelting furnace could consume as much timber in 40 days as to deforest an area of 1 mile (1.6 km) radius. It had become necessary by 1640 to stop iron production altogether to avoid the total destruction of forests in Britain, a country which depended on the availability of timber for ship construction. In consequence of this desperate measure, iron had to be imported from Sweden, Norway and even from Russia.

Coal had not been used in ironmaking because of its sulfur content, and the difficulty in forcing the air-blast through the caked mass of coal in the furnace. In the period from 1621 to 1638, Dud Dudley (1599-1684), a son of the Earl Dudley and a military engineer of King Charles I in the Civil War against Cromwell, almost solved the problem of using coal as the fuel in iron smelting by first converting the coal into light and porous coke, but the opposition to his method by both the ironmakers and the charcoal workers, the Civil War, and other practical difficulties left him in debt and in financial ruin without having been able to perfect his method of making iron with coal as the fuel.

In the period from 1707 to 1713, a Quaker ironmaster of Coalbrookdale at the Severn river, Abraham Darby (1677-1717), together with his son-in-law Richard Ford, devised a method to smelt large quantities of iron, from 5 to 10 tons a week, in which the fuel consisted of charcoal supplemented by coke. This coke smelting process was useful only for cast-iron; wrought-iron made by coke was brittle because of the many impurities left by the coke. Darby also invented the method of casting thin-walled iron pots in dry sand without loam. Despite this significant development, the iron production in Britain was still insufficient for domestic use, and iron had to be imported from Scandinavia to meet the demand of the domestic market for iron.

In 1745, Abraham Darby II (1711-1763) and his son-in-law Richard Reynolds, used coke regularly as the fuel in their blast-furnaces in which the blast was provided by the largest water-driven bellows to obtain high-grade cast-iron. Already in 1732, Abraham Darby II had used a Newcomen engine to raise water on top of a waterwheel which drove the bellows. Introduction of coke removed the restrictions the charcoal had placed on the size of the blast-furnace because coke was able to carry a much increased load of iron ore. By 1760, Reynolds was able to convert pig-iron into malleable iron by this process, thereby establishing volume production of pig-iron and malleable iron. In 1767, Reynolds used cast-iron rails in his colliery tramways supported on 3¾ by 1¼ inch (9.5 by 3.2 cm) sleepers spread 2 feet (0.61m) apart.

In 1760, John Smeaton constructed a large blowing-engine powered by a waterwheel to provide a blast powerful enough to pass through large quantities of pig-iron in the blast-furnaces which made large-scale iron production possible.

In 1775, Abraham Darby III (1750-1791) had so much improved the coke-fired furnace and the iron smelting process, that he was able to cast in 1779 the first iron bridge of 100-foot (30.48m) span over the Severn river in Colebrookdale.

An imperious ironmaster, an outstanding industrial talent and production engineer, John Wilkinson (1728-1808), who made wet-sand casting a standard method of iron casting, is the first known cast-iron enthusiast who made great contributions to the production engineering and industry, who anticipated the machine-tool industry, who pioneered in applying steam-power to iron production, who made the first seamless lead tubes by casting them around a steel bar, and who probably became one of the first British millionaires. In 1762, Wilkinson standardised the mass-production of iron pipes. In 1774, he patented his celebrated robust cylinder boring mill for accurate boring of cannon barrels which introduced the guide principle by means of a guided boring-bar into machine-tooling as an important improvement over Smeaton's supported boring-bar. This patent of Wilkinson marked the birth of the machine-tool industry.

In 1782, Wilkinson erected the first rotative steam-engine of Watt to work his forging hammer which made it possible for the tilt-hammer to do 9 times as much forging than before the use of steam power. In 1792, he erected a steam-driven rolling and slitting mill which weighed about 8 tons, and in which only the top roller was powered. Wilkinson began to use steam-powered blast to smelt iron in 1794, which made Darby's ironmaking techniques practical and thereby gave the foundry an advantage over the forge. In 1787, he launched the first iron boat, The Trial, a 70-foot

(21.34m) long iron vessel made of rivetted iron plates at Willey Warf. He built iron chairs, iron vats, issued iron currency to his workers, and even made an iron coffin in which he intended to be buried. In Wilkinson's opinion everything was better if made of iron, and as a result of his iron-clad conviction of the superiority of iron, his contemporaries called him 'iron mad'. It is not too far-fetched to say that as concerns his stalwart character and his keenest personal interests, John Wilkinson was truly a 'man of iron'. Wilkinson's brother, William, who was also an ironmaster, established iron industries not only in France but also in Russia. Unfortunately, the two brothers were implacably inimical to each other, and, therefore, were not able to collaborate in ironmaking.

About 1746, a Doncaster maker of clocks and watches who was of Dutch stock, Benjamin Huntsman (1704-1776), rediscovered and improved the ancient method of making wrought-iron into steel, an alloy of iron with carbon, by means of the crucible process. The Sheffield steel industry was founded on Huntsman's secret method, which was literally stolen from him.

However, the actual founder of British iron industry is Henry Cort (1740-1800) of Farenham in Hampshire, a Royal Navy paymaster who purchased in 1775 an ironworks near Plymouth with his 10-year savings and borrowed finances. Cort was able to improve and combine ironmaking processes invented by others into a successful and efficient continuous industrial iron production process. In 1784 Cort patented an improved version of the reverberatory furnace [also called the 'fining furnace'] which posited metal in separate furnace chambers to produce malleable iron, a furnace which had been invented in 1766 by the brothers Thomas and George Cranage, ironmasters of Colebrookdale. More heating is required in this furnace, in which the fire reflects, or reverberates, from the roof of the furnace down onto the metal. However, this malleable iron in its red-hot state crumbled in forging because it contained too much carbon and sulfur. To improve this condition, Cort introduced in 1784 the 'puddling process' invented by a Welsh ironworks foreman of Merthyr Tydfil, Peter Onions, a method which broke the bottleneck in iron smelting. Onions had devised in 1783 a coal-fired reverberatory 'puddling furnace', in which the bloom was stirred, or 'puddled' with rods, called 'rabbles', that eliminated the excess carbon and produced as good a 'coke iron' as 'charcoal iron'. Moreover, he replaced heating and hammering of cast-iron by 'rolling' to remove impurities and produce wrought-iron. Onions patented his combined 'puddling and rolling' process in the same year. The Cort furnace had a sandy bottom of the hearth, and the iron in it was heated to a pasty state. It was called the 'dry puddling' process. The bloom was then shingled by Cort under the power hammer and put through a rolling mill outfitted with grooved rollers, which pressed out the dross such as silicon, phosphorus and other impurities in the form of slag, and produced the typical corrosion-resistant wrought-iron in the form of square and hexagonal bars, round and oval rods, and flats. In this process various symmetrical cross-sections, which were of practical value to the construction and manufacturing industry, could be easily produced from the puddled iron in a single operation without additional heating.

The remarkable aspect of Cort's continuous production of iron in an 'integrated mill' was that all the component processes of it had been invented by others. Cort only combined these processes into a highly efficient and successful continuous industrial production process. Even the use of rolling mills had a precedent in Sweden. For instance, a London toy maker, John Pickering, had already introduced in 1766 the method of drop-forging with the trip-hammer. In 1746, the brilliant Swedish mining engineer Kristofer Polhammer (1661-1751) [after ennoblement, Christopher Polhem], who was a pioneer of mass-production and probably the first modern engineer who combined science with engineering in engineering practice, designed and constructed a series of mills, among them, power-driven rolling mills to produce wrought-iron and malleable iron at Stjernsund in Sweden. The rollers, which were handmade from wrought-iron or cast-iron with steel-hardened surfaces, were up to 7 inches (17.7 cm) in diameter. Two high-speed rollers were used to press impurities out of an iron bar passing between them. Rollers were also cut with grooves of various patterns for the shaping of the iron. The rolling mill of Polhammer, which was powered by waterwheels, produced up to 15 tons (13.6 metric tons) of wrought-iron in 12 hours. Polhammer's complicated production method was discussed in detail in his treatise, <Political Testament>. The rolling mill was actually invented in the Mediæval Age, for rolling mills are known to have existed in Germany before A.D. 1400 for the rolling of lead bars for windows. However, much greater power was needed for the rapid rolling of iron. Since Polhammer had to use waterwheels, a limited power source, he was only able to roll smaller iron shapes, whereas Watt's powerful steam-engine was available to Cort which enabled the latter to roll large iron shapes of great industrial utility. Polhammer's rollers working at high speed processed about 20 times more wrought-iron than the contemporary hammering method. In 1745 he built a rolling mill which refined and shaped wrought-iron in a single process. On account of this invention and the ready availability of high-grade iron ore, Sweden became the world's leading producer of wrought-iron in the 18thcentury. The 7 waterwheels at Stjernsund drove more than 9 drop-hammers, as well as various shearing mills and rolling mills of Polhammer's design. Polhammer invented and designed sawmills, flour mills, nail works, clockworks, dock works, tin-plating shops, wheel-cutting and file-cutting machinery, hoists, conveyors, textile machines, steel presses, mechanical devices and scientific instruments. Many of these inventions were far in advance of the contemporary technology, and were later reinvented by others. He manufactured a staggering variety of items such as interchangeable clock gears, pots, pans, mortars, plates, tankards, bowls, spoons, knives, forks, turnspits, army cots, sheet-iron, roofing, gutters, screws, bolts and ornamental brass work by his mass-production methods. Polhammer, who recognised the value of division of labour, also introduced the idea of 'unit operations' into the method of mass-production by devising a series of drop-hammers lifted by cams on a horizontal shaft turned by a waterwheel: a cast-iron drop-hammer shaped and moulded the white-hot metal blank which was then finished by a machine of 5 smaller drop-hammers in a sequential operation, 3 drop-hammers to finish the edges of the plate and 2 drop-hammers to finish the interior of the plate. He produced 15 dozen dishes a day by this mass-production method.

Polhammer also designed large public works, particularly canals. He was a graduate of the University of Uppsala in Sweden, and he combined science and engineering in a nearly balanced manner in his professional work, which makes him one of the most prominent universal engineers of all ages. He declined very attractive and remunerative invitations from the Russian and the English sovereign to resettle in their respective countries.

Cort's integrated industrial process of iron manufacture increased iron production 16 times whilst the labour and energy spent in the manufacture of iron remained nearly the same. Cort as a technological entrepreneur established the British iron industry upon its modern foundation, which made Britain the foremost iron-producing country in the world within 10 years. The iron production in Britain increased from 18,000 tons a year to 4,000,000 tons a year in 1820, all because of the Cort's integrated iron production process. Cort invested his entire 10-year savings as an agent of the Royal Navy into the expansion and development of his ironmaking process, and to increase his capital he went into partnership with Samuel Jellicoe, whose father Adam Jellicoe was the Deputy Paymaster of the Royal Navy. Adam Jellicoe invested about £50,000 in Cort's ironworks and was entitled to receive in compensation half of the profits earned by the partners. Although his transaction was sanctioned by the British Government, Adam Jellicoe had peculated £27,000 of his capital from the Royal Navy accounts which was discovered in 1789 upon his sudden death. Cort's patents were confiscated by the British Government and his ironworks was confiscated by the Royal Navy, since Cort had surrendered his patent rights to Adam Jellicoe as a security for his loan. Cort was deprived of his patents, as well as of his industrial property that was worth £25,000. Curiously, Samuel Jellicoe, the son of the peculator, was put in charge of the operation of the confiscated ironworks of Cort. This shabby treatment by the British Government bankrupted Cort, and left him, his wife and his eleven children to the mercy of a measly annual pension of £160 fit for a pauper which was provided by the Treasury. In 1800, Cort died in disgrace.

Remarkable improvements of industry had been brought about in the period from 1730 to 1790. Darby had given the method of producing large quantities of pig-iron, and the wet-sand iron casting method. French scientist René Antoine Ferchault de Reaumur (1683-1757) had given a method to make strong malleable iron castings. Cort had established an integrated production method to make large quantities of iron for industry, Huntsman had discovered how to make good steel which could be forged into tools necessary for precision machining, and Watt's steam-engine supplied the great power needed by the industry. Therefore, the great industrial revolution that began in the 1760's was, in a sense, made possible by the improvements in metallurgy.

In 1828, the manager of a Glasgow glassworks, James Beaumont Neilson (1792-1865), devised his hot-blast stove, which produced more economical hot-blast to the blast-furnace and, thereby, reduced the fuel consumption by 60 %. Neilson's preheating stove was a cast-iron vessel incorporating air passages enclosed in brickwork, and at first fired by solid fuel burning on a grate. After 1834 the hot air in the hot-blast stove as a heat-transfer equipment was provided by the recirculated hot waste-gases taken from the top of the blast-furnace, and a cup and cone device for the closing of the top of the blast-furnace was installed to save fuel, and allow charging of the furnace. If the blast air was preheated to 600° F (316° C), 3 times as much iron could be produced with the same amount of fuel. Iron ore which had been useless with the cold-blast could now be used with the hot-blast, since the hot-blast burnt out more impurities than the cold-blast which made possible the use of coal or anthracite as fuel that was extremely important to the Scottish economy.

In 1838, Joseph Hall, a Staffordshire ironmaster, improved Cort's 'dry puddling' process, by lining the furnace with iron oxide and heating the pig-iron to a high temperature in the puddling furnace. Since the oxygen combined with carbon in the pig-iron to produce carbon monoxide, the carbon content of the pig-iron reduced as a consequence, and the melting point of the iron rose so that it was changed from a liquid to a pasty state. In this so-called 'boiling process', it was possible to reduce the phosphorus and the silicon content of pig-iron by means of the oxygen contained in the lining of the furnace, thus resulting in a superior and purer iron. The oxides of silicon and phosphorus produced by this process passed into the slag, and the slag and cinder were more easily separated from the metal in rolling mills.

The so-called 'boiling process' of Hall could have been used to produce mild low carbon steel, which would have required temperatures above 2,700°F (1,500°C), but the difficulty was to provide strong enough furnaces with refractory material that could withstand such high temperatures, and remain unaffected by the chemical reactions with the slag and oxides in the furnace as the 'boiling out' process continued to remove the carbon from the pig-iron. These difficulties slowed down the progress in the production of steel as a replacement for wrought-iron, and they were finally overcome in the 1860's, which marked the beginning of the 'Age of Steel'.

New Metal Elements:

During the 18th and 19th centuries many metals were isolated as new elements, particularly in Sweden: *phosphorus* was isolated by the German alchemist Henning Brand in 1669; *cobalt* by the Swedish physician Georg Brandt in 1742; *zinc* by the German chemist Andreas Sigismund Marggraf in 1746 [Marggraf was the first chemist to use a microscope in chemical research by means of which he discovered sugar in beets]; *platinum* by the Spanish naval officer Antonio de Ulloa in 1748; *nickel* was isolated by the Swedish mineralogist Axel Fredrik Cronstedt in 1751 [Cronstedt, a pupil of Brandt, was the first modern mineralogist who classified minerals according to their elements] but first refined by the Swedish mineralogist Torbern Olof Bergman in 1775; *bismuth* by the French chemist Claude Francois Geoffrey in 1753; *metallic manganese* by the Swedish chemist Johann Gottlieb Gahn in 1774; *molybdenum* by the Swedish chemist Peter Jakob Hjelm in 1778; *tellurium* by the Austrian mineralogist Franz Josef Müller in 1782; *tungsten* [a.k.a., *wolfram*] by the Spanish chemist Don Fausto de Elhuyar together with his older brother Don José in 1783; *titanium* by an English clergyman William Gregor in 1791; *chromium* by the French chemist Louis Nicolas Vauquelin in 1798; *magnesium* by the English chemist Humphry Davy in 1808; *calcium* by the German chemist Friedrich Stromeyer in 1817; *silicon* by the outstanding Swedish chemist Jöns Jakob Berzelius in 1824; *aluminium* by the Danish physicist Hans Christian Ørsted in 1825, and the German chemist Friedrich Wöhler in 1827; *thorium* by the Swedish chemist Berzelius in 1829; *vanadium* by the Swedish chemist Nils Gabriel Seftström, a pupil of Berzelius, in 1830; *zirconium*, *titanium* and *tellurium* independently isolated by the German chemist Martin Heinrich Klaproth in 1794.

In the 19th century it was discovered that steel alloyed with these metals singly or in combinations acquired most valuable properties for technological purposes. Many of these metals would later become extremely important in the production of alloy steels for machines, metal cutting tools, electrical apparatus, automobile and aircraft parts, transformers, generators, motors, magnets and utensils.

Age of Alloy Metals:

Painstaking research on ferrous alloys was done by Michael Faraday (1791-1867), a self-taught scientist, in England, from 1819 to 1834. He made nickel and chromium alloy steels experimentally in 1819, which marks the beginning of the modern age of alloys. Nickel became one of the most important metals after Faraday had successfully coated another metal electrolytically with it in 1845. At his death, Faraday left 79 alloy steel specimens he had produced in his research on alloy steels, which were much later carefully tested by the English metallurgist Robert Hadfield (1858-1940), head of a foundry at Sheffield. In 1822 Faraday wrote in an article in Philosophical Transactions that in the future alloy metals will become even more important than the metals themselves, a prescient prediction which came true.

Industrially, nickel steel was made by Fischer in 1824; manganese steel by Mushet in 1845; tungsten steel by Köller in 1855; chromium steel by Holtzer in 1877; vanadium steel by German metallurgists in 1914; and stainless steel as a chromium-nickel alloy steel by the American metallurgist Elwood Haynes in 1919, after the English metallurgist Harry Brearley of Sheffield had noticed in his experiments with alloy steels for gun barrels in 1913, among many samples which he had thrown aside as unsuitable a sample alloy steel containing 14 % chromium did not rust on a scrap heap outside. It was later found that even a more corrosion and heat resisting steel with up to 20 % chromium and 10 % nickel were useful in the chemical industry, in the surgical scalpels, and in cutlery. Alloy steels with higher chromium and nickel content are useful in chemical plant equipment, furnace parts, and pump components. The growth in the production of stainless alloy steels is more than any other ferrous alloys, about 10 % per annum or more.

Already in 1799, William Reynolds, an ironfounder in Coalbrookdale, had taken out a patent for mixing manganese oxide with pig-iron in order to improve the quality of iron castings. In 1839, J.M. Heath, an associate of Robert Mushet's father, had demonstrated that the quality of Huntsman's crucible steels can be remarkably improved if some manganese were added to the metal bath before it was poured. Heath used a mixture of manganese oxide and coal tar which he placed in the steel-smelting crucible. Robert Forrester Mushet (1811-1891), a talented Scottish experimental metallurgist, recognised that manganese in steel not only improves the forgeability of steel, but also cleanses the steel from iron-oxide before casting. He patented his idea of the manganese alloy steel in 1856, before steel was commercially produced by Bessemer.

In 1868, Mushet began to manufacture a new steel alloy out of pig-iron rich in manganese and containing an added portion of finely pulverised wolfram ore (tungsten). It produced a self-hardening, high-carbon, tungsten-manganese alloy steel with superior cutting qualities for machine-tools. The superiority of Mushet's alloy steel was so great that its success was immediate and worldwide. Mushet, after an unfortunate experience with his manganese steel patent in 1856, very wisely it seems, did not patent his alloy steel but kept it such a well-guarded secret that the precise details of his production techniques have remained unknown till this very day.

CHAPTER XIII

POWER TECHNOLOGY: From Wind and Water Power to Steam Power

Ancient Sources of Power

Man's first power source was some other living creature. The neolithic man domesticated some wild beast which he captured and tamed by virtue of his superior intelligence and willpower, and he made the tame animal to haul heavy loads for him. By domestication of animals man had discovered a new source of power which he could use for his own benefit. The next new source of power for man was a slave-gang when enemies captured in tribal wars were set to work instead of being slaughtered as was the immemorial custom. The human slave in comparison with the beast had less strength, but this inherent disadvantage was compensated by man's superior intelligence: a human slave could do work that the beast was incapable of doing. The lack of strength of the human slave relative to the beast was compensated by grouping slaves, or workmen into gangs (called *corvées* during the rule of the Holy Roman Emperor Charlemagne in Europe) in order to combine their individual strengths into a much greater collective power. Ancient Egyptians employed this method of creating power in their monumental construction projects, and they further increased the efficiency of their working gangs by using such mechanical devices as the inclined plane, the lever and the roller.

After the Sumerians, Egyptians were the first people to harness the power of the wind in their sailboats and to combine it with the collective power of the oarsmen. The wind-power could propel their sailboats across, and even against the current, and the water-power was harnessed by sailing with the current down the Nile river.

However, it was much more difficult to harness the power of wind or water on land. In the first century B.C. the first primitive waterwheels were developed by the Hellenistic engineers for grinding corn and raising water for irrigation of the fields. The first waterwheels were horizontal to accommodate the horizontal millstones.

A Roman engineer Vitruvius Pollio in the 1st century B.C. described a vertical waterwheel the movement of which was transmitted through the right angle by means of wooden Archimedean gears, known as 'cog and lantern gears' [also 'lantern and crown gears'] which had a gear ratio of 1– to –5. This waterwheel together with its gearing and millstones, now called the 'Vitruvian Mill', was used to grind corn, and it constituted the first known practical solution to the problem of continuous power transmission.

The first use of air as a source of propulsive power was made about 400 B.C. by the ancient Greek mathematician and engineer Archytas of Taras (428-c.347 B.C.) in his wooden toy-dove which was propelled through the air by the reaction to a jet of compressed air and steam expelled from the toy-bird. The application of steam as a source of power was made by the founder of Hellenistic mechanical engineering, Ktesibios of Askra in Alexandria, in the 3rdcentury B.C.. According to Vitruvius, Ktesibios constructed a gadget called *aiolopyle* [the ball of *Aiolos*, the Greek god of winds] consisting of a spherical bronze boiler containing air which upon heating produced a blast of compressed hot air through a pipe controlled by a stop-cock. The *aiolopyle* was used throughout the Roman Empire to provide air-blast for small furnaces or fires. Military engineer Philon of Byzantion (*floruit* c.250 B.C.), a student of Ktesibios, invented about 240 B.C. the hydraulic reaction-turbine for fountains which had the form of the modern rotative lawn sprinkler with outlet tubes bent into the shape of *swastika*, the venerable Indo-European symbol of wellbeing.

Another important Hellenistic engineer Heron of Alexandria (*floruit* c.130 B.C.), headmaster of a school of mechanics in Alexandria, compiled a book of ingenious devices for the Temple of *Serapeis* many of which functioned by steam-power. One of these devices was a small steam-turbine which was operated by the reaction to the jet of steam expelled from a bronze boiler containing water. This design of Heron demonstrates a knowledge of physics and mechanics that was not equalled in the Occident for almost 2 millennia.

During the Mediæval Age the water and the wind power came into general and wide use in the Christian feudal society in such a precipitous manner which can be called 'revolutionary', yet muscles of horses, oxen and men still remained important sources of power for many common tasks.

After Ghengis Khan had invaded China and made Chinese rockets weapons of offence, Roger Bacon added such a proportion of sulphur to the mixture of saltpetre and carbon to make it a highly explosive gunpowder. The Byzantine weapon consisting of a cylinder fitted with a piston which shot liquid 'Greek Fire' was developed into a cannon with the use of gunpowder. The cannon represented the first heat-engine that functioned by internal combustion for it produced effective power under controlled conditions. Such a cannon is first shown in a manuscript written in 1327.

The first recognition of the power of steam in Europe was by the Italian Renaissance mathematician Girolamo Cardano (1501-1576), whose father was a friend of Leonardo da Vinci. Leonardo believed that Archimedes of Syracuse had invented a steam-gun, and in his notebook he had a sketch of a heated cylinder containing water and fitted with a piston. Cardano, who knew the contents of Leonardo's notes, mentioned the power of steam and the method of producing a vacuum by condensing steam in his book, < On the Nature of Things > (*De Rerum Natura*), published in 1557.

In 1601, the Italian mystic and speculative natural philosopher, Giovanni Battista Della Porta (c.1535-1615), described in his book on natural magic, < Three Books on the Spirit > (*Tre Libri de'Spiritali*), published in 1606, 2 gadgets, one expelling liquid through a pipe by steam pressure, and the other drawing up water through a pipe by means of a vacuum created in a vessel by condensation of steam. Porta correctly explained the action of condensing steam as producing a vacuum in an apparatus which is subsequently filled up with water. In this explanation Porta came close to making a positive distinction between steam and air. However, Porta made no other practical application of these devices.

Steam Power

Another great British invention was the steam-engine without which the so-called 'Industrial Revolution' could not have been possible. Its historical development in Britain has a slender beginning in the book < Elements of Artillery > (*Les Elements de l'Artillerie*) published by the French military engineer David Rivault in 1608, in which he describes the bursting of bomb shells that are filled with water and heated to explosion. It created a renewed interest in the power of steam among British engineers.

A well-known Norman fountain-builder and engineer, Salomon de Caus (1576-1626), came to England where he resided from 1609 to 1613. He built in England a waterwheel-driven pumping installation and some fountains for the Hatfield House at Greenwich Palace, and for a house of Henry, the Prince of Wales, at Richmond. Salomon de Caus declared that steam is evaporated water, and he contrived a small steam-operated water-lifter during his stay at Richmond, which was quite similar to the simple ***aiolopyle*** of Ktesibios. Prince Henry, the eldest son of King James I, took great interest in Salomon de Caus' devices, but he died suddenly in 1612. Salomon de Caus published a book on perspective in the same year in London, in which he called himself *Ingénieur du Serenissime Prince de Galles*, the first man known to call himself an 'engineer'. Henry's sister, Elizabeth, married the Elector Palatine Frederick V, and in 1613 Salomon de Caus left England with her for Heidelberg in Germany where Frederick V appointed Salomon his engineer-architect.

In 1615 Salomon de Caus published in Frankfurt, Germany, his most celebrated book, both in French, and in German, under the title < The Reasons for the Moving Forces in Different Machines ...> (*Le Raisons des Forces Mouvantes, avec Diverses Machines tant utiles que Plaisance par Salomon de Caus, Ingénieur et Architecte du Roy*). In the first part of this book he makes a number of propositions that represent a great advance upon all previous accomplishments on gaseous phenomena in which air was distinguished from steam, and the potential usefulness of steam-pressure in practice as distinct from air-pressure was stressed. He described a small device for raising water by the power of steam, and he wrote about a device of raising water by condensing steam, but apparently the last device was never made. He also described a steam-engine of his own design consisting of a kind of impulse wheel with paddles, which rotates when the steam issuing from a pipe strikes the paddles. The rotating impulse wheel drove a gearing system. In this book

Salomon de Caus is the first person to use the term 'work' for the mechanical effect of a force moving over a distance.

In 1629, the Italian engineer, Giovanni Branca (1571-1640), described in his book < Diverse Machines of Mr. Giovanni Branca > (*Le Machine Diverse del Signor Giovanni Branca*) a kind of simple steam impulse turbine in which a jet of steam rotates a horizontal wheel by impinging on its vanes. This particular apparatus, meant to power a stamping-mill by means of triple gearing, was apparently never constructed. Branca also described a vacuum pump drawing wine up a pipe, but it did not use condensation of steam to create the vacuum.

At the time when Salomon de Caus was employed at the court of King James I, 2 men at the court were inspired by the ideas of Salomon de Caus on steam-driven gadgets: a Scot, David Ramseye (died about 1653), a member of Prince Henry's household, and the young Edward Somerset, the second Marquis of Worcester (1601-1667), whose grandfather was the Keeper of the Privy Seal. First Ramseye, and then Somerset attempted to apply Salomon's ideas to urgent practical problems of pumping.

In 1630, after the Statute of Monopolies had been passed in 1624, David Ramseye, who was a prolific early patentee, obtained a patent from King Charles I for various inventions, one of them was a machine < To Raise Water from Lowe Pitts by Fire >. Unfortunately, no design details are available on how the machine of Ramseye functioned to raise water from low pits by fire, and there is no evidence that it ever was given practical execution, yet it was one of the first instances in which steam was used as a source of power in practical application.

Marquis of Worcester and 'Water-Commanding Engine':

Edward Somerset, Marquis of Worcester, wrote a pamphlet in 1659 entitled, < Century of the Names and Scantlings of Such Inventions as at Present I Can Call to Mind to Have Tried and Perfected >, which was published in 1663. In this pamphlet Somerset described 100 diverse ideas, and one of them was his so-called 'water-commanding engine' for which he obtained in 1663 a 99-year monopoly from the Parliament. It appears that he built the first steam-operated engine in his Raglan Castle in 1630, and the second one in a cannon foundry at Vauxhall with his collaborator, a German mechanic Kaspar Kaltoff, but these 'water-commanding' engines were probably not practical successes for the promised complete account of them never appeared. It seems reasonable to assume that his hydraulic engine consisted of one boiler connected to 2 receiving tanks from which water could be forced by the pressure of steam controlled by cocks in a continuing operation. It appears to have been an enlarged and refined form of the della Porta apparatus.

Morland and Packed Plunger Force-Pump:

Samuel Morland (1625-1695), a gentleman 'Master of Mechanicks' of King Charles II educated at Oxford, bought the Vauxhall foundry, and he is said to have filled the house and grounds of it with ingenious devices. Morland invented the 'packed plunger force-pump' consisting of a solid cylindrical plunger which moved in a pump barrel through a stuffing box on top of the barrel. The leakage in the stuffing box was prevented by a rather ingenious leather gland made of cupped leather seals. This arrangement did not require a snug fit between the plunger and the pump barrel, and it became a standard feature of most reciprocating pumps and engines. The ancient Græco-Roman plunger pump had no packing gland and, therefore, it could only function when submerged. Morland did invent a steam-powered pumping engine which probably was an improved version of Worcester's 'water-commanding engine'. The precise nature of it is not known, but it appears likely that the steam from a separate boiler acted through a forked pipe on 2 moveable heavy plungers, or pistons, which lifted up the plungers in sequence. The falling plungers apparently operated a toothed rack-and-pinion gear which turned a wheel that opened and closed the steam valves automatically, making this steam pump the first self-controlled machine. King Charles II issued a patent to Morland in 1682 after witnessing a demonstration of his steam pump. Morland wrote a small booklet entitled < Raising of Water by all Kinds of Machines > (*Elevation des eaux, par toute sorte de machines*), published in 1683 in France, after he had been sent to France by Charles II to advise and assist the French King Louis XIV with the problem of supplying water to the new royal palace in Versailles. In it he gave a table on the relation existing between steam-pressures and temperatures in a cylinder which measured up to 10 feet (3.05 m) in length and 6 feet (1.83 m) in diameter. He also wrote that water after evaporating into saturated steam occupies a volume that is 2,000 times that of the water.

It is possible that Morland failed to develop his engine because the waterworks he proposed for the Versailles Palace was not successful after £ 2,500 had been spent on his design, and as a result of this failure he fell into disgrace and on difficult times. The huge Versailles waterworks at Marly, were finally built from 1681 to 1685 for the French King Louis XIV by the Flemish engineer Arnold de Ville (1653-?), and the Flemish carpenters Renniquin Sualem (1644-1708) and his brother Paul. Abbé Edmé Mariotte (1620-1684) replaced Morland as the consulting engineer for this huge waterwheel installation popularly known as the 'Machine of Marly' which initially generated about 80-horsepower at a low efficiency from 6% to 7%, but the performance of this hydraulic power generator declined rapidly because of neglect.

Discovery of Atmospheric Power

When a suction-pump at the Villa of the Grand Duke of Tuscany failed to draw water from a 50-foot (15.24 m) deep well in 1641 by drawing water no more than 32 feet (9.75 m) by suction, Galileo Galilei (1564-1642) was consulted, but he was unable to find an explanation for this failure. His amanuensis, Evangelista Torricelli (1608-1647), was the first scientist to explain the suction-pump phenomenon by demonstrating that Earth's atmosphere had weight, and that the column of water *in vacuo* in the suction pipe of the pump would rise only so high until its weight equalled that of the atmosphere. Torricelli proved his theory with his famous mercury column experiment, in which mercury column rose only to a height of 29 inches (0.74 m). Torricelli's mercury column was the first barometer.

In 1650 the German physicist and engineer, Otto von Guericke (1602-1682) of Magdeburg, invented the first vacuum-pump by means of which he evacuated the space under a tightly fitting piston in a cylinder open on top, and as a result of it the piston was driven down by the great force of atmospheric pressure. This experiment of Guericke revealed that the atmosphere can be a source of great power. It remained to find other more practical means than the air-pump to create vacuum under the piston.

In 1678 a French Abbé, Jean de Hautefeuille (1647-1724), proposed a pumping system for Versailles consisting of 2 cisterns into which water could be drawn by vacuum created by exploding in alternate cisterns small charges of gunpowder.

Christiaan Huygens (1629-1695) experimented with the same idea, but he used Guericke's metal cylinder open on top and tightly fitted with a piston in his experimental gunpowder engine, or machine. Huygens attempted to create vacuum under the piston by exploding a small charge of gunpowder which expelled the air under the piston through a non-return valve, but this method of producing power was not only dangerous but also inefficient since the vacuum under the piston created by the explosion was incomplete. In this gunpowder machine the atmospheric pressure produced the power-stroke by driving the piston into the cylinder. In these experiments, Huygens was assisted for 2 years, from 1673 to 1675, by Denis Papin (1647-c.1713). Huygens believed after the moderately successful experiments with the gunpowder engine that if the gunpowder engine can be perfected it could be used for the propulsion of vehicles, ships, and even aircrafts on account of its light weight relative to the power it can produce.

Papin and Invention of Steam-Engine:

In 1675, Denis Papin, a Huguenot of Blois, was sent by Huygens to England, where he remained until 1687, to escape the religious persecution of Calvinist Protestants in Catholic France. Papin was a physician by training but physicist and technologist by vocation. In England he worked to improve the air-pump for Robert Boyle (1627-1691) and, later, he became the assistant to Robert Hooke (1635-1703) at The Royal Society in London.

In 1679, during his stay in London, Papin invented his celebrated pressure-cooker ['*marmite*'] which incorporated a safety valve, and a double-acting air-pump. His air-pump was powered by a waterwheel, which operated from a distance by means of a pipeline, and a piston-cylinder type of atmospheric engine. This apparatus of Papin was the first installation able to transmit power over a considerable distance.

Since Huygens' gunpowder engine presented difficulties for any practical development, and experimenting with gunpowder was a dangerous pastime, it occurred to Papin to evacuate the cylinder by the condensation of steam rather than by the explosion of gunpowder, a method which had been explained correctly by della Porta in 1606, and already used by Huygens in 1666.

In 1688, Papin emigrated to Germany where he was appointed professor of mathematics at the University of Marburg by Carl-Augustus, Landgrave and Elector of Hesse-Cassel. Papin obtained financial assistance from the Landgrave to construct a simple small-scale atmospheric engine similar to Huygens' gunpowder-engine. Papin tested both the gunpowder and the steam for the evacuation of the cylinder under the piston, thus letting the atmosphere execute the downwards power-stroke of the piston and raise a weight of 60 pounds (27.2 kg) by means of a cord and a pulley. The gunpowder experiment was not successful since about 20% of the air still remained under the piston after the explosion. His experimental model of the atmospheric engine weighing only 5 ounces (141 grams) used steam to produce vacuum in a bronze pipe of 2½-inch (6.4 cm) bore and leather strap valves. It took about a minute to heat the cylinder to produce steam under the piston, and more than a minute to cool it down and condense the steam. Papin discovered that about 20% of the water remained in the cylinder after the power-stroke. Papin had great difficulties in the making of sufficiently accurate cylinders to fit the piston because the tools and workmanship of the time were inadequate for this task. He placed a layer of water on top of the leather membrane disk which covered the piston to minimise leakage due to lack of fit between the piston and the cylinder. The waterlogging in the cylinder, in his opinion, made the utilisation of his method for large-scale atmospheric engines questionable. He published his conception of the atmospheric engine in 1690, < New Method of Obtaining Very Great Power at a Low Cost > (***Nova methodus ad vires motrices validissimas levi pretio comparandas***), in ***Acta Eruditorum***. Papin's atmospheric engine was vigorously criticised by Robert Hooke for the slow movement of its piston [one stroke a minute], and the necessity of moving the fire from under the cylinder to produce cooling. This criticism discouraged Savery, and therefore he did not use Papin's cylinder-and-piston system in his pumping engine patented in 1698.

Papin turned his attention to the use of his atmospheric steam-engine in a modified form for the propulsion of a model steam-carriage and a boat, but their mechanical details became even more complicated. Therefore, he turned his attention to the steam-pump similar to the one Savery was working on. Papin's new steam-engine was the first model of a thermally powered machine the boiler of which was separated from the steam-cylinder. In this steam-operated reciprocating piston-engine the steam produced both the vacuum under the piston and the power-stroke of the piston. It was the first direct-acting, high-pressure experimental pumping steam-engine without condenser, and it consisted of a separate copper boiler controlled by a safety valve, and a pumping cylinder fitted with a free-floating, hollow piston. The cylinder had a 20-inch (50.8 cm) bore in which the free-floating, hollow piston separated the hot steam from the cold water under the piston which had to be pumped. The piston had a stroke of 16 inches (40.6 cm). The cylinder was provided with a hollow cylindrical pipe along its axis which was accessible from the outside so that hot irons could be placed into it to provide a kind of primitive 'superheating'. The cylinder was a closed vessel into which steam was introduced from the boiler. The steam-pressure drove the free-floating piston down into the cylinder thereby driving the cold water under the piston into a tank by means of steam-pressure. Cooling of the cylinder created a vacuum on top of the cylinder which pulled the piston up and filled the cylinder under the piston with the new supply of water by suction. This process represented Papin's pumping cycle. Therefore, Papin's steam-pump worked on the two principles demonstrated in della Porta's book of 1606. Papin called his steam-pump, 'Elector's engine', in honour of the Landgrave of Hesse-Cassel who supported and financed his invention. Remarkably, Savery, an English inventor, worked on a steam-pump at the same time in England, and finally succeeded in producing a prototype pumping engine in 1698.

Papin gave a demonstration of his pumping steam-engine to the Landgrave of Hesse-Cassel, but this trial was not a complete practical success for the soldered seams of the steam-pump tended to leak and, in consequence, Papin was able to pump water to a height of only 70 feet (21.34 m), although his steam-pump was an improvement upon that of Savery because of the free-floating piston. Papin wrote a memoir < The New Art of Raising Water Efficiently with the Aid of Fire > (*Ars nova ad aquam ignis adminiculo efficacissime elevandam*) in 1707, in which he described the construction of his steam-pump. Unfortunately, Papin's efforts did not meet with general approval, and because of his difficult personality he was ridiculed in public as a 'steam-pot mathematician'. He died in poverty and in such obscurity in England, that even the year of his death is not definitely known. Yet his contributions to the principles of operation of the steam-engine are fundamental, and his accomplishments in the experimental steam-engine construction are important. In fact, Papin is the first inventor of the steam-engine which he provided with all the necessary technical details to make it work. Unfortunately, the tools, the materials, and the workmanship of the period were unequal to the requirements of precision in the manufacture of Papin's steam-engine.

Papin had other inventions to his credit. He designed the first known centrifugal pump, called the 'Hessian pump', which was used for land draining near Marburg, Prussia. Papin also built a submarine and tested it successfully near Cassel in 1692.

First Practical Industrial Steam-Pump

Thomas Savery (c.1650-1715), a prolific British inventor and a shrewd businessman, patented his pumping engine in 1698. Savery's pumping steam-engine was based on the ideas of Marquis of Worcester and Morland, and it consisted of a boiler 2½ feet (0.76 m) in diameter and a single receiving vessel of about 50 gallons (189.3 liters) capacity. Savery's pumping engine had no moving parts except the non-return valves and stop-cocks, and it operated alternately by the indirect work of suction, caused by a vacuum created in the vessel through the condensation of steam, and by steam-pressure. In 1699, Savery demonstrated the new version of his pump which consisted of 2 receiving vessels and a 'wheelwork' which opened and closed the cocks automatically. In his final pump design, the interconnected steam cocks had been replaced by a single 'sector-valve' which was mechanically actuated. This sector-valve was the forerunner of the later 'slide-valve'. Savery provided in this version of his pump a second boiler which supplied the main boiler with steam through a clack-valve by steam-pressure. Savery's steam-pump made about 5 cycles a minute and delivered about one horsepower. The cocks and pipework of his pumping engine were of brass, and the receiving vessels and the boilers were of beaten copper.

Jean Théophile Desaguliers (1683-1744), a French physicist living in London, proposed an improvement to Savery's steam-pump by condensing the steam inside the receiving vessel by means of a jet of cold water introduced through an injection-cock as proposed in his book in 2 volumes, < A Course of Experimental Philosophy > in 1744. In this treatise Desaguliers displayed his ignorance of the working principle of Newcomen's engine, and his lack of understanding the operation of its ingenious valve-gear.

Papin was aware of Savery's steam-pump for Gottfried Wilhelm Leibniz (1646-1716) had sent him detailed information of it in 1705. In his own pumping steam-engine, Papin had eliminated the most serious defects present in Savery's steam-pump. What Savery had accomplished in his pumping engine was to make a small-scale model into a practical large-scale steam-operated pump capable of pumping 3,000 gallons (1135.5 liters) of water an hour. Surprisingly, Savery's engine had no safety features and it operated with 30 up to 50 psi (207 to 345 kPa) steam-pressure. The materials from which his steam-pump was constructed were in general of poor quality and, therefore, its performance was erratic for it frequently became waterlogged. 20 feet was the maximum practical length of the suction-pipe through which the water could be forced into the evacuated receiving vessel by the atmospheric pressure, and the forcing lift of the water by the steam-pressure was limited to a height not substantially exceeding 20 feet because that was all the pressure Savery's boiler could withstand without exploding. For this reason, Savery's pumps would have had to be installed in a vertical series at about 50-foot (15.24 m) intervals in order to pump water from deep mines, which commonly were several hundred feet underground. This arrangement would have been obviously too expensive, impractical, and also too dangerous because of Savery's unsafe boilers. Owing to these shortcomings, Savery's pumping engine which delivered about one horsepower was solely used for pumping to heights under 20 feet (6.09 m) throughout the 18thcentury.

Although the inefficiency of Savery's pump could be tolerated, it failed in one important respect: it could not be used to pump deep mines. The pumping of mines was the most important economic problem of the period, because most of the cost of deep mining was the cost of pumping water out of the mines.

Newcomen's Atmospheric Pumping Engine

The first practical atmospheric pumping engine was finally designed by a nonconformist engineer, ironmonger, blacksmith and lay Baptist preacher of Dartmouth, England, Thomas Newcomen (1663-1729), and constructed in 1712 by his associate, a plumber and glazier, John Calley (or Cawley) (?-1717), after a decade of diligent effort. Newcomen was an acquaintance of Robert Hooke, a brilliant experimental physicist in London. Newcomen atmospheric pumping engine consisted of a heavy horizontal working-beam pivoted at the center, with the steam-cylinder of 21-inch bore, 7 feet and 10 inches high, located beneath one end of the beam and the lifting pump-rod suspended by a chain from the other end. At rest, the pumping end of the working-beam was down and the engine end was up because of the heavy weight of the pumping-rod. The piston within the cylinder was suspended by a chain from the cylinder end of the working-beam. The top of the cylinder was open to the atmosphere and the piston was made tight by hemp-rope packing, a disk of leather and a layer of water on top of the piston. When the cylinder was filled with steam and then condensed by a cold water-spray injected into the cylinder through the so-called 'snifting-valve', vacuum was created under the piston, and the atmosphere acting on top of the piston drove the piston into the cylinder thereby lifting up the pumping-rod end of the beam and producing the working-stroke. Then, by letting the steam into the cylinder, the vacuum was broken and the weight of the pumping-rod lifted up the piston end of the working-beam by lever-action. This cycle could be repeated 12 times a minute.

Newcomen designed, improved and constructed his immensely successful atmospheric pumping engine over a span of 10 years. Newcomen, besides being familiar with scientific work, was also a very skilled mechanic, and he had the great talent of selecting the right components and methods of functioning in his development of the atmospheric pumping engine which he had successfully completed by 1710. One of the earliest Newcomen atmospheric pumping engines located above the ground was built in 1712 for the colliery at Dudley Castle in Worcestershire near Wolverhampton, which was powerful enough to pump large quantities of water from any reasonable depth. This pumping engine made 12 strokes a minute, and each stroke lifted 10 gallons (38 liters) of water 153 feet (46.63 m) through tiers of pumps. Yet the efficiency of this Newcomen pumping engine of about 5½-horsepower was very low, about 0.6 % for it consumed a huge amount of coal for the small amount of power it delivered. The excessive inefficiency of Newcomen's atmospheric pumping engine stemmed from the necessity of having to cool down the large cylinder with cold water for each power-stroke. Since the cylinders of Newcomen's early pumping engines were rough-cast, and not bored, and then polished manually with abrasives, there existed much friction between the cylinder and the piston. The variations in the diameter of the cylinder frequently caused the piston to seize and the vacuum to fail. The cylinders of this engine had to be made thin and of brass because of the thermal requirement for its rapid cooling and heating during the complete cycle of operation of the engine. An old brewer's kettle was fitted with valves to serve as the boiler, and located underneath the steam cylinder. The working of the Newcomen pumping engine was slow because it operated with the atmospheric pressure or less, and the expansive power of the steam was not used for the working-stroke. First calculations of the power of Newcomen's atmospheric pumping engine were made by Henry Brighton (c.1686-1743), a Fellow of the Royal Society who was also the editor of 'Ladies' Diary'.

James Brindley (1716-1772), the canal builder, tried to reduce the coal consumption of a Newcomen atmospheric pumping engine by placing the furnace and the flue inside the boiler instead of underneath it, and he even experimented with wooden cylinders in his efforts to drain mines. Brindley patented his improvement of Newcomen's atmospheric pumping engine in 1758.

However, the longevity of the Newcomen atmospheric engines is impressive: the last such engine, which had no history of serious breakdowns was still in use in 1934 at Parkgate in Yorkshire, England,

Savery's patent, extended by special act of Parliament to expire in 1733, was so inclusive that Newcomen, who had invented an entirely different atmospheric pumping engine along the lines of Papin's design of 1698, had to join the Savery company in order to have the legal right to produce his very different pumping engine. Upon Savery's death in 1815, Newcomen founded his own company by taking over the remaining patent rights of Savery and, therefore, his company controlled the manufacture of steam-engines in England until 1733. The packing of the piston and valves of the Newcomen atmospheric pumping engine was a great problem at the time, for leaky valves and a leaky piston made its operation somewhat uncertain, and the condensation of steam under the piston tended to limit the number of strokes of the engine because of waterlogging. Newcomen overcame the leaky piston like Papin by placing a layer of water on top of the leather membrane disk which covered the piston, and by introducing the 'snifting valve' for draining the condensed water from the cylinder. The early Newcomen atmospheric pumping engines were operated manually, but after 1712 all Newcomen atmospheric pumping engines were provided with a self-acting valve-gear which made their operation automatic, and they reigned supreme in pumping for 60 years until James Watt introduced his atmospheric steam-pump which was a Newcomen atmospheric pumping engine fitted with a separate steam condenser. The basic patents under which the Newcomen company built their atmospheric pumping engine expired in 1733.

Smeaton and Experimental Method in Engineering Technology:

In 1767, John Smeaton (1724-1792), assumed a methodical approach to the steam-engine, and began painstaking and detailed experimental studies of the Newcomen pumping engine. He applied statistical and experimental methods in his systematic scientific investigation of the Newcomen pumping engine, and constructed an experimental model of the engine for the development of detailed design criteria that improves the relative proportions of the engine parts. In these studies, Smeaton carried out altogether 130 experiments. In 1772 he tabulated his results for the optimum dimensions for the bore of the cylinder, stroke, number of strokes per minute, boiler size, feed of water, and coal consumption of Newcomen engines ranging from 1 to 78 horsepower. To compare the performances of different Newcomen engines he introduced a unit of measure for the power of the engine which he called 'duty'. Smeaton defined duty as the number of foot-pounds of work done by the engine per hundred-weight (a bushel) of coal, or the number of pounds of water the engine can raise one foot for one bushel (84 pounds) of coal burnt. Unfortunately, this measure of power did not consider the quantity of heat used. Smeaton found that Newcomen engines average a duty of 5,590,000 pounds (2,535,624 kg) of water raised one foot (0.30 m) for each bushel of coal consumed. Smeaton's changes in the Newcomen engine nearly doubled its performance to 9,450,000 pounds (4,286,520 kg) of water, but the efficiency of the engine was still very low, about 0.8 %, and it was still wasteful of fuel. Smeaton did not introduce any fundamental improvements to the Newcomen atmospheric pumping engine but merely perfected its original design concepts. The duty of Newcomen atmospheric pumping engine was about 4.3 million in 1718, 7.4 million in 1767, and 12.5 million in 1774. After his experimental research, Smeaton designed Newcomen atmospheric pumping engines of unparalleled size and power. In 1774, he designed a 6 ½-ton Newcomen engine with 3 boilers of 15-foot (4.57 m) diameter, a cylinder of 6-foot (1.83 m) bore and 9.5-foot (2.90 m) stroke, which had an efficiency of 1.4 %.

Since these large piston cylinders had to be accurately bored, Smeaton designed a boring-machine for such machining in the Carron Iron Foundry which produced a true circular bore, but this great advance in machining did not guarantee a parallel bore. A true parallel bore became possible only in 1774 with the guided boring-mill of John Wilkinson. In workmanship Smeaton had to tolerate manufacturing errors of about ½ an inch in a cylinder of 28-inch (1.27 cm on 71 cm) bore for Newcomen engines. As late as in 1830, a fitter who could work to 1/16th of an inch (0.16 cm) accuracy was considered a skilled craftsman.

In 1775 Smeaton was commissioned by the Tsarina Catherine II of Russian Empire to build one of his largest Newcomen atmospheric pumping engines to pump out the water from the dry-docks at the fortress of Kronstadt, in Russia. This was a monumental undertaking which formerly required a year of pumping with 100-foot (30.48 m) high windmills. Smeaton's Newcomen atmospheric pumping engines pumped the water out of Kronstadt dry-docks within two weeks!

Smeaton was an incessant and talented experimenter and one of the most prominent engineers of his age. Smeaton's papers on his excellent experiments on the windmill, the waterwheel and the Newcomen atmospheric pumping engine are classics, and the pioneering rational engineering experiments he carried out were the first model experiments in power engineering which exerted great influence in France as well as in Britain. Smeaton's investigations on waterwheels were awarded the Copley Medal, the highest Blue Ribbon award of The Royal Society in London. Smeaton demonstrated that the overshot waterwheel had a much higher efficiency, about 63%, than the undershot waterwheel the efficiency of which was about 22%. Although he did not test the breastfed waterwheel, he indicated that the same principles must be applied to the breastfed, as to the overshot and to the undershot waterwheel. His experiments opened the era of rational experimenting and laboratory testing in engineering.

Smeaton built many industrial waterwheels in the design of which he was guided by his experimental results. The largest one was an undershot waterwheel 32 feet (9.75 m) in diameter and 15½ feet (4.72 m) wide, which he built in 1768 for the London Bridge Waterworks on the Thames river. This waterwheel was reversible since it was in almost constant operation by the tidal flow. From 1582 to 1820 reversible waterwheels were in use at the London Bridge waterworks to operate force pumps which raised water from the Thames river. Wherever possible, Smeaton preferred overshot waterwheels.

Smeaton's efforts to improve the Newcomen atmospheric pumping engine came at the end of the era of the atmospheric pumping engine, for just about that time James Watt introduced his revolutionary condenser into the steam-engine design that nearly trebled the efficiency of the Newcomen type of atmospheric steam-engine.

Savery, Newcomen and Smeaton were the important progenitors of Watt in the development of the practical steam-engine as an entirely British creation.

Steam-Engine of Watt

Whilst the Newcomen atmospheric steam-engine was being used for the pumping of mines, waterwheels and windmills were used as power sources for other industrial purposes, because they were able to produce rotative action instead of reciprocating action of the Newcomen engine. Smeaton's efforts to improve these two sources of power have been mentioned above, but there were important technical improvements made to windmills by other inventors.

In 1745 a British millwright Edmund Lee invented the so-called 'fantail', which is a device that keeps the sails of the windmill automatically facing the wind. The fantail as an auxiliary windmill consisted of a set of vanes set at right angles to the sails, the rotation of which through a system of gears brought the sails automatically back into the wind when the wind direction changed. It is one of the earliest examples of automatic self-correcting guidance-control in machinery.

In 1751, an ingenious Scottish millwright, Andrew Meikle (1719-1811), invented the sails of windmills consisting of unequal, spring-controlled pivoted shutters that were able to spill any excessive wind. He developed the fantail-gear for the small auxiliary windmill of Lee consisting of a wooden worm-gear and wheel, and pinion-and-rack that was able to provide a gear-reduction of 3000-to-1. Meikle is also reputed to have invented the first threshing-machine in 1788, which not only threshed the sheaves but also automatically blew away the chaff.

In 1787, a British millwright Thomas Mead patented the application of the centrifugal governor, a double conical pendulum, to maintain a specified gap between the rotating and the stationary millstone in a windmill, and to regulate the sails of the windmill. It represented the first 'automatic feedback' control system in a machine. It is known that centrifugal governors had been in use before 1787. Stephen Hooper patented another version of the centrifugal governor in 1789.

These ingenious inventions dealing with the automatic control of machinery had a great influence on the Scottish engineer and instrument-maker James Watt (1736-1819) who invented the first practical working steam-engine. Watt came from a family of teachers, mathematicians, instrument-makers and contractors. He had a thorough classical education, and in mathematics he excelled particularly in geometry. At 17 years of age he went to Glasgow to work as an apprentice at mathematical instrument-making. When he was 21 years of age, Watt attempted to open his own shop, but was prevented from doing so by the Corporation of Hammermen because Watt had not served his proper period of apprenticeship. A family friend helped Watt to obtain an appointment as mathematical instrument-maker of the Glasgow University. The university faculty respected Watt and treated him as an equal. He quickly made friends with professors of natural philosophy and particularly with the Professor of chemistry, Dr. Joseph Black (1728-1799), who was engaged in an experimental study of heat, because Watt's interest in science included chemistry. He also became a close friend of Joseph Priestley (1733-1804). Watt studied both the German and the Italian language to be able to read theoretical mechanics and original technical papers in both languages. When he was 48 years of age he was elected a Fellow of the Royal Society in Edinburgh, and at 49 years of age a Fellow of The Royal Society in London. In 1806 Watt was awarded the LL.D. degree from the University of Glasgow.

When James Watt was the mathematical instrument-maker to the university in Glasgow, Scotland, a demonstration model of Newcomen's atmospheric engine used in the physics department of the University had broken down, and was sent in 1759 to a prominent instrument-maker Jonathan Sisson in London for repair but he was unable to restore its proper functioning. In 1764 Professor John Anderson gave the task of restoring the model Newcomen engine to Watt who succeeded to make it function again, but the engine ran out of steam after 5 strokes and had to remain inactive for a period to build up new steam to continue functioning. This made Watt curious about why the Newcomen engine required so much steam although the atmospheric pressure did all the useful work of the engine. After a careful study of the Newcomen engine, he discovered that the process of generating steam continued to absorb heat although there is no rise of temperature once the boiling point of water was reached. Conversely, when the steam is condensed it gives up this quantity of heat. Watt realised that most of the heat potential of the steam in the Newcomen atmospheric engine was wasted in cooling the cylinder, which was mainly responsible for the huge consumption of fuel.

Joseph Black measured heat energy quantitatively in 'calorics'. He showed that when hot steam is injected into cold water it raises the temperature of about 6 times its weight of water to the boiling point. This large hidden quantity of heat is liberated from steam by condensing steam to water. Black called the quantity of heat energy liberated, or absorbed, by a unit of mass of a substance, such as water, in its change of state to steam at constant temperature, the 'latent heat'. Boiling water at 100°C (212°F) into steam absorbs about 540 cal./gram (972 Btu/lb). The same quantity of heat energy is released when steam is condensed to water at this temperature. Black informed Watt about his experiments with 'latent heat', after Watt had done experiments on boiling water into steam. In 1765 James Watt conceived his revolutionary idea to separate the hot cylinder from the 'condenser' of steam for the atmospheric steam-engine and to keep the cylinder as hot as possible by enclosing it with a steam-jacket, which made the atmospheric engine for the first time a really practical power source for it nearly doubled the efficiency of this pumping engine to 2.7 %. He built a model of his engine that worked perfectly, and was granted a patent for it in 1769 which was in force until 1782. Watt transformed the crude atmospheric pumping engine of Newcomen into a relatively efficient prime mover capable of supplying power to work machinery in factories, textile mills, and even to propel river boats and ships. From 1770 to 1774 when technical and financial problems with his experimental steam-engine made it necessary for Watt to engage in the practice of civil engineering in London, Watt was fortunate to meet and form a partnership with a progressive manufacturer and a capable entrepreneur, Matthew Boulton (1728-1809), son of a Birmingham engineer who not only inspired Watt to improve his steam-engine but also made valuable suggestions for several technical improvements of it.

In 1760's Boulton had wanted to expand his factory in Soho near Birmingham, but the waterwheels which supplied power to his factory where not able to drive additional machines in the factory. Boulton discussed the problem of power with Benjamin Franklin (1706-1790), who was in London for a few years to represent American colonists in the negotiations with British government, particularly how to improve the Newcomen type of fire-engine, and subsequently Boulton built a model of an improved fire-engine on the advice of Franklin. Boulton, who was technically competent about steam-engines, was responsible for making the Boulton & Watt steam-engine a huge business success. Without Boulton, Watt who suffered depressions from frequent technical disappointments would have probably ended up a poor man like many other inventors who had neither drive nor business acumen. The energetic Boulton had both insight and foresight about the commercial possibilities of steam-power in an industrial civilisation.

In Watt's atmospheric steam-engine the vacuum was created under the piston by letting the steam escape and expand into a condenser, thus avoiding the need to cool down the hot cylinder itself like in the Newcomen engine. Besides a steam-jacket around the cylinder, Watt added an air-pump to his engine to maintain a vacuum in his condenser by pumping out its contents of air, steam and water. Until 1769, when Watt was still associated with Dr. John Roebuck of the Carron Iron Works as his partner, his engine was still an atmospheric engine because the power-stroke was executed by the atmospheric pressure. In this form the engine was useful only for pumping. From 1770 to 1774 Watt had to earn his living by surveying a number of canals in Scotland and by improving surveying instruments, because work on his steam-engine had come to a virtual standstill. Whilst working in partnership with Roebuck, the construction of Watt's full-scale engine in an outbuilding adjacent to Dr. Roebuck's Kinneil House had been plagued by the poor quality of workmanship available in Scotland at the time which made him despondent.

The firm Bolton & Watt was created in 1775. In 1776, during Watt's new partnership with Boulton, the first Watt pumping steam-engine was installed. Fortunately for Watt, John Wilkinson, who was a friend of Boulton, was able to bore large, up to 70 inches (1.78 m) in diameter, cylinders for Watt's atmospheric steam-engine accurately enough with his large guided boring-mill. The maximum error made by Wilkinson's boring-mill was (1/1000) of an inch per inch of the cylinder's diameter, an achievement in machining accuracy which was quite remarkable for its time. By 1900 the boring-mill accuracy had reached (1/2000) of an inch per inch. In 1795, by the time Boulton & Watt built their own boring-mill in their Soho factory, Wilkinson had supplied virtually all of the several hundred cylinders required for Watt's atmospheric steam-engines.

In 1781, Watt patented the first genuine steam-engine, which was single-acting and used steam to execute the power-stroke. In 1782, Watt patented his double-acting rotative engine in which the expansive force of the steam-pressure was applied alternately to the opposite sides of the piston, although expansive working was not of great advantage with the low steam-pressure used by Watt.

In 1783, the first large Watt engine was installed by Wilkinson to operate the trip-hammer in Wilkinson's foundry. It had 4 times the power of a Newcomen atmospheric pumping engine, an efficiency of about 5 % and it worked with 5 psi up to 7 psi (34.5 to 48.3 kPa) steam-pressure. This was a rotative single-acting engine, instead of a reciprocating engine, which used a heavy flywheel to overcome the dead-center, and the sun-and-planet gearing to change the reciprocating motion of the piston into a rotary motion of the output shaft. Wilkinson always took an active part in the construction of Watt's steam-engines.

In 1782, Watt invented a parallel motion mechanism for his double-acting steam-engine to convert the rectilinear motion of the piston into a circular motion at the end of the rocking-beam. In all, Watt patented 5 different mechanisms capable of producing parallel motion in 1784. In the same year Watt also patented his 'trunk engine' in which the connecting-rod was directly linked to an annular piston and fitted with a trunk. However, the trunk-engine had certain disadvantages such as increased friction, larger size of the cylinder, heavier piston, difficulty of access, and loss of heat from the trunk. John Penn (1805-1878) patented a double-trunk steam-engine much later, in 1845, and installed one in the first armoured iron warship, H.M.S. Warrior, in 1861 in England.

In 1784, Baron von Kempelen (1734-1794) from Austria patented a steam-turbine, which temporarily aroused the concern of Watt. Von Kempelen's steam-reaction-engine worked on the same principle as the Barker Mill, a steam version of Philon's hydraulic reaction device. Watt calculated its efficiency by using the theory of waterwheels of Antoine Parent (1666-1716) in a generalisation of water-power theory to encompass also a heat-engine, and found it to be low.

In 1788, on the suggestion of Boulton, Watt introduced the centrifugal governor of Mead to control the steam-throttle of his steam-engine. John Rennie (1761-1821), a Scottish engineer and former apprentice of Andrew Meikle, had used centrifugal governors in the Albion Flour Mill at Blackfriars where upon the invitation of Watt & Boulton he had designed and erected the millwork, and installed 2 Watt steam-engines which supplied power by means of almost exclusive use of iron shafting and gearwheels to the machinery in the mill between 1784 and 1788. This feedback-method made Watt's steam-engine self-regulating, and not merely automatic as the Newcomen atmospheric engine. The celebrated Albion Mills, the first large steam-powered flour mill in the world, although it burnt down within its first year of operation in 1791 owing to an overheated hardwood bearing in the opinion of Rennie, revolutionised the flour-milling industry.

Watt produced a mercury steam-gauge, a glass water-gauge, and a poppet-valve to admit and release steam. Watt's draughtsman and assistant, John Southern (1758-1815), improved the engine indicator which automatically drew the 'indicator diagram' [P-V diagram], an instrument devised by Watt, for the mechanical measurement of the power output of his steam-engine, and used for the automatic setting of the steam-valve. Watt had obtained the idea for the indicator from the mathematical analysis of the expansive power of the steam-engine given as evidence in the trial of Hornblower versus Watt in 1791.

Rotary motion to drive machinery was customarily produced by making a Newcomen atmospheric steam-engine lift water on top of the waterwheel which turned the machinery by its rotary motion. In 1779, Watt had built an experimental model of his steam-engine in the Soho Works at Birmingham which incorporated a crank, 1- to -2 gearing and a counterweight to produce rotary motion. In the same year, Matthew Wasbrough (1753-1781) of Bristol patented mechanisms to convert reciprocal motion into rotary motion without the intermediation of a waterwheel. Wasbrough's mechanism consisted of pulley-wheels driven by chains and controlled by pawl-and-ratchet, or of toothed-racks which were similar to the mechanisms of Papin and Hulls. He employed a flywheel to even out the rotary motion. In 1779, Wasbrough outfitted his second Newcomen atmospheric engine for a mill at Snow Hill, Birmingham, owned by a button manufacturer, James Pickard, with the pawl-and-ratchet mechanism to produce rotary motion but the result was not satisfactory. In 1780, Wasbrough substituted a single connecting-rod, a crank and a flywheel for the pawl-and-ratchet mechanism, and it produced quite satisfactory rotary motion. James Pickard patented the crank and connecting-rod mechanism in late 1780, although this mechanism had been known for centuries in the literature. Watt suspected that Wasbrough had obtained the crank and connecting-rod idea from Watt's experimental steam-engine, but he was loath to break Pickard's patent for fear that it might provide a precedent for others to try to break Watt's patent, which is what happened to Arkwright's spinning-frame patent in 1785. Therefore, Watt adapted, for the duration of Pickard's crank patent, the 'sun-and-planet' gearing to his steam-engine which also converted reciprocal motion into rotary motion. In the sun-and-planet gearing the planet-wheel which was rigidly attached to the connecting rod was made to rotate by toothed gearing round the perimeter of the sun-wheel which was keyed to the driving shaft to be rotated. If the two wheels were of equal size, the driving shaft rotated twice for every double stroke of the steam-engine. Even shortly before and after the expiry of Pickard's crank patent, all Watt's double-acting steam-engines incorporated the flywheel to carry the crank and the crankshaft over the 'dead center' at the end of each stroke.

Watt's resident engineer in the Cornish mining district from 1779 to 1798, the Scot William Murdock (1754-1839), who had entered the service of Boulton & Watt in 1777, was behind many improvements of Watt's steam-engine such as the 'sun-and-planet' gearing. He invented the D-shaped slide-valve to replace Watt's poppet-valve and the eccentric [a crank of small radius]. He designed many machine-tools in Boulton & Watt Works in Soho, and constructed the first steam-engine with oscillating cylinder. In 1785, Murdock introduced the practice of cementing two iron surfaces with a mixture of iron filings and sal-ammoniac as a cement where machining proved to be too expensive. He also improved pumping by introducing a plunger-type pump, which was later made popular by Trevithick. He introduced a worm-and-wheel drive in 1799. Murdock constructed a compressed-air motor to drive through gearing the pattern-making lathe at 225 rpm in Soho Works. It was a reciprocating engine with a cylinder of 12-inch (30.5 cm) bore and 18-inch (45.7 cm) stroke. Compressed air for this motor was provided by the blowing-engine of the foundry cupola. Later, in 1819, he proposed the use of compressed-air instead of steam to drive locomotives.

In the Soho machine shop Murdock drove every machine independently by its own vacuum piston engine individually connected by pipes to the vacuum-pump system in which the vacuum was maintained by a central exhausting pump. It was probably the first successful application of such a central system of power-transmission, which had been invented by Papin. After the retirement of Boulton and Watt in 1800, Murdock became the manager of the Boulton & Watt Works which were now under the direction of the sons of Boulton and Watt. Murdock resumed Lebon's experiments on gas lighting, and found that cannel coal produced excellent gas as illuminant of gaslighting. In 1798, Murdock installed the first practical gas lighting in Boulton & Watt's foundry in Soho, Birmingham, with a cast iron retort and a long system of tubing connecting it to the fishtail gas burners. His illumination of the factory with gaslights created wide public interest in gaslight, and in 1808 he was awarded the Rumford's Medal for his gaslighting system.

Watt introduced a new power unit, the 'horsepower', which corresponds to 30,000 foot-pounds of work performed a minute (40,674 Joule) in terms of which he was able to compare the performance of various steam-engines. The indicator [pressure-volume] diagram was a valuable instrument to determine the actual power produced by the steam-engine.

All of Watt's steam-engines worked under relatively low pressure, about 5 up to 10 psi (34.5 to 69.0 kPa) above the atmospheric pressure. All throughout his life, Watt did not approve of the use of high steam-pressure in his steam-engines which he considered a dangerous practice on account of the possibility of boiler explosion for he had seen Savery's pumps explode at London waterworks as a young man. Watt's original atmospheric steam-engine had a thermal efficiency of 2.7 %, whereas his expansive steam-engine had a thermal efficiency of 4.5 %.

The major difficulty in the production of steam-engines in this period was the maintenance of small tolerances and effective packing for the piston since this was of cardinal importance to the efficient operation of the steam-engine. To increase the efficiency of his steam-engine Watt was urged by Boulton to use the expansive power of steam by cutting off the steam before the end of the power-stroke. Watt followed Boulton's advice and introduced the 'steam cut-off' method of operation of the steam-engine consisting of shutting off the steam supply at about ¼ of the power-stroke, thereby letting the remaining ¾ of the cylinder to be filled by the expansion of the steam already in the quarter space of the cylinder. Watt expected to save considerable amount of steam by the expansive working of the steam.

Hornblower and Compound Steam-Engine:

The compounding of steam and expanding it in a second cylinder in a power-stroke rather than in a condenser was patented by the enginewright, Jonathan Carter Hornblower (1753-1815), who came from a family of distinguished engineers and had a good formal education. In 1782 he built the first compound steam-engine of 11.5 horsepower, the next advance in steam-engines, in which the steam released from the first cylinder of 19-inch (0.48 m) bore and 6-foot (1.82 m) stroke was further expanded in another 8-foot (2.44 m) working-stroke under lower pressure in the second larger cylinder of 24-inch (0.61 m) bore, at Radstock Colliery near Bath. Compounding did away with the condenser and increased the overall efficiency of the steam-engine. From 1784 to 1791, Hornblower erected a number of compound steam-engines in Cornwall. In 1791, when Hornblower applied for an extension of his patent, Watt sued him for violation of Watt's patent, and Watt subsequently won the long and involved litigation in 1799 which broke Hornblower financially. In this litigation, Mr. David Giddy [who later changed his name to Gilbert] (1767-1839), an Oxford mathematician and a Member of Parliament who later became President of The Royal Society in London, gave evidence on Hornblower's behalf in the course of which he presented a mathematical analysis of the expansive operation of the steam-engine. Giddy's analysis of the power of the steam-engine suggested to Watt a mechanical method to measure the power-output of his steam-engine. Watt's patent, which was extended by the Parliament in 1775 to the year 1800, effectively stopped all alternative steam-engine developments for the duration of Watt's patent. The compound steam-engine was actually no real threat to Watt's steam-engine as long as it functioned by low steam-pressure. In 1803, after the expiry of Watt's patent in 1800, the first compound steam-engine with a condenser was patented and installed in Cornwall by Arthur Woolf (1776-1837), who formerly had worked as a millwright under Hornblower. Woolf's compound steam-engine saved 50% of the fuel used by the Watt steam-engine for the same power output. The first compound steam-engine which was built by Jonathan Hornblower in 1780 at a Radstock Colliery, had been declared by the court an infringement of Watt's patent in 1799. In 1805, Hornblower wrote a description of a 'new steam-wheel' engine similar to a steam-turbine, but he never exploited this idea. The growing opinion of the Cornish mine operators was that the monopoly of Watt's steam-engine was oppressive, and retarding the development of the steam-engine. They encouraged a number of contemporary steam-engine builders such as Hornblower, Bull and Trevithick to develop an alternative to Watt's steam-engine.

In 1790, Edward Bull, a former steam-engine erector for Boulton & Watt, erected his type of the pumping engine in which he had eliminated the cumbersome rocking-beam by mounting the steam-cylinder vertically above the pumping-rod to which it was connected, but he still used a separate condenser in an infringement on Watt's patent. In the erection of Bull's pumping engines, he was assisted by Richard Trevithick. Between 1790 and 1794, 10 pumping steam-engines of the Hornblower and the Bull type were erected in Cornwall, England, but the court forced the owners to pay Watt royalties for the infringement on Watt's patent.

Watt, a stubborn man with self-esteem, was a Scottish individualist who believed that genuine personal worth is obtained only through personal achievements which needs no official approval. In his view, to accept honours granted by others is degrading to a man of genuine dignity since it makes his personal worth appear to depend on the opinion and whims of others rather than on the personal values created by the individual himself through his singleminded pursuit of personal excellence. Mainly for this reason, and to spurn servility to the English nobility, Watt declined a baronetcy offered to him by the Prime Minister of Great Britain.

During Watt's 'monopoly' period, Boulton & Watt were unable to satisfy all the market demands for steam-engines. Probably less than half of the steam-engines built in the period from 1775 to 1800 were built under Watt's patent. Wilkinson built more than twenty Watt's steam-engines, and Batemen & Sherrat in Manchester built close to 40 Watt's steam-engines without obtaining licences from Boulton & Watt in a direct piracy of Watt's steam-engine. Many Newcomen and Savery pumping engines were supplied to customers who could not obtain Watt's steam-engines, or who preferred less efficient but less expensive steam-engines. By 1800, Boulton & Watt Company had 496 engines in operation: 164 were pumping water back on the waterwheels which provided rotary motion to move machinery, others were mostly rotative steam-engines that directly moved other machinery, and 24 steam-engines were blowing engines producing blast-air for blast-furnaces. The average power of Watt's steam-engines ranged from 15 to 16-horsepower; the largest engines produced 50-horsepower.

The doubly-compounded steam-engine came into its own only when high-pressure steam came into use. The triply-compounded steam-engine in which steam produces power-strokes in 3 stages in 3 different cylinders fitted with pistons before exhausting into a condenser, was patented in 1871 by the French marine engineer Charles Benjamin Normand (1830-1888). He built the first such engine in 1873. The more economical use of fuel with steam-pressures above 180 psi (124 kPa) required a quadruply-compounded steam-engine, for compounding greatly improved thermal efficiency of the steam-engines. A Newcomen atmospheric pumping engine used 30 pounds (13.6 kg) of coal for every horsepower hour, Watt's steam-engine used 7½ pounds (3.4 kg), and the customary doubly-compounded steam-engine in the 1850's used 2½ pounds (1.1 kg).

High Pressure Steam-Engine
Verbiest, Grimaldi and Model Locomotive:

Steam-pressures used in steam-engines, which were much higher than the atmospheric pressure, were initially used in model locomotives. The first such model locomotive known is the toy-cart which was built by a Flemish Jesuit Ferdinand Verbiest (1623-1688) in his spare time in far-off China, where he was a Christian missionary. This toy-locomotive which had three wheels was about 2 feet long and its impulsive steam-wheel was geared by means of cog-wheels to the axle of the two rear driving-wheels. The steam-jet impinging on the impulse-wheel (ball of *Aiolos*) was fed from a centrally located pear-shaped boiler which was heated by live coal. This model was able to move with appreciable speed around a circular track for more than an hour. Verbiest was apparently inspired to construct this model after reading a description of such an impulsive steam-turbine by Salomon de Caus in 1615, and the depiction of it by Giovanni Branca (1571-1640) as a power-plant in a stamping-mill in Branca's book, < The Machine > (*Le Macchine*), in 1629. The steam in this impulse-turbine was under quite high pressure. Verbiest published an account of his model vehicle in his book, < European Astronomy > (*Astronomia Europaea*), in 1687.

About 1730, another Jesuit priest, P.M. Grimaldi, copied Verbiest's design and built another model cart powered by the impulsive steam-turbine. He entertained as well as instructed the young Chinese Prince Hang-Hi with this self-propelled cart.

Papin and Model Self-Propelled Vehicle:

Denis Papin made the second known model of a self-propelled vehicle about 1698, which was powered by his model reciprocating high-pressure steam-engine consisting of a boiler, a cylinder and a piston, and a transmission by ratchet-wheels. Papin intended to build a prototype vehicle for military transport, for which this small locomotive was a model, in order to show another use for his steam-engine but it was not a complete success and, therefore, Papin had to abandon this project.

Even the ancient reaction principle of Archytas of Taras in mechanics provided the idea of a steam-carriage in a form of a steam-boiler on wheels which was to propel itself by the reaction to a jet of steam issuing backwards from the boiler. It was provided with steam-controlled levers and steering to be operated by the 'coachman'. A model of this jet-propelled carriage was first made in 1680.

Leupold and High-Pressure Steam-Engine:

In 1720, the German engineer Jacob Leupold (1674-1727), who was Councillor for Mines under the Prussian Government and a member of the Berlin Academy of Sciences, designed a high-pressure, non-condensing, single-acting, pumping steam-engine of two cylinders working two lift-pumps, which followed Papin's steam-engine concept. Leupold was unable to construct a sufficiently strong boiler for it, owing to the generally inadequate state of workmanship at the time, a problem that had frustrated Savery a quarter of a century earlier. In 1725 Leupold described his steam-engine in < Theater of Hydraulic Machines > (*Theatrum Machinarum Hydraulicarum*) published in 2 volumes in 1724 and 1725 of his 9 volume encyclopædic treatise on mechanical engineering written in German language with Latin titles,< Theater of General Machinery > (*Theatrum machinarum generale*), published from 1724 to 1739. The last 2 volumes of this treatise, including the very last supplementary volume on the machinery of mills by Johann Matthäus Beyer, appeared in print posthumously. James Watt, who had learnt to read German, studied Leupold's work with great care when he was creating his own steam-engine. Leupold also wrote a volume on bridge engineering < Theater of Bridges > (*Theatrum pontificiale*), and described his mechanical calculator in the volume < Theater of Arithmetical Geometry > (*Theatrum arithmetico-geometricum*), published after his death in 1727, as part of his comprehensive treatise of 1,764 pages and 472 full-page copper engravings in 9 volumes. This voluminous treatise gave the first systematic treatment of mechanical engineering and the first systematic treatment of kinematics of mechanisms. Leupold was a practical engineer of eminent repute in Europe.

Cugnot and Fardier:

In 1769, another unsuccessful locomotive was designed and built by the Swiss engineer living in Brussels, Nicolas Joseph Cugnot (1725-1804). It was a 3-wheel ratchet-driven steam-carriage, or *fardier*, intended to haul artillery. Cugnot put this full-size steam-propelled vehicle on the road in 1769, and he was the first to do so. Regrettably it was unstable, had poor directional stability, and crashed into a wall on its very first trial run. Cugnot constructed his own steam-engine based on Papin's piston and high-pressure steam principle, and he seems to have been influenced by Leupold's design of the high-pressure steam-engine. Cugnot improved his design in 1770, and powered the single front wheel by using 2 vertical cylinders of 13-inch (0.33 m) bore straddled over the single front wheel in which the pistons were moving in opposite directions and delivered their motion by pawls to ratchets on the front wheel. Steam was alternately introduced into the cylinders through valves and exhausted at the base of the cylinders. A rocking-beam connected the 2 piston-rods to give the pistons their opposite motions. Unfortunately, the firebox allowed a great deal of heat to escape and the boiler was woefully inadequate for the task. This arrangement rapidly exhausted the steam-pressure provided by the cupcake-shaped copper boiler suspended in front of the single front-wheel. This vehicle, which was only able to make 4 miles (6.4 km) an hour, had to stop frequently to raise the steam pressure: after each ¼ hour of travel it had to stand for another ¼ hour to build up the steam. The vehicle was, moreover, unstable on turns and difficult to steer by means of its small tiller. Cugnot's larger version of the steam-carriage is still extant and on exhibition in the *Conservatoire des Artes et Métiers* (*Conservatory of Arts and Crafts*) in Paris. It has a high quality workmanship, but it appears to be unfinished. Cugnot's engine is the only large-scale high-pressure steam-engine which was actually built in the 18th century.

Murdock and Model Steam-Carriage:

Watt's assistant, the Scottish mechanic William Murdock (1754-1839), designed and constructed a model of a high-pressure steam-engine, but Watt, who was highly circumspect, discouraged Murdock to proceed with this project because Watt considered high-pressure steam-engines dangerous despite the urging of Boulton to increase steam-pressure in their engines. In 1784, Murdock constructed a 3-foot high model of a steam-locomotive fitted with a steam-cylinder of 0.75- inch (1.9 cm) bore and 1.5-inch (3.81 cm) stroke, but no condenser, which he also successfully road-tested. It ran at a speed of 8 miles (12.9 km) an hour. Watt considered such models dangerous toys, and he seems to have only pretended to collaborate with Murdock in its development in which he went as far as to take out a patent on the locomotive in 1784 as part of his general patent in order to save expenses for Murdock. In his patent Watt had sketched a rather incomplete proposal for a steam-carriage machinery after repeated urging of physician and poet Erasmus Darwin (1731-1802), and other friends of Watt to produce a steam-carriage. Watt did provide specifications for steam-carriages which included rocking-beams, sun-and-planet gearing, a hooped wooden boiler with internal firebox, and a three-speed variable-ratio transmission of great merit. The constant-mesh 3-speed system of Watt had spur-gears so arranged that those on the second-motion shaft could revolve freely on it, or be selectively 'clutched' to it as required by a sliding feather, or elongated key, actuated by a suitable collar and forked yoke. This type of constant-mesh gearing was revived for motor-car transmission about 1900. Watt's efforts seem to have been mainly devoted to making certain that if he did not produce a steam-carriage nobody else could. Regrettably, as a result of Watt's obstructive attitude, Murdock finally had to put the locomotive idea aside, but only after having built a larger locomotive model in 1792 with a steam-cylinder of 1-inch (7.54 cm) bore, which was large enough to transport him to the mines and back. The second locomotive was fired with gas fed from a reservoir. In 1819, Murdock proposed to use compressed-air to drive locomotives, but apparently nothing came of it.

Read, Kinsey and Fitch, and Self-Propelled Steam-Carriage:

In 1790, American teacher and apothecary Nathan Read (1759-1849), who invented in 1791 the multitubular water-tube boiler, built a small-scale steam-carriage and drove it through Warren, a town in Massachusetts. It was the first vehicle known to have used a steering-wheel and high-pressure steam. It has also been reported that about the same time, Dr. Apollos Kinsey, a very ingenious man of his time, drove a steam-waggon of his own design in Hartford, Connecticut. In 1796, the American pioneer steamboat builder, John Fitch (1743-1798) is supposed to have built a small self-propelled steam-carriage with fitted flanged-wheels.

Symington and Direct-Acting Marine Steam-Engine:

The actual pioneer of direct-acting steam-engines was William Symington (1763-1831), a Scottish engineer. He first worked on a model carriage driven by steam-power in 1786, but he was unable to finance the construction of a full-scale carriage despite the encouragement from the faculty of University of Edinburgh. In 1787, Symington patented an atmospheric engine fitted with two cylinders of 4-inch (10.2 cm) bore open on top and 18-inch (45.7 cm) stroke, and supplied with separate jet-condensers. In 1801, Symington designed a direct-acting engine of 10-horsepower without a working-beam, and he applied it to the propulsion of a tugboat, the Charlotte Dundas, on the Forth and Clyde Canal. This steam-engine which incorporated a condenser, had a horizontal double-acting cylinder of 22-inch (0.56 m) bore and 4-foot (1.22 m) stroke, and it drove the crankshaft of the paddle-wheel directly through a connecting-rod. This engine worked by a steam-pressure which was considerably higher than the atmospheric pressure.

Trevithick and First Steam-Powered Locomotive:

The important developers of the high-pressure steam-engine without a condenser were the English mining engineer, strongman and wrestler, Richard Trevithick (1771-1833), and the American mechanical genius, Oliver Evans (1755-1819). The two men worked independently of each other.

Trevithick's father was the general manager of the Dolcoath coal mine, and his son Richard had a good education in science and mathematics. Trevithick received his engineering training by serving an apprenticeship under Hornblower and also under William Murdock, the best of Watt's assistants. Trevithick constructed a noncondensing double-acting steam-engine, called a 'puffer', about 1798, before Evans built his steam-engine. At first he placed his cylinder upright inside the boiler, which served as a steam-jacket, and then, he eliminated the working-beam by connecting the piston-rod to an overhead crosshead which acted through connecting-rods on both sides on the crankshaft beneath the boiler. This steam-engine was compact, self-contained and portable as a unit. He changed it into a direct-acting steam-engine by setting the cylinder horizontal and connecting its piston by means of connecting-rods to the crankshaft. Trevithick introduced two safety valves regulated by weights, one beyond the control of the operator to avoid mistakes. He also introduced a fusible plug of lead in the boiler to avoid too high heat in the structure of the boiler when the water-level falls too low in the boiler. He later introduced also a steam-gauge after one of his boilers had exploded. In 1802 Trevithick proved in a test that high-pressure steam-engines were inherently more economical than low-pressure steam-engines, and that the efficiency of a steam-engine improved with the increase of operating steam-pressure. His pumping engine at Coalbrookdale despite its small size produced great power. It had a cast-iron boiler 1½ inches (3.8 cm) thick and 4 feet (1.22 m) in diameter and a cylinder of 7-inch (17.8 cm) bore and 3-foot (0.91 m) stroke. It worked with the unprecedented steam-pressure of 145 psi (1000 kPa), 10 times the atmospheric pressure. The great power this small-size steam-engine developed, astonished everyone who saw it. In 1803 Trevithick constructed his original high-pressure steam-engine with horizontal cylinder built inside of the dome-topped boiler. In 1805 he built a high-pressure steam-engine with vertical cylinder and internally fired return-flue boiler. By 1804, Trevithick had built nearly 50 high-pressure steam-engines which drove sugar-mills, ground corn, pumped water and rolled iron.

The steam-engine for a locomotive had to be powerful, compact and light, and Trevithick was the first engineer to realise fully that only high-pressure steam-engines can satisfy these requirements. Trevithick began to work on the steam-propelled vehicle almost at the same time when he worked on his steam-engine. He built a model locomotive in 1798, and had developed it by 1801 into a full-size steam-carriage, after having seen Murdock's model steam-carriage in 1794, which he planned to use for passenger service. He obtained a patent in 1802, and with his partner and cousin Andrew Vivian he shipped his steam-carriage to London in 1803, and ran it in the streets. It was an adaptation of his puffing-engine, which had a cylinder of 5½-inch (14.0 cm) bore partly enclosed in the boiler and 2½-foot (0.76 m) stroke, to serve as the engine of a mobile vehicle. The steam-engine operated with 30 psi steam pressure, and made 50 strokes a minute. It had 3 wooden wheels with wrought-iron tyres and a primitive two-speed transmission. This vehicle had a successful trial, but was left to boil dry in the celebration of success, and after this misfortune it had a road accident which left the frame of the carriage twisted. Trevithick lost interest in steam-carriages because of these mishaps, and thenceforth he restricted his attention to locomotion on rails.

Trevithick demonstrated the first clearly successful locomotive engine in 1804, when his locomotive hauled ten tons of iron, 70 men and 5 cars, 9½ miles (15.3 km) with a speed of 5 miles (8.0 km) an hour from a Welsh Pen-y-daren Iron Works, in South Wales, to a seaport with his single-cylinder steam-locomotive. It had a single-cylinder horizontal engine of 8¼ -inch (21.0 cm) bore and 4½-foot (1.37 m) stroke. The piston-rod, by means of the connecting linkage, turned a large flywheel 7-foot (2.13 m) in diameter which, in turn, was connected to the wheels through an intermediate idler-gear. Later in 1804, Trevithick modified his locomotive with a major improvement by exhausting the steam through the smoke-stack above the damper thereby increasing the draught in the furnace of the boiler which in turn improved combustion. He demonstrated this locomotive in 1805 by hauling 25 tons of iron on a wooden-track railway at Wylam, near Newcastle. The contemporary cast-iron tracks were inadequate for his locomotive, which was too heavy for the tracks, and soon the locomotive derailed because of the failure of the track. After the derailment Trevithick put this locomotive to stationary work as he had done with the overturned steam-carriage in London. The steam-pressure in these locomotives may have been between 40 and 50 psi (276 and 345 kPa).

Trevithick shipped a new locomotive to London in 1808 which he exhibited to the public for an admission fee of one shilling on a circular railroad track laid down near Euston Square. It had a cylinder of 14½ -inch (0.37 m) bore and 4-foot (1.22 m) stroke. He advertised it as a 'Racing Steam Horse' and as the 'Catch-Me-Who-Can'. This locomotive developed a speed of 12 miles (19.3 km) an hour on the small circular track, whereas on the straight track it travelled with 20 miles (32.2 km) an hour. He coupled all four wheels of his locomotive by gears to make them all driving-wheels. Finally, this locomotive also ran off the track and overturned, and since Trevithick had exhausted his funds, he could not put it back on rails.

In 1812, Trevithick built a modified Watt-type of steam-engine which used high-pressure steam, but also incorporated a condenser. This steam-engine was a prototype of what became known as the 'Cornish engine' which initially worked with 50 psi (345 kPa) pressure and the steam cut-off at (1/9)th of the stroke, and had 3 times the thermal efficiency of Watt's steam-engine, about 17%. The efficiency of these steam-engines was progressively improved by Arthur Woolf (1776-1837) until in 1842, Trevithick's Cornish engine achieved a duty of 107 million. These steam-engines were highly reliable and remained in use throughout the 19th century. Trevithick's internally fired return-flue boiler was essential for the high-pressure expansive operation of steam-engines in which the exhaust steam preheated the water in the boiler. His 'Cornish boiler' consisted of a cylinder heated by an inner furnace tube, called the 'flue', which carried the combustion-gases through the boiler, and the internal fire-plate. The flue was very effective in increasing the efficiency of the boiler. This type of boiler is still in use.

In 1815, Trevithick patented his well-known plunger-pole pumping engine which pushed the rods and pumped water at the down-stroke. It worked with 120 psi (827 kPa) pressure, had a plunger of 33-inch (0.84 cm) bore and 10-foot (3.05 m) stroke. All of Watt's pumping steam-engines had bucket pumps which lifted water on the up-stroke. In the same patent he also claimed his 'whirling-engine', an *aiolopyle* reactive steam-engine, which consisted essentially of a 15-foot (4.57 m) diameter wheel mounted on a shaft and having hollow radial arms. The steam escaped with the nozzle speed of 200 feet (60.96 m) a second at a pressure of 100 psi (690 kPa) tangentially to the wheel through a small hole in each arm. The maximum rotational speed he could obtain was between 250 and 300 rpm which was far too slow and produced only one-fifth of the power potentially available in the steam. Therefore, this noncondensing, high-pressure reactive steam-turbine which was to be mounted on the shaft of an Archimedean screw for marine propulsion was not a practical success, and Trevithick did not develop it any further, although he liked the machine for its remarkably light weight.

Trevithick was the most inventive steam-engine builder of his time and had he patented his device of exhausting steam through the smoke-stack in order to provide adequate draught in the boiler it would have given him similar financial control over locomotive development as the separate condenser had given Watt over the steam-engine development. Unfortunately, Trevithick was not a shrewd businessman, and by this oversight he missed a great opportunity for financial success.

Trevithick was a restless man, and after all the mishaps and business failures he turned his attention to dredging, tunnelling under the Thames river, steam-powered threshing-machines, cultivators, tugboat engines, screw-propellers, and improving his stationary steam-engines. When these efforts also failed as business ventures, he left for Peru in 1816 to pump out the silver mines, where after some initial success his incredible string of bad luck continued when a revolution broke out in South America. He returned destitute to England in 1827 with the assistance of Robert Stephenson, and upon his death in 1833 he was buried in a pauper's grave at Dartford.

Evans, Automatic Flour-Mill, 'Columbian Engine' and 'Orukter Amphibolos':

Oliver Evans was a farmer's son who attended country school until he was 14 years of age, then he apprenticed himself to a wainwright where his mechanical ingenuity immediately manifested itself. He was determined to obtain a scientific education on his own, and bought books on technical and theoretical subjects, particularly mathematics and mechanics, which he mastered by teaching himself during his free hours.

In 1787 Evans, who was the pioneer of automation, built the first completely automatic mill for mass-production of flour in which 5 inventions consisting of power-driven roller mills, cylindrical bolters with various sieve meshes, vertical bucket elevators and horizontal screw-conveyors allowed shifting, grinding, bagging and loading of flour without the intervention of man. Since Evans was self-educated and had no scientific prestige, his inventions had to face frequent public ridicule despite the fact that his ideas were sound.

When Evans was 17 years of age, he had the opportunity to study Watt's new steam-engine design with its separate condenser. In 1802, Evans produced a direct noncondensing, double-acting, high-pressure steam-engine, called the 'Columbian engine', which was a remarkably sound design that used the 'elastic power' of high pressure steam [about 50 psi (345 kPa)] after its admission into the cylinder had been cut off which would raise the power and efficiency of the steam-engine so that it could be used in small sizes to propel land vehicles as Evans maintained. The upright double-acting cylinder of this steam-engine had a mere 6-inch (0.15 m) bore, only an 18-inch (0.46 m) piston stroke and a wooden flywheel 7½-foot (2.29 m) in diameter operating at 30 revolutions per minute. Evans designed a horizontal cylindrical copper boiler with fire-plate and an internal flue running through the center of the boiler before connecting with the chimney, a boiler which was also a sound improvement that later led to the design of firetube boilers. The steam-valves and the exhaust-valves in the form of 3-way cocks of Evans' steam-engine were operated by pins on the flywheel. Evans used his steam-engine to grind plaster of Paris, to saw marble blocks, to drive boring machines, and to propel boats. He had devised an ingenious rectilinear linkage for the steam cylinder, later known as 'Evan's rectilinear linkage', which was located at the same end of the beam as the crankshaft, thereby solving the rectilinear motion problem for the piston-rod and drastically reducing the size of the working-beam which now had a fulcrum at the head rather than in the center. With this novel arrangement Evans was able to translate the rectilinear motion of the piston rod to the rotary motion of the shaft in a much simpler design than Watt's rectilinear mechanism. It was called the 'grasshopper' engine, a name suggested by the peculiar loping motion of the rocking links at the end of the 'rocking beam' when the steam-engine was operating. The grasshopper design eliminated the lengthy rocking beam and expensive slide-bars, and required only the relatively inexpensive pin-joints for the guidance of the crosshead in a rotative steam-engine. It was the first mobile prime mover. The construction of the small high-pressure steam-engine cost Evans his entire financial resources, $3,700. The 'grasshopper' type steam-engine was also patented in England in 1803 by the English mechanic William Freemantle, who used this type of steam-engine in several locomotives including the well-known 'Puffing Billy' of William Hedley (1779-1849).

Evans immediately recognised the potential use of his steam-engine for the driving of carriages such as his 'steam waggons' which he had patented in 1797, and boats. In 1805, Evans built his well-known 'Both-Way-Digger' (*Orukter Amphibolos*), a wooden dredging scow on wheels, which moved on water and as well as on land because it was equipped both with wheels and paddle-wheels. It moved in a 3-mile (4.8 km) trial run on land with a speed of 15 miles (24.1 km) an hour, and with a speed of 16 miles (25.7 km) an hour on the Delaware river, but despite this successful trial, Evans found no financial support for the production of his dredger on wheels. In 1807 Evans established the Mars Iron Works in Pittsburgh for the manufacture of his steam-engines. In 1816, Evans built his first ship named <u>Oliver</u> <u>Evans</u> for the Mississippi River service, but it was later involved in a boiler explosion causing human fatalities.

In 1815 Evans built a steam-engine fitted with a cylinder of 20-inch (0.51 m) bore and 5-foot (1.52 m) stroke which was supplied by four boilers designed to deliver steam at 200 psi (1379 kPa) boiler pressure for the Fairmount Waterworks of Philadelphia. However, the operation of this high-pressure steam-engine was so costly that the steam-powered plant was replaced in 1822 by a hydraulically powered plant. Evans had built 50 steam-engines by the time of his death in 1819, and his influence on the use of high-pressure steam-engines in United States and in Germany was decisive: both prominent extremely high-pressure steam-engine builders, Jacob Perkins of Philadelphia, and the German eye surgeon Dr. Ernst Alban (1791-1856), considered themselves to be disciples of Evans, for Evans had published influential pioneering handbooks < The Young Millwright and Miller's Guide > and < The Young Steam Engineer > in which he advocated the scientific method in engineering as the 'true path of invention'. Evans admired Smeaton's scientific experimental approach in the improvements of Newcomen's atmospheric pumping engine, the waterwheel and the windmill. Evans, who was one of the greatest engineers of his age, was apparently the first engineer to propose in 1805 a closed cycle mechanical refrigeration machine.

Both Trevithick and Evans used about 50 psi (345 kPa) high steam-pressure in their boilers, but they did not realise that as a result of abandoning the condenser and exhausting the steam at relatively high pressure and temperature into the atmosphere, they lost a great deal of the thermal efficiency they had gained from higher boiler pressure.

The high-pressure steam-engines were made possible by the invention of the interchangeable metallic packing by Edmund Cartwright in 1797, which encouraged Trevithick to begin his experiments on high-pressure steam-engines in 1798. The metallic packing, which in 1816 was replaced by metallic piston rings invented by the Englishman John Burton, allowed small tolerances, and avoided serious leaks in steam-engines, which made high-pressure, multicylinder, noncondensing compound steam-engines possible.

Murray, Side-Lever Steam-Engine, Epicycloidal Mechanism and Textile Machines:

Matthew Murray (1765-1826) of Leeds was the first inventor to improve the steam-engine after James Watt, and to develop new applications for steam-power. He introduced new techniques of forging iron and new machines for the British textile industry, and was mainly responsible for making the city of Leeds a major manufacturing center. In 1790 he devised and built a new flax-spinning machine, and soon after the 'wet-spun' process for flax which produced more silky and stronger yarn.

He was the first to make the steam-engine a self-contained unit by supporting everything on a single bed-plate, which eliminated the possible interference of the local millwrights with the efficient operation of the steam-engines through unskilful erection. He invented the **D**-slide valve by bringing the admittance and exhaust ports together in a steam-chest [or steam-box] and the screw-feed. Murray devised an epicycloidal mechanism to convert reciprocal motion of his steam-engine into rotative motion which consisted of a gear-wheel rolling along the inside of a gear-ring of twice the diameter of the gear-wheel with the piston-rod of the steam-engine being attached to a pin on the rim of the gear-wheel. This particular epicycloidal mechanism which converts a rotary motion into reciprocal rectilinear motion is based on a theorem in Ptolemaic astronomy proved by the Persian astronomer and mathematician Nasir Eddin at-Tusi (1201-1274) at Maragha observatory established by Hulagu Khan, grandson of Genghis Khan and brother of Kublai Khan: "If a point moves with uniform circular motion clockwise around the epicycle whilst the center of the epicycle moves counterclockwise with half this speed along an equal deferent circle, the point will describe a straight-line segment." In its technological application Nasir Eddin's theorem predicated that if a circle rolls without slipping along the inside of a coplanar circle having twice the diameter, a point on the circumference of the smaller circle will move along the diameter of the larger circle. Nasir Eddin's astronomical theorem was used by Nicolaus Copernicus (1473-1543) in his heliocentric astronomy, but it is doubtful that Murray had any knowledge of this theorem. This epicycloidal mechanism, which is still used in certain types of machinery, was published by John White as his own 'invention', and is known as 'White's Parallel Motion'.

Through all his improvements, Murray was able to make the steam-engine considerably smaller, lighter, more efficient and easier to construct. Murray's steam-engines were, therefore, cheaper and less expensive to operate than Watt's steam-engines, which aroused the enmity of Boulton & Watt, and led to the first known case of industrial espionage when James Watt, the younger, and William Murdock, the manager of Watt's factory, attempted to discover the manufacturing secrets of Murray, particularly his superior casting methods.

In 1811, Murray collaborated with Trevithick in building a high-pressure steam-engine and boiler to outfit a captured French privateer as a packet boat. This engine was later used in another ship named Courier.

Murray invented a much larger, 2-cylinder steam-engine, called the 'side-lever engine'. In this steam-engine the beam was mounted in a shallow-well below the cylinder instead of above it, thereby reducing the overall height of the engine. It became a popular marine engine and the standard source of power in Mississippi paddle-steamers, which gives Murray an important place in the early history of steam-engines. Murray's factory known as the 'Round Foundry' became an important manufacturer of locomotives after his death. Many men trained at Murray's Round Foundry spread Murray's tradition far and wide. For instance, Murray's only son trained at the Round Foundry went to Russia where he established a successful engineering business in Moscow. He died there in 1835. Murray's own firm, Fenton, Murray & Jackson, went out of business in 1843.

Krupp and First Integrated Industrial Empire:

Many engineers who served their apprenticeship in Murray's 'Round Foundry', which was operated after Murray's death by Smith, Beacock & Tannett, became prominent in industry. The remarkably talented and enterprising German industrialist and engineer Alfred Krupp (1812-1887) and his brother served their apprenticeship in the 'Round Foundry'. Krupp was one of the gold medal winners at the Great Exhibition of 1851 in the Crystal Palace in the Hyde Park, in London, England, for his design of a 6-pounder cannon of cast-steel. Krupp also displayed at the Great Exhibition a huge flawless block of cast-steel weighing several tons produced with a new method of casting steel that he had perfected. Krupp's industrial exhibits in the Crystal Palace in the Hyde Park made him famous worldwide, and established his international reputation. He also succeeded in developing a seamless railway wheel that was soon widely adopted. Krupp became the leading industrialist in Europe by establishing the giant **Krupp Werken** (*Krupp Works*), the imposing steel, armament and machinery works in Essen, in Germany, and after acquiring shipyards, coal and iron mines in the 1860's he created the huge integrated industrial empire of Krupp. In 1854, **Krupp Werken** in Essen, Germany, built the first rolling mill which rolled flat strips from solid blocks of steel. By 1887 Krupp steel works were the foremost producers of steel in the world.

Krupp was also responsible for establishing the first industrial social welfare system which provided his workers and employees with showers and dressing rooms, a company hospital, company benefits, a company pension fund, company banking, a company store, a company kindergarten, a company school, company housing, and a company church. The pioneering social welfare system of Krupp served as an exemplary industrial model which demonstrated how a large industrial enterprise can provide reasonable social welfare for its employees, and much of Krupp's popular welfare system was adopted on a national scale by the Chancellor Otto von Bismarck (1815-1898) in the first State sponsored pension scheme in the world instituted between 1883 and 1887 in the united German Empire.

In 1871, another giant industrial enterprise, the ***Thyssen & Co. KG***, was founded at Mülheim in the Ruhr Basin in Germany by August Thyssen (1842-1926), son of a banker and manufacturer who was educated at the ***Polytechnikum*** (*Polytechnical Institute*) in Karlsruhe and at a business school in Antwerpen in Belgium. After 1878, the ***Thyssen & Co. KG*** acquired, in addition to their iron works and rolling mills for iron flats and bars, also industrial works for the manufacture of pipes, as well as the ***Maschinenfabrik Thyssen*** (*Machine Factory Thyssen*) for the manufacture of steam-engines and steam-turbines. The ***Thyssen Co.***, one of the largest mining concerns in the world before the First World War, owned and operated extensive coal fields. After the First World War, ***Thyssen Co.*** also owned 26% of the ***Vereinigten Stahlwerke*** (*United Steel Works*).

Both industrial giants, ***Krupp*** and ***Thyssen***, are still prominent international corporations in modern Germany.

Woolf and Edwards, High-Pressure Compound Steam-Engine with Condenser:

The Hornblower compound steam-engine had little practical advantages when used with low steam-pressures, but when high steam-pressure was used and the 2 cylinders were properly proportioned to the expansion of the steam in the ratio of 1–to–5, then it had a considerable advantage over the Watt's steam-engine. The millwright Arthur Woolf (1776-1837), who had formerly worked for Hornblower, constructed a rotative high-pressure compound steam-engine with a condenser that was considerably more efficient and economical than Watt's steam-engine. The addition of the condenser was Woolf's fundamental improvement of Hornblower's compound steam-engine. Woolf used 50 psi (345 kPa) steam-pressure and it was about 50% more efficient than Watt's steam-engine by achieving a thermal efficiency of 7.5%. Both Trevithick and Evans had not realised that the great advantage of a condenser is the increase of thermal efficiency and the increase in the output of work for each pound of steam generated. Woolf and his collaborator Humphrey Edwards, a millwright of Lambeth, perfected the high-pressure compound steam-engine by using an eight-fold or nine-fold expansion, but despite its fuel economy, it did not become popular in Britain, where coal was relatively inexpensive, because of the greater initial cost of the compound steam-engine and its mechanical complexity. Woolf built no compound steam-engines after 1824 because they could not compete with Trevithick's high-pressure Cornish engine. However, Humphrey Edwards, who went into business for himself in France where he became successful, made the compound steam-engine popular in continental Europe despite its high initial cost and complicated operation, mostly because fuel for the steam-engine was quite expensive in France.

One measure of the performance of the steam-engine is its thermal efficiency, something that the pioneers of the steam-engine overlooked in their measures of the steam-engine efficiency. The thermal efficiencies of the various reciprocating steam-engines were:

1718, Newcomen (duty,4.3), thermal efficiency 0.5% ;1767, Smeaton (duty,7.4), thermal efficiency 0.8%;1774, Smeaton (duty,12.5), thermal efficiency 1.4%; 1775, Watt (duty, 24.0), thermal efficiency 2.7%; 1792, Watt expansive steam-engine (duty,39.0) thermal efficiency 4.5%; 1816, Woolf compound steam-engine (duty,68), thermal efficiency 7.5%; 1828, improved Cornish steam-engine (duty,104), thermal efficiency 12.0%; 1834, improved Cornish steam-engine (duty,149), thermal efficiency 17.0%.

Perkins, Uniflow-Engine and Mechanical Refrigerator :

In 1827, American engineer and engraver, Jacob Perkins (1766-1849) of Philadelphia, a fellow Philadelphian of Oliver Evans, experimented with ultrahigh-pressure steam-engines which worked with steam-pressures as high as 1,400 psi (9,653 kPa). He devised a horizontal high-pressure steam-engine, later called the 'uniflow-engine', in which the steam entered at both ends of the steam-cylinder through slide-valves, and flowed to the middle of the cylinder where it was exhausted through a ring of ports. The piston having a length equal to the 12-inch (30.5 cm) stroke less the clearance space acted as the exhaust valve. The ends of the cylinder in the Perkins engine were kept hot and the central outlet cool at the exhaust temperature, which made the 10-horsepower steam-engine thermally efficient, but it was difficult to construct. Perkins' steam-engine worked with a steam-pressure of 514 psi (3545 kPa) and a boiler temperature of 392° F (200° C). It was only 1/5th the size of a comparable Watt steam-engine. Such a uniflow-engine was much later patented by Todd in 1885, but it became a successful steam-engine only after the German engineer Johann Stumpf, professor at the **Technische Hochschule** (*Technological Institute*) in Charlottenburg, a suburb of Berlin, patented a new valve-gear for it in 1908, which made the uniflow-engine the ultimate refinement of reciprocating steam-engines.

In 1834, Perkins also invented a mechanical refrigerator operating by a process of closed ether-cycle in which volatile liquid ether was first boiled in an evaporator at low temperature and pressure to freeze water, and next the resulting vapour of ether was compressed and condensed back into a liquid at a higher temperature and pressure. Then, the liquid ether was throttled through an expansion valve back into the lower-pressure evaporator where the temperature returned to its initial condition and thus completed the cycle. Perkins was granted a U.S. patent, and also an English patent for his machine when he moved to England, where his first compression machine was built by John Hague. Perkins' ice machine was mechanically successful and stood out as a remarkable achievement for the period because its design was based on sound physical principles. Although it was the first power-driven machine capable of manufacturing ice in sufficiently large quantities, Perkins failed in his attempt to commercialise icemaking with his refrigeration machine in England, yet it was urgently needed in European dairy and brewing industries as well as for importation of perishable foods.

It is remarkable that the art of refrigeration in the artificial production of ice has an ancient history. Ancient Egyptians in the South, ancient Indians in the Far East, and ancient Estonians in northern Europe were able to produce small quantities of ice by an artificial method which depended upon the idea of simultaneous loss of heat through evaporation and radiation in the cool dry air of a cloudless night.

Dr. Gorrie, Cold-Air Machine, Refrigeration and Airconditioning Industry:

The first commercially successful refrigeration machine was designed in 1844 by a Florida physician Dr. John Gorrie (1803-1855) of Apalachicola which worked on a closed air-cycle system and produced ice and cold air for his patients suffering from malaria and high fever. Gorrie's cold-air machine, which in some respects was similar to an air-cycle heat pump, served as a model for the design of many different air-cycle refrigeration machines. After Gorrie obtained a U.S. patent in 1851 for his cold-air machine, the refrigeration and airconditioning industry in Europe began with Gorrie's cold-air machine and rapidly grew to an immense size, and it is still growing today.

Dr. Alban, Ultrahigh-Pressure Steam-Engine and Superheated Steam:

Another self-declared disciple of Oliver Evans, the German eye surgeon, Dr. Ernst Alban, who had spent some time in Britain, built a 30-horsepower ultrahigh-pressure steam-engine in 1840 which worked with 650 psi (4482 kPa) steam-pressure. It had a centrifugal governor, and a cylinder which oscillated about the crankshaft fitted with a flywheel thereby eliminating the need for a connecting-rod. This ultrahigh-pressure steam-engine operated the textile machinery in a textile mill in Plau, Germany, for 60 years. Alban rarely used condensers since he found alternative uses for spent steam, such as heating the buildings. He anticipated the use of superheated steam, but considered its use not practical in his own time. Superheating of steam produces 'dry' steam at temperature up to 300° F (149°C) above the boiling temperature of water, and superheated steam owing to its elevated temperature has much higher energy content than saturated steam. Alban published an excellent treatise on high-pressure steam-engines in 1848.

McNaught and Compounded Watt's Steam-Engines:

In 1845, John McNaught of Bury, a Scot living in Lancashire, was able to compound the existing Watt's rotative beam steam-engines by adding a small short-stroke high pressure cylinder, which had its own parallel motion, to act on the beam about half-way between the central fulcrum and the end driving the connecting-rod. Many existing Watt's engines were thus compounded to operate with high-pressure steam, whereas the 'McNaught' cylinder was added to the new engines operating with steam-pressures varying from 120 psi to 150 psi (827 to 1,034 kPa). This practice was called 'McNaughting' the steam-engine, and it obviated the great expense of replacing the old Watt steam-engine.

Neither the compound steam-engine of Woolf, nor Trevithick's plunger-pole steam-engine long survived in heavy pumping work, for they were unable to compete with the high-pressure Cornish steam-engine of Trevithick which dominated the field almost to the end of the 19th century without serious rivals. The efficiency of the Cornish steam-engine was gradually improved until in 1840 the Taylor's Cornish pumping steam-engine achieved a duty of 107 million. Its steam cylinder of 85-inch (2.16 m) bore and 11-foot (3.35 m) stroke developed 220-horsepower at 7½ double strokes a minute whilst pumping 500 gallons (1,892.7 liters) of water a minute. The skill of workmanship in the construction of this pumping engine for its time was quite remarkable.

Maudslay and Direct-Acting Table Engine:

The first direct-acting steam-engine which came into extensive use in industry was the table engine of Henry Maudslay (1771-1831), the founder of modern machine-tool industry, which he patented in 1807. In one form or another this compact table steam-engine, in which the cylinder was placed on a table and its crankshaft mounted on the floor, survived in a simplified form as the industrial source of power until the end of the 19th century.

Maudslay's son, Joseph, designed an oscillating cylinder steam-engine, an original invention of William Murdock, for the propulsion of ships. A combined pair of such steam-engines providing upwards of 20-horsepower were installed in a small steam-vessel, S.S. Endeavour.

In 1845, Frederick Ellsworth Sickels (1819-1895), inventor of a steam-powered steering system for ships, devised an improved valve for steam-engines. He noted that there was some throttling in the slow-closing variable cut-off valve as steam passed through the partly closed valve [called 'wire-drawing']. He invented a 'trip' lifted by a 'wiper', which dropped a 'poppet-type' valve quickly to its seat cushioned by a hydraulic 'dashpot'.

Giant Cross-Compounded Steam-Engines of Corliss:

In 1849, the American engineer George Henry Corliss (1817-1888) invented a novel quick-acting rocking valve as a replacement for the slide-valve which was at the time in general use in stationary steam-engines. Corliss' valve gear consisted of rapidly closing cylindrical rocking valves which were moved partly by an eccentric, and then tripped to make the valves close suddenly under the action of springs. The centrifugal governor controlled closely the variable cutoff of the steam without 'wire-drawing' of the steam as the valves were closed, and to maintain the free-exhaust by keeping the exhaust valves fully open for the entire return-stroke. Corliss' valve afforded a great saving in steam as well as improved the speed regulation. In 1856 Corliss founded the Corliss Steam Engine Company, and he soon became the leading international authority on steam-engines. His last wrist-lever valve gearing of the 1880's was his best design.

In 1876, Corliss built a gigantic steam-engine, the largest steam-engine at the time, for the Centennial Exhibition in Fairmont Park, Philadelphia. This giant steam-engine, consisting of two separate, single cylinder steam-engines of the beam-type and a common flywheel of 30-foot (9.14m) diameter, produced an output of 2,500-horsepower that was used to run several thousand machines in the 1,500 sections of the Machinery Hall of the Exhibition. The Main Hall of the Centennial Exhibition in Philadelphia was the largest building in the world. This colossal, 44⅓ feet (13.51 m) high, beam-type steam-engine weighing 600 tons had two high-pressure cylinders of 40-inch (1.02 m) bore and 10-foot (3.05 m) stroke at 36 rpm. The giant Corliss steam-engine, which also worked the pumps that produced a spectacular 33-foot (10 m) high and 36-foot (11 m) wide waterfall carrying 30,000 gallons (136,380 liters) of water a minute at the Centennial Exhibition in Philadelphia, was ceremonially started by the Emperor of Brazil, Dom Pedro de Alcantara.

In 1881 Corliss built a 4,000-horsepower cross-compounded Corliss steam-engine which had a cylinder of 50-inch (1.27 m) bore for high-pressure, a cylinder of 84-inch (2.13 m) bore for low-pressure, and a common 8-foot (2.44 m) stroke. It worked with 100 psi (690 kPa) boiler-pressure. The Corliss compounded steam-engine had a duty of 150 million pounds of water raised one foot, and a thermal efficiency of 17.2 %.

Corliss sold his steam-engines on the bases of fuel saving. For instance, his 180-horsepower steam-engine saved about $20,000 a year compared to other steam-engines of equal power. Corliss steam-engines were particularly popular in textile mills, where the regularity of the engine speed was of paramount importance.

Vertical High-Pressure Steam-Engines
Nasmyth and Steam-Hammer Engine:

About 1850, Scottish machine toolmaker James Nasmyth (1808-1890) invented a vertical high-pressure steam-engine similar in configuration to his steam-hammer, which had the steam-cylinder on top and the crankshaft below in place of the anvil of his steam-hammer. Such engines for years were called 'inverted engines', which later became its generally accepted name when the 'inverted engines' were adopted for the duties of pumping, blowing the blast-furnaces, and turning the shafts of screw-propelled steamships. The comparatively small space this 'inverted steam-engine' occupied, its compactness, its easy accessibility, and its uniform wear of the cylinder brought the 'steam hammer engine' into almost universal use. In 1851 the steam hammer engine was awarded the first prize at the Great Exhibition in the Crystal Palace in London.

High-Speed Steam-Engines

When in the late 19th century the steam-engine's prominence as the best prime mover began to be challenged by steam-turbines, many different types of high-speed steam-engines were devised which were more efficient than the standard-speed steam-engines.

Elder, Brotherhood, Normand, Willans, Kirk and Multiple-Expansion Steam-Engines:

The first sea-going vessel powered by a compound steam-engine was the British ship S.S. Brandon, which had been fitted in 1854 with the double-expansion steam-jacketed engine built according to sound thermodynamic principles by John Elder (1824-1869), a close friend of the famous Scottish engineer-scientist-poet and thermodynamics expert William John MacQuorn Rankine (1820-1872). In 1862, Elder patented the triple-expansion and quadruple-expansion compound steam-engines. Some of the Elder's steam-engines were the most efficient steam-engines of the period, consuming about 3½ pounds (1.1 kg) of coal per kilowatt hour of power. The first ocean-going vessel to have double-expansion compound steam-engine was the steamship S.S. Holland in 1869. More economical use of fuel when steam-pressures at the boiler were higher than 180 psi (1241 kPa) required more than double-expansion of steam. Such multiple-expansion steam-engines were first introduced in 1871.

In 1871, English mechanic Peter Brotherhood (1838-1902) patented his multicylinder steam-engine which had three cylinders inclined at 120 degrees to one another in the vertical plane and directly drove the machinery at considerably higher speeds than the customary steam-engines. His steam-engines ranged in size from 1¼-horsepower at 100 rpm, to 55-horsepower at 500 rpm. Brotherhood finally designed compound steam-engines of this type with cranks 180 degrees apart, which developed an output up to 1,000-horsepower at 250 rpm.

In 1871 a French engineer Charles Benjamin Normand (1830-1888) patented a triple-expansion compound steam-engine in France, and built the first such steam-engine in 1873.

In 1874, the first triple-expansion steam-engines in Britain were fitted by A.C. Kirk of John Elder & Company in the steamship S.S. Propontis of 2,083-ton displacement. This steam-engine worked with the boiler steam-pressure of 150 psi (1,034 kPa).

One of the most successful high-speed steam-engines developed was the 'central-valve steam-engine', patented by Peter William Willans (1851-1892) in 1884. The cylinders of this single-acting steam-engine were co-axially arranged above one another in which the highest steam-pressure was exerted on the top. The interconnected piston valves were all mounted on the same hollow piston-rod with ports cut into its wall. The piston-valves were concentric with the engine's pistons and were worked by an eccentric in the middle of the crank pin. The pistons were attached to the hollow piston-rod which rotated the crankshaft by means of 2 connecting-rods. This steam-engine was splash-lubricated and its speed was controlled by a spring-loaded centrifugal governor located at the opposite end of the crankshaft to the flywheel. In this single-acting steam-engine there was no reversal of the load on the bearings and, therefore, its operation was quiet. Willans' steam-engines were compounded so that the steam expanded from the high-pressure cylinder which was on top, down to the pressure in the condenser in 2, 3 or 4 stages. Willans introduced the triple and the quadruple expansion in his steam-engines in 1871 and 1875 respectively to achieve a better operating efficiency. Willans introduced 9 steam-engines the steam cylinders of which ranged from 5-inch (12.7 cm) bore at 750 rpm, to 20-inch (50.8 cm) bore at 350 rpm. All of his steam-engines could be erected as simple, or as compounded steam-engines with double-expansion or triple-expansion and incorporating single, double, or triple cranks. Willans was also a pioneer in efficient working procedures, and he standardised most of the parts of his steam-engines of different size. Therefore Willans' steam-engines required high-precision workmanship. It took a long time for steam-turbines to become more economical to operate than the central-valve steam-engines of Willans. Willans' first steam-engines were designed to drive directly the electric current generators of his friend, Rookes Evelyn Bell Crompton (1845-1940), a self-taught electrical engineer with whom he collaborated. It was the first time that a direct-current generator was coupled to a steam-engine to make both work as a power-unit. The steam-engines of Willans were so effective that only at the close of the 19th century the reciprocating steam-engine was being gradually superseded by the steam-turbine and the internal combustion engine.

Dr. Alban, Brown, Sulzer and Superheated Steam:

In 1870's, a German textile mill owner, Gustav Adolf Hirn (1815-1890), in Alsace, concluded after examining steam-engines for their thermal efficiency that their heat-loss would be considerably reduced if superheated steam were used as the fuel, since the cylinder condensation, the largest heat-loss in steam-engines, would then be drastically reduced. In 1866, the English engineer Charles Brown (1827-1905) built the first superheated steam-engine in collaboration with Heinrich Sulzer of the firm **Gebrüder Sulzer** (*Sulzer Brothers*) in Winterthur, Switzerland.

Superheating first suggested by Dr. Alban was at first used in locomotives, yet its greatest advantage was in stationary power plants using steam-turbines which by the mid-20th century operated with temperatures over 1000° Fahrenheit (540° Celsius) and steam-pressures over 3,000 psi

(20,685 kPa) which gave turbines a thermal efficiency that ranged from 30 % to 40 %. Superheating of steam produces the so-called 'dry steam' at temperatures far above the boiling temperature of water. Superheat is defined as an increment of heat to produce vapour at a higher temperature than the saturation temperature, which normally varies from 50°F to 300°F (10°C to 149°C). Superheaters installed in boilers are heat-transfer surfaces of tubular form in which saturated steam received at the inlet of the small diameter tubes is delivered as superheated or 'dry steam' at the outlet of tubes. Heat is received by the superheater from the combustion of fuels by direct heat radiation to the tubes, by convection of hot furnace gases flowing over the tube surfaces, or both. The less erosive character of 'dry steam', and the lower heat losses from pipes carrying 'dry steam' have made superheating very desirable and, therefore, superheaters are installed in most modern steam boilers.

Steam-Engine Boilers
Waggon-Boiler, Cornish Boiler, Multiple-Firetube Boiler, Lancaster Boiler and Scotch Boiler:

Boilers had to be capable of providing an adequate rate of steam at a proper pressure for the steam-engine. First boilers of Newcomen steam-engines were adaptations of the brewer's kettle which were made of copper. The furnace was made of brickwork constructed under and around the boiler so that flames and hot gases of the furnace swept around the boiler before entering the chimney. A little later, only the bottom of the boiler was made of copper, a very expensive metal, whereas the beehive top of the boiler was beaten out from lead sheets. Still later, by 1725, rivetted iron sheets were used in the place of lead.

The early boilers were very inefficient; it took a long time to heat the water to the temperature at which it turned into steam, and most of the heat applied went directly out of the chimney. Watt's boilers, which were externally fired, had the form of a water tank with a half-cylindrical top and a concave segmental cylindrical bottom.

By 1820, more steam was required by larger steam-engines and, therefore, the heating surfaces of the boilers had to be increased. For this reason, boilers were made long and boxlike with flat-sided flues and internal fireboxes. Such boilers were called 'waggon boilers'. However, these boilers were not stiff and they were unsatisfactory for steam-pressures higher than 5 psi (34.5 kPa) which is the pressure that Watt's steam-engine used. Therefore, 'waggon boilers' tended to leak. Both Evans and Trevithick realised that cylindrical boilers were much stiffer and deformed less than 'waggon boilers'.

After 1800, Evans built a 15-foot (4.57 m) long cylindrical boiler made of copper, which incorporated an internal firetube called a 'flue'. The heads of this cylindrical boiler were made of cast-iron and held together by longitudinal wrought-iron rods. Almost at the same time, Richard

Trevithick also made a cylindrical Cornish boiler, 6-foot (1.83 m) in diameter, with an inner flue of 3½-foot (1.07 m) diameter and two cast-iron boiler-heads bolted to the cylindrical shell. The cylindrical boiler with a flue soon became standard practice for steam-engines. The weak components of such boilers were the flat or rounded boiler-heads. Trevithick placed the fire on a grate inside the flue and, thereby, invented the Cornish boiler. Evans initially used the smaller cylinder for water with fire in the larger cylinder, but he soon adopted the same design as in the Cornish boiler.

George Stephenson (1781-1848) employed 2 flues in the boiler of one of his earlier locomotives, but in 'The Rocket', his son Robert (1803-1859) designed a multiple-firetube boiler and exhausted steam from the cylinders of the steam-engine through a pipe in the smoke-stack to increase the draught in the furnace, a method George Stephenson had learnt from Trevithick. Marc Séguin (1786-1875) in France designed a similar multiple-firetube boiler at the same time for his locomotive which had a separate blower to force hot gases through the firetubes.

In 1845, William Fairbairn used in the construction of his 'Lancashire boiler' the double flue of George Stephenson and also double furnaces at both ends of the boiler, which were fired alternately to prevent formation of smoke. The 'Lancaster boiler' was in continuous use until 1900.

American Nathan Read (1759-1849) invented the multi-watertube boiler in 1789. John Fitch (1743-1798), James Rumsey (1743-1792) and John Stevens (1749-1838) invented multi-watertube boilers by realising that the customary contemporary 'waggon boiler' used by Watt was incapable of producing sufficient quantities of steam for marine propulsion. In 1828, Goldsworthy Gurney (1793-1875) also came to the same conclusion in the design of his steam-powered road-carriages. The multi-watertube boilers were not only difficult to clean and expensive to construct, but the watertubes easily clogged up with deposits causing the boiler tubes to burn-out in the absence of water circulation. Such multi-watertube boilers became really practical only in the late 19th century when steel replaced iron as construction material, and the technological means in general had appreciably improved.

The traditional use of sea water in marine boilers created an acute corrosion problem in the watertubes. Expensive copper tubes were often used to resist corrosion caused by salt water. This problem was finally overcome by the invention of the surface-condenser by Samuel Hall (1781-1863) in 1834, which supplied distilled water to the boiler in a continuous operation. After 1835, firetube marine boilers were also introduced by following the practice in locomotive engineering. The firetube marine boilers were of the 'box-type'.

In 1840 Dr. Ernst Alban, a German eye surgeon, built the first reliable ultrahigh-pressure steam-tube boiler which was capable of attaining pressures in the range from 50 to 80 atmospheres. Alban built his furnace with multiple alternating layers of smaller and larger watertubes in which water flowed in the opposite direction to the movement of the furnace gases, much like the watertubes in the non-radiative part of a modern furnace boiler. Water entered the lower tubes first, then the larger tubes on top, and so forth. This design was excellent for conductive heat-transfer, but less so for radiative heat-transfer. Alban did not rely on relief valves for safety, but designed his boiler to survive the rupture of boiler tubes.

For compound steam-engines, which used boiler pressures of 60 psi (414 kPa) or higher, stronger cylindrical boilers and cylindrical flues were introduced. The well-known 'Scotch marine boiler' was first built in 1862 which had a cylindrical shell and furnaces, and internal firetubes.

The most famous and successful watertube boiler is that of Babcock and Wilcox, designed by Stephen Wilcox (1830-1893) and patented in the United States in 1867. This boiler consisted of a staggered set of inclined watertubes, a design first used by the versatile German engineer Carl Anton Henschel (1780-1861) in 1837, in which the water rose up by natural convection and by steam bubbles in the water due to the heating. The water was separated from the steam in the top drum and returned to the lower end of the inclined tubes to rise again. This boiler produced steam much faster because the quantity of water in it was smaller than in other boilers. It was also called a safe boiler for an explosion of an internal watertube was never a catastrophe in Babcock-Wilcox boiler which was cylindrical in shape. Henschel designed and built his explosion-proof high-pressure boiler, good from 6 to 10 atmospheres overpressure in 1837, for which he was awarded in 1844 a gold medal and 6,000 Francs first prize by the *Société d'Encouragement pour l'Industrie National* in Paris. Many other types of watertube boilers were invented and used in practice.

Boiler-Feed Pump of Worthington and Steam Injector of Giffard

In steam-engines the boiler had to be supplied by water at the proper rate, a problem that had presented great difficulties ever since the invention of the steam-engine. If a pump was attached to the steam-engine itself, it ceased to operate when the steam-engine was shut off, and therefore, it became necessary to resort to manual pumping.

The first person to solve this difficult problem was an American mechanical engineer, Henry Rossiter Worthington (1817-1880) who invented a small, simple, and portable steam-pump which effectively solved the boiler-feeding problem. It was an independently operated pump of novel design having opposing cylinders with pistons connected to a common piston-rod. Steam admitted to the larger cylinder moved the piston of the pump cylinder by direct action, which is the reason why it was called a direct-acting steam-pump for boiler-feeding. Automatic reversing of the steam-valve was at first accomplished by means of various kinds of tappets and springs, but later they were replaced by steam-thrown valves and a boiler-float controlling the admission of steam to the pump, which made the operation of the pump completely automatic. Worthington's pump soon found other applications in industry.

In 1858, Worthington's boiler-feed pump was replaced by a more ingenious device, the steam injector, invented by a versatile, self-educated French engineer and aeronaut, Henri Giffard (1825-1882). It was a simple but ingenious device which used the moving power of the steam jet from the boiler to inject the feed water at the proper rate into the same boiler. It consisted basically of an ingenious arrangement of three nozzles, which were convergent for subsonic flow of steam. This steam injector was theoretically conceived by Giffard by using fluid dynamics and thermodynamics available at the time. It was generally adopted almost immediately after its invention. In 1869 divergent nozzles were incorporated into the steam-injector for supersonic flows of the steam by a German engineer Schau. Giffard was awarded prizes and honours for this invention in 1859, 1863, 1867 and 1876. He was the first to install a compact steam-engine in a lighter-than-air steam-powered dirigible of his own design. It was a 3-horsepower steam-engine weighing 99 pounds (45 kg). Income from the steam-injector patent made Giffard a millionaire.

CHAPTER XIV

BRITISH ENGINEERING: Industrial Technology and Engineering

British engineers were self-taught engineers who relied on their native talent and apprenticeship. It is no exaggeration to regard the Yorkshire engineer, John Smeaton (1724-1792), a well-educated son of a lawyer who became through self-study and his apprenticeship to the ingenious clockmaker and instrument-maker Henry Hindley (1710-1771), the most prominent engineer of his time, and the founder of British engineering as a profession. Smeaton acquired his rational engineering knowledge from his careful study of the influential French engineering books written by Bernard Forest de Belidor (1693-1761), and his practical engineering background from his firsthand examination of important engineering works during his visits to the Low Countries. Smeaton surpassed his French colleagues in his efforts to replace the old intuitive ***modus operandi*** of skilled craftsmen with more precise scientific procedures through the use of systematic experimental methods in engineering, and by demonstrating how such systematic scientific experiments can throw light on engineering problems. He was eminently successful in combining the practical skill of craftsmen and the empirical knowledge of his contemporary engineers with the technique of precise measurement that was an essential skill of instrument-makers, and the experimental method of British scientists.

Smeaton's systematic experimental studies were largely responsible for the general adoption of rational methods in practical engineering, and ushering in a new era of experimental and laboratory testing in engineering. Smeaton asserted that experiment was the only way of coming at the truth in mechanical enquiry. Smeaton, who became the leading exponent of the scientific experimental method in engineering, was a pioneer. Even the more scientifically trained French engineers found it necessary to follow Smeaton's rational approach to engineering problems by using scientific experiments and statistical methods to find the most efficient engineering design criteria, and to translate Smeaton's memoirs on experimental investigations in engineering into French. Many years later, the American self-taught genius Oliver Evans (1755-1819) considered Smeaton the only author in engineering who joined practice and experience with theory. This high praise of Smeaton was entirely justified, for Smeaton made a determined effort to improve anything that he undertook to build. Therefore, it is not surprising to learn that beginning with Smeaton engineering became a respected profession in Britain. Smeaton soon became a consulting 'civil engineer', a title which he coined for himself, meaning an engineer who did all kinds of engineering work that was not military engineering.

Any significant engineering project which was of public importance required the passage of a bill by the British Parliament. The passage of any such bill required the presence of the engineer responsible for the project in order to defend the bill against any opposition on the floor of the Parliament. Therefore, many prominent engineers came to London when the Parliament was in session. In 1771, Smeaton and 5 other eminent consulting engineers agreed to form themselves into an informal society which met at King's Head Tavern in London once in a fortnight whilst the Parliament was in session for the express purpose to discuss engineering matters. This first private engineering society in Britain, the Society of Civil Engineers, became popularly known as the 'Smeatonian Society', the guiding maxim of which was adopted from the wisdom of King Solomon in < Apocrypha > : "All things in measure, number and weight."

Smeaton's first important engineering project was the building of the celebrated Eddystone Lighthouse on the Plymouth Rock located in a rough sea 14 miles (22.5 km) off the Plymouth coast. British scientific engineering may be said to begin with this structure. Smeaton designed the lighthouse in the form of a bole of an oak tree, and constructed it with interlocking and fitted granite and limestone blocks. In this construction Smeaton carried out an extensive series of experimental tests on samples of limestone and sand obtained from all parts of England, and mixed lime, clays, lias, and volcanic trass in various proportions into mortars which were tested for their setting and strength. These experiments of Smeaton, in which every variable affecting the cement was investigated, are classics in the history of cement, and resulted in the production of the first completely successful hydraulic cement since Roman times. He concluded that a cement capable of setting under water must consist of limestone containing an appreciable amount of clay. With this series of experiments, Smeaton became a pioneer of 'factorial experimentation', which is fundamental to multivariable technological problems. The construction of the Eddystone Lighthouse, a structure which gave a great boost to Smeaton's professional reputation, began in 1756 and was finished in 1759. Smeaton's Eddystone Lighthouse was replaced by a new lighthouse built by James Douglass from 1878 to 1882, because by that time, the rock on which the lighthouse stood had been undermined by the sea.

Smeaton's resident engineer in the construction of the Eddystone Lighthouse was a millwright, Josiah Jessop. After Jessop's death, Smeaton took charge of the education of Jessop's son, William Jessop (1745-1814), and he trained the young lad for 10 years as his engineering apprentice and assistant. William Jessop became the most prominent British engineer of his generation, and through his personal influence upon younger engineers such as Rennie and Telford, Jessop became the most important link between Smeaton and the subsequent generations of British professional engineers.

Smeaton did every kind of engineering required in his time. He collaborated with the millwright James Brindley in the planning of the Grand Trunk Canal, and designed and constructed the Forth and Clyde Canal in 1768 with his resident engineer Robert Mackell. He designed the bridges at Hexham (1756), Perth, Banff, Coldstream (1763) and Edinburgh, and cast the first iron beams. The Perth bridge is 900 feet (275 m) long, and it crossed the Tay river at Perth on 7 arches. Hexham bridge was destroyed in a great flood in 1782, but was rebuilt as a replica of Smeaton's bridge by Robert Mylne on a new site with the original masonry. The other bridges of Smeaton are still standing. Smeaton built the road from Markham to Newmark in the valley of Trent in 1768, an event which raised many eyebrows since road building in Britain before 1800 was considered to be below the calling of an engineer.

At the same time, Smeaton was involved in designing canals, highways, harbours and other civil engineering works. In 1774, Smeaton was appointed engineer of the Ramsgate harbour, his last major work, in the design of which he created the first well-functioning artificial harbour that was free of silt and safe. The Ramsgate harbour was formed by 2 piers projecting 1,300 feet (395 m) into the sea leaving a 200-foot (60 m) opening to the harbour. He used the tide and sluices on the inner basin to flush the silt out of the main harbour into the sea at the ebb.

The Swedish engineer Kristofer Polhammer [after knighthood, Christopher Polhem] (1661-1751) had performed the first model experiments to measure the efficiency of waterwheels about 1700, but his results were not very reliable because he failed to take the 'scale factor' completely into account. In 1752, Smeaton carried out the first reliable model experiments on waterwheels to test the theoretical conclusions on the efficiency of waterwheels given by the French theoretical engineer Antoine Parent (1666-1716) in 1704. Smeaton demonstrated that the 'undershot' waterwheel could develop much more power than Parent's theory predicted. He also investigated the 'overshot' waterwheels, and proved the 'overshot' waterwheels to be about twice as efficient as the 'undershot' type. Smeaton also established experimentally the optimum speed of rotation for the 'undershot' waterwheel for its best performance. He did not investigate the efficiency of the 'breastfed' waterwheels.

In 1759, The Royal Society in London sent Smeaton to Holland and Flanders to examine the engineering of Dutch windmills. Subsequently, he made the multisailed windmill popular, and constructed an influential 5-sail windmill at Leeds. Smeaton published his celebrated memoir on his experimental enquiry on waterwheels, windmills and other power machines in 1759, which made him famous worldwide. His memoir was translated into French by Pierre Simon Girard (1765-1836), and it exerted great influence upon the French engineering practice. Smeaton's memoir was reprinted in 1760, 1796, 1814 and 1826 which indicates its immense professional importance. Smeaton built many industrial waterwheels guided by his experimental results, the largest one of which was his undershot waterwheel for the London Bridge Waterworks built in 1768. It was 32 feet (9.75 m) in diameter and 15 feet (4.57 m) wide. Wherever possible, he preferred the 'overshot' to the 'undershot' waterwheel.

Smeaton was also an expert mechanical engineer: he designed boring mills, blowers for smelters, and atmospheric pumping engines. Smeaton analysed the performance of 15 Newcomen atmospheric pumping engines in great detail, since the Newcomen atmospheric pumping engine he had designed for the New River Company in London had failed to reach the expected performance. He carried out a great number of scientifically controlled experiments over a 4-year period on a small steam-engine which he had built for experimental purposes on the grounds of his property. On the basis of his experimental investigation, in which he changed one variable at a time that was possible in his equipment, Smeaton was able to build Newcomen-Smeaton atmospheric pumping engines which had more than twice the efficiency of the conventional Newcomen engine, about 1.5% compared to 0.6% of the conventional Newcomen engine. Smeaton improved the accuracy of the steam cylinders by designing a boring mill in the Carron Iron Foundry which produced a true circular bore of the cylinder to fit the piston, although it did not assure a parallel bore which prevents seizing of the piston in the cylinder.

Sometime between 1768 and 1771, when Richard Reynolds of the Coalbrookdale Iron Works replaced the wooden rails of his colliery tramway to Ketley with cast-iron rails which had inner flanges, Smeaton suggested to Reynolds to transfer the flange from the rail to the wheel, a design concept which became universal in the railroad era. He was the first to use cast-iron girders in buildings when he supported a factory floor on such girders in 1755.

After the retirement of Smeaton in 1787, William Jessop became the foremost British engineer of his generation. Jessop designed canals, harbours, horse-drawn colliery railways, and was the first promoter of horse-driven railways. He invented the type of rail cross-section still used today, and the flanged wheels for the horse-driven railroad at Loughborough in 1789. His most famous engineering works are the Grand Canal of Ireland, the West-India Docks in London, and the Bristol Docks. Jessop helped to shape the professional development of 2 outstanding Scottish engineers, John Rennie (1761-1821), whom he advised, and Thomas Telford (1757-1835) with whom he collaborated in the design of the Ellesmere Canal in 1793. Both Scots became the leading British engineers of their generation. Jessop actively assisted younger men in the profession, never begrudging them the fame they won or sought to claim. Success never went to his head, never made him arrogant or pompous, a common fault of many other prominent professional men of his time.

The 6 ½-foot tall Scottish giant, John Rennie, was educated by John Robinson (1739-1805), who was the first professor to institute lectures for engineering students at the University of Edinburgh. Rennie was unique among British engineers of the period in that he attempted to fuse theory of mechanics with his engineering practice. He had studied the treatises of Bernard Forest de Belidor, the memoirs of Antoine Parent, of Johann Heinrich Lambert (1728-1777) and of Smeaton, and he held theoretical knowledge in high regard. Though Rennie is mostly remembered for his bridges, canals and harbours, he was basically a mechanical engineer who was active both in the design and in the manufacture of machinery throughout his career. In his youth, Rennie had apprenticed in the workshop of the very able Scottish millwright Andrew Meikle (1719-1811), a training which had made Rennie an expert practical mechanic. Rennie's first important engineering work was in the famous Albion Flour Mill at Blackfriars, the first steam-powered flour mill in the world, where he demonstrated his all-round abilities. Rennie was hired by James Watt (1736-1819) and Matthew Boulton (1728-1809) to design and erect the millwork in the Albion Mill between 1784 and 1788, and to install 2 Boulton & Watt steam-engines. In the design of Albion Mill, probably on the advice of Rennie, modern ideas of foundation design were used: the soil was excavated to the level where the soil weight equaled the weight of the building to avoid excessive settlement of the building. In the Albion Mill, Rennie designed and erected many centrifugal governors for the automatic control of the flour mills, which inspired Boulton to advise Watt to use a centrifugal governor as the automatic means for the control of the quantity of steam injected into Watt's steam-engine. Although most of the machinery in the Albion Flour Mill were made of iron, there were also some cogwheels with cogs made of hard wood. Rennie suspected that an overheated wood bearing may have caused the fire in 1791 which totally destroyed this technologically important flour mill, since it was the first mill in which power from 2 central steam-engines was exclusively transmitted to the machinery by a transmission mechanism made almost entirely of iron. The shafts and axles of Rennie's power transmission mechanism were made of iron but the bearings were of brass.

During the construction of London Docks in 1801, Rennie began to use the steam-engine for the operation of various mechanical devices such as pile-drivers, water-pumps, and so forth. He built an enormous mile (1.61 km) long breakwater at the entrance to the Plymouth Sound for the protection of ships from the southwesterly winds. 3½ million tons of stone were used to construct this immense jetty.

Rennie expressed his belief that the professional ethics in engineering required the engineer to keep aloof of the financing of engineering projects under his direction when he stated: "Engineers should be entirely independent of these connexions – not dabblers in shares and fee alike of contractors and contracts." John Rennie was the first British engineer to assume responsibility for all aspects of an engineering project.

Rennie designed and built the 3 famous bridges in London: the London Bridge, the Waterloo Bridge, and the Southwark Bridge, which made him famous. He was not only an engineer but also an architect for architecture was an organic component, and not only an ornament of his structures.

If Rennie had any shortcoming in his engineering, it was in his design of canals: they were peculiarly deficient in water supply.

John Rennie, a man with an exceptional sense of personal worth and self-esteem, refused to accept the knighthood which was offered to him by the Prime Minister of Britain after his successful completion of the Waterloo Bridge. Rennie complained that he had a "devil of a time to escape this 'honour'."

A contemporary engineer of Rennie was Thomas Telford (1757-1835), who deservedly can be considered one of the greatest civil engineers of all time. Telford, another tall Scot, was a self-made man who rose to engineering prominence the hard way from being a poor sheepherder in his youth and later a stonemason. Telford's early ambition was to make a name for himself as an architect and a literary man, and a great deal of his self-education was devoted to this end. He studied chemistry, general science, and other subjects only so far as they could be applied to the art of building. He became a cultured man of letters and his artistic temperament was always apparent in his engineering works. According to his artistic tastes, Telford remained always a little contemptuous of theory, particularly any mathematical theory, an attitude in which he was quite a contrast to Rennie, and he remained adamant in regarding practical training of young engineers to be an absolute necessity. In fact, Rennie and Telford soon were at professional odds with each other, and it appears that Rennie resented Telford, perhaps because Rennie considered Telford an 'upstart'. He seems to have resented Telford's surprising success in engineering which had begun to rival that of Rennie, who was considered at the time the most prominent engineer in Britain. Perhaps, Telford's rapid rise in professional importance aroused Rennie's ire because the growing engineering reputation of Telford began to threaten Rennie's preeminence in engineering. Shortly before Rennie's death, the 2 Scottish engineers were locked in a keen rivalry for several important commissions in British engineering. It is reasonable to assume that Rennie particularly resented the appointment of Telford as the engineer to the Commissioners for Highland Roads and Bridges despite the fact that Rennie considered road building below the dignity of a professional engineer of his stature. However, many bridges had to be built on the Highland roads, and he most certainly thought that he had a much better claim for the job of a bridge builder than Telford owing to Rennie's considerably more extensive experience and technical background in bridge construction.

Telford built 920 miles (1,480 km) of new roads during the 18 years he spent in the Highlands, the most famous road among them was the Holyhead road. The method of construction of the really hard-surfaced Telford road was quite similar to the practical and economical method of road construction of the French engineer Pierre Trésaguet (1716-1796) who first proposed such a road surface in 1775. He also rebuilt 280 miles (451 km) of old roads. Among the pioneer road builders in Great Britain was a former Scottish horsedealer, John Metcalfe (1717-1810), who had been blind since the age of 6, yet he built 180 miles (290 km) of turnpike roads in Yorkshire. Metcalfe laid a foundation of large stones, covered with road material having a cambered surface and drained the surface water into large ditches at both sides of the road. In road building Metcalfe preceded Telford.

Another contemporary Scot of Telford, John Louden MacAdam (1756-1836), who had made a small fortune in America as a businessman before his return to Britain, became an expert in the economical construction of roads. MacAdam invented the first inexpensive road surface by eliminating the expensive Telford foundation for roads and letting the native soil carry the traffic. He claimed that the native soil in a dry state will carry any weight of traffic without sinking. He relied on adequate drainage and adequately cambered top surface of the road, and very importantly, on the impervious wearing surface. His road surface consisted of 3 cylindrical layers of 2-inch (5 cm) stones the top surface of which was ground by the iron-tyred wheels of the vehicles into small chips and packed by the traffic into the surface of the road, which made the road surface hard and impervious. It was called 'traffic-bound macadam road surface'. MacAdam was a prolific writer, and mainly through the influence of his writings by 1830 Britain alone had 160,000 miles (257,488 km) of 'macadam' roads, and by 1900 about 90% of the roads in Europe had been 'macadamised'.

Altogether, Telford built 1,117 bridges, the most famous of his bridges is the celebrated suspension bridge built from 1819 to 1826 which carries the Holyhead road over the Menai Straits between the Holyhead Island and Wales, located about 1½ miles (2.4 km) southwest of Bangor. This bridge which is still in service, has a 579-foot (176.48 m) span. Its unstiffened timber deck was wrecked by a storm in 1839, and replaced by a heavier timber deck that lasted till 1893 when Benjamin Baker (1840-1907) reconstructed it in steel, and its original wrought-iron chains were replaced by high-tension steel chains in 1940. In 1826, Telford built the Conway Castle suspension bridge in Wales which has a span of 417 feet (127 m), and is also still in service.

Telford, moreover, engineered numerous important canals including the 130-mile (209 km) Ellesmere Canal which he designed in collaboration with William Jessop, the Caledonia Canal, and the famous Göta Canal between Stockholm and Gothenburg in Sweden, all still in service today.

Telford had about 120 engineering contracts over a span of 18 years, which indicates the great scope of his engineering activity. He had a tendency for the practical, and was zealous in his duties. Telford was a very shrewd, mechanically minded, skilled planner and organiser of engineering works. However, since Telford was an architectural engineer, his work was not merely a product of cold calculations and rigid formulae, but also incorporated his æsthetic taste in his designs.

Telford, who was a proud Scot, said that he wanted to be remembered for his own works produced by his own efforts, and not by 'honours' conferred on him by others. Unfortunately, he was too late to decline the knighthood conferred on him by the Swedish monarch after he had completed the construction of the Göta Canal.

Telford's motto was: "Stirring the lake to prevent it from stagnating," a professional attitude he maintained in all of his engineering activities. Some of his bolder engineering proposals did not materialise, such as the huge iron bridge spanning 600 feet (183 m) across the Thames river in the heart of London, but he took great pride in his own idiosyncracies: "I am a queer creature, and am not ashamed of being thought singular." Despite Telford's great talent as an engineer and organiser, he had a professional shortcoming typical of his time: his cost estimates for his engineering projects were often grossly underestimated. For instance, his estimate for the cost of the Göta Canal was several times too low, so much so that the Swedish government, because of the unforeseen excessive cost of this canal, incurred serious financial difficulties that delayed the development of railroads in that country for several decades.

Jessop, Rennie and Telford worked in the so-called 'Canal Era', which lasted from 1760 to 1830, an era which had begun with the Worseley Canal from Worseley to Manchester built by James Brindley (1716-1772). Brindley had collaborated with Smeaton, which makes the professional succession from Brindley to Smeaton to Jessop to Rennie to Telford a direct line of apprenticeship. It was Telford who established in Britain the professional engineering system consisting of the chief engineer, the resident engineer, and the contractor for an efficient execution of any engineering project. Telford's assistant engineers, such as John B. MacNeill (1793-1880), who became professor of engineering at Trinity College in Dublin, William Cubitt (1785-1861), who became a prominent railroad engineer and harbour builder, and James Meadows Rendel (1799-1856), who became the most prominent harbour engineer of his time, were the leaders of the next generation of British engineers, and so it continued and spread. Rennie's son, John Rennie Jr. (1794-1874) became a prominent engineer who collaborated with Telford on a number of engineering projects. The assistant engineer to John Rennie Jr., Charles Blacker Vignoles (1793-1875), became the chief engineer of Irish railroads, the designer of the notable suspension bridge at Kiev in Ukraine, and, in 1844, the first professor of engineering in England. Rennie's most famous employee was William Fairbairn (1789-1874), who became a world-famous engineer and industrialist. Fairbairn was elected a Fellow of The Royal Society in London, and a foreign member of the French Academy of Sciences – a rare distinction for a British engineer.

In great contrast to French professional engineering which was acquired through formal education, in Britain engineering was learnt essentially through self-study and apprenticeship. The native talents in British engineering, beginning with Smeaton, soon snatched the initiative and leadership in engineering from the French professional engineering establishment.

Telford and his works were soon forgotten after his death, when England became 'railway mad'. Rennie's sons continued their father's enmity towards Telford by trying to diminish Telford's stature as a leading engineer by promoting Smeaton as the 'Father of civil engineering' in Britain instead of Telford.

The first canals which were built quite inexpensively in the 1760's, earned such large dividends for their owners that they created in the period between 1790 and 1794 a 'canal mania' of promotion and speculation. The new canal companies had to build their canals in a period of rapid inflation brought on by the speculation and the Napoleonic Wars and, therefore, they never had the financial success of their predecessors.

George Stephenson's railroad ended the 'Canal Era' in 1830, and gave rise to a new kind of engineer, the railroad engineer, who was at the same time a civil as well as a mechanical engineer. Simultaneously, the inventive craftsmen made themselves into 'makers of machine-tools', engineering specialists who were absolutely essential to the 'industrial revolution'.

Technological and Scientific Revolution

The so-called 'low technology' includes all the crafts that men of every culture have used to build their civilisation such as to construct houses, roads, water supplies, to make clothes and pottery, and to grow food.

The so-called 'high technology' consists of particularly sophisticated crafts and manufactures, which in some way are intimately associated with sciences by means of the use of scientific theories, instruments and techniques to increase the knowledge and technical competence of the craftsman. It is the development of the 'high technology' begun by the ancient Greeks that primarily distinguishes Western civilisation from all other civilisations, past or present.

In the middle of 17th century, the so-called 'scientific revolution' which spread out of Florence, Paris and London had a profound influence upon the technological practitioners of science: the instrument-makers, the teachers of navigation and surveying, and the authors of books on the useful scientific crafts. They influenced a large group of amateurs of science who were the customers of the technological practitioners of science. As a consequence of this association, the artisans made every effort to popularise and promote the use of mathematical and other scientific instruments in technology.

The important movement of industrialisation taking place in England, the so-called 'industrial revolution' which brought about a drastic change in the world economy, had its origin partly in the making of fine machines, and partly in the crafts of the millwrights, the wheelwrights, and the mining engineers. The most dramatic changes that particularly made this revolution possible came in the ingenious new designs based on the crafts of the instrument-makers and clockmakers.

Ever since the Mediæval Age the clockmakers had maintained in their craft skills and qualities of artisanship that were well-balanced between sciences and crafts, and which proved to be of crucial importance to the 'industrial revolution' and to the Modern Age of 'high technology'. In its widest perspective, the craft tradition of clockmakers led to such diverse technological designs as the magnetic compasses, calculating machines and computers, as well as automata and robots.

CHAPTER XV

TOOLS OF INDUSTRY: From Manual to Machine Tools

Machine Toolmakers

It was quite evident to most shrewd observers at the time that it was impossible to sustain a brisk and rapid industrial growth and progress without machine tools. The construction of machines in the Renaissance Age had been by tradition the responsibility of millwrights. Millwrights used in their construction of machines simple measuring devices such as foot-rules, squares, compasses and calipers, and hand-and-foot-driven lathes, but these tools were not adaptable to heavy metalwork. Even as late as when James Watt (1736-1819) was active, there existed no large-scale power-tools in Boulton & Watt's Soho Works at Birmingham.

French mechanic Charles Plumier stressed in his monograph, < The Art of Turning >(*L'art de tourner*), in 1701, that the lathe has to be solidly constructed, and be sufficiently stiff to offer resistance to large forces acting between the machine tool and the workpiece. In the 1770's, the ingenious French mechanic and maker of automatic robots, Jacques de Vaucanson (1709-1782), constructed a metal lathe on an iron bed made of two 1½ x 1½ inch (3.8 x 3.8 cm) square bars set on edge so that the carriage slid on two faces inclined at 45 degrees, thereby providing a much greater degree of precision than any previous lathe. A brass saddle casting had a screw-operated cross-slide. A square-threaded screw fitted to make the saddle traverse the length of the bedways was an improvement upon previous lathes. The entire machine tool was braced by a framework of bolted square iron bars, which rendered the machine very stable and stiff. This lathe had a sliding tool-carriage advanced by a long lead-screw which ran parallel with the workpiece. The sliding tool-carriage, called the 'slide-rest', had been used by German craftsmen as early as about 1480. Vaucanson also built a horizontal drilling-machine in which the positioning, and the advance of the drilling bit were controlled quite like in the modern drill-press. Vaucanson's lathe served as an exemplary model for the further development of machine tools over the following decades.

Another lathe made in 1795 by a French mechanic Senot had besides the sliding tool-carriage a system of gears, the so-called 'change-gears', which connected the spindle of the lathe with the lead-screw, and thereby controlled the rate of advance of the tool-carriage in a way that facilitated the cutting of helical screws. David Wilkinson in United States had built similar screw-cutting lathes in 1798 but his lathes were only capable of producing screws of the same pitch as the lead-screw. The use of the lead-screw in combination with the change of wheels in a screw-cutting lathe had already been proposed by Leonardo da Vinci about 1500. The lead-screw had subsequently been used in the *fusee* engine, a special clockmaker's tool, and described in 1741 by the French clockmaker, Antoine Thiot.

One of the important contributions of the British industry to the progress of European civilisation was the creation of the machine tool industry, which was essential to rapid industrial growth.

A friend of the brilliant young English astronomer Jeremiah Horrox (1619-1641), the English astronomer William Cascoigne (c.1612-1644) who died in the Battle of Marston Moor, had used the optical theory of the German astronomer and mathematician Johann Kepler (1571-1630) in the optical design of his astronomical telescope which used 2 convex lenses that produced a real image in the ocular where measuring wires could be used for reference. In 1637, Cascoigne built the astronomical telescope of Kepler that incorporated a screw-operated filar micrometer of his own design as a component of the eyepiece in the telescope which made possible the precise measurement of angles as small as a few seconds of an arc.

The English instrument-maker Henry Hindley (1710-1771) invented in 1741 a gear-cutting machine for clocks which incorporated a system of differential indexing about 150 years before the same feature appeared in milling machines. The English scientific instrument-maker, Jessie Ramsden (1735-1800), invented a screw-cutting lathe with a lead-screw and gearing about 1770 which facilitated the mass-production of standardised screws, and his linear ruling engine which made the manufacture of Vernier scales for the first time rapid and inexpensive. The machine tool industry on a large scale began in earnest with John Wilkinson (1728-1808), who patented in 1774 his boring mill with an automatic guide and a hollow boring bar. Unfortunately, Wilkinson's boring mill was similar to the boring mill already installed in the Woolwich Arsenal in 1771, and therefore, his patent was later revoked.

However, the first important British machine toolmaker was Joseph Bramah (1748-1814), a Yorkshire farmer's son who invented an improved water-closet in 1778, a type of patent tumbler-lock in 1784 that is still manufactured, a screw-propeller in 1785, a rotary gear pump (known since 1636) in 1790, a hydraulic press and a fluid-powered control system in 1796, a method to extrude lead pipes that was further improved by a plumber Thomas Burr in 1820, an automatic device for printing the numbers on banknotes, a suction beer-pump, a crude form of fountain pen the patent for which he bought from Bryan Donkin (1768-1855), a spring-winding machine for the manufacture of springs of varying pitch that was a forerunner of the screw-cutting machine made by his assistant Maudslay, and many other mechanical devices. Bramah, who had altogether 18 patents, set new standards of craftsmanship when precision in manufacture became important in mechanical engineering. In 1802, Bramah invented a planing machine for wood with a rotating cutter consisting of a disk fitted with gouges and planing irons turned by a steam-engine at 90 rpm which functioned with an unprecedented precision. This wood-planer had 28 cutting tools. The wood in this planer was moved back-and-forth on carriages by means of an endless chain, and 40-foot (12.19 m) long rails above the planer held the turning disk fitted with cutting tools. The axis of the principal shaft of Bramah's planing machine for wood was supported by a piston set in a vessel of oil, a pioneer hydraulic bearing which considerably reduced friction. Bramah's planing machine for wood remained in use in the Woolwich Arsenal for half a century. The first known planing machine for metal had been built in 1751 by the French mechanic Nicolas Focq for the planing of the pump barrels in the Marly pumping plant, but it was a relatively small machine. Bramah was a pioneer designer of machine tools, and he can be regarded as the true progenitor of British machine tool industry.

In 1812, Joseph Bramah proposed, but never carried out in practice, a system of hydraulic 'mains' in every street of a town which could be used to operate hydraulic machinery and to fight fires. This hydraulic system of pipes incorporated an accumulator to store up the hydraulic power in order to balance the supply of, and the demand for power. This principle was first put into practice by William George Armstrong (1810-1900), a pioneer builder of hydraulic machinery who later became a prominent armament manufacturer, with his installation of hydraulic cranes on the Tyne Quays at Newcastle. At first, these cranes were powered by the pressure of the Newcastle water mains, but by 1858, there were 1,200 hydraulic cranes in operation which were supplied with power from weight-loaded accumulator systems (invented by Armstrong in 1851) provided by 125 steam-engines producing a total of 3,000-horsepower. The simplest accumulator consisted of a vertical cylinder containing a weighted piston, the weight and size of the piston was proportioned to equal the hydraulic pressure in the water main. The accumulator afforded to operate hydraulic cranes, jacks and lifts, and other devices even when the pumps were not active, and to maintain a constant pressure supplied to the hydraulic machinery even when the pumps were working. Armstrong was also a pioneer builder of hydraulic motors which he invented in 1838. The swing-bridge carrying the road over the Tyne at Newcastle is still operated by the hydraulic motor with an oscillating cylinder of Armstrong's design.

For many years, Bramah offered a prize of 500 British pound sterlings for any person who can pick his patent tumbler-lock. Only 67 years later, 37 years after Bramah's death, was Alfred C. Hobbes, an American locksmith and salesman for Day & Newell Bank Lock, able to pick Bramah's lock after 51 hours of intensive work, and after finding it necessary to make a number of tools to accomplish this task. Unfortunately, Bramah's complex design of his tumbler-lock made its mass-production impossible.

Bramah's assistant, Henry Maudslay (1771-1831), a young blacksmith from the Woolwich Arsenal, helped Bramah in the design and building of machinery for the manufacture of his complex tumbler-lock. Maudslay brought the accuracy of machining to a new precision, to 0.0001 of an inch by means of his bench-micrometer invented in 1805, and he became the founder of the British precision machine tool industry. In 1797, Maudslay improved the slide-rest lathe by fitting it with an accurate lead-screw and standardised interchangeable gear-wheels. Maudslay's lathe was the first industrial lathe with a cross-feed. He invented an improved screw-cutting lathe in 1800, and developed the standard range of screws with respective taps and dies, and interchangeable bolts and nuts. Maudslay as Bramah's young assistant was probably responsible for the introduction of the self-tightening leather-cup washer, first conceived by Samuel Morland (1625-1695) and later used by John Smeaton (1724-1792), for Bramah's hydraulic press without which the operation of the press would have been neither reliable nor satisfactory because it leaked. Later Maudslay invented a punching-machine for boiler plates.

Maudslay's signal contribution to machine tool industry was making crude machines highly accurate, and producing perfect screws. Beginning with Maudslay, precision mechanics began to enter the manufacture of machine tools.

Maudslay built the models of the 44 blockmaking machines designed by the French engineer Marc Isambard Brunel (1769-1849) which Brunel used to convince the British Admiralty that his ideas and his concept for the mass-production of blocks used in naval rigging were sound, and such a full-scale plant was ordered by the British Admiralty for the Plymouth Dockyard with the active support of Samuel Bentham (1767-1831), the Inspector General of British Naval Works. Maudslay launched his own machine-building workshop with the contract to manufacture the 44 single-purpose process machines for the mass-production of blocks of naval rigging for the Portsmouth Dockyard, which took him 5½ years to complete. By 1809, all 44 mass-production machines were in full operation in the Plymouth Dockyard.

In 1810, Maudslay founded his engineering works in Westminster Road, Lamberth, which soon began to surpass in reputation the Soho Works of Boulton & Watt and Murray's Round Foundry. Since Maudslay was also a talented steam-engine designer and constructor, his company Maudslay, Sons & Field, founded together with his partner Joshua Field (1787-1863), a former draughtsman of the Portsmouth Dockyard, became an important manufacturer of marine steam-engines. In 1815 he designed a compact table steam-engine which drove the crank directly from the crosshead. This steam-engine was used in factories until electric motors forced it out of the market by 1900. Maudslay was a pioneer builder of marine steam-engines of about 400-horsepower which worked by the side-lever principle of Murray that gave this steam-engine greater stability than the conventional marine steam-engines.

Many of the important second generation machine tool builders in Britain had at one time been working under Wilkinson, or Bramah, or Maudslay. Most of them had their apprenticeship under Maudslay. Matthew Murray (1765-1826), who was the first and the most formidable rival of Boulton & Watt in the business of building steam-engines after the expiry of Watt's patent in 1800, designed and constructed the first steam-engine operating on the side-lever principle, an engine which saved in head-room. He also improved the hydraulic press, and became a pioneer builder of the hydraulic press for tensile testing when in 1814, he proposed to the Royal Navy Board the acquisition of a 1000-ton hydraulic machine for the tensile testing of chain-cables, a machine he finally built in 1826. Murray's testing machine had a 16-inch (0.41 m) ram and a 4-foot (1.27 m) stroke, in which the ram and the cylinder weighed together about 5 tons.

Later, brothers James and Joseph Tangye concentrated on the development of hydraulic jacks, and it was the Tangye brothers who provided all the hydraulic jacks for the launching of Isambard Brunel's colossal iron ship, The Great Eastern. In 1862, James Tangye adapted his hydraulic jacks for the punching and shearing of iron and steel plates. In 1871, R.F. Twedell designed an hydraulic rivetting press which was constructed by Fielding & Platt. This manufacturing company, and others, built powerful and massive hydraulic presses for the purpose of shaping keel and boiler plates for ships, which virtually revolutionised shipbuilding and boilermaking.

Murray was also the most consummate craftsman of his era, and an important inventor of machine tools. His most famous machine tool was his planing machine which he used to plane the surfaces of D-valves for his steam-engines. It was probably the first British planer ever devised and built, and it existed already in 1814. In the planing machine the workpiece moves forwards and backwards under a stationary cutting tool. Murray had no connexion with Wilkinson, or Bramah, or Maudslay, for he was essentially a self-taught mechanical engineer, and machine toolmaker.

Joseph Clement (1779-1844), a son of a Westmorland weaver, in 1814 began to work as the chief draughtsman and superintendent of the workshop for Bramah. After Bramah's death in 1814 he worked as the chief draughtsman for Maudslay, Sons & Field where he was responsible for the designs of the first marine steam engines. He struck out on his own in 1817, and soon thereafter he became one of the best early machine toolmakers in Britain. In 1820 Clement invented one of the first, and the best, hand-driven planing machines in existence capable of planing surfaces up to 6-foot (1.83 m) square. He produced an improved version of his planing machine in 1824 which brought him most of his income. Clement devoted special attention to self-acting tools, particularly to the slide lathe. In 1827 Clement designed and built his facing lathe that incorporated an infinitely variable speed drive. This ingenious and complex machine tool far surpassed any existing machine tool by its ingenious design and excellence in workmanship. The Society of Arts awarded Clement a gold metal for his remarkable facing lathe. He was one of the first toolmakers after Bramah to use revolving cutters. He continued Maudslay's standardisation of threads, and began routine manufacture of taps and dies for screw-cutting. Moreover, Clement was probably the best skilled draughtsman of the last century as his drawings of the complicated digital calculating machine of Babbage, the construction of which was Clement's most admired work, demonstrate.

Another early machine toolmaker was James Fox (1787-1859) of Derby, a former butler of a country parson, who came independently to machine toolmaking without having apprenticed under Wilkinson, Bramah, or Maudslay. Fox invented one of the earliest planing machines in 1814, almost at the same time as Murray. Another well-known planing machine was produced in 1820 by George Rennie (1791-1866), a son of the famous Scottish engineer John Rennie (1761-1821). Fox, who produced a variety of lathes, built his first lathe in 1817, and he also invented machines for index-cutting of gears. Fox mainly produced all sorts of textile machinery, such as lace-making machines, in his workshop.

In 1817, William Fairbairn (1789-1874), who became a celebrated engineer, and his partner, James Lillie, produced one of the first slotting machines.

The most versatile and prolific machine toolmaker and inventor was the ingenious Welshman, Richard Roberts (1789-1864), who had first worked for John Wilkinson at the Bradley Ironworks and, then, for Maudslay. Roberts invented his lathe with the back-geared headstock in 1817 which was the first in the great spate of his machine tool inventions: the first planing machine for metal, a variety of drilling machines among which was a multiple-spindle drilling machine, a self-acting hole-punching machine controlled by a Jacquard punch-card mechanism (1848), the first satisfactory power-loom (in 1820, and in 1830), an automatic self-acting spinning mule (1825) as the final step in the complete mechanisation of the textile industry, and a combined self-acting machine for shearing iron, and punching both webs of an angle-iron or T-iron simultaneously to any required pitch.

Robert's versatility was demonstrated in 1845, when he built the most powerful electromagnet then in existence, designed the first differential gear for a steam-carriage in 1833, the first successful gasmeter, and produced many other diverse inventions. For all his native ingenuity, Roberts was a poor businessman, and despite the great number of his inventions, he finished his days a poor man.

Joseph Whitworth (1803-1887), son of a minister and schoolmaster in Stockport who had been an apprentice to both Maudslay and Clement, became the foremost machine tool manufacturer of the 19th century after he set himself up as one in 1833. Whitworth made machine tools for sale which required the use of self-acting machine tools in his own works to manufacture accurate machine tools for sale in large quantities. In 1858, following the example of the remarkable Swiss mechanic and inventor Bodmer who had worked in his factory for some time, Whitworth installed an overhead travelling crane in his factory. After the Conway and Menai tubular bridges had demonstrated the strength of box sections in his day, Whitworth adopted the hollow box sections extensively in his machine tool design. Whitworth gave his machine tools both the unprecedented strength by using hollow box frames, and great precision in performance. He raised precision in machining to an extremely high accuracy by inventing in 1856 his millionth-measuring machine based upon the micrometer capable of detecting differences of 0.000001 part of an inch which made high accuracy in end measurement in machining possible. He introduced his well-known standard screw-threads in 1841, and invented the first hollow-box lathe in 1835 which drove and controlled both the cross-feed and the longitudinal-feed automatically by using a single lead screw. He also invented a variety of other machine tools such as a shaping-machine with quick return, a slotting-machine, a planing machine with unidirectional cut in combination with quick-return motion of the carriage, and a key-seater. Whitworth, more than anybody else, brought into general practice the principles of modern production engineering which were based on accurate end measurements, precision tools, interchangeability of machine parts, and standardisation. In Whitworth's Manchester factory each machine tool was driven by a separate steam-engine by means of a flexible shaft mounted with a coiled spring, a flexible means of power transmission invented by Nasmyth. Instead of being an inventive genius, Whitworth rather possessed personal standards so exacting that even the best was not good enough for him. He ferreted out with unerring judgement the best features of contemporary designs, improved upon them, and consciously put them together in a masterful synthesis. He introduced many improvements of details in machine tools.

In his Manchester Works, Whitworth specialised in the production of machine tools to such an unprecedented extent that for the first time manufacturers could promptly obtain a machine tool of the highest quality at reasonable cost. The less specialised tool builders on account of their multifarious activities were not able to compete with Whitworth neither in rapid delivery of a machine tool nor in the price. The business venture of Whitworth was so immensely successful, that by 1850 his machine tools dominated the workshops of the world.

The last successful apprentice of Maudslay, who also became a prominent toolmaker of machine, was the Scot, James Nasmyth (1808-1880), son of the celebrated Scottish painter and a gifted amateur engineer Alexander Nasmyth of Edinburgh, who was a skilled artist himself. In 1829 James went to London with a model steam-engine he had built as evidence of his mechanical ability in the company of his father, and paid a visit to Maudslay, who hired James as personal assistant in Maudslay's private workshop. At the time Maudslay's firm was building a very large marine engine, and Nasmyth helped Maudslay to make a model of this engine which is still exhibited in the Science Museum in London. Nasmyth invented during his model-making activities his flexible shaft of coiled spring steel intended for drilling holes in inaccessible places, an idea he did not patent. This flexible drive shaft soon became commonplace in many different applications, such as in the flexible shaft dental drill, but its original inventor was already forgotten in Nasmyth's lifetime.

In 1830, Nasmyth invented his famous index milling machine. The only earlier industrial milling machine could have been that of the American inventor Eli Whitney (1765-1825) which might have been built in 1820, but it is not certain that Whitney's machine at that time had the form of his extant machine. In 1835, Nasmyth invented his shaping machine which could produce any small surface formed of rectilinear elements. In the shaping machine the workpiece is held stationary and the cutting tool is reciprocating. In 1839, he conceived and built his famous 'steam hammer' for the purpose of forging the 30-inch diameter shaft of the paddle-wheel of Isambard K. Brunel's steamship, The Great Britain, a forging job which was subsequently not carried out because the paddle-wheel of the ship was replaced by the screw-propeller.

In 1839 he designed one of the simplest and most reliable valves used in hydraulic machinery, a stop-valve in the form of a double-faced wedge-form sluice-valve which is the stop-valve still in common use today. In 1843, Nasmyth built his steam-powered pile-driver to drive piles in the construction of Robert Stephenson's Newcastle railroad bridge, and in harbour construction in general. The steam-powered pile-driver speeded up pile driving appreciably by working at the rate of a blow every second instead of a blow every few minutes delivered by the traditional gravity pile-drivers. He also invented a hydraulic punch, a grinding machine, a ball-and-socket joint, and a ventilating fan for mines. In 1862, Nasmyth invented a slot-milling machine which eliminated one of the most tedious jobs in metal working consisting of drilling a series of holes and then chipping out the metal between the holes with the chisel. As mentioned earlier Nasmyth invented a device in his model-making days to make rotary power turn corners, a device which consisted of a spiral coil of spring steel encased in a flexible sleeve. This device immediately became an important means of power transmission in workshops, and it was later used by Whitworth to operate all the machine tools in his factory. Nasmyth also designed a teeming ladle operated by worm-gearing in the smelting of iron which prevented serious accidents in dangerous pouring operations of hot metal.

In retrospect it is quite obvious to any observer that without these self-taught, ingenious craftsmen who were interested in mechanics and science, the Industrial Revolution in Britain would not have been possible. The Industrial Revolution was brought about by such self-made, and self-taught men, and not by scholars and formally educated professionals. Their approach to machine design and production engineering was empirical insofar as they improvised as they went along.

In 1862, an American mechanic, Joseph Rogers Brown (1810-1876), produced the first successful universal milling machine as a replacement for a great number of difficult manual operations, and the first universal grinding machine, both of which were important advances in machine tools. The first, but different, universal milling machine had been invented in 1852 by the American mechanic, Frederick W. Howe, who worked for Brown at the time.

Automatic Machine Tools

Copying-machines of various kinds were devised in the latter part of the 18th century. Even James Watt during his years of retirement made some improvements upon machines of this kind then in use. The copying lathe came into general use in 1818 with the patented copying lathe invented by the American machinist Thomas Blanchard (1788-1864) for the reproduction of gunstocks which operated automatically by the pantograph principle. Blanchard's copying machine for wood was so accurate that for its time it surpassed the precision of all metal working machines. It is the progenitor of all later copying and profile-turning lathes. His other automatic copying machines which mass-produced axe handles and shoe lasts were highly profitable to him, although he was unable to capitalise on his greatest invention, the gunstock copying lathe.

The first automatic machine tool processing metal instead of wood was made in 1873 by the American engineer Christopher M. Spencer (1833-1922). It was a turret lathe the camshaft and pulley of which not only moved the levers and fed the workpiece but also changed the tools. This automatic turret lathe was designed to mass-produce parts for a sewing machine. He founded his own company for the building of automatic machine tools, called 'automats', that mass-produced nuts, screws, cogwheels and other parts of various machines. Spencer's 'automats' exerted an immediate influence on the machine tool industry.

American System of Manufacturing and Mass-Production

Mass-production by means of power-driven machinery consisting of special purpose machines and based on the principle of interchangeable parts, automation and assembly-line came to be called the 'American System of Manufacturing'. This name was coined by Englishmen after seeing at the Great Exhibition in Crystal Palace, in London, England, in 1851 the American exhibits of the McCormick reaper, the Day & Newell safe and, particularly, the Colt revolver, and the rifles produced by the firm of Robbins & Lawrence by such mass-production method. Americans exhibited about 500 products among which McCormick's reaper, David Dick's antifriction punch press, Charles Goodyear's vulcanised rubber products, William C. Bond's astronomical clock and Gail Borden's dried-meat biscuit were granted Council awards, the Exhibition's highest honour, besides 102 other American products which were awarded medals. In general, many American manufactured products received generous public acclaim at the Great Exhibition in London. The huge success of the Great Exhibition in London in 1851 led to other similar industrial exhibitions in New York City in 1853, in Paris in 1855, in London in 1862, in Paris in 1867, and in Vienna in 1873.

Actually the basic ideas of the system of mass-production went through a long, tortuous, and sporadic development throughout the history of Western civilisation. The Sumerians were the first people in history known to have recognised the commercial importance of division of labour, and the first to mass-produce commercial wares such as pottery after Sumerians had invented the pottery wheel. Ancient Greeks also came to appreciate division of labour in order to increase production of marketable items, since ancient Greeks were the first modern traders who operated on monetary economy without which no real progress in mass-production is possible. Moreover, the Hellenistic Greeks in the first century B.C. were the first technologists to attempt to use power-driven machinery in production. Mithridates (132 B.C.-63 B.C.), the Parthian king of Pontus, ordered his Hellenistic engineers to build in his palace at Kabeira a horizontal watermill to grind corn. Since this mill had no transmission gearing, it was far too slow for efficient grinding of corn. The Hellenistic engineers were the first technologists capable of taking advantage of natural power to drive machinery, since they had discovered the usefulness of both the impulse-wheel and the reaction-wheel as natural sources of power.

However, when it came to mass-production, Romans were the first people to attempt to use such methods, since philosophically at least, Rome was an Empire of common men. Large Roman farms certainly made efforts to mass-produce grain by using farm machinery which were worked by animal power. They also made the first conscious attempt to implement standardisation of manufactured products such as water-spouts and water-outlets. Romans were also the first to apply power-driven machinery to the mass-production of flour during the last centuries of existence of the Western part of the Roman Empire. Romans had installed a battery of hydraulically powered flour mills fed by an aqueduct on the Ianiculum hill (modern Vatican hill) to supply flour to the Roman citizenry. Earlier, in A.D. 308, Roman engineers had built a battery of 16 cascading overshot waterwheels turning 16 millstones at Barbegal near *Arelate* (modern Arles) in *Gaul* (modern France), to mass-produce 28 tons of flour in 10 hours for Rome and the Roman army. There existed other such hydraulically powered mills utilising the efficient Vitruvian mill, which incorporated a continuous power-transmission over the right-angle by means of lantern-and-crown gearing with a gear ratio of 1– to – 5 that produced an efficient grinding speed of millstones.

The first more elaborate mass-production method was used in the Arsenal of the Republic of Venice, an oligarchic republic governed by a merchant aristocracy like in the ancient Carthage, and ruling over a large overseas dominion consisting of Greek islands taken from the Byzantine Empire in A.D. 1204. About the middle of the 15th century, the Venetian shipwrights not only led the world in the best design of ships, but also in the best methods of shipbuilding. The shipwrights of the Arsenal of Venice, an arsenal which employed about 16,000 technicians and shipwrights, made a concerted effort to standardise the parts of the galleys, their warships, thereby anticipating the modern manufacturing principle of interchangeable parts. They also used the assembly-line manufacturing system by assembling calibrated and standardised parts to outfit the mass-produced galleys. The Venetian shipwrights had to use some kind of standardised measurements with calipers, which were essential to such assembly-line manufacturing method of calibrated parts. In an emergency, the Arsenal of Venice could outfit a war galley in about a day by assembling standard, prefabricated parts of the galley. The Venetian shipwrights could work within fairly wide tolerances of their interchangeable parts, but by the 17th century the manufacturing tolerances had become much finer in military weapons, and to check on the precision of the caliber of a weapon inside and outside calipers were employed, an idea that led much later to limit-gauges, an important method in interchangeable-parts manufacture of high precision. Despite their excellent shipbuilding method, Venetian naval construction began to wane as their sea power began slowly to decline and their timber supplies, particularly oak for the frame and hull planking, became scarce.

It is probably justified to consider the ingenious Swedish mining and mechanical engineer, Kristofer Polhammer [after ennoblement, Christopher Polhem] (1661-1751), the first European engineer who successfully combined science with engineering, to be the actual founder of modern mass-production by inventing automatic machinery for use in a new process of production which he organised in such a way that most of the human labour as well as the detailed human control of the process was reduced to a bare minimum. Polhem substituted power-driven machine labour for human labour wherever possible. He was an ardent advocate of replacing human power by machine power, and he fully recognised the value of division of labour. Polhem reduced the entire manufacturing process to unit-operations, and wherever possible designed a power-driven machine for each unit-operation. His power source was the waterwheel. Polhem justified mechanisation in mass-production in the following way: "Nothing increases the demand as much as a considerable lowering of the price of a product. That is why we need machines to take over the heavy and time-consuming labour by hand. The result will be a saving from 100% to 1,000% in cost."

In the 1720's, Polhem began to manufacture standardised gears for clocks, a pioneering step in the mass-production of standardised, interchangeable parts. One of the first automatic precision machine tools he designed was the cogwheel-cutting machine in 1729. His mortising machine employed a clone-clutch, probably a novel feature in machines. Polhem manufactured lathes made of iron and powered by waterwheels in which the tool-holder was fed by a lead-screw driven by a gear-wheel. Such lathes were in operation in his Stjärnsund Works by 1700. Therefore, Polhem's contributions were not only important in metallurgy, and factory design, but also in toolmaking.

In 1699 Polhem built a series of mills in his industrial plant near a tumbling stream running between two lakes at Stjärnsund, in Sweden. The more than 300-foot (914.40 m) long main building of this industrial plant was a huge smithy, and a factory filled with various mills. It had 7 waterwheels driving more than 9 drop-hammers as well as a number of shearing mills and rolling mills of Polhem's design. He appears to have been one of the first technologists to use grooved rolling mills to press out the dross from the metal. His wrought-iron or cast-iron rollers of diameters as large as 7 inches (17.7 cm) were carefully made by hand, and had steel-hardened surfaces. These heavy rollers, which could be cut in various patterns to shape the iron to any desired form, could operate at high speed and process up to 20 times more wrought-iron than the traditional hammering-method. A tilt-hammer could forge at most a ton of wrought-iron bars in 12 hours, whilst the rolling mill produced up to 15 tons of superior quality wrought-iron bars in the same time. Polhem's rolling mill produced up to 15 tons (13.6 metric tons) of wrought-iron in 12 hours. In 1745 he designed a rolling mill which could refine and form wrought-iron in a single process. Polhem invented the so-called 'three-high' rolling mill which permitted the passing of the rolled metal back-and-forth without reversing the rotation of the rollers. The rollers had been used before Polhem in metal working, but never as extensively, and as efficiently as by him. In England, John Payne had obtained a patent for grooved-rollers for manufacture of bars in 1728. Polhem's mass-production methods helped Sweden to become the leading producer of iron in the 18th century. Other buildings in his industrial plant housed a saw-mill, a flour-mill, a granary, an ironworks, a nailworks, a clockworks, and a tin-plating shop. In this remarkable industrial complex, Polhem manufactured a surprising variety of products, mostly produced by power-driven machines: pots, pans, plates, tankards, spoons, forks, knives, bowls; hand-tools for the carpenter, blacksmith and sculptor; clocks, locks, screws and bolts; sheet-iron roofing and gutters, and ornamental brass-work; steel-presses, turn-pits, mortars, and army cots. For instance, to fashion sheet-iron dishes, Polhem devised a sequence of well-organised unit-operations by means of a series of drop-hammers lifted by cams on a horizontal shaft which was turned by a waterwheel. The drop-hammers had different shapes for a sequence of unit-operations: a cast-iron drop-hammer shaped and moulded the white-hot metal blank, which was then finished by a machine incorporating 5 smaller hammers: 3 to finish the edges of a plate, and 2 to finish the interior. By this remarkable mass-production machinery, Polhem was able to produce 15 dozen dishes a day. In the manufacture of gears for his clockworks, Polhem used automatic machinery and jigs, and pioneered in precision measurements with limit-gauges to ensure interchangeability of standard parts. Polhem gave up his Stjärnsund Works in 1735.

Polhem designed a number of machine elements. He built a series of 80 models of machine elements, including all the machine elements known in his time and a few of his own invention, which he called 'mechanical alphabet', and he intended to use it in the teaching of machine design. He believed that an engineer could use this 'mechanical alphabet' to design useful machines much like a poet uses letters to compose poems. The 'mighty 5 elements', the so-called 'simple machines', represented the vowels of his alphabet, and the rest of the elements represented the consonants. In 1697, he set up his *laboratorium mechanicum* as an experimental workshop for the development of new machinery. After Polhem, the German engineer Jacob Leupold (1674-1727) also separated machine elements systematically from the machines in his< Theater of General Machinery > (*Theatrum machinarium generale*),1723-1739, the first systematic treatise of mechanical engineering in 9 volumes. Polhem built his renowned **Machina Nova**, an ore-hoisting machine powered by an overshot waterwheel for the Falu copper mine at Blankstöten in Sweden. It contained an elaborate power-transmission mechanism consisting of oscillating rods and levers, two gear-wheels, an endless chain, rods and hooks. This machine was able to hoist 22 barrels of ore in an hour, whereas the conventional machines were able to hoist only 16 barrels in an hour. Polhem's *Machina Nova* was the 17th century counterpart to Juanelo Turriano's **Artificio** built from 1565 to 1573 in Spain. Polhem's power transmission devices conveyed water power from a waterwheel across a rough countryside over a distance of 1½ miles (2.5 km).

Polhem's ideas of mass-production and his machines were far in advance of his time, and the main reason why he could not mass-produce on a larger scale was the limitation of his power source, the waterwheel, for the steam-engine as a prime-mover did not yet exist.

Polhem, moreover, was an important civil engineer, for he was responsible for the construction of the docks at the naval base in Karlskrona, and the construction of a canal connecting the Baltic Sea with the city of Göteborg (also **Gothenburg**). The construction of this canal was suspended when the sea-route was reopened after the Treaty of Nystadt in 1721. This canal was finally finished by the Scottish engineer Thomas Telford (1757-1834) a century later, and is called the 'Göta Canal'.

For a period of time, Polhem worked in the mint and mining works of George I in Hannover, Germany. Both George I, who became King of England, and the Russian Tsar Peter I tried in vain to persuade Polhem to come to their respective countries. In 1715, the well-known Swedish philosopher-scientist, Emanuel Swedenborg (1688-1772) was Polhem's student at Stjärnsund. He wrote on many philosophical ideas which had interested Polhem. Polhem was elected President of the Swedish Academy of Sciences.

Concepts of production engineering began to appear in the 18th century because of the attempts by some engineers to increase productivity of men in engineering construction or in manufacture by studying the efficiency of engineering and manufacturing operations. The French engineer from Switzerland, Jean Rodolphe Perronet (1708-1794), who perfected the construction of stone voussoir bridges, realised in 1760 that quantitative measures could be devised to integrate men, materials and equipment in such a way that a higher productivity per man-hour could be achieved. At the time he was particularly concerned with the means to optimise the manufacture of common pins used in construction, and he carried out extensive time-and-motion studies to set 494 pins an hour as the standard for production. Perronet appears to have been the first pioneer in 'production engineering' as a new discipline of engineering.

The idea of standard, interchangeable parts in the manufacture was also implemented by a French General of artillery, Jean Baptiste de Gribeauval (1715-1789), who invented a bronze cannon, and suggested in 1765 to standardise cannon carriages in production. Since the limits of tolerances in the production of such carriages were quite generous, Gribeauval was able to carry out satisfactory mass-production of cannon carriages on a small-scale. In 1770, General Gribeauval was also entrusted with the task of reporting on the military value of the self-propelled steam-truck called the *fardier* of the Swiss engineer Nicholas Joseph Cugnot (1725-1804), for the transportation of cannon carriages.

The first instance of a complete automation of an entire industrial plant, a flour mill, was designed and built about 1780 by the ingenious American millwright, Oliver Evans (1775-1819) of Philadelphia, a pioneer of automation. Evans designed this first completely automated plant on the basis of a sequence of 'unit-operations' and a moving conveyor system. The entire automated flour-mill was operating with the following inventions: vertical bucket-elevators, horizontal screw-conveyors, a 12-foot (3.66 m) revolving rake which continuously stirred the grain and guided it to the central chute and the millstones, and a moving belt on rollers made to move under the weight of grain. None of these inventions were original, but the method by which they were combined into one continuous, automatic mechanical operation and powered by a single waterwheel was entirely original. The grain entered the mill and after a sequence of automatic processing and handling came out in sacks as flour without the manipulation, control or judgement of a human operator. The entire automated flour mill of Evans was operated by one man, and it remained a singular achievement in a completely automatised industrial plant until modern times.

In 1785, a French gunsmith, Honoré Le Blanc, who worked in various French government arsenals, had used 'interchangeable parts manufacture' of muskets, which required rather high accuracy in the production of interchangeable musket parts. Le Blanc published a memoir in 1790 on his method of interchangeable parts manufacture in which he demonstrated his profound understanding of this method of production and its probable effects on manufacture in general. Thomas Jefferson (1743-1826), a self-educated engineer and technologist who had visited Le Blanc's works and tested the Le Blanc system of manufacture by assembling a gunlock from the mass-produced parts himself, reported in 1785 about the advantages of the interchangeable parts method of manufacture, and described Le Blanc's work with this method in his letter to John Jay, Secretary of Foreign Affairs and the first Chief Justice of the United States. Le Blanc did not succeed with his method because he did not use limit-gauges in his system to size the components, something that was readily recognised by the American gunsmith right from the outset. The principle of establishing the margin of error, both plus or minus, which are tolerable for critical dimensions, and enforcing them in the workshop by means of suitable gauges such as limit-gauges, is essential to the success of the interchangeable parts method of manufacture.

Samuel Bentham, younger brother of the famous political philosopher and founder of the utilitarian philosophy, Jeremy Bentham (1748-1832), was an 'efficiency engineer' and a specialist in efficient organisation of production methods. Whilst he was serving in the Navy of the Russian Empire of Catherine the Great, he was placed in charge of Potemkin's large, but grossly mismanaged shipyard and its factories, at Krichen in Southern Russia. Bentham was able to reorganise the operations of this Russian navy yard much more efficiently whilst conserving whatever labour skills were available in that country. In general, Russian labour was plentiful but utterly unskilled. Bentham was knighted by the Tsarina for this outstanding achievement, and appointed brigadier-general and a commander of a Russian flotilla. He led the attack against the Turks in a naval battle where shells were fired for the first time in naval history. He returned to England in 1791 and soon took out a patent for woodworking machinery which he attempted to introduce to the British Royal dockyards. He greatly extended his patent designs in 1793. Samuel Bentham was appointed Inspector-General of the British Naval Works in 1794, a post which he held until 1812, and he immediately developed a scheme for a complete mechanisation of the manufacture of naval ship-blocks. A single navy frigate needed about 1,500 blocks for its rigging, and the British Admiralty needed about 130,000 ship-blocks a year as replacements for damaged or worn-out ship-blocks.

In 1801, Bentham met the French engineer and inventor, Marc Isambard Brunel who fled France in 1793 during the French Revolution, and had spent 12 years in United States as a surveyor, engine-builder and gunsmith. Brunel became a United States citizen in 1796 and served as the chief engineer of New York City. Brunel had also developed a scheme for blockmaking by the interchangeable parts method which was similar to that of Bentham. Bentham, a man of integrity, readily recognised that Brunel's manufacturing scheme was superior to his own, and he immediately adopted it. The Brunel's manufacturing scheme based on unit-operations required altogether 44 separate special purpose machines, which performed unit-operations such as sawing, boring, mortising and scoring for 3 different sizes of pulley blocks. This interchangeable parts manufacturing scheme was officially adopted in 1802 after Brunel had convinced the Admiralty of the soundness of his manufacturing scheme with the 44 models of his special purpose machines made by Henry Maudslay and with the resolute support of Bentham. Brunel immediately engaged Henry Maudslay, who had recently left Bramah's employment to establish his own workshop. Maudslay built all the special purpose machines for Brunel, a task which took him 5½ years to accomplish. These blockmaking machines were installed from 1802 to 1807.

Brunel's machines, which were driven by two 30-horsepower steam-engines, one manufactured by Boulton & Watt and the other by Murray & Wood of Leeds, did everything but the final assembly and polishing in changing the elm logs into finished pulley-blocks. Steam-power was used to turn iron shafts, wheels, gears and feed-screws of the block-making machines. Some of these machines remained in use for more than a century after Maudslay's death. This mass-production scheme replaced a team of 110 skilled craftsmen with 10 unskilled men. The 10 men produced 130,000 pulley-blocks a year by Brunel's mass-production method, which equalled the total output of the 6 largest British naval dockyards using conventional methods of production.

In 1798, American mechanic Eli Whitney of Connecticut secured a government contract to produce 10,000 Charleville muskets by the manufacturing process of interchangeable parts. It appears, however, that Whitney, who is popularly regarded as the founder of the 'American system of manufacture', does not deserve this honour because his production method was not as advanced as that of Honoré Le Blanc in France. In 1807 another American gunsmith, Col. Simeon North (1765-1852), produced parts of his pistols also by the method of interchangeable parts.

Capt. John A. Hall, who had installed his machines in the Harper's Ferry Armoury by 1817 to build his rifle for the United States government by the interchangeable parts manufacturing techniques, appears to have been the technician who is mostly responsible for the success of this type of mass-production. His methods seem to have been quite superior to the techniques used by Whitney, and appear to have been influential so that at the beginning of 1819 two major arsenals in United States adopted the interchangeable parts of manufacture as their normal method of production. At the same time, clockmaker Eli Terry (1772-1825), also adopted the interchangeable parts manufacture of shelf-clocks in large quantities, the wheels of which were accurately stamped with dies from sheets of brass. From 1828 forwards, the Hubbard rotary-gear pump was also manufactured by the interchangeable parts method. In 1830, Chauncey Jerome began to produce a clock a day by the interchangeable parts manufacture.

The automatic draw-loom was developed in France, a country where figured weaving was traditional and popular. The 'shedding mechanism' for selecting the threads to be drawn in the loom was also invented in France. In 1725, a French weaver Basile Bouchon of Lyon, whose father was an organ-builder familiar with the cylinders with pegs that controlled the melodies played by carillons, devised a 'control-mechanism' based upon the concept of carillon cylinders. He prepared a roll of paper perforated to the weaving-pattern in place of the pegs for the selection of the threads of the warp which he suspended from the loom, and when pushed towards the selection box made the needles meeting the unperforated paper slide along and carry their threads with them, whilst the other needles were made to pass through the perforated holes and remain stationary. The selected threads of the warp were then drawn by a foot-operated comb followed by the throw of the shuttle, and then the perforated paper was manually moved to the position of the next series of perforations when the same operation was repeated. In 1728, another French weaver Henri Falcon improved this weaving mechanism by increasing the number of needles to several rows and by replacing the roll of perforated paper, which created some problems in operation, with a long train of perforated rectangular cards, but the operation of this vertical draw-loom still remained manual.

The talented Frenchman, Jacques de Vaucanson, was a mechanical genius, and the best but not the only builder of automata since Hellenistic times. Since Vaucanson had studied music, anatomy and mechanics, he organised this knowledge to devise and construct some of the most remarkable and mechanically sophisticated robots ever made: the robotic androids in the humanlike form of a flutist and a drummer, as well as the mechanical robotic duck. The robotic android in the form of a sitting flutist exhibited in 1738 at Paris, was 5½ feet (1.68 m) high. It had movable lips, movable tongue functioning as the air-flow valve, and movable leather-tipped fingers that opened and closed the openings in the flute to play 12 tunes. The robotic android in the form of a drummer, also 5½ feet (1.68 m) high, was built on the same principle as the flutist except that the drummer simultaneously played 20 tunes on a 3-hole shepherd's pipe. The most sophisticated robot of Vaucanson was the mechanical robotic duck, which could waddle and swim, had wings that imitated the motion of the wings in beating the air in all of its details, could wag its head, quack and pick up the grain in swallowing movements, and grind up the grain and expel it in an imitation of a real duck. The mechanical robotic duck of Vaucanson was widely admired in France as a mechanical wonder. He had tried out almost 300 different versions of design of the robotic duck before he was completely satisfied. Vaucanson became quite prosperous from exhibiting his automatic robots in Europe. He was responsible for creating a new era in the design of self-propelled mechanisms by breaking new ground in the theory and practice of automation.

No less famous than Vaucanson's robots was the robotic android of the Swiss theologian Pierre Jaquet Droz (1721-1790). In 1760 Droz built a robotic android in the form of a sitting child which wrote with crayon every letter of the alphabet in an almost perfect imitation of the motion of human hand.

Since 1747, Joseph Gallmayr, a German court mechanic in Munich, built a number of curious robots such as a bouncing and singing robotic canary, a life-size android in the humanlike form of a flute player, and a robotic dog performing canine acts. In 1776, he made the first known self-winding pocket watch using an eccentric spring winder.

Apparently, similar automata had been produced already in the previous centuries. The very learned, remarkably versatile and very well-informed scientific authority of his time, the German Jesuit Pater Athanasius Kircher (1601-1680), was keenly interested in automata and robotics of various kind. He built automatic mechanical robots such as a talking head, singing birds, humanlike androids playing musical instruments, and automatised theaters for gardens which he described in a book published in 1650 with the collaboration of another Jesuit Pater, Kaspar Schott (1608-1666).

Kircher was professor of ethics and mathematics at the University of Würzburg in Germany, but before he retired in 1643 to devote himself entirely to archaeological research, he was professor of physics, mathematics and Oriental languages at the **Collegio Romano** in Rome, where his remarkable collection of antiquities is still stored.

In 1741, Vaucanson was appointed to the post of the inspector of the silk manufacturers by Cardinal Fleury, the real power behind the French throne at the time, which raised Vaucanson's interest in the mechanisation of textile manufacture in which craft he was able to make a number of significant mechanical improvements in spinning and weaving. He was appointed director of the State silk mills, after which he abandoned building automata. However, he did invent a machine to make rubber hoses and was the first to use a rubber hose made of caoutchouc.

In 1745, Jacques Vaucanson built an automatic draw-loom for the weaving of figured silk cloth which was operated by water-power and represented an improvement of the draw-loom of Bouchon and Falcon. He located the selecting-box, which operated directly on hooks attached to the neck-threads of the warp, above the loom on a chassis. The hooks passed through needles and were raised by a strong metal bar. The needles were selected by a sliding and rotating cardboard drum perforated to the weaving pattern which automatically controlled the pattern of weaving that was repeated after each completed revolution of the drum operated by water-power. The heddles were automatically raised and lowered at each throw of the shuttle when the perforated drum moved back-and-forth on the chassis and rotated through a small angle into the position of the next series of perforations which controlled the lifting of a different selection of warp threads according to the weaving pattern for the next throw of the shuttle that was automatically executed by hydraulic power. The length of the repeated weaving pattern in the Vaucanson's automatic draw-loom was limited by the perimeter of the perforated drum which was also its major shortcoming. Since Vaucanson's automatic draw-loom put many weavers out of work it was not popular among the weaving trade. Fortunately, Vaucanson's automatic draw-loom was saved from oblivion by Vandermonde, who found all the parts of a disassembled draw-loom of Vaucanson after the Revolution.

In 1756, Vaucanson established the first textile plant in the modern sense with his silk factory at Aubenas near Lyons which he designed to its minutest details. Every component of the textile machinery down to the very reels, which were ingeniously supplied by thread from the cocoons submerged in a bath and fitted with twisting-frames that spun the thread, were carefully designed. Every machine in the textile-mill was supplied with power from a single overshot waterwheel. The 3-storeys high factory buildings (for he soon built a second factory) were planned with the utmost attention to sound physical conditions of a factory: ample windows covered with oiled paper provided throughout the well-lit halls of the factory the soft lighting necessary to operate the textile machines, and a primitive ventilating system maintained the temperature and moisture required for the spinning of silk. Unfortunately, Vaucanson's pioneering effort in the mechanisation of the textile industry in France came to nought, like many other good ideas of French engineers, because the highly centralised Catholic France was not a fertile soil for an enterprising and ingenious inventor-entrepreneur like Vaucanson.

French mathematician, Alexandre Théophile Vandermonde (1735-1796), made experiments on what might be called production engineering. In 1793 he was given a laboratory in a Benedictine priory by the Revolutionary Government, and in 1794 *St. Martin des Champs* was placed at his disposal. Vandermonde was keenly interested in automation, and he collected every automatic device he was able to find, particularly the robotised automata built by Jacques de Vaucanson, and housed them in the facilities placed at his disposal. This collection became known as the *Conservatoire des Arts et Métiers* (*Conservatory of Arts and Crafts*). A Lyon weaver and inventor, Joseph Marie Jacquard (1752-1834), who was employed by Vandermonde to work in the *Conservatoire des Arts et Métiers* and develop further completely automatic devices for fancy weaving looms, reassembled the surviving parts of the disassembled Vaucanson's automatic draw-loom, which he had not seen before, at the request of Vandermonde. Jacquard began to construct an improved version of the automatic draw-loom of Vaucanson which was not restricted to the limited size of the repeated weaving pattern of the perforated cardboard drum of Vaucanson's loom. Jacquard modified Vaucanson's automatic draw-loom by replacing the perforated cardboard drum, the perimeter of which was responsible for this shortcoming, with an endless train of Falcon's perforated control-cards attached to a mechanism which slid back-and-forth on the same chassis whilst turning the next perforated control-card into position with each throw of the shuttle. This relatively minor, yet important improvement of Vaucanson's automatic draw-loom, made it possible to weave any size of repetitious weaving pattern by merely adding more perforated control-cards, so-called 'punched cards', to the endless train of perforated control-cards. In 1805 Jacquard perfected his automatic draw-loom which wove repeated patterns by means of an endless train of punched-cards that automatically controlled the lifting of different selections of threads of the warp to form the pattern of weaving at each throw of the shuttle. Jacquard automatic draw-looms which were programmed by punched-cards to control 400 warp-threads became almost immediately a huge business success. Although Jacquard could not patent his minor improvement of the Vaucanson's automatic draw-loom as an employee of the *Conservatoire des Arts et Métieres* (*Conservatory of Arts and Crafts*), he became very famous for his loom, and was granted a yearly pension in France. After the death of Jacquard, his 57-inch (1.50 m) square portrait, which looks like a fine engraving, was woven on a Jacquard-loom controlled by a train of more than 24,000 punched-cards. By 1812 there were more than a 1,000 Jacquard-looms in operation in France alone. In the 19th century

Jacquard-looms were operated by steam-power. Early modern computers in the 1940's still made use of the well-developed technology of information-carrying punched-cards of Jacquard.

The Swiss mechanic, inventor and a former child prodigy, Johann Georg Bodmer (1786-1864), came to England in 1824 where he worked on steam power and textile machinery, and soon became a pioneer of the assembly-line and interchangeable parts method of manufacture. Already in 1806, he made guns and, particularly, locks by the interchangeable parts method in Switzerland. In 1824, Bodmer patented his improved designs of cotton spinning machinery, and built a cotton spinning factory in collaboration with an English engineer at Bolton in which he had integrated processing and handling in manufacture much like Oliver Evans had done in a flour mill in America, which allowed the output of one machine to be fed directly into another machine for the next stage of the manufacturing process without human intervention. In 1828 he returned to Switzerland for reasons of health but returned to Bolton in 1834. For a time Bodmer worked in the factory of Whitworth in Manchester as a mechanic and technologist who was an expert on the integration of handling and processing in the manufacturing industry. In 1834 he took out a patent on steam-engines and boilers, and in 1835 he received a patent for a collection of improved spinning machines for cotton and wool. In 1839 he was granted a patent on 40 different improved machine tools among which was the first vertical facing lathe that became popular in America where it was developed into multi-spindle and completely automatic machine tool. In 1840 he obtained an important patent on the standardisation of various different machine elements. He laid out a textile factory, Charlton Mills, in Manchester, in which the processing and handling were completely integrated. Bodmer was a true pioneer of the assembly-line consisting of the moving crane, the endless belt, and the travelling grate [a quasi-automatic stoker], all of which he put into functional operation in Lancashire in the 1840's where he was busy as a consulting technologist on factory design. In the factories of his design particular attention was devoted to the functional arrangement of machines to minimise the transport of products in their manufacturing process within the factory by employing such devices as conveyor belts and overhead cranes. Bodmer's model installation was described in his lengthy memoir published posthumously in the Transactions of the Institution of Civil Engineers in 1868. Bodmer left England in 1848, and never returned. Following Bodmer's example, the workshop of James Nasmyth by 1860 was set up with mass-production machinery which allowed him to employ unskilled labour in the production of machine tools for sale. In England, Bodmer was later called the 'uncrowned king of machine tool technology'.

Bodmer's 'assembly line' system was utilised in the mass-production of biscuits in 1833, when the British Admiralty built a bakery in which machines for kneading dough, and other machine operations were sequentially organised for the baking of biscuits in large quantities for the Royal

Navy. The dough was passed from one workman to another on trays transported on power-driven conveyors.

American inventor Col. Samuel Colt (1814-1862), who devised his handgun with the revolving drum in 1835, the first completely successful revolver, gave a great boost to mass-production by interchangeable parts. The first government order for his revolver was manufactured at Whitney's Armoury under the direction of Eli Whitney Jr., in 1847. But Colt was not entirely satisfied with Whitney's manufacturing method, and he built his own plant at Hartford, Connecticut, in which the manufacture of his revolver was broken down to a great number of unit-operations each one done by a special device, a machine or a tool designed solely for a specific single job, a manufacturing process reminiscent of the method of Polhem and Brunel. In 1849, Colt hired Elisha K. Root (1808-1865), a most ingenious mechanical engineer, to be his factory superintendent. Root acquired, designed and built about 1,400 special-purpose production machines even for the minutest manual operation such as burr removal, and provided special jigs, fixtures, and limit-gauges required for the mass-production of the revolver. Interchangeable parts of the revolver were assembled on an assembly-line, one man fitting one or two parts at a time. It was Root, more than Colt, who was directly responsible for the manufacturing system of Colt's revolvers. Colt's factory was a marvel of the modern system of mass-production by interchangeable parts employing assembly-line, which makes Colt and Root largely responsible for establishing the 'American system of manufacturing' as a modern mass-production system. Right from the beginning, American gunsmiths recognised the value of Polhem's limit-gauge method of measuring in interchangeable parts manufacture, which they promptly adopted to their mass-production of guns.

Almost simultaneously with Colt and Root, Richard Lawrence (1817-1892) developed a factory for mass-production of rifles by the interchangeable parts method. In 1851, the exhibits of Colt's mass-produced revolver, and of Robbins & Lawrence's mass-produced rifles at the Great Industrial Exhibition in Crystal Palace in London brought the American system of mass-production to the attention of the world and, particularly, to the attention of the British, who established in 1853 a special commission headed by Whitworth and Nasmyth to examine what the commission called the 'American system of arms manufacture'. In 1851, Robbins & Lawrence were contracted by the British government to equip the Royal Small Arms Factory at Enfield with 157 American machine tools including 74 milling machines for the mass-production of rifles. It marked the very beginning of the dissemination of the 'American system of mass-production' in Europe.

The American gunsmith had mechanised the manufacturing process so completely by breaking it down into a sequence of very simple unit-operations that, in effect, eliminated at the same time all the satisfaction a skilled craftsman experienced in his work. The division of labour and mechanical unit-operations also divided the interest of the skilled workman and, thus, made every job in mass-production almost totally uninteresting.

In conclusion it is apposite to stress the point that in spite of the improved lathes and other machine tools that were invented by the ingenious craftsmen of the Industrial Revolution, the workshop practice still remained to a great extent a practical art in which the precision of workmanship involving the accuracy of finish and fit essentially depended upon the skill of the individual craftsman rather than upon applied science. However, when the engineers were called upon to produce a series of identical machines the traditional methods of workshop practice were found to be inadequate because wherever interchangeability of parts was required, the traditional skill of hand and eye, however magnificent, was not good enough. This type of manufacture required a rather complete mechanisation of the production of machines as pioneered by Polhem, Marc Brunel, and the American gunsmiths.

Continuous Papermaking by Machines

Paper had been produced by manual methods for countless centuries. At the beginning of the 19th century the demand for paper had increased so much that manual methods of production of paper had become totally inadequate. Books and newspapers were in great demand by the public, and the famous London newspaper, The Times, which now appeared daily, had a circulation of 10,000. It was mostly through the influence of the owners of newspapers that mechanical papermaking came into existence.

In 1762 the French scientist René Antoine Ferchault de Réaumur (1683-1757) had suggested to make paper from wood, after observing that certain wasps made paperlike material out of wood fibre, but he never put his idea into practice. A German pastor, Jacob Schaffer, later proved quite conclusively that paper can be made from wood fibre.

The first practical method of making paper in continuous length from rags by mechanical means was invented by a young Frenchman, Nicolas Louis Robert (1761-1828), a former soldier turned printer. Robert devised a papermaking machine in 1798, which he patented in 1799 for 15 years, whilst he was a printer at the Publication House Didot (Saint-Léger) in Essonnes. Robert attempted to manufacture his machine commercially after he had been awarded 3,000 francs by the *Conservatoire des Arts et Métiers*, but he lost a great deal of money, both his own and his backers, when this business venture failed. Disheartened, Robert retired to become a poor schoolmaster, and he sold his patent rights to his employer Didot, whose brother-in-law, John Gamble, a stationer, improved the machine and brought a crude model and some rough drawings of it to England. Gamble persuaded the Fourdrinier brothers, Henry (1766-1854) and Sealy, to finance the development of Robert's papermaking machine. The 2 brothers bought the patent rights of Robert's papermaking machine in 1804 and employed a young and talented English technician, Bryan Donkin, to develop such a machine. The first partially successful papermaking machine able to produce a continuous web of paper was made by Donkin in 1804. It had an appearance of a washtub supplied with a wringing machine. Paddles stirred up the pulp and splashed it onto an endless wire sieve passing over rollers and transporting a web of wet paper to roller-presses covered with felt which squeezed out the remaining water and consolidated the paper web. Then the paper web was cut into sheets and hung up to dry as had been the earlier custom. The first machines used in papermaking had only replaced the initial stages of papermaking: the hand bruising and hydrating of cleaned, bleached and shredded cotton to form a fibrous pulp. By 1810 the mechanical process of papermaking was technologically ready to be marketed, but by that time the Fourdrinier brothers were bankrupt, mostly because of their difficulties in establishing their patent.

Donkin introduced additional steam-heated drying cylinders, which had been patented by T.B. Crompton in 1821, into continuous papermaking that speeded up the manufacturing process a great deal. Donkin continued improving his papermaking machine so rapidly that by 1830 he had brought the continuous papermaking machine to a state of practical perfection in which the machine was able to produce continuous rolls of paper up to 40 feet (12 m) in length and over six feet (1.8 cm) in width, a machine that revolutionised the papermaking industry. In the perfected machine the water- and-pulp mixture was fed to a continuous moving roll of fine wire mesh, the thickness of the paper being controlled either by the speed of the roll or the rate of feeding. The wet pulp from the wire mesh roll was picked up by felt-covered rolls which squeezed out the excess water in the paper pulp, after which the pulp fell onto a moving felt web which carried it through a second pair of rolls drying the paper further. The continuous sheet of damp paper was then wound on a drum and placed on the racks to dry. Donkin manufactured about two hundred of his papermaking machines, and one such machine won the gold medal at the Great Exhibition in London, in 1851. In the course of this work Donkin, like his friend Henry Maudslay, made important contributions to the precision and accuracy of workshop practice, which were the *sine quo non* to the development of mass-production. By 1830, Donkin had built such high precision into his machine tools in his workshop that he was able to supply perfectly fitting interchangeable parts to his papermaking machine on a moments notice.

Donkin's machine made paper mechanically, and papermaking from rags remained a purely mechanical process until about 1860, when the demand for paper had increased so prodigiously that soon a general shortage of rags forced the industry to search for new raw materials in papermaking, such as mechanical wood pulp, which had been patented in 1845 by Keller of Saxony in Germany, esperato grass, and chemical pulp consisting of wood chips cooked in calcium sulphate, a process which was patented in 1867 by Tilghmann of Philadelphia in United States. Introduction of chemical treatment of pulp since 1880 made papermaking a chemical industry, which it still is today. These 2 methods, as well as chemically treated waste paper are used today, and the finished paper up to 20 feet (6.11 m) wide is produced with a speed of 20 miles (32.19 km) an hour on the modern papermaking machines. A modern paper mill may produce as much as 700 tons, or more, paper a day.

Donkin was a versatile engineer and technologist. In 1814, Donkin, together with the mathematician and engineer Peter Barlow (1776-1862), assisted Thomas Telford to perform more than 200 experiments on an improved kind of chain-cable invented by William Brunton (1777-1851), which Telford used for the suspension chains of his famous suspension bridge over the Menai Strait. About 1854, Donkin also built the duplicate of the Difference Engine of Per Georg Scheutz (1785-1873), which was used in the mechanical computation of a life expectancy table.

Metal-Cutting by Machines

By 1880, the mass-production methods had become quite commonplace in various manufacturing industries of United States. At this time, the foreman of the machine shop of the Midvale Steel Company, Frederick Winslow Taylor (1856-1915), became convinced that the output of the machine tools was far too low in the production of locomotive tyres and railway axles made of manganese alloy steel in which his company specialised. In his opinion, the machining of the products which involved metal cutting was not fast enough for mass-production. Taylor realised that the cutting tools, and the cutting techniques used in machining were not good enough for the rate of modern production, and he began a programme of careful experiments with cutting tools that lasted twenty-six years. His aim was to provide the machine tool designers with mathematical formulae by means of which the performance of the machine tool could be predicted in terms of tool pressure, the speed of cutting, and the feed. He distinguished between twelve variable factors in metal cutting by machines, and he proceeded to examine their effect on metal cutting by changing one variable at a time, as John Smeaton had done in his famous experiments. Taylor carried out 50,000 methodical experiments in the course of which he cut 800,000 pounds (362,880 kg) of metal with his experimental cutting tools. Taylor learnt that the Mushet tungsten-manganese round-nosed steel tools can be run at much higher speeds than the old diamond-pointed tools, and that coarse cuts at low speeds remove more metal in a given time than fine cuts at high speeds. In his experiments on metal cutting Taylor distinguished between the depth and the thickness of the cut. A constant stream of water directed at the point of the chip removal allowed an increase in cutting speed so that the output of his experimental machine increased from 30% to 40 %. In order to prevent rusting of the metal, the cooling-water was saturated with carbonate of soda to form suds. Taylor used low-alloy carbon steel, the so-called 'high-speed steel', which was alloyed with tungsten, chrome, vanadium and molybdenum that gave the tools a high resistance to wear. He found that 0.7 % vanadium in the alloy steel markedly improved the wear resistance of the cutting tool.

In 1900, at the Paris International Exhibition, Taylor, who became the leading American machine toolmaker and efficiency expert in production methods, gave a remarkable exhibition of metal cutting in which the metal shavings were peeling off at blue-heat whilst the tip of the cutting tool operated at a visible red-heat of about 1550°F (843°C).

Beginning in 1894, Taylor used an electric motor to drive his experimental machine with high-speed cutting tools, and it is quite certain that without the electric motor he would have been unable to carry out his experimental programme. In fact, the improvement of the electric motor was promoted by its importance to drive individual industrial machines in factories and workshops. Taylor noted that as much power was required to feed the machine as to drive the cut when the tool had become dull.

Since 1889, Taylor had continued his experiments mostly at the Bethlehem Steelworks with the metallurgist Maunsel White with whom he treated the composition of his steel-tools and the degree of heat applied to them before permitting them to harden in the open air. The best comparative basis of tool performance was established as the exact cutting speed in feet a minute which could cause the tool to be ruined after a 20 minute run, all other factors being maintained constant. Mushet's original cutting steel had contained 7% tungsten, 2% carbon and 2.5% manganese. Taylor and White were able to prove that it is the manganese content which gave the alloy steel its self-hardening property.

In 1892, Henri Brustlein began producing chromium tool-steel at the steelworks near Unieux, in France, where Jacob Holtzer had earlier practiced pioneering metallurgy in using chromium in alloy steels. In 1894, Brustlein sent samples of his chrome-tungsten steel to be analysed by Taylor and White, who discovered that the self-hardening property of the manganese could also be obtained with chromium, which further improved the performance of tools. Then White increased both the chromium and the tungsten content, and added some silicon to improve the shock resistance of the cutting tools. The lathe used by Taylor at the Paris Exposition was fitted with the Taylor-White chrome-tungsten tools. The Mushet tungsten-manganese steel tool began to crumble when overheated, whereas the chromium-tungsten steel tool in cherry-red heat failed at 30 feet (9.14 m) per minute in 20 minutes but when heated to 2000°F (1093°C), which is just below the fusion temperature, the cutting speed could be increased up to 90 feet (27.43 m) per minute before failure. After 1900, Taylor and White experimented with other alloys such as molybdenum and titanium, but these alloy steels were too expensive. They discovered that the addition of 0.3% of vanadium improved the cutting tool, an amount which they gradually increased to 0.7% with further improvement in the cutting ability of the tool. The results of Taylor's extensive tests were published in his pathbreaking memoir in 1906. Based on his test, Taylor was able to devise a kind of a slide-rule for production engineers which replaced the rule-of-thumb methods customarily employed by machine tool operators with scientific control. The working life of a cutting tool depends largely upon the speed with which it is driven: the carbon-steel tool of 1850 had the maximum speed of 40 feet (12.19 m) a minute; the self-hardening steel alloy containing tungsten, vanadium and manganese invented by Mushet in 1865 increased the cutting speed of the tool to 60 feet (18.29 m) a minute; the Taylor-White cutting tools could work up to 100 feet (30 m) a minute. Thus Taylor's experiments gave the industry high-speed tool steel which from 1900 to 1910 had increased the allowable cutting speed 4–fold.

The direct impact of Taylor's experiments on the machine industry was to make the existing machine tools rapidly obsolete. The machines had to be completely revamped, for they had to be much more robust for the new cutting tools made of high-speed steel. The feeding-power and the driving-power of the machine had to be considerably increased, which, in turn, required the replacement of the customary cast-gears of the machines by hardened steel-gears in both drives, and the lubrication method as well had to be improved. Moreover, the speed-range for both drives had to be much more extensive than before, and as nearly infinitely variable as possible to achieve optimum cutting speeds for various distinct materials, for different size of workpieces, and for different thicknesses and depths of cut. All this meant that machines had to be designed according to the Taylor-White high-speed cutting tools by the scientific method conceived by Taylor.

These new requirements imposed by the high-speed cutting tools of Taylor made the old problem of machine tool drives, and particularly the feed-drives in the lathe, a problem which had existed since 1827 when Joseph Clement (1779-1844) brought out his revolutionary lathe with infinitely variable speed-drive, into a central technological problem in machining which needed an immediate solution.

The pioneering solutions, which were provided by Murdock with his vacuum motor and Nasmyth with his steam-powered motor to drive individual machines in the workshops, were occasionally used in the factories but these motors were not sufficiently versatile and amenable to great improvements. The steam-engines as central prime-movers drove networks of overhead belts in most factories, which provided little control of the driving-power at each individual machine. American blacksmith Thomas Davenport (1802-1851), was the first to design electric motors using magnetic induction to operate a lathe and a drill in his workshop in 1837, but his power source, the battery, was far too uneconomical to make it practical.

A Belgian inventor, Théophile Gramme (1826-1901), by accident, had demonstrated at the Vienna Exhibition in 1873 that an electric generator can be driven like an electric motor by another generator. An Italian physicist, Antonio Pacinotti (1841-1912), had demonstrated already in 1863 that an electric generator run in reverse becomes an electric motor. In 1874, Gramme installed an electric motor, which drove the line-shafting for various machines in his workshop in Paris, and it seems to have set a precedent. However, Gramme's electric motor still served as a central power source by replacing a steam-engine. Yet the great potential of the electric motor in the machine shop was not properly recognised until 1894, when Taylor used electric motors in his experiments to drive his experimental machine tool.

The first attempts to combine the electric motor and the machine tool into one unit were made in 1901. At first it was attempted to replace mechanical variable-speed gears by variable-speed, direct-current electric motors but this arrangement was seldom successful for only a motor of excessive size was able to produce sufficient power for heavy cuts at low speeds. Experience soon proved that the combination of a variable-speed electric motor and a mechanical form of gear change was the best, since it gave a wide range and a closely stepped range of speed. The quick-change gearbox for machine tools invented by W.P. Norton was introduced in United States in 1892, which enabled the speed of the spindle to be varied at will and over a wide range of thread cutting by 1910. In contemporary machine tools, the mechanical steps and electrical steps are automatically shifted by electrically operated clutches. The auxiliary electric motors were introduced to power feed-motions of machine tools before 1914.

By 1900 the automatic lathe and the turret lathe, which were provided with electric drive, chip pans, coolant sumps, preselected spindle speeds, and fast traverse properties, had become commonplace. By 1914 the universal cross-slide had been introduced which permitted multiple simultaneous cuts. The automatic lathes used in various industrial productions had many such features incorporated into their design. By 1920 rolling contact bearings, hardened and ground precision guides and ways were in general use.

In 1911, Taylor had published his landmark book, < Principles of Scientific Management >, which dealt with the efficiency of industrial processes, and it initiated the general trend to build the mechanical skill into the machine tool, and, even into the whole production process. The auxiliary electric motor eliminated the major portion of the complex mechanism in machine tooling, and since the appearance of Taylor's monograph, the skill in production became the skill in designing, building and maintaining the automatic machine tools. Therefore, Taylor was also a modern pioneer in production engineering.

After the first 'revolution' in machine tools brought about by Taylor, sophisticated, and yet rugged machine tools were developed at surprising speed beginning in the 1920's. About 1915, Edward Haynes produced non-ferrous, high-temperature alloy cutting tools. In 1928 Krupp Company in Germany produced the tungsten-carbide tools in which finely ground tungsten carbide were sintered together with a cobalt binder. The brittle properties and the difficulties of grinding the tungsten-carbide tools were overcome in the 1930's with the result that by 1939 the tungsten-carbide machine tools for cutting which admitted higher cutting speeds were in general use in Europe and North America. Since 1940, the ceramic cutting tools were developed from the abrasive materials of the earlier grinding-wheels, which were capable of heavy and large cuts instead of the many rapid fine cuts of the grinding wheel. The ceramic cutting tools require very rigid machines with great power. The most recent developments in machining technology are the robotised machine tools with powered control-systems and electronically controlled machines to which instructions are fed by punched-cards, tapes or, in general, by computers.

In 1926 the next important advance in the machine tools which came from Germany were the tungsten-carbide tools of tremendous cutting power, the first introduction of hard metal as cutting material marketed under the trade-names in United Kingdom as 'Widia' and 'Widmet', and in America as 'Carboloy'. This hard metal is an alloy of chrome, tungsten, titanium or molybdenum with their carbon compounds, or carbides. These alloys can be either cast or sintered. In sintering the metallic, or mineral elements in powdered form, the elements are heated to the temperature at which one, or more of the metals melt and cement the rest together into a hard metal alloy. As a result of this development, a number of hard but expensive metals were commercially marketed.

Since the tungsten, tantalum, titanium and other carbides are very hard, but very low in tensile strength as well as high in cost, the new carbide tools were made as composite tools consisting of a small cutting tip of hard metal brazed onto a steel shank.

In 1970, the cutting properties and durability of hard metals were improved further by new metallurgical processes, and by the application of solid-state physics. Tools made of hard metals coated with a layer, frequently no more than a thousandth of a millimeter in thickness, of titanium carbide and aluminium oxide, created another 'revolution' in metal cutting which could reduce the time of machining done by uncoated tools by as much as hundred times.

Metal Milling

The milling machine since the days of Eli Whitney had become quite a sophisticated machine tool by 1900. A great deal of improvements in the milling machine had been made by A.L. DeLeeuw and Fredrick Holtz since 1908, such as the constant speed geared-drive, and the mutually independent spindle-drive and feed-drive. The constant speed drive was first developed by John Parker in 1900, which made feasible the use of constant-speed electric motor. Since 1910 a great number of automatic controls of different types were devised by DeLeeuw, Hazelton and Solomon Einstein. Hydraulic drive had been introduced to a milling machine by 1926. Massive and powerful milling machines which had wide cutters and worked rapidly began to replace the traditional shapers and planers because the new machine tools produced better surface finishes than the older types. The emerging automobile industry needed specialised milling machines by 1912, and it required high precision, high speed and high production-rates from such machines, as well as greatly reduced labour cost. It led to scientific study of the cutting tools, and to larger, more rigid, more powerful, more complex, and, ultimately, more automatic machines.

Metal Grinding

The development of precision cutting tools of hardened steel made it necessary to develop precision grinding machines to dress the cutting tools. The basic development of grinding machines was done in United States. An early cutter-grinder had been used in 1783 by the French mechanic Réhé. In 1834 Wheaton of Providence, Rhode Island, made a grinding machine which had many features of modern grinders. James Nasmyth made a grinder in 1845 which had two cast-iron wheels mounted with abrasive stones, about 6½ feet (c.2 m) in diameter. The real development of grinding began with the silicate grinding-wheel made in 1872 by Hart. In 1877, Pulson made a successful grinding-wheel from emery, clay, and feldspar, with which precision grinding became possible. In the 1870's, Francis A. Pratt and Amos Whitney from the Pratt & Whitney Company, and Joseph R. Brown and Lucien Sharpe from the Brown & Sharpe Company developed grinding lathes, and plane and surface grinders for precision machining. In 1876, Brown & Sharpe produced the first universal grinding machine which had most of the modern features, such as provisions for form-grinding.

The early use of grinding was more or less restricted to grinding hardened gears to obtain a good surface finish. It was a slow and expensive operation. As the metal grinding-wheels wore away quite rapidly, an accurate compensation for the grinding-wheel wear in precision grinding work became essential. In 1908 Ward and Taylor invented a device which automatically corrected for the wear on the formed grinding-wheel by means of two diamond points which compensated for the wear of the formed grinding-wheel and maintained the correct form automatically. The first special machines to grind hardened gears for automobiles was the one designed by J.E.R. Reinecker in Germany, and another one produced by the ***Maschinenfabrik Oerlikon*** (*Machine Factory Oerlikon*) in Switzerland .

When the automobile industry began to require parts in their cars which had to be made of hardened alloy steels that had to have certain precision of manufacture as well as a satisfactory surface finish, the old methods of grinding based on manual skills were no longer acceptable because of the great costs involved. This new industrial requirement called for new and inexpensive grinding-wheels which had to be available in great quantities, and heavy, powerful, and accurate grinding machines capable of producing large cuts at low cost and at large production rates. This new challenge posed by the automobile industry was first met by the mechanic Charles A. Norton of United States.

The grinding-wheels suitable for such a large production-grinder was made possible by the invention of two artificial abrasives. The silicon carbide, known as 'carborundum', was a fortuitous invention of Edward Goodrich Acheson (1856-1931) resulting from his attempts to produce synthetic diamonds. Since carborundum improved the cutting power and durability of grinding-wheels, Acheson recognised its importance to production grinding, and the carborundum grinding-wheels appeared on the market as early as in 1898. However, they were quite expensive at the time, and it took about a decade to overcome its various shortcomings. In 1897 Charles B. Jacobs invented artificial corundum, known commercially as 'alundum', which was acquired by Norton Emery Wheel Company to carry out large-scale experiments with it. In 1906 Norton grinding-wheels were no longer made from emery and natural corundum but rather from aluminium oxide. Further research at the Norton Company with bauxite of extra high quality made it possible for them to market in 1910 solid grinding-wheels of various grades and types, and with uniform and given characteristics, as well as information on their use in precision grinding. All that was lacking was a powerful grinding machine capable of precision production in large quantities. Such a machine was designed by Charles A. Norton, who was not associated with the Norton Company, but was hired by this company to develop such a large grinding machine for them, after Norton's extensive experience with small grinding machines in the firm of his previous employers, Brown and Sharpe, who had ridiculed Norton's idea to make an accurate micrometer out of a large and powerful grinding machine. By 1900, Charles A. Norton had built two such large precision grinding machines which proved to be huge successes as industrial grinding machines. From that time on Norton devoted his efforts to designing and building grinding machines which were able to meet the particular production requirements of the automobile industry. By 1905, Norton's crankshaft grinder did in 15 minutes what had formerly required 5 hours of skilled labour. By 1910, he had produced a large production grinder for camshafts of automobiles made of hardened alloy steel.

In their efforts to solve the precision grinding problem in mass-production, Norton Company had established one of the first industrial research laboratories. In 1934, Norton Company made a diamond-bonded wheel which was able to dress the tungsten-carbide tools of Krupp Company, as a direct result of the research done in its research laboratory. The Norton Emery-Wheel Company formed in 1885 in United States changed the precision grinder into a heavy-duty production tool without which the automobile and the aircraft industry could not have come into existence.

Charles A. Norton's production grinding machine influenced other machine designers to produce special purpose production grinders for industry, such as James A. Heald's planetary cylinder-grinder and piston-ring grinder in 1904 and in 1905 respectively, automatic ball-bearing grinders, automatic thread grinders, and L.R. Heim's machine for centerless grinding in 1915 which came into industrial use in 1925. The science of grinding and the grinding-wheel came mostly from the efforts of James J. Guest in 1915.

Precision Shop Measurement

Joseph R. Brown's vernier caliper in 1851 and his micrometer caliper in 1867 had made precision measurements in the workshop possible by 1890. However, it was quite clear by the turn of the century that shop precision-standards used together with mechanical drawing techniques to manufacture interchangeable parts were necessary for mass-production. The Swedish technician, Carl Edvard Johansson (1864-1943) had recognised the need for precision-standards for use in workshop calibration of micrometers. He made an outstanding advance in metrology in 1908 when he introduced his block and slip gauge system consisting of 81 small rectangular blocks of hardened steel varying in length from 0.05 to 4 inches for the end measurement method. By means of the combination of the spirit-cleaned blocks together can give any desired gauge length in 0.0001 inch steps. By 1907, Johansson had scientifically linked his block gauge system to the standard meter at Paris and to the physical properties of the steel from which his blocks, that required absolutely parallel plane surfaces of high surface finish and the precise dimensions indicated, had been produced by the skill of his hands. Using this ingenious set of blocks any dimension over a wide range could be obtained. Ever since mechanical means of producing the so-called 'Joe blocks' was developed by Major Hoke of the U.S. Bureau of Standards in 1918, these blocks still stand as the standard of precision measurements in the workshop despite the electronic, pneumatic, and hydraulic measuring methods developed since 1907.

CHAPTER XVI

SHIP PROPULSION: From Steam-Engine to Diesel-Engine

Anticipation of Navigation by Natural Power

The oldest known mechanical marine propulsion system not operated by manpower was the paddle-wheel turned by oxen. Some of the boats transporting Roman armies under the command of General Claudius Caudex were equipped with such a propulsion system. There existed a bas-relief from about A.D. 527 depicting a war galley fitted with 3 pairs of paddle-wheels, each driven by a pair of oxen. Warships with paddle-wheels operated by manpower is reported to have been used in China in the 7th century A.D..

In the West, Roberto Valturio of Rimini reproduced in his book < On Military Matters > (*De re militari*), Veronae, 1472, (in Libri XI, page 2), boats fitted with paddle-wheels, the concept of which he attributed to Matte de Pastis.

A Spanish sea captain, Blasco de Garay, proposed for ship propulsion the use of paddle-wheels fitted to both sides of the ship. In 1543, this method of marine-propulsion was tried successfully in the harbour of Barcelona, where the ship named <u>Trinidad</u> of 209-ton displacement was propelled by paddle-wheels with a speed of 3½ miles (5.6 km) an hour, but it seems the paddle-wheels were turned by 25 sailors rather than by steam-power. In England, propulsion of ships by paddle-wheels was first proposed by William Bourne (died 1583) in 1578.

Salomon de Caus (1576-1626), after studying the experiments with bomb shells pressurised by steam which were performed by the French military engineer David Rivault and recorded in his book < Elements of Artillery > (*Les elements de l' artillerie*), Paris, in 1608, appears to have been the first to recognise the great power of steam as the motive force capable of moving vehicles on land and ships at sea. He probably had the first idea of a steamship.

David Ramsey (died 1653), a member of the English Royal Court at the time when Salomon de Caus was attached to the English Court as the Royal Engineer, obtained patents in 1618 and 1630 to raise water by steam-power, and to propel boats by steam, but his ideas were still far too impractical.

In 1661 and 1662, two Englishmen, Thomas Toogood and James Hayes introduced the idea of propelling boats by the reaction to the water expelled from the boat. Their propulsion system consisted of bellows which forced jets of water below the water level through the bottom or sides of the ship. This proposal represents the first known application of the reaction principle of the ancient Greek mathematician and mechanical engineer Archytas of Taras (428-c.347 B.C.) to the propulsion of naval vessels.

Edward Somerset, Marquis of Worcester (1601-1667), another member of the English Royal Court who was acquainted with Salomon de Caus and Ramsey, carried out the first practical experiments using steam as the prime mover. He also suggested the use of paddle-wheels turned by the force of the river to wind a rope upstream round a trommel with the other end of the rope attached to a boat downstream which is thereby moved upstream. This method of propulsion, however, was not original, it was known since 1450.

Navigation by Steam-Power

It is certain that the first definite proposal for a steamboat was given by the French physicist and physician Denis Papin (1647-1714), a native of Blois who invented the principle of steam-engine. Papin proposed to use his atmospheric steam-engine, in which the piston was forced down into a cylinder by the atmospheric pressure thereby condensing the steam under the piston and communicating this reciprocal motion by a transmission mechanism as rotational motion to the paddle-wheels of the boat. In 1695, Papin specified the toothed-rack which drove a pinion on the shaft of the paddle-wheel as the mechanism of transmission of motion. His atmospheric steam-engine was designed to have from 3 to 4 pistons. In 1707, Papin built a boat with paddle-wheels, and tried it out on the Fulda river in Germany. In this trial the paddle-wheels were turned by manpower, since his steam-engine was not yet connected to the paddle-wheels. Unfortunately, Papin's boat was destroyed by the boatmen who considered it a menace to their livelihood. In 1708 Papin prepared a new proposal to construct a steamboat and submitted it to The Royal Society in London. All mechanical details were worked out by Papin and given in the proposal of his steamboat, the steam-engine of his steamboat drove a bank of oars. This makes Papin, besides being the conceptual inventor of the steam-engine, also the conceptual inventor of the steamboat.

After the successful introduction of the Newcomen atmospheric steam-engine, Dr. John Allen (1660-1741) revived the reaction propulsion principle in 1729. He proposed to use Newcomen atmospheric steam-engine to pump water through the stern of the boat under the surface of water. Allen made a successful trial of the reaction principle of propulsion with a canal boat, but the pumps in his trial boats were operated by manpower rather than by steam-power.

Jonathan Hulls (1699-1758) obtained a patent in 1736 for a steam-powered tugboat propelled by a paddle-wheel driven by a Newcomen atmospheric steam-engine with a vertical single-acting steam-cylinder of 30-inch (0.76 m) bore. The power-stroke of the engine was the downwards stroke of the piston produced by the atmosphere which raised a counterweight equivalent to ½ of the effective pull. The descent of the counterweight due to gravitation produced the working return-stroke so that the entire operation of the engine was similar to a double-acting engine. Hulls used ratchet wheels driven by ropes from the piston of the engine, and the counterweight as his motion transmission mechanism 32 years before Nicolas Joseph Cugnot (1725-1804) built his ratchet-driven tractor. Although Hulls supplied considerable details for his tugboat, he was unable to obtain capital for the building of an experimental version of his tugboat. It appears that Hulls' tugboat, had it been built, would not have been successful for the Newcomen engine was far too heavy for the boat.

In 1753, Daniel Bernoulli (1700-1783) submitted 2 proposals in a competition given by the French Academy of Sciences in Paris for the best propulsion system of boats which does not use the power of wind. Bernoulli's first proposal for such a propulsion system was a reaction system which consisted of a vertical tube fitted with a funnel on top into which water is being continually pumped. This water is discharged under gravity in the stern of the boat beneath the surface of water and the boat is propelled by the reaction to the expelled water. Bernoulli's second proposal consisted of a 6-foot (1.83 m) diameter wheel with inclined iron blades which were to be mounted on parallel shafts on both sides as well as at the stern of a 100-ton boat. These propellers acting as screws were to be operated by steam pumps, horses, or men, but he gave no details for the steam pump.

In 1755, a French canon from Nancy, Joseph Gautier (1717-1776), undertook a comparative study of the steam-powered and the man-powered propulsion of boats. He designed a Newcomen steam-engine incorporating a steam-cylinder of 31-inch (0.79 m) bore and 6-foot (1.83 m) stroke and executing 15 strokes a minute. According to Gautier's calculations this steam-engine exerted a total force of about 11,000 pounds (48,895 N). A chain and rachet wheel mechanism mounted on a shaft carrying rotative oars at its terminus converted the reciprocal piston motion into the rotary motion of the shaft which propelled the boat. The oars were idle on the return stroke of the piston. Gautier had designed the steam-power to operate the pumps, to heave the anchor, and to heat the furnace, and even to cook the food. This boat was never built.

A different system of propulsion was devised in 1759 by a Swiss pastor, J. A. Genevois of Bern. His design employed compressed springs to move the paddles resembling weblike feet, and he proposed gunpowder, or Newcomen atmospheric steam-engine to compress the springs. His proposal of 1760 to the British Admiralty to build an experimental boat of this type was turned down.

An American mechanic, William Henry (1729-1786) of Lancaster, Pennsylvania, who had visited Britain in 1760 and had become interested in applying steam power to boat propulsion, built a paddle-wheeled boat about 1763, but it foundered before he was able to make a trial run with it. Henry also designed an *aiolopyle*-type reaction steam-turbine based on the principle of Heron of Alexandria (*floruit* c.150 to 100 B.C.), but he never got around to making a model of it. It appears that Henry's work on the steamboat influenced John Fitch, and also other American steamboat designers John Stevens and Robert Fulton.

All these pioneering efforts in building a steamboat were doomed to failure, mostly because a more efficient steam-engine than the atmospheric steam-engine of Newcomen did not yet exist. The steamboat became a practical proposition only after James Watt (1736-1819) invented the first genuine steam-engine with a separate condenser in 1770. Watt himself proposed a 'special oar', a screw-propeller, for the propulsion of canal boats by steam-power, but he subsequently did not take a keen interest either in steamboats or in steam-carriages, and, therefore, he did not promote steam-powered navigation.

Practical Steamboats

The first pioneer builders of practical steamboats were the French. The first outstanding attempt to build an experimental steamboat was made by Count J. B. d'Auxiron (1728-1779), a retired French army officer. He decided to develop Papin's concept of the steamboat, and the design he prepared in 1771 was granted a 15-year concession in 1772 by the French Finance Minister, A.L.J.B. Bertin, Count de Bourdeilles (1719-1792). D'Auxiron formed a company, the *Société d'Auxiron*, in partnership with Marquis Claude de Joffroy d'Abbans (1753-1832), who later became the first successful steamboat builder, and Comte Charles M. de Follenay (1734-1814) for the special purpose of constructing his steamboat. The construction of his steamboat was subject to the approval of the French Academy of Sciences, and Jacques C. Périer (1742-1818) was charged by the Academy to report on the progress of the construction to this august society. Construction of the steamboat began in 1771, and in 1772 an atmospheric steam-engine was fitted to the boat. Unfortunately, the counterweight of the beam of the engine fell through the bottom of the boat because of sabotage, causing the steamboat to founder at Meudon.

After this disaster, Jacques C. Périer took it upon himself to construct in 1775 the first practical steamboat at Chaillot, near Paris. Périer fitted his small boat with a model steam-engine which had a steam cylinder of 8-inch (0.rier20 m) bore to turn 2 paddle-wheels fitted to each side of the boat. Unfortunately, his small steam-engine delivered only one horsepower, which was insufficient to propel his small craft up the Seine river. It was calculated that 15 to 20 horsepower would have been necessary to move the boat with the speed of 4½ knots (8.3 km an hour) up the river Seine. Périer, after unsuccessful experiments with mechanical devices of propulsion other than paddle-wheels, abandoned his steamboat experiments and, thenceforth, concentrated on land applications of steam-engines. However, Périer's experiments inspired Abbé Étienne d'Arnal (1733-1801), Canon of Alais Cathedral, to make small-scale experiments with steamboats beginning in 1788, but he was no more successful than Périer.

One of d'Auxiron's partners, Marquis de Jouffroy d'Abbans took up steamboat experiments after Périer had failed, and he became the first steamship pioneer whose steamboat propulsion experiments were crowned with some success. His first trial in 1778 on the Doubs river was carried out with a 43-foot (13.11 m) long boat which had a beam of 6 feet 5 inches (1.95 m), and a displacement of 9 tons. This boat was outfitted with 2 inclined copper steam-cylinders of 22 feet 5 inches (0.57 m) bore which were reinforced with iron hoops. The pistons had a 64-inch (1.63 m) stroke, and produced a reciprocating movement of weblike paddles, which were 32 feet (9.75 m) wide and 1 foot 7 inches (0.49 m) deep, on both sides of the boat. Nevertheless, this experiment ultimately failed because the propulsive mechanism was insufficient to move the boat upstream.

His second steamboat, which had a wooden hull and was called <u>Pyroscaphe</u>, was built in 1783 at Écully near Lyons in collaboration with de Follenay. <u>Pyroscaphe</u> was driven by a double-acting steam-engine which incorporated a steam cylinder of 25 ⅝-inch (0.65 m) bore and a piston stroke of 77 inches (1.96 m) enclosed within the boiler. The rotative motion of the paddle-shaft was produced by a piston-rod and double-ratchet mechanism similar to Papin's mechanism. The 2 paddle-wheels of this steamboat were 13 feet 1 inch (3.99 m) in diameter and each had 8 radial floats 6 feet 5 inches (1.95 m) long and 2 feet 1 inch (0.64 m) wide. Regrettably, the boiler used was insufficient to maintain the required steam-pressure for the steam-engine, yet despite this shortcoming the steamboat was able to make at least a partially successful 15-minute trip up the river Saône near Lyons, the first steam-propelled boat ride in history, after which the steam power failed. <u>Pyroscaphe</u>, which displaced 182 tons, had an overall length of 148½ feet (45.26 m), 25-foot (4.57 m) beam, 28-foot (8.53 m) width over paddle-wheels, and a draught of 3 feet 2 inches (0.98 m). The machinery of <u>Pyroscaphe</u> was built by *Frérejeon et Cie*. in Lyons.

In 1784, Jouffroy made a 1-to-24 scale model of his steamboat which is still extant at the *Musée de la Marine* in Paris. Unfortunately, the French Revolution put a stop to Jouffroy's steamboat experiments. Long after the Revolution, in 1816, Jouffroy patented and built another steamboat, named <u>Charles-Philippe</u>, at the river Seine, but by that time other steamboat builders had surpassed Jouffroy both in the steamboat design and construction.

In 1775, about the time Jouffroy began his experiments on steamboats, an American inventor, David Bushnell (born c.1742), a contemporary of John Fitch, devised and built the first successful submarine, which he named <u>The American Turtle</u>. Bushnell discovered, and proved experimentally, the 2 principles of submarine: firstly, that the submerging can be carried out by means of a floating tank, and secondly, that the gunpowder can be detonated under water with great force which does not dissipate itself as it does in the air. Bushnell's submarine was 6 feet (1.83 m) high, and it accommodated one man. It had a tank for submerging which could be emptied by a pump operated by a foot-treadle. Screw-propellers were provided to move the submarine forwards, sideways and upwards. It was the first marine craft to make practical use of screw-propellers for the propulsion of any boat. His submarine had a conning tower but no periscope. Bushnell's Master of Art thesis at Yale College contained a description of his submarine.

The first completely successful steamboat experimenter was the American inventor John Fitch (1743-1798) of Windsor, Connecticut, who in 1785 made his model steamboat which was to be propelled by an endless chain of paddles fitted to the port side of the model and equipped with a leeboard on the starport side. However, this chain-and-paddle device failed to move the boat. In 1786, Fitch made a successful test with his first steam-powered boat on the Delaware river. It was propelled by 12 reciprocating vertical oars of 12-foot (3.66 m) length, 6 on either side of the boat, moved by a chain-sprocket and driven by Fitch's own steam-engine at a speed of 3 miles (4.83 km) an hour. His engine was a low pressure, condensing steam-engine fitted with one horizontal double-acting cylinder of 12-inch (0.30 m) bore and 3-foot (0.91 m) stroke of the piston. The displacement of Fitch's boat was 9 tons, its overall length was 34 feet (10.35 m), its beam was 8 feet (2.44 m), and its depth was 3½ feet (1.07 m). Fitch, and his fellow mechanic, Henry Voight, a Dutchman, were able to develop and build their double-acting steam-engine in a relatively short time, a task which had taken Watt many years. In 1787 Fitch installed a singly wound watertube boiler in a better, but smaller steamboat, which he demonstrated to farmers in Philadelphia. It did work, but not for long since after a brief operation of the steam-engine the watertube in the boiler clogged up. In 1788, Fitch launched another boat, 60 feet (18.29 m) long, which was propelled in a similar manner. This boat carried 30 passengers about 20 miles (32.2 km) from Philadelphia to Burlington in 3 hours and 10 minutes at an average speed of 5½ knots (10.2 km an hour).

In 1790, Fitch completed his 3rd ship equipped for its propulsion, as recommended by Oliver Evans (1755-1819), with paddle-wheels at the stern, by means of which Fitch established a ferry service for passengers and freight between Philadelphia and Bordertown. This steamboat ran 80 miles (128.7 km) a day at 8 miles (12.9 km) an hour – much faster than Fulton's ship <u>Clermont</u> in 1814. This steamboat was propelled by a beam-engine fitted with one steam cylinder of 18-inch (0.46 m) bore. This particular steamboat of Fitch ran altogether from 2,000 to 3,000 miles before bankruptcy forced Fitch to close down his ferry service. The basic reason for the bankruptcy was that the steam-engine was still too bulky and occupied so much space on the boat that there was little room left for the cargo and passengers.

In 1793 Fitch built another boat, named <u>Perseverence</u>, of 25-ton displacement for use on the river Mississippi, but it was wrecked before completion. After an unsuccessful promotional trip to France, Fitch returned to United States, and in 1796 he constructed another steamboat which was 18 feet (5.49 m) in length and had a 6-foot (1.83 m) beam. It was driven by a beam-type steam-engine with 2 cylinders and propelled by a screw with 3 complete turns. The boiler of this steamboat was an iron vessel with a thick plank lid held tight by a transverse iron bar. The steam-engine's cylinders were barrel-shaped and made of wood slats tightened with iron hoops. The 2 piston-rods worked upwards and were attached to the ends of the beam which was supported by an iron column, and the connecting-rod turned a crank of the forward end of the propeller shaft. Fitch demonstrated this steamboat on the Collect Pond in New York City, but despite his successes he was unable to raise capital for further practical development of his steamboats. Fitch was so discouraged from his string of failures to attract financial support to his projects that he ultimately committed suicide.

At the time Fitch constructed his singly wound watertube boiler in 1787, an Italian scientist Serafino Serrati of Florence built a small model of an air-propelled boat which was fitted with a vertical funnel, the mouth of which was bent towards the stern. Air heated under the funnel escaped through the funnel and propelled the model boat by reaction. It was later reported that Serrati propelled a full-size boat by this reaction device on the Arno river in 1789, but no documentary evidence of this event has survived. The reaction principle of marine propulsion was first proposed by Thomas Toogood and James Hayes in 1661, and later revived by Dr. John Allen and Daniel Bernoulli. In 1776, Arthur Donaldson made a similar proposal to the Council of Pennsylvania. The practical development of this idea was promoted in 1775 by Benjamin Franklin (1706-1790) in his memoir on maritime affairs to the American Philosophical Society in which he proposed to pump water manually from the bow of a small boat and force it out under the surface as a water-jet at the stern in order to propel the boat by the reaction to the water-jet, since Franklin had become convinced that propulsion of boats by paddles is inefficient.

The reaction principle of propulsion for steamboats was first carried to practical success in 1785 by the American inventor James Rumsey (1743-1792) of Virginia, when he travelled with his steam-driven water-jet vessel which had a wooden hull down the Potomac river to Stephertown. After his initial success Rumsey constructed another steamboat which he tested in 1787 and successfully moved it against the current of the Potomac River with 2 tons of cargo at 4 miles (6.5 km) an hour. This steamboat was about 18 feet (5.49 m) long, had 6-foot (1.83 m) beam, 2 feet 8 inches (0.82 m) depth, and a displacement of 3 tons. His steam-engine was designed on Newcomen's atmospheric steam-engine principle, and operated with approximately 2 psi (13.8 kPa) steam-pressure. The 2 cylinders of the engine were on top of each other and had a 12-inch (0.30 m) bore. The pistons, which were brazed around by 15-inch (0.38 m) wide strips, had a common 26-inch (0.51 m) stroke. The valves of the engine were made of brass. The boiler of this experimental steamboat was an iron pot which leaked so badly, that Rumsey replaced it in 1786 by a pipe boiler of his own design which was 3½ feet (1.07 m) high. The furnace was 2 feet (0.61 m) square inside, and it incorporated a 120-foot (36.58 m) long coiled iron pipe of 2-inch (5.1 cm) bore. The steam-engine and the boiler occupying an area of about 4-foot (1.22 m) square in the boat weighed 810 pounds (363 kg), and consumed about 4 bushels of coal in 12 hours. In this project Rumsey was financially supported by General George Washington (1732-1799).

Rumsey went to England where he obtained a patent for his reaction-driven steamboat. He obtained financial backing in England, and built in Dover an experimental vessel, named <u>Columbian Maid</u>, of 101-ton displacement. It was fitted with a steam-powered pump of 2-foot (0.61 m) bore, the piston of which executed from 20 to 22 stroke a minute. Rumsey collaborated with James Watt in the design of the steam-engine of his last boat. Unfortunately, Rumsey died of a stroke in the middle of a conference with his backers in late December of 1792. After his death, in early 1793, a successful trial of Rumsey's steamboat was made on the Thames river in which the boat reached an average speed of 4 knots (7.4 km an hour). As late as in 1865, the Royal Navy launched an armoured sloop, named <u>Waterwitch</u>, which had a reaction propulsion system similar to that of Rumsey's steamboat.

The next important development of steamships was taken in 1787 by a Scot, Patrick Miller (1731-1815) of Dalswinton, who was a banker in Edinburgh. Miller designed a ship with triple-hull, named <u>Edinburgh</u>, which was built and launched in 1786. He provided this ship with manually operated paddle-wheels as an auxiliary propulsion system. The length of this ship was 73 feet 4 inches (22.34 m), and its width was 22½ feet (6.86 m). The vessel was outfitted with 2 paddle-wheels, 6 feet (1.83 m) in diameter and 4 feet (1.22 m) in width. Miller also had a ship with double-hull and a paddle-wheel built to his design which was 60 feet (18.29 m) long, and 14½ feet (4.42 m) wide. The paddle-wheel operated by a capstan was located between the 2 hulls. It took 5 men to operate the capstan to achieve a speed of 4 miles (6.4 km) an hour, but it proved to be too exhausting for the crew of the ship to maintain. Therefore, Miller contacted a Scottish inventor, William Symington (1763-1831), a son of a Scottish mining engineer who had constructed and patented the first direct-acting steam-engine, to build a steam-engine to power Miller's double-hull ship. Symington became the British pioneer of steamship propulsion by constructing a marine steam-engine which had 2 cylinders of 4-inch (0.10 m) bore, open to the atmosphere at the top, and outfitted with jet-condensers. Each piston had an 18-inch (0.46 m) stroke, and was connected by 2 chains with a drum which alternately turned in opposite directions. The 2 horizontal shafts of the two paddle-wheels each had 2 loose pulleys with ratchet-teeth round their inner flanges which were turned by the chains from the drum in opposite directions. Between each pair of loose pulleys was a disk fitted with 2 panels and keyed to the shaft. The teeth of the ratchet wheels alternately meshed with the panels to turn the paddle-wheel in one direction. The two parallel wheels were located between the 2 hulls, one in front of the other. This marine steam-engine, which is still extant, was separated from its boiler. This steamboat was 25 feet (7.62 m) long, 7 feet (2.13 m) wide, had a draught of 2 feet 2 inches (0.67 m), and displaced 5 tons. In 1788 Miller's steamboat made a successful trial voyage on Dalswinton Lake reaching an average speed of 5 miles (8.0 km) an hour. Among the passengers were the poet Robert Burns and the painter Alexander Nasmyth, father of the mechanical engineer and toolmaker James Nasmyth. Symington claimed that his engine, which was an atmospheric engine with a separate condenser (and for this reason an infringement on Watt's patent) was 25% more efficient than a comparable Watt's engine. Miller did invite Watt to cooperate in an effort to provide an effective steam propulsion system for boats, but Watt declined.

In 1801, Lord of Dundas of Kerse (1742-1811) had a steam-driven tugboat built which was propelled by a recessed paddle-wheel at the stern to carry out experiments on the Forth and Clyde canal for the use of steam-propelled tugboats as a replacement for horse-towing of vessels. The tug, named <u>Charlotte Dundas</u>, was fitted by Symington with a 10-horsepower steam-engine and the associated power transmission mechanism. This steam-engine had a horizontal double-acting steam cylinder of 22-inch (0.56 m) bore and 4-foot (1.22 m) stroke of the piston. The piston rod was sliding in guides with the connecting rod acting directly on an overhung crank of the paddle-wheel shaft, a mechanism patented by Symington in the same year. The steam-engine was located on the port side, and the boiler on the starboard side of the boat. The condenser and vertical air pump worked by a bell-crank mechanism from the crosshead were placed below the deck, and the paddle-wheel was located at the stern of the boat. In 1802 this steam-driven tugboat successfully towed 2 loaded vessels, each carrying a cargo of 70 tons, over 19½ miles (31.4 km) in 6 hours against a strong head wind in the Fourth and Clyde Canal, but the wash from the paddles damaged the banks of the canal and as a consequence of this discouraging experience the tugboat was rejected. The length of this tugboat was 56 feet (17.07 m), its beam was 18 feet (5.59 m), and its depth was 8 feet (2.44 m).

Francis, the Third Duke of Bridgewater (1736-1803) ordered 8 similar tugboats for his Lancashire canal but, unfortunately, the Duke died before the order was filled, and the committee which took over the canal after his death did not honour the Duke's agreement, and cancelled the order. The cancellation of the order for additional 8 tugboats forced Symington into financial bankruptcy.

In 1773, American mechanic, businessman and inventor, Oliver Evans became convinced that horseless carriages can be built by using high-pressure steam. In 1802 Evans had an 80-foot (24.38 m) wooden hull built for which Evans himself constructed a 15-horsepower high-pressure steam-engine. This steamboat was intended to run between New Orleans and Natchez on the Mississippi river. Tragically, the hull was wrecked in transit by a hurricane and, therefore, its steam-engine was installed on the shore to saw timber in a lumber mill.

In 1804 Evans constructed near Philadelphia, his famous amphibian craft, a wooden scow, 30 feet (9.14 m) long and weighing 15 tons, to demonstrate that crafts can move on land as vehicles and on water as vessels by the power of steam. It was the first amphibious steam-driven vehicle, and one of the first vehicles to use successfully a high-pressure steam-engine. The steam-engine developed about 5-horsepower with a steam cylinder of 5-inch (0.13 m) bore and a 10-inch (0.25 m) stroke of the piston packed with hemp under 30 psi (206.9 kPa) steam pressure. The steam-engine drove either the paddle-wheel at the stern of this amphibious craft, or the 4 wheels to run the craft as a vehicle on land. Under normal working conditions the engine made about 30 revolutions a minute. The steam at 50 psi (344.8 kPa) pressure was supplied from a horizontal iron boiler 2½ feet (0.75 m) in diameter and 10 feet (3.05 m) long set in brickwork. This amphibious craft Evans named 'Both-Way Digger' (*Orukter Amphibolos*) since it was meant to be a wooden dredging scow. Evans drove this amphibious craft 3 miles (4.83 km) from the Central Square in Philadelphia to the Schuylkill river where he removed its wheels, and steamed the scow down the Schuylkill to the Delaware river, and then up the Delaware river to Philadelphia. The average speed of the *Orukter Amphibolos* was 15 miles (24.14 km) an hour. An operating model of the *Orukter Amphibolos* in 1– to – 8 scale is still in the collection of the Franklin Institute in Philadelphia. The length of the full size *Orukter Amphibolos* of 15½-ton displacement was 30 feet (9.14m), breadth 12 feet (3.66 m), and draught of 1 foot 7 inches(0.49m).

In 1802 in England, Richard Trevithick (1771-1833), and his cousin and collaborator Andrew Vivian, obtained a patent for a high-pressure steam-engine, and they performed experiments in 1804 with a steam cylinder of 8-inch (0.20 m) bore and 54-inch (1.37 m) stroke.

In 1805 Trevithick fitted a canal boat of 70-ton displacement with a steam cylinder of 10-inch (0.25 m) bore driving side paddle-wheels, each with 6 radial float-boards 26 inches(0.66 m) by 14 inches (0.36 m). The speed reached by his canal boat was 7 miles (11.27 km) an hour. In 1806, Trevithick constructed a steam-driven bucket-ladder dredger mounted on the wooden hulk of a dismantled ship. Trevithick's high-pressure steam-engine of 6-horsepower had a steam cylinder of 12-inch (0.30 m) bore and 2½-foot (0.76 m) stroke of the piston. Its cylindrical iron boiler was 4 feet and 10 inches(1.46 m) in diameter and 8 feet (2.44 m) in length, and it supplied steam at about 40 psi (275.8 kPa). The boiler was supplied with water by means of small force-pump driven by the steam-engine. 180 tons of spoil could be raised in a tide of 7 hours from the depth of 14 to 18 feet (4.27 to 5.49 m). The length of the dredge was 860 feet (24.38 m), the breadth of the hull was 18 feet (5.49 m), the depth was 11½ feet (3.51 m), the draught was 10 feet (3.05 m), the depth of the dredge was from 14 to 18 feet (4.27 to 5.49 m). This dredger remained in service about 10 years.

Trevithick and Robert Dickenson of Middlesex obtained a patent in 1808 for a paddle-wheel propulsion [later used by Maudslay, Sons & Field without much success], and in 1809 for an iron ship construction in which both the deck and the hull were to be made of iron plates. John Urpeth Rastrick (1780-1856), an ironmaster, constructed Trevithick's machinery.

In 1815, Trevithick proposed an *aiolopyle*-type reaction steam turbine for ship propulsion based on the principle of Heron of Alexandria's which was to be mounted on the same shaft as an Archimedean screw-propeller. The boiler for it was vertical, the top of the flue in contact with steam was intended to produce some dry superheat. It was meant to be a towing engine.

The Archimedean screw had been known since antiquity, but the first to propose and test its use for marine propulsion was the French engineer Pierre Bouguer (1698-1758) in 1746. The first British patent of the screw-propeller was taken out by Samuel Miller in 1775. American David Bushnell was the first marine engineer to use it for the propulsion of his submarine in 1775. Fitch and John Stevens made trials with screw-propellers but without much success.

In 1829, Austrian engineer Joseph Ressel (1793-1857) built a screw-propeller into a boat of 5-ton displacement, which was operated by 2 sailors. Pierre Louis Frederic Sauvage (1785-1857), a French naval engineer, recognised that a screw-propeller of a single turn was more efficient than a screw-propeller of several turns and he patented a propeller with a single helicoidal blade in 1832, but his propeller was rejected in comparison with other designs in 1843. The first efficient screw-propellers were invented by Francis Pettit-Smith (1808-1874) of Hendon, Middlesex, England in 1837, and by a Swedish engineer John Ericsson (1803-1889) in 1836, but it was not until the 1860's that the superiority of the propeller over paddle-wheels was clearly established. As late as in 1858, paddle-wheels and screw-propellers were still used in combination to propel Isambard Kingdom Brunel's mammoth steamship of 27,000-ton displacement, The Great Eastern, the largest ship of the century. The first screw-propelled steamer was the British ship Archimedes which displaced 237 tons.

In 1789, an American inventor and judge, Nathan Read (1759-1849) of Salem, Massachusetts, invented a compact and safe multitubular boiler for steamboats, which contained 76 vertical watertubes inside the boiler, and also a new type of high-pressure steam-engine similar to the one later constructed by Richard Trevithick in England.

The first commercially successful steamboat was built by the American artist, civil engineer and inventor, Robert Fulton (1765-1815). Fulton was a very systematic inventor, a brilliant professional engineer, and a gifted businessman. He made himself familiar with all the previous work on steamboats, and thus avoided the mistakes made by other inventors. Fulton was able to use well-proven design components and experimental results of Col. Mark Beaufoy (1764-1827) for the best hull proportions. His interest in commercial steamboats had been raised by Robert R. Livingston (1746-1813), President Jefferson's American ambassador in Paris who negotiated the purchase of the Louisiana territory from Napoléon Bonaparte (1769-1821). Livingston, who had experimented in 1793 with a boat propelled by hydraulic reaction in collaboration with Marc Isambard Brunel (1769-1849) and John Nisbet in New York, and with a paddle-wheel steamboat in 1798 in collaboration with his brother-in-law, Col. John Stevens (1749-1838), provided the financial backing for Fulton to build a steamboat.

Fulton, who worked in Paris, constructed an experimental steamboat which had a length of 66½ feet (20.27 m), beam of 10 feet 7 inches (3.23 m), depth of hull of 3⅓ feet (1.00 m), draught of 1 foot 8 inches (0.55 m), and displacement of 25 tons. He acquired the steam-cylinder of Jacques

C. Périer which had a vertical double-acting cylinder of 17.7-inch (0.45 m) bore, a piston stroke of 31½ inches (0.80 m), and 8-horsepower. The rest of the machinery was built by Étienne Calla (1760-1835). The boiler was an early type of watertube boiler invented for steamboats by his friend Joel Barlow (1754-1812), a poet and American ambassador in Paris, which was 7 feet (2.13 m) long, 5 feet (1.52 m) wide, 5 feet (1.52 m) high and worked at a low steam pressure of about 2 psi (13.8 kPa) above the atmosphere. The condenser was fitted with a vertical air pump of 6-inch (0.15 m) bore, and 15-inch (0.38 m) stroke. Just before the trial run of the steamboat, when the boat was still moored at the Seine river, a violent storm broke the hull of the boat in half. This misfortune forced Fulton to build a new hull 74 feet 7 inches (22.73 m) long, with 8 feet 2 inches (2.50 m) beam, and 3 feet 2 inches (0.97 m) depth. This boat was tested in late 1803, and it proved to be an outstanding success. The boat attained a speed of 4½ miles (7.24 km) an hour against the current of the Seine river. After further experiments on the Seine river, Fulton achieved in 1807 the first commercial success with his paddle-wheel steamboat named The Clermont. He fitted this boat with a Boulton & Watt beam-type steam-engine, which had a cylinder of 24-inch (0.61 m) bore, and 4-foot (1.22 m) stroke. Its boiler was 20 feet (6.10 m) long and 8 feet (2.44 m) wide. The Clermont was 150 feet (45.72 m) long, had an 18-foot (5.49 m) beam, 7-foot (2.13 m) draught, and displaced 110 tons. It took up service in the Livingston & Fulton Shipping line on the Hudson river between New York and Albany, and made a profit of $16,000 in the first year of its commercial operation. Fulton had carried out experiments in the tow tank for the best design of The Clermont's hull. Fulton had taught himself engineering mathematics, and he always took great care to calculate mathematically the practicability of his engineering ideas. He was an excellent draughtsman and always worked out his plans to the minutest details.

Fulton had been a painter, a civil engineer and inventor before he took up submarine design in 1800. He designed a submarine, The Nautilus, which he built for the French in 1801, when Napoléon Bonaparte became the First Consul and war broke out between France and England. The Nautilus as a submarine was a slight advance over the first one-man submarine, The American Turtle of David Bushnell, but Fulton's submarine did possess all the basic features of a modern submergible. It had a 212-cubic foot (6.00 m^3) capacity which contained enough oxygen to support 4 men and 2 small candles for 3 hours. It was capable of submerging to the depth of 25 feet (7.62 m), and it had a conning-tower, ballast-tank with pumps, diving-planes and, later, also a compressed-air reservoir. The submarine was propelled by a screw-propeller operated by manpower, like in the submarine of David Bushnell. Fulton also designed his own torpedoes, and gave a successful demonstration of their effectiveness. Fulton and his crew chased the blockading British vessels, which were ordered not to fight his submarine, for a few years.

Fulton also constructed the first naval steamship, The Demologos, in 1814 for the United States Navy. This ship was a floating fort which had a paddle-wheel between its double wooden hulls, and was outfitted with twenty cannons. The ship had a displacement of 2,475 tons,156-foot (47.55 m) length, 56-foot (17.07 m) breadth, 20-foot (6.10 m) depth, and 10-foot (3.05 m) draught. It had a steam cylinder of 48-inch (1.22 m) bore and a 5-foot (1.52 m) stroke producing 120-horsepower. Its copper boiler measured 22 feet (6.71 m) in length, 12 feet (3.66 m) in width and 8 feet (2.44 m) in height. Its paddle-wheel had a 16-foot (4.88 m) diameter, ten radial 14-foot (4.27 m) by 4-foot (1.22 m) float boards, and made 15 revolutions a minute. The battleship The Demologos, which reached 5½ miles (8.85 km) an hour average speed, was never tried in battle because the war ended before it was launched.

Fulton built 15 steamships for various other lines and countries before he died in 1815. His last steamship, Chancellor Livingston, launched after Fulton's death in 1816, was the first American steamship to use coal as fuel. It had a 60-horsepower engine and developed a record speed of 9.2 miles (14.81 km) an hour. Fulton's great contribution to steam navigation was his ability to coordinate the hull design with the steam-power plant to make the steamboat economically successful.

Col. John Stevens (1749-1838), a brother-in-law of Livingston who had made him interested in steamboat experiments, had keen scientific interest and had followed the steamboat experiments of Fitch and Rumsey. Stevens and Livingston had met Fitch and witnessed some of Fitch's trial runs. They readily realised that steam navigation had an important future, and that Fitch failed mainly because he had insufficient financial resources. Livingston, who was a politician, immediately bought up the unfulfilled patent grant to Fitch by the New York State, a grant which allotted monopoly rights for regular commercial boat service on the Hudson river between Albany and New York. Stevens, being mechanically minded, began his own steamboat experiments about 1787, and secured his first steamboat patent in 1791. In 1798, Stevens collaborated with Livingston and Nicholas J. Roosevelt, a skilled mechanic, in the construction of an experimental steamboat on the Passaic River, but the experimental steamboats built by Stevens were too small for Livingston's charter, and since the charter's term of existence was soon running out, Livingston formed a partnership with Fulton in Paris, France.

Right from the outset, Stevens appreciated the high-pressure steam-engine which Oliver Evans had developed before Stevens began his steamboat experiments. Stevens recognised from the beginning that in high-pressure steam-engine the weak link was in the boiler design, and he set himself to work on an improved boiler design but his progress was too slow for Livingston, who was worried about his charter running out. Steven's experimental steamboats were 20 feet (6.10 m) long and 6 feet (1.83 m) wide scows travelling at 4 miles (6.4 km) an hour.

When Livingston was in France, Stevens solved the boiler problem by designing a multitubular boiler consisting of 2 flat cylinders with water flowing from one cylinder to the other through multiple tubes which were directly in the furnace, thus permitting a rapid heating of water into steam. Stevens who was able to operate his boiler at least at 50 psi (344.8 kPa), aimed to use 100 psi (689.5 kPa) boiler pressure.

Stevens became one of the first pioneers of practical screw-propulsion of ships after David Bushnell, when he built a twin-screw launch named the Little Juliana, the first successful screw-driven steamboat, in 1803. This boat with a wooden keel, had an overall length of 24 feet 8 inches (7.53 m), beam of 6 feet 1 inch (1.86 m), depth of 2⅓ feet (0.70 m), and displacement of about 5 tons. The boat was fitted with an iron rudder. The vertical steam-cylinder of the engine had a 4½ inch (0.11 m) bore, and the stroke of the piston was 9 inches (0.23 m), but like the steam-engine of Evans it had no condenser. The reciprocal motion of the crossbeam was transmitted through connecting-rods and 2 cranks to twin propeller shafts which were in phase, but rotating in opposite directions to preserve directional stability of the boat. Each screw-propeller had 4 blades about 18 inches (0.46 m) in diameter. His multitubular boiler 1½ feet (0.46 m) in length contained 28 copper watertubes of 1½-inch (3.8 cm) bore which supplied steam at 50 psi (344.8 kPa). Unfortunately, his boilers occasionally exploded. In 1806 Stevens built another screw-driven scow which had a steam-cylinder of 10-inch (0.25 m) bore and 2-foot (0.61 m) stroke. The steam-engine of this scow turned a 4-blade screw-propeller. The multitubular boiler of this steamboat contained 100 tubes of 2-inch (5.1 cm) bore.

In 1808, Stevens abandoned screw-propulsion and built a commercial steamship, The Phoenix, which was driven by paddle-wheels and became the first steamship to venture sailing the open sea. He sent this steamship under the guidance of his son, Robert L. Stevens (1787-1856), unto Atlantic Ocean, around the Cape May and then up the Chesapeake to serve the Stevens Steamship Line on the Delaware river, which became as important as Livingston & Fulton Steamship Line on the Hudson. Robert L. Stevens made a number of improvements in steamships and increased their speed to 15 miles (24.1 km) an hour. Initially The Phoenix had 2 steam-cylinders each of 16-inch (0.41 m) bore and 3-foot (0.91 m) stroke. Steam was supplied at 3 psi (20.7 kPa) from a cylindrical boiler with the return flue set in brickwork to turn the paddle-wheels which were about 13 feet 10 inches (4.21 m) in diameter. Later this steamship was fitted with one steam cylinder of 24-inch (0.61 m) bore and 3-foot (0.91 m) stroke. The paddle-wheels were provided with pivoted vertical floats. The Phoenix had the overall length of 103⅓ feet (31.49 m), breadth of 16 feet (4.88 m), depth of 6 feet 10 inches (2.07 m), and displacement of 95 tons.

After 1812, Stevens became also a pioneer in railroad locomotion, and in the 1820's he built his own locomotive to run on a great circular track on his large estate. The sons of John Stevens became the most successful pioneers of American railroads.

In 1811 Nicholas J. Roosevelt built, at Pittsburgh, a steamboat, The New Orleans, and ran with it a steamship line on the Ohio and the Mississippi river until the ship was wrecked on a snag at Baton Rouge in 1814.

Capt. Henry M. Shreve (1785-1851), who broke the monopoly of Livingston and Fulton in 1819, built The Washington, the first of the Mississippi steamboats, which had a paddle-wheel at the stern. He abandoned the condenser, introduced two horizontal stationary cylinders of 24-inch (0.61 m) bore and 6-foot (1.82 m) stroke of the pistons. The pistons were connected with cranks by connecting-rods, and the cranks were at right angles to obviate the dead-center problem. The ship's 4 boilers, which operated under high pressure, were horizontal, and had flues in them like the boilers of Evans, and like the boilers of Evans, the boilers of his ship suffered many explosions. He finally introduced a cam-operated shutout valve for steam in his steam-engine to take full advantage of the expansive power of steam.

The first European steamboat in commercial service was The Comet of Henry Bell (1767-1830), which was put into a daily steamship service on the Clyde in 1812. The Comet made a trip three times a week and travelled 27 miles (43.5 km) in 3½ hours. The steam-engine of The Comet was similar to that of Fulton's engine having an output of 4-horsepower. This steamship had an overall length of 51 feet (15.54 m), hull width of 11¼ feet (3.43 m), depth of 5 feet 7 inches (1.71 m), mean draught of 4 feet (1.22 m), and a displacement of 28 tons.

In 1818, the steamboat Rob Roy, which was designed, built and fitted with a 30-horsepower side-lever steam-engine by David Napier (1780-1869), began regular service between Greenock and Belfast under the guidance of William Denny. The vertical steam cylinder had 20-inch (0.76 m) bore, and the piston had a 3-foot (0.91 m) stroke. The steam pressure from the boiler was about 2 psi (13.8 kPa) above atmospheric pressure. The paddle-wheels were 10 ½ feet (3.20 m) in diameter, each fitted with 10 radial float-boards 4¼-foot (1.30 m) by 1½-foot (0.46 m), and making 30 revolutions per minute. The speed of this steam-powered vessel was 7 knots (12.97 km an hour). This ship had an overall length of 80 feet 11 inches (24.66 m), hull width of 15 feet 10 inches (4.82 m), depth of 9 feet (2.74 m), draught of 5 feet 9 inches (1.75 m), and the displacement of 88 tons.

Although John Wilkinson (1728-1808) had built the first iron vessel, a barge, for traffic on the Severn river as early as 1787, only in 1821 was the first iron steamboat, The Aaron Manby, launched. It was built by Aaron Manby (1776-1850), master of the Horseley Ironworks, in collaboration with his son Charles and Charles Napier (1786-1860), who promoted the use of iron steamships on the Seine river in France. The hull of this ship was constructed of ¼ inch (0.64 cm) thick iron plates with a flat bottom and square stern. The fore and aft seams were lapped and rivetted, and ribs were angle irons. The keelson was wood, and wooden beams were supporting his patented steam-engine in 1821 that rated 30-horsepower, and consisted of 2 oscillating cylinders of 27-inch (0.69 m) bore and 3-foot

(0.91 m) stroke. Manby soon discovered that oscillating cylinders had been used by William Murdock (1754-1839) in 1785, which made his patent invalid. Manby's ship was prefabricated and assembled at the Survey Canal Dock in London. It had 2 flat-sided iron boilers under 2 psi (13.18 kPa) pressure, 5 feet (1.2 m) wide, 6 feet (1.83 m) high, and 12 feet (3.66 m) long, each containing 2 furnaces 4½ feet (1.37 m) long and 1½ feet (0.46 m) wide with return flues uniting in a 47-foot (14.3 m) funnel of 3-foot (0.91 m) diameter. The funnel was stayed from the bowsprit, stern and sides, and served as a mast. It had one pair of paddle-wheels of 12-foot (3.66 m) diameter and 2½ -foot (0.75 m) width. The steamboat dimensions were: 120 feet (36.58 m) in length, 17 feet 2 inches (5.24 m) in breadth, 7 feet 2 inches (2.19 m) in depth, 3½ feet (1.07 m) in draught, and 116 tons displacement. It was the first iron steamer to cross an open sea, the English Channel, from London to Paris, with a speed of 7 knots (12.95 km an hour).

In the oscillating steam-engine, which had been first proposed by William Murdock in 1785, the piston-rod was directly connected to the crank which eliminated the connecting-rod. This type of connexion was first used in the first iron-built paddle-steamer, The Aaron Manby. In 1827, Henry Maudslay (1771-1831) adapted this arrangement of the steam-engine after providing it with an efficient form of valve-gear, and patented it. In 1828 the Thames steamer Endeavour was the first vessel to be fitted by Maudslay with oscillating cylinders. The oscillating steam-engine became popular only after John Penn (1805-1878) of Greenwich introduced his improvements to this engine. However, with the advent of the high-pressure steam, the oscillating steam-engine could not be easily adapted to compounded expansion and, therefore, was superseded by the inclined direct-acting steam-engine first patented in 1822 by Marc Isambard Brunel (1769-1849).

The British founders of marine engineering were the cousins David Napier (1790-1869) and Robert Napier (1796-1876). David Napier built the marine engines for the first Glasgow-Belfast ferry. In 1823, Robert Napier set up a small works at Camlachie in the east end of Glasgow on the Clyde for the building of marine engines.

Samuel Cunard (1787-1865), a successful Canadian businessman of Nova Scotia who was engaged in banking, lumbering, shipping, and shipbuilding enterprises, founded the British & North American Royal Mail Steam Packet Company in 1839 as the first regular steamship service between the continents, which was the beginning of the renowned Cunard Company Line between the American and the European continent. 15 of Robert Napier's iron vessels, the so-called 'Atlantic Greyhounds', were built for the Cunard Company fleet numbering about 40 steamships. Robert Napier also built iron vessels for the French shipping companies **Pereire** and **Ville de Paris**. Napier's iron steamships established new records in Atlantic crossing.

Rise of Internal Combustion Engine for Ship Propulsion

High-speed steam-engines came into prominence in the middle of the 19th century. John Elder (1824-1869) launched the first seagoing vessel, S.S. Brandon, in 1854, which was fitted with a double-expansion, steam-jacketed compound steam-engine built according to sound thermodynamic principles. In 1862, Elder patented a triple-expansion, and a quadruple-expansion compound steam-engine, since more economical use of fuel under steam pressures higher than 180 psi (1,241 kPa) necessitated more than the double expansion of steam. Such multi-expansion steam-engines were first introduced into marine engineering practice in 1871, when the French marine engineer, Charles Benjamin Normand (1830-1888), patented his triple-expansion compound steam-engine, and built the first ship with such a steam-engine in 1873. In 1874, the first triple-expansion steam-engine in Britain was built by A.C. Kirk of John Elder & Company in the steamship S.S. Propontis of 2,083-ton displacement, which worked with the steam pressure of 150 psi (1,034 kPa).

After the use of superheating was advised for the increase of the thermal efficiency of steam-engines by the German eye surgeon Dr. Ernst Alban (1791-1856) in 1845 and demonstrated by a German textile manufacturer in Alsace, Gustav Adolf Hirn (1815-1890), steam-engines operating with 'dry' superheated steam were used in ship propulsion. The first superheated steam-engine had been built in 1866 by the English engineer Charles Brown (1827-1905) and Heinrich Schulzer of the firm **Gebrüder Schulzer** (*Schulzer Brothers*) in Winterthur, Switzerland. The first steam-turbine for ship propulsion was built by Charles Algernon Parsons (1854-1931) in 1897 when he outfitted the vessel Turbinia of 44-ton displacement with his reaction steam-turbine. The steamship Turbinia attained an unprecedented speed of 34½ knots (63.9 km an hour) for its time.

The internal combustion engine began to threaten the dominance of the steam-engine as the prime mover of ships after the turn of the century. Otto's internal combustion engine was not powerful enough to replace steam-powered engines in ships, but after Rudolf Diesel (1858-1913) demonstrated his first successful oil engine in 1898, which could be developed into a powerful prime mover, the days of steam-engines and steam-turbines as the prime movers of ships were numbered. The first marine Diesel-engine was built in 1902 by the German engineers Adrien Bocket and Friederich Dyckhoff, and installed in a French canal boat in 1903. It was still a relatively small horizontal Diesel-engine of 20-horsepower which required a reversing gearbox. **Gebrüder Schulzer** in Winterthur, Switzerland, built another marine Diesel-engine which was fitted with an electric transmission. In 1907, Diesel himself worked on the design of his marine engine for ship propulsion. Since the submarine warfare in the First World War necessitated the construction of efficient and safe Diesel-engines which saved weight and space, and had more than 10 times the power of contemporary Diesel-engines, by 1914 the engineers were able to build Diesel-engines of 500-horsepower to propel submarines. Engineers had developed the 'direct' or 'solid' injection as a substitute for Diesel's compressed-air injection by 1927, which increased the power of a single cylinder Diesel-engine to a 1000-horsepower. By that time, the Diesel-engine had almost completely replaced the steam-engine as the source of power for ships.

CHAPTER XVII

AGE OF HEROIC ENGINEERING: Advent of Railroads

Railroads

Modern railroad had its beginning in the horse-drawn colliery tramways. The original railroads were the rut-roads of antiquity along which the horse-drawn carts travelled. In more recent times, the mining industry gave the impetus for modern development of railroads. The rough conditions of the floors of early mining tunnels made the hauling of carts by men or beasts extremely difficult and slow. Since it was not practical to keep smoothing the floors of the tunnels because they rapidly deteriorated, it was much simpler to lay down wooden boards, or rails, in the ruts worn in the tunnel floor by the wheels of the carts. This solution eased the traction difficulties to a great degree. Such wooden tracks, or 'waggon-ways' were used in the early 15th and the 16th century in the German mining tunnels of the Erz and the Harz mountain, and in the 16th century in England. Many such 'rail-ways' have been found in the British coal districts of the north, in the latter part of the 17th century, where they became known as 'Newcastle roads'. The earliest known British Railway Act stems from 1758, issued for a 'waggon-way for conveyance of coal to Leeds'. Horse-drawn railways spread rapidly in the mining districts in Wales and northern England, but they were private and virtually all were used for hauling coal. In 1800, a Mr. Thomas of Denton, proposed to build roads in Newcastle on the principle of 'coal-waggon ways' for the general carriage of goods. In 1801, Dr. Anderson suggested, in his < Recreations in Agriculture >, to lay down tramways operating with horse-drawn wagons in places where canals could not be constructed. He proposed a development of an organised network of such tramways similar to the public road system.

The first such horse-powered public tramway was built by William Jessop (1745-1814) between 1801 and 1803. It was called the Surrey Iron Railway from Wandsworth on Thames, London to Croydon, with a branch to Carlshalton. This railway was 9 miles (14.5 km) long and intended mostly for the transport of agricultural produce to London, and the return of the manure to the country. Jessop designed this railway with cast-iron rails laid on wooden sleepers at a gauge of 4½ feet (1.3 m), but it ceased operation in 1846. During the next 25 years, 29 such railway companies had sprung into existence, among which the most important railroad lines were: the Kilmarnock and Troon (1808), the Gloucester and Cheltenham (1809), the Plymouth and Dartmoor (1819), the Stratford and Moreton (1821); and the Bolton and Leigh, the Canterbury and Whitstable, the Cromford and High Peak, the Nantile, and the Rumney (all incorporated in 1825).

Historically the most important of these horse-drawn railways was The Stockton and Darlington Railway, which was intended to connect the coal fields of Durham with the navigable part of the Tees river. In the late 18th century, a canal had been proposed for this purpose, but this idea was abandoned when John Rennie's survey for it was made known in 1813. In 1818, a tramway was proposed for it, and the chief engineer of the company, Thomas Overton, prepared the plans for a tramway operated by horse-traction, which was given Royal assent in 1821. Upon the recommendation of George Stephenson, who succeeded Overton as the chief engineer, the new Act of 1825 included the locomotive as a means of traction.

Locomotive as *Primum Mobile*

Railroad had its origin in the horse-drawn colliery tramways. In 1804, Richard Trevithick (1771-1833) built the first locomotive which he successfully tested on Penydaren tramway in South Wales. However, a complete success was denied to Trevithick because his heavy locomotive broke up the cast-iron rails of the tramway. Later experiments of Trevithick with his locomotives at Gateshead and in London also did not succeed because of the inferior track and road conditions.

Despite Trevithick's failure, the owners and the Leeds agent John Blenkinsop (1783-1831) of the Middleton Colliery thought that steam-power might help the company out of its economic difficulties. Blenkinsop realised that the locomotive must be much lighter than Trevithick's locomotive if the cast-iron track on which it travels is to last. Therefore, Blenkinsop devised a type of rack-drive to increase the tractive power of a light locomotive which he patented in 1816. Since Blenkinsop was not an engineer, he trusted the design and construction of the locomotive to Matthew Murray (1765-1826), the best engine-builder at the time, and the first serious competitor of Boulton & Watt. Murray has been called the 'Father of Leeds engineering' since his many inventions helped to establish Leeds as an important manufacturing center. Although Murray followed Trevithick's locomotive design and paid a royalty for the use of principles in Trevithick's patent, he did incorporate a number of technological improvements of his own in his locomotive.

Trevithick had employed a single horizontal cylinder in his locomotive engine which probably made its action irregular despite the heavy flywheel. Murray's locomotive had 2 double-acting cylinders set vertically on the top of the barrel of the single-flue boiler driving 2 cranks at right angles so they could not stop at 'dead center'– a mechanical feature followed in all subsequent locomotive designs. The twin crankshafts, which were mounted on the wooden frame of the locomotive and driven by the pistons by means of crossheads and long connecting-rods, turned the pinion shaft by spur-gearing. The pinion located on the near side of the engine, had 38¼-inch (0.97 m) diameter and 20 teeth which engaged the teeth of the rack cast on the outside of one of the running rails. The Blenkinsop locomotive was tested on new cast-iron rails laid on stone sleeper-blocks at a gauge of 4⅛-feet (1.26 m), and it proved to be a complete success. Murray's locomotive design was superior to that of Trevithick except that his locomotive tended to run short of steam because he did not exhaust the steam through the smoke stack, as Trevithick had done in his locomotive, which drastically increased the draught in the furnace. Upon the successful trial of the first locomotive, 3 more locomotives were built by Murray, one of which worked for a period of time on the Kenton & Coxlodge tramway on Tyneside, where it was seen by George Stephenson (1781-1848), a semiliterate Northumbrian enginewright. Murray's locomotives, which hauled coal from a mine at Middleton to Leeds until 1835, proved to be too slow for general freight and passenger transport and, therefore, they occupied an intermediate stage in the locomotive development between Trevithick's locomotive which failed to run on tracks, and Stephenson's locomotive which was capable of general railway use. However, Murray's locomotive was the first steam locomotive to operate with commercial success on a railway.

The Blenkinsop-Murray locomotive, which used the pinion-and-rack drive, did operate with profit and saved the company from its financial difficulties. However, it was somewhat unsteady, and the strain put upon the rack attached to the rail tended to pull the rails apart, and on some occasions even the boiler blew up. This locomotive was able to haul 90 tons at a speed of 3½ miles (5.63 km) an hour on a level track, and its rack-and-pinion drive enabled it to climb grades up to 5½ %. A practical limit of speed had to be imposed on the rack-and-pinion locomotive, which, as was mentioned above, eliminated it from competition on regular lines since locomotives became larger and more powerful.

The Scottish engineer, William Brunton (1777-1851), one of the pioneers in steam transport, particularly steamships, and an employee of Boulton & Watt, believed that locomotive engines could not grip the railroad tracks except on level ground. In 1813, he designed his 'mechanical traveller', a veritable 'iron horse', a steam locomotive fitted with 'feet', that is, it was attached to humanlike 'legs' with articulated 'knee joints'. These 'iron legs', that were moved by the piston of the steam engine by means of connecting-rods attached to the 'knee joints' of the 'legs', took 6-foot (1.83 m) long 'steps', which made the locomotive move jerkily at 2 miles (3.2 km) an hour. Brunton's locomotive worked through 1814, but it blew up in 1815 when its boiler exploded.

Another engineer who was in doubt about the adhesion of locomotive wheels to the rails was William Chapman (1749-1832), and, therefore, he designed a locomotive which hauled itself along a chain-cable laid between the 2 rails. Chapman was the first to see the need to flex the wheelbase of a vehicle on rails, and in 1812 he patented his 'bogey', a swivel-truck which moved along the curved track whilst the vehicle itself moved along the chord of the arc.

William Hedley (1779-1849), inspector of Wylam Colliery, built with the assistance of the enginewright, Timothy Hackworth (1786-1850), his first successful locomotive 'Wylam Dilly' in 1813. Hedley had examined the locomotives of Trevithick and Blenkinsop before he designed his locomotive. In 1814 he built his new locomotive, the first to be fitted with the grasshopper-beam type steam engine patented in 1803 by Freemantle. He had conducted traction tests, and patented his new locomotive fitted with the grasshopper-type steam engine, called 'Puffing Billy', in 1813. The boiler of this locomotive had a return flue, and a silencer for the venting of steam in the smoke stack. 'Puffing Billy' locomotive ran between the Wylam colliery and the Tyne river. Since the locomotive ran on plate rails, which had flanges inside, the locomotive wheels were flangeless. After 1825, the plate rails at Wylam were replaced by 'edge rails', since the locomotive wheels now had flanges. Surprisingly, the 'Puffing Billy' locomotive remained in continuous service until 1862.

The next contribution in locomotive design came from George Stephenson, who more than anybody else was responsible for the rapid development of railroad in Britain. He began working at an early age, and did not learn to read and write until he was 18 years of age. He was a naturally talented mechanic, rapid learner, shrewd, brashly self-confident, dedicated and tough. He was soon promoted superintendent of the collieries at Killingworth and elsewhere. He suggested a number of improvements of the engines and railroads for the collieries, and by 1813 he had become a 'self-educated civil engineer'. After examining the locomotive of Blenkinsop, Stephenson decided to build a better 'travelling engine'. In 1814 Stephenson constructed his first locomotive, which he called 'Blücher' in honour of the Prussian general then campaigning with Duke of Wellington against Napoléon Bonaparte, for Sir Thomas Liddell. He followed Murray's design by setting 2 vertical cylinders in the boiler along the center line of the boiler, one for each axle. Each cylinder had an 8-inch (0.20 m) bore and a 2-foot (0.6 m) stroke. The pistons drove upwards 2 crossheads from which the power-stroke was directed downwards on both sides of the boiler by means of connecting-rods to cranks attached to the outside of the wheels. The cranks of one axle were set at right angle to the cranks of the other axle in order to avoid the 'dead-center' effect on the engine. The axles were connected by chains on sprocket wheels, so that all wheels operated together like in the locomotive of Trevithick. Stephenson's cylindrical boiler, 2 feet 10 inches (0.86 m) in diameter and 8 feet (2.44 m) long, was made of wrought-iron and had a single 20-inch (0.51 m) flue running along its entire length. He exhausted the steam through a nozzle into the smokestack, like Trevithick, which increased the draught in the furnace. Stephenson was able to haul with his locomotive 30 'waggons' at 4 miles (6.44 km) an hour at the Killingworth colliery, but his locomotive consumed an excessive amount of fuel.

Stephenson built a second locomotive in 1815, but again it did not meet his expectations. Stephenson kept on building new, and better locomotives, but by 1818, he realised that he needed to know much more about the resistance the locomotives have to overcome: the 'rolling resistance', the friction of the axle, and the force of gravity on grades. He devised an instrument to measure the resisting force of the train against the pull of the locomotive with Nicholas Wood (1795-1865), a colliery engineer, and he found this friction to be constant at all speeds. The resistance of a train of cars on straight and level track was approximately 10 pounds (44 N) per long ton, but a rise of one foot in 100 feet required an additional pull of 22.4 pounds (97 N) per long ton to overcome the force of gravity. Therefore, he found that a gradient of 1% would add more than twice as much resistance as on the level track and, therefore, required 3 times the tractive effort from the locomotive than on level track. Thenceforth, Stephenson insisted that railroads should operate on as low grades as possible.

The colliery railways of Hetton and Killingworth were the testing grounds for a great number of Stephenson's experiments. The Hetton railway was laid out by George Stephenson, and built by his 19-year-old son Robert Stephenson (1803-1859) in 1822. It ran from the colliery to the Wear river and was about 8 miles (12.9 km) long. The Hetton railway was worked partly by locomotives, and partly by stationary steam engines hauling wagons by cable over a 300-foot (91.44 m) high hill.

In 1822, Stephenson began to build his first railway connecting the coal field of Darlington with the port of Stockton-on-Tees. It was not yet possible to make a nearly level track for this railway and, therefore, the road had several inclined sections with stationary steam engines for hauling by cable, and in some sections horse-drawn transport was still used. By 1825, when this railway was opened, Stephenson had improved his locomotives to such an extent, that his locomotive, called 'Locomotion No. 1', which weighed 6 ½ tons, hauled a load of 90 tons consisting of 36 cars with about 600 passengers, and some freight at a speed of 12 miles (19.3 km) an hour. By that time, Stephenson had built altogether 55 engines, 16 of which had been locomotives. During the various stages of the construction of this railway, Stephenson adopted the double-headed rails rolled by a new process invented by John Birkinshaw in 15-foot (4.57 m) length from wrought-iron for the main running line, whilst cast-iron rails were still used for passing loops because of their lower cost.

The first suspension bridge carrying a railroad was built on the Stockton-Darlington Railway line, when in 1830, Capt. Samuel Brown (1776-1852) built his unstiffened suspension bridge to carry the Stockton-Darlington Railway across the Tees river. Two coal waggons passing over this suspension bridge deflected it about 12 inches (0.30 m) and sent a downwards wave ahead of the waggons, and an upwards wave behind the waggons in the bridge deck which made the bridge unstable and unsafe. The bridge deck was propped up with piles in 1831 to stabilise it, and, finally, in 1844, Robert Stephenson decided that the suspension bridge was not stiff enough to carry a railroad, and replaced it with a cast-iron bridge.

The Stockton-Darlington Railway, which marked a transition between the colliery tramway and the railroad proper, attracted worldwide attention and created some controversy among engineers. The networks of canals and metalled roads, which at the time represented a prodigious capital investment in transportation, had paid rich dividends to canal companies, turnpike trusts, and coach-line proprietors. By using the horse in transporting mail and passengers on the new roads, and by hauling barges on the canals, people regarded the horse as the only proper means of overland locomotion. It was not an easy task to convince the public that the new 'iron horse', the steam-locomotive, is superior to the horse. The powerful vested interests, such as the turnpike trusts, coach-line proprietors, landowners, and canal companies, to arrest the building of railroads, exaggerated the dangers of steam-powered travel in the Parliament and in public.

George Stephenson, and a minority of merchants and manufacturers, had realised that roads, and particularly canals, had been doomed by their own commercial success because they had helped the commerce to expand to the point where it had outgrown them. These powerful vested interests felt threatened by the new railroad of Stephenson and, therefore, called upon their engineers to make belated improvements to the existing canal systems, to promote new canal schemes, and to oppose and ridicule Stephenson's opinions. Since these engineers were members of the Institution of Civil Engineers, Stephenson and the public came to view this Institution as the stronghold of opposition to railroads. On account of their ill-advised opposition to railroads, Stephenson developed a profound professional contempt of the 'London engineers' as he called them, but this professional assessment was not well-advised, because it led him to assume that they were inferior to him in overall engineering experience, which they were not. Stephenson came to regard them as the 'horse-age engineers'.

George Stephenson was an incredibly persistent and stubborn man, who had an almost infinite confidence in his own judgement. On account of his obstinacy he sometimes refused to accept constructive criticism, but on many other occasions it helped him to reject views which would have been very detrimental to the success of railways.

Facing such powerful opposition, the coming of railroads would have not succeeded in 1830 had it not been for the incredibly stubborn confidence of George Stephenson in the future of railroads, and for the finances provided by a handful of merchants and manufacturers who believed that canals were doomed by their own success.

The immediate success of the Stockton-Darlington Railway encouraged a dozen new railway companies to be chartered, among which was the Liverpool-Manchester Railway that became the decisive success for the future of railroads. George Stephenson was appointed the chief engineer of this company in 1826 but he had to work without the assistance of his very able son Robert, who had left for Columbia, South America, to develop ports and railways for the mines of that country.

Stephenson was faced with the difficult task of administering the organisation of the construction of the railroad which involved many much more serious problems of civil engineering than he had encountered in the past. The majority of the directors of the Liverpool-Manchester Railway tended to favour stationary steam engines and cable haulage for the entire 30-mile (48.3 km) length of the line. Stephenson intended to use cable-haulage and stationary steam engines on three inclined planes on the line, each one of which was more than 1 mile (1.61 km) long, and to use his locomotives on the level track, a judgement which was questioned by the majority on the board of directors of the company. Over the protest of a minority of directors led by the treasurer Henry Booth, well-known engineers James Walker and John Urpeth Rastrick (1780-1856) were called to re-examine the proposed railway. Walker and Rastrick, after their study of the proposed railway, reported in favour of haulage. In their scheme, each section of haulage operated by a stationary steam engine would be about 1½ miles (2.42 km) long, but Stephenson took strong exception to the reports of Walker and Rastrick.

Stephenson had concentrated all his efforts on the construction of a floating railroad track on timber sleepers over a treacherous stretch of land, a huge peat bog called the Chat Moss, which he began in early 1826, against the judgement of all the 'London engineers', who ridiculed him for attempting it. Unfortunately, Stephenson lost control of the administrative side of his duties as the chief engineer of the Company owing to his unbusinesslike approach. When financial difficulties in securing a new loan for the project occurred after Thomas Telford's report to the Commissioners of the Bill, which criticised Stephenson for ignoring sound business practice in his method of construction, and the lack of clear policy by the Company for the system of transportation, the Company's directors decided to arrange a competition to find the best locomotive for the line upon the urging of Stephenson.

In the meantime, Robert Stephenson had returned to England in late 1827, and had taken charge of the locomotive building business at Newcastle which his father had founded. Robert Stephenson readily realised that the success of the Liverpool-Manchester Railway, and railways in general, depended upon the success of the locomotive as a long-distance transport vehicle of both freight and passengers. He set up a programme of mechanical engineering for the development of the locomotive, and in a surprisingly short time of 3 years he was able to build 4 locomotives: the 'Lancashire Witch', the 'Rocket', the 'Northumbrian', and the 'Planet'.

The celebrated locomotive of Robert Stephenson, the 'Rocket', took part in the first competition between machines, called the 'Rainhill Trials', arranged by the Liverpool-Manchester Company in October, 1829, on the Manchester side of the Rainhill Bridge at Kenrick's Cross, about 10 miles (16.10 km) from Liverpool. The course in the competition was 1½ miles (2.41 km) long each way with an additional 660 feet (201.17 m) at each end for stopping. Each competing locomotive was required to make 20 round trips in one day, equalling the distance from Liverpool to Manchester. The main 4 conditions placed on the locomotives were:

1. the locomotive should 'consume its own smoke',
2. the locomotive should weigh at most 6 tons and draw a 20 ton load, including the tender and water tank, at 10 miles (16.10 km) an hour,
3. the total boiler pressure should not exceed 50 psi (344.75 kPa),
4. the cost of the locomotive should not exceed 550 British pound sterling.

The first prize in this competition was 500 British pound sterling.

Originally 10 locomotives were scheduled to compete, but only 5 showed up on the competition day. One locomotive, 'Cycloped' of Mr. Brandreth was not admitted to the competition, and another, 'Perseverance', of Timothy Burstall was unable to exceed 6 miles (9.66 km) an hour speed and withdrew from the competition. The light, elegant, and fast locomotive, named the 'Novelty', of John Braithwaite and John Ericsson (1803-1889), a noted Swedish engineer and inventor, and 'Sanspareil', the locomotive of Timothy Hackworth (1786-1850), an engineer of the Stockton-Darlington Railway, competed against the rather awkward looking 'Rocket' of Stephenson.

'Novelty' had some mishaps with its bellows on every trial run, whilst 'Sanspareil' broke down repeatedly on every trial, and had to give up before it had travelled half the required distance. Only the 'Rocket' fulfilled all the conditions of the competition, and ran the required distance at an average speed of 15 miles (24.14 km) an hour, reaching on the last lap a speed of almost 30 miles (48.28 km) an hour.

The first day of competition was attended by a crowd of nearly 15,000 spectators.

The 'Rocket' was essentially designed by Robert Stephenson. Its multitubular boiler suggested by Henry Booth was equipped with 25 copper firetubes of 3-inch (7.6 cm) bore which greatly increased the thermal efficiency of the boiler and its capacity to make steam. The firebox at the rear end of the boiler was surrounded by water compartments to conserve heat, an arrangement which was not essentially improved upon in later times. It operated with 40 psi (275.8 kPa) steam-pressure and the draught in the furnace was controlled by exhausting it into the smoke stack through a variable opening. The cylinders were inclined at an angle, and the piston was connected directly to the pair of spring-mounted driving wheels. The 'Rocket' together with its tender weighed only about 7¼ tons.

The 'Planet', a prototype of the modern steam locomotive, designed by Robert Stephenson, which was delivered to the Liverpool-Manchester Railway in 1830, had its cylinders between the frames and beneath the smoke-box, driving a cranked axle, and its steam was provided by a multitubular boiler. All these locomotives were fired with coke because they had to 'consume their own smoke'. The Rainhill Locomotive Trials, which was the first competition between machines, and the 'Planet' convinced all skeptics that steam-powered locomotion was here to stay and, thereby, inaugurated the 'Age of the Railroad'.

Although Robert Stephenson was the first to use the multitubular boiler in 1829, it had been patented in France by the French engineer Marc Seguin (1786-1875) in 1828. Seguin had visited the Newcastle works of Stephenson in 1828, and ordered locomotives which were delivered in late 1828 and early 1829. Seguin, a suspension bridge builder, constructed the first railroad in France between St. Etienne and Lyons which was completed in 1832. Seguin wanted to increase the speed of the Stephenson's locomotive, and for that purpose he introduced a multitubular boiler in which the firetubes passed through the water in the boiler. In order to produce a draught in the boiler and to force the hot furnace gases through the firetubes, he installed axle-driven centrifugal fans on the tender which acted like bellows by impelling the hot furnaces gases through the firetubes rather than exhausting them directly through the smoke stack. This design increased the heating surface in the boiler and made the boiler thermally much more efficient. Seguin employed his multitubular boiler later on his own railway, but it was first used in practice by Robert Stephenson in the 'Rocket' apparently independently of Seguin, because his design of the boiler was suggested to him by Henry Booth, the treasurer of the Liverpool-Manchester Railway. The multitubular firetube boiler was an important improvement in locomotive engines.

The Newcastle works of Stephenson was the first locomotive factory in the world, and orders for locomotives came from many countries which had embarked on building their own railway system. By 1840, Stephenson had delivered locomotives to France, United States, Belgium, Austria, Italy, Germany and Russia. Soon other countries began to manufacture their own locomotives, like Seguin had done in France.

George Stephenson performed remarkable engineering feats in constructing the Liverpool-Manchester line. To meet his criterion for railway grades, he drove a 2-mile (3.2 km) long cutting through the Olive Mount, made of solid rock, near Liverpool. Some 48,000 cubic yards (36,700 cubic meters) of rock were removed, and the railroad track at the maximum point ran 80 feet (24.38 m) below the surface. He ran his railroad track supported on a floating raft over the 4-mile (6.4 km) swamp called Chat Moss which a man could cross without sinking only with boards tied to his feet. Stephenson built 63 bridges and 2 tunnels to keep his track level. The famous Kilsby Tunnel provided his most difficult engineering problem on this line, since it had to run through a 400-yard (1.1 km) pocket of quicksand pouring ground water into the tunnel. It took 13 pumping-engines, pumping 1,800 gallons (8,138 liters) of water a minute, a year and a half to drain the water so that construction could continue. The Liverpool-Manchester Railway cost 50,000 British pound sterling per mile (1.61 km), more than twice the initial estimated cost by Stephenson. In comparison, the Grand Junction Railway built some years later by his assistant Joseph Locke cost 18,846 British pound sterling per mile.

After the completion of the construction of the Liverpool-Manchester Railway it became an instant commercial success, and as a consequence of it George Stephenson's reputation as a railroad engineer, and the 'Father of Railroad Age' was made unassailable.

After a major mine explosion in 1815, George Stephenson invented a miner's safety lamp; a similar lamp was also devised almost at the same time by the prominent English chemist Humphrey Davy. All the credit for this invention was accorded to Davy by the influential, educated circles of the society, because they considered a formally uneducated man who spoke a lower class brogue, such as George Stephenson, unequal to the ingenuity of an educated man such as Davy, a viewpoint expressed by Davy himself. Stephenson never forgot this arrogant and snobbish injustice. He declined the knighthood offered to him on several occasions, an invitation to the Fellowship of The Royal Society in London, and a membership in the Civil Engineers Society, because as he said: "... I objected to these empty additions to my name." However, he did accept the Presidency of the newly organised Institution of Mechanical Engineers. Upon the retirement of George Stephenson, his son Robert succeeded him to the Presidency of this Institution.

The huge success of the first railroad lines led to public enthusiasm and gullibility concerning railroads, which culminated in the 'railway mania' of 1845-1847. During this time the public rushed to buy shares in any railway company, no matter how improbable, as soon as its prospectus was published, even before the Parliamentary bill was passed for the surveyed route. The earlier 'canal mania' seems insignificant in scale in the face of the flood of new railway bills introduced in the British Parliament. It was a period of public mania in which charlatans flourished. By the end of 1844, there were already 2,234 miles (3,595 km) of railways built in Britain. In 1845, Parliament authorised the construction of further 2,170 miles (3,492 km) of railways at an estimated cost of 500 million British pound sterling. By early 1846, Parliament was considering 477 more railway bills at an estimated cost of 200 million British pound sterling. The equipment and labour necessary for such a massive construction programme was, of course, not available, and the whole promotion was unrealistic. It was a time when Britain was still an open society for the authority of the State was quite limited, the newly created 'police' had no more power than the citizens had rights, and the individual could still enjoy his personal profits and property in peace and safety.

The steam locomotive was improved in 1854 when John Ramsbottom invented the metallic piston-rings to replace the hemp packing previously used. His split piston-rings were rolled from strip brass. In 1855 David Joy devised his double-coil piston-rings turned from a cylinder of cast-iron which showed considerable reduction of wear in comparison with the brass piston-rings.

Ramsbottom introduced his safety valve that could not be tampered with by the engineman in 1856, and the steel boiler for locomotives in 1873.

Steam-Carriages

At the time the Stephensons were developing their locomotives and the railway as a means of transportation, the railroad was not without competition from another form of steam-powered transportation – the steam-carriage. In fact, Robert Stephenson adopted some valuable details from Gurney's steam-carriage to design the 'Rocket'.

After Trevithick's pioneering road locomotive had failed to attract public support in 1802, a Czech engineer, Josef Bozek (1782-1832) built a ½ horsepower steam engine in 1815 which he installed in a carriage and in a paddle-wheel steamboat. The limited boiler capacity and low power of his engine required frequent stops of his steam-carriage in order to build up steam. In 1817, Bozek organised a public demonstration in the Stromovka woods near Prague, and had a collection made to cover the debts he had incurred in building the steam-carriage and in perfecting the engine. Unfortunately, somebody stole the collected funds during a sudden thunderstorm, and in a mood of discouragement, Bozek destroyed his steam-carriage and the 2 steamboats he had built.

The next steam-carriage builder was an English inventor, Medhurst, who produced his first steam-coach in 1819. It was a single-seater, and it developed a speed of 5 to 7 miles (8.0 to 11.3 km) an hour on its trial run between Paddington and Islington. Encouraged by his first experience, he built then a 4-seater steam-carriage which covered 7 miles (11.3 km) an hour without refueling. But for some reason Medhurst abandoned any further development of his steam-coach.

After Medhurst, a number of steam-powered road vehicle builders, such as Timothy Burstall, Goldsworthy Gurney (1793-1875), Col. Francis Maceroni (1788-1846), John Scott Russell (1808-1882), Walter Hancock (1799-1852), and William Henry James (1796-1873) were connected with steam-coach lines such as Hill & Burstall, Church & Goldworthy, Steam-carriage Company of Scotland & Russell, Squire & Maceroni, London and Paddington Steam-carriage Company & Hancock, and Anderson & James with various success, but ultimately all were forced off the road by the combined opposition of the stagecoach operators, turnpike trusts, and, most of all, by the new railways, which soon after the beginning of the reign of Queen Victoria, forced even the stagecoach lines out of business in a keen competition.

In 1821, David Gordon took out his first patent on a steam-carriage which was driven by a locomotive mounted inside a 9-foot (2.74 m) drum, 5 feet wide. David Gordon believed like Brunton, that wheels are not capable of gripping the road on inclines and, therefore, he developed a set of 'walking feet', a system of mechanical feet, 3 on each side, operated by an ingenious system of cranks and connecting-rods. He used such 'walking feet' in 3 or 4 steam-coaches he built from 1823 to 1830. Gordon discovered that although the motion of his steam-carriages was far less jerky than that of Brunton's bilegged locomotive, the wear and breakdowns were excessive if speeds more than walking speed were attempted. Finally he came to the conclusion that the power of the engine should be applied to the wheels, and he gave up his work on steam-carriages.

Goldworthy Gurney and Steam Coach:

The most well-known steam-carriage builder, but not the most successful, was Goldworthy Gurney, a well-educated and wealthy surgeon from Cornwall, who gave public lectures at the Surrey Institution on his chemical researches which were attended by the English physicist and chemist Michael Faraday (1791-1867), and later developed the limelight, the Bude light, the Gurney stove, and the central heating by radiators. Gurney, who had been impressed as a child by Trevithick's steam-carriage and had appreciable knowledge of mechanical engineering, began his work on a mechanically driven common road vehicle in 1823. He constructed what was probably the first ammonia-engine, and he used it to drive a small locomotive. Then he turned to steam-power. Surprisingly, Gurney also thought that wheels are not capable of gripping the road surface, as he had been told by the most competent engineers of the period. Therefore, Gurney's first steam-carriage in 1827 relied entirely on 'mechanical legs' for propulsion, but they proved very troublesome. In the rebuilt vehicle the wheels were connected to the 'propellers', his name for the 'mechanical legs', but he soon discovered that it was never necessary to use the 'propellers', and he eliminated them altogether in 1828.

The design of the steam-carriage involved packing the required power into the small available space, and this meant that quite high steam pressures had to be used. At the time when 25 psi (172.4 kPa) was considered 'high pressure', the steam-carriages were using pressures from 70 psi (482.7 kPa) to 250 psi (1,723.8 kPa). Although the materials and methods of construction were not up to the necessary standards at the time, boiler accidents were relatively few. The boilers for steam-carriages had to be light, compact, and steam-efficient. Gurney designed a watertube boiler for his steam-carriages which minimised the possibilities of explosion. His boiler consisted of U-tubes, filled with water, turned at right angles and connected at their termini to 2 horizontal drums. The steam drawn from the upper drum representing the steamchest of the boiler passed into 2, or more, vertical receivers serving as separators which contained water in their lower part. The separators were connected to the lower drum of the boiler. Applying heat to the lower leg of the U-tubes created circulation of the water which avoided the interference between the up and the down currents in the standard cylindrical boilers. The steam was drawn from the upper drum serving as the steamchest, and passed into 2, or more, vertical receivers which served as separators to ensure dry steam. Gurney's boiler evaporated about 5 pounds (2.27 kg) of water to each pound (0.45 kg) of coke incinerated. The steam pipes from the tops of the separators were connected to a main feed pipe, which passed over the fire to reach the control valve and engine, thereby drying and partly superheating the steam. This boiler was quite safe against bursting pipes, but it did have the priming problem. Gurney's boilers, which had been tested in their cold state to withstand 200 psi (1,379 kPa) pressure, produced steam pressures from 70 psi (482.7 kPa) to 120 psi (827.4 kPa). Although his boilers were modified in some details over the 8 years he built his steam-carriages, in fundamental design his boilers remained essentially the same. Gurney's steam-carriages weighing from 1½ tons to 2½ tons, had four wheels of 4-foot (1.22 m) diameter with iron tyres about 4 inches (0.10 m) wide.

In the steam-engine, Gurney used a fixed cut-off, which was controlled by throttling or 'wire-drawing'. Since the steam pipes were rather small compared to the cylinder and valve-port areas, the steamchest pressures were about 15% to 25 % lower than boiler pressures. This design was probably used to reduce the risk of priming, and give the advantage of flexibility to the engine that was useful for the single-ratio transmission employed.

In 1828, Gurney built his 14-seater steam-coach, which was a rather well-designed vehicle for its time that incorporated a servo-assisted steering. Most steam-carriages had mechanical bellows or fans driven by the engine to create a draught in the furnace. Gurney installed a small auxiliary engine capable of driving the blower and water-pump even when the carriage had stopped. Gurney's main engine with an output of 14-horsepower had 2 cylinders with 9-inch (0.23 m) bore and 18-inch (0.46 m) stroke, and its pistons cranked the rear axle with two 5-foot (1.52 m) diameter wheels giving the coach a speed over 10 miles (16.1 km) an hour. It consumed 10 gallons (45.5 liters) of water and 20 pounds (9.1 kg) of coke for every mile (1.6 km) travelled. In his early work on steam-carriages Gurney was associated with Col. Francis Maceroni, who also had engineered a steam-coach but soon found himself in debt and in jail. A basic disagreement between the 2 men in 1829 dissolved this contentious collaboration.

Gurney took the longest motor trip in history with his 14-seat steam-coach when in 1829 he drove his steam-coach at 15 miles (24.1 km) an hour from London to Bath and back, a 200-mile (322 km) trip, but it was not a complete success. The steam-coach and its passengers were attacked and stoned at Melksham by a mob of hostlers, horse dealers, drivers and agricultural workers who saw the steam-coach as a threat to their jobs. Gurney was seriously injured in this attack, and his fireman was beaten unconscious. After this experience with mob-violence, Gurney built a 'steam-drag' in 1829, which was a road locomotive designed to pull the horse-drawn vehicles. In the steam-drag the auxiliary engine was eliminated and the exhaust blast was used to induce a draught in the furnace. The boiler was modified by using a single horizontal separator in place of a vertical one. Both wheels were free to turn round the axle, and were driven by T-arms fixed to the shaft, and bearing on removable pins or bolts in the felloes of the wheels. Usually only one wheel was driven, but when more adhesion was needed, bolts were put into the second wheel. By setting the driving-pins in the felloes the spokes were relieved of the driving stress. Gurney built about 6 or 7 steam-drags, and he gave a demonstration ride to Duke of Wellington and Robert Stephenson in a barouche drawn by Gurney's steam-drag. Stephenson took notice of some good design features in Gurney's steam-carriages which he used with profit in his own locomotives.

In 1831, Charles Dance established a steam-carriage service between Cheltenham and Gloucester with 3 of Gurney's steam-drags which Gurney sold to Dance, and which were fitted with a modified Gurney boiler. This service ran regularly four trips a day except on Sundays, from late February to the late June without an accident. The passengers were transported in vehicles resembling horse-drawn coaches which seated 16 passengers. The total distance covered was 3,640 miles (5,858 km) in 396 trips and more than 4,000 paying passengers were transported. The average running time for the 9-mile (14.5 km) trip was 50 minutes, but on a few occasions it was reduced to 40 minutes. The turnpike trustees had large stones strewn across the road to a depth of 18 inches (0.46 m) to stop the operation of the steam-carriage line, and after the steam-drag had struggled through the deliberate stone obstructions 3 times and finally broken the axle, this enterprise came to an abrupt end. At the same time the renewed Cheltenham Road Bill authorised a heavy toll on steam-carriages. On some roads the tax on steam-carriages was 6 times the tax on horse-drawn coaches.

After the failure of the line, Dance had the boiler of one of the Gurney drags redesigned by Joshua Field (1787-1863) of Maudslay, Sons & Field, which led to a new patent in the name of Field and Dance. This boiler was developed by Field into the successful Field Boiler with hanging tubes, which provided steam-power for fire-engines, river launches and other power equipment.

Gurney himself gave up his steam-carriage development after having spent about 30,000 British pound sterling on it, and turned to his other more lucrative projects such as the limelight, from which he made a great deal of money.

A fate similar to Dance's steam-carriage line was suffered by the Glasgow to Paisley steam-carriage service run by the Steam-carriage Company of Scotland for 7 months in 1834. This company used the 6 steam coaches, each seating 26 passengers, designed by John Scott Russell. Russell's steam coaches were superior to those of Gurney, for in Russell's coach the motion of the pistons in the 2 vertical cylinder engine was transmitted by crosshead and external connecting-rods to 2 crankshafts, and from there over 2–to–1 spur-gearing to the line-axle which was spring-mounted. The line-axle was guided by a rather ingenious linkwork so that the functioning of the spur-gearing was not affected by the motion of the axle on springs. Each steam-carriage also transported a 2 wheel trailer carrying enough fuel and water for the next trip and accommodating 6 more passengers. Its boiler was similar but inferior to that of Hancock, the most successful steam-carriage builder of the period. Often 45 passengers were carried on a single trip. Since Russell's steam-carriages were popular, the trustees of the turnpike had heaps of loose stones thrown on the road to obstruct the steam-carriage. When one of the steam-carriages struggled through a heap of stones, its wheel collapsed, its boiler burst and killed 5 people, in the second fatal motor accident in Great Britain. The Court of Session interdicted the whole fleet of steam-carriages, and a subsequent damage action against the turnpike trustees was finally settled out of court. The service, however, was never reopened because of the damage done by the harmful publicity. The sabotage had done its work.

Maceroni and Steam Carriage:

Francis Maceroni designed several steam-carriages which he built in his Paddington workshop, but he had incredibly bad luck with his financial backers and associates, who cheated him out of his just dues. He designed in 1841 another steam-carriage which had an extremely efficient boiler with 81 vertical watertubes arranged in nine rows in a horseshoe form round the firebox. The bottom water connexion between the rows of tubes served at the same time as a fire container, and by keeping them cool, the water helped to prevent clinkering. A steam-dome was instrumental to prevent priming. The boiler was very quick in steaming and was regularly worked from 150 psi to 200 psi (1034 kPa to 1379 kPa), much higher than in Gurney's boiler. For this reason Maceroni's 7½ by 15½ inches (0.19 by 0.39 m) steam engine gave as much power as Gurney's 9 by 18 inches (0.23 by 0.46 m) steam-engine. It had a modern looking chassis and unsprung driving axle, but it suffered from crank axle breakages like the other steam-carriages. The automatic engine lubrication of Maceroni's steam-carriage was much in advance of his time. Maceroni's steam-carriage had a trial run in London where it achieved an average speed of 16 miles (26 km) an hour with 17 passengers aboard, and it climbed the steep Shooter's Hill at 8 miles (12.9 km) an hour. Owing to the high cost of his engine, his financial arrangement collapsed.

Hancock, Steam-Omnibus and Wedge-wheel:

The most successful steam-carriage builder was Walter Hancock, who patented in 1824 a novel type of steam-engine working by a rubber bag in place of a cylinder and piston. The inflation and deflation of the bag under low steam pressure moved the connecting-rods and cranks of this inexpensive and simple 4-horsepower engine, which worked for a number of years in his workshop at Stratford. In 1825 he decided to use this engine to power a steam-carriage, but it was not suitable for high-pressure steam-engine. Hancock built altogether 9 steam-carriages, really omnibuses, which used very high steam-pressure of 200 psi (1379 kPa) for its time, and they proved to be remarkably safe and reliable.

His first steam-carriage was a 3-wheel carriage with the single front wheel that acted both as a driving and steering wheel turned by 2 oscillating cylinders one on either side of the forked frame. Since the flexible steam connexions were unsatisfactory, and the boiler was inadequate, the steam-engine was removed whilst the rest of the vehicle was scrapped. Hancock realised that the major weakness of the steam-carriages of his time lay in their inadequate boilers. In 1872, he designed, and patented, an ingenious boiler with embossed and very flexible iron plates rivetted together which looked like a modern radiator. The thin walls of the water-chambers bulged under high steam pressure and touched adjacent chambers thereby receiving lateral support and increasing the surface area exposed to the hot gases. The furnace had a hopper-feed for the coke, and a rack-and-pinion mechanism which removed the clinkered fire bars and fed in fresh sets without slackening the fire. He went on to develop boilers with forced draught supplied by an engine-driven fan, a dog-clutch which permitted the fan and water pump to be operated whilst the steam-carriage was stationary, and other improvements. In the last 3 of his steam vehicles, Hancock set the furnace slightly to one side of the water chambers, and admitted part of the exhaust steam to pass over the fire in order to produce the draught for the promotion of combustion. Hancock's boilers in general were twice as efficient as those of Gurney. In 1832, Hancock also invented his' wedge-wheel', which is the forerunner of the automobile wheel of the 20th century. It was particularly designed for use in steam-omnibuses since the failures of their wheels proved to be one of their greatest hazards. The wedge-wheel was exceptionally strong, both vertically and laterally, and on turns all the spokes delivered the effort from the hub to the felloe. Hancock's wedge-wheel was so much stronger for its weight, that it was promptly adopted for gun carriages, ammunition carts and other heavy duty military vehicles, and became known as the 'artillery' wheel. Hancock's wedge-wheel was the most common automobile wheel in the 1920's, and some American cars were still equipped with it in 1932.

His first rebuilt steam-carriage, named the 'Infant', which seated 10 passengers, had 4 wheels and an exposed oscillating engine carried on an outrigger-frame behind the rear axle, was driven by a single chain-transmission consisting of a link-chain passing over grooved and spiked pulleys. The 'Infant' was soon completely rebuilt to seat 14 passengers, its exposed oscillating engine was replaced by an enclosed, fixed-cylinder vertical engine with a place for the engineman. The single chain transmission was supplied by a screwed radial arm to maintain the correct chain tension and protected from the road by an undershield.

The subsequent vehicles of Hancock incorporated many improvements, and given various types of body-work, they essentially followed the design of the rebuilt 'Infant'. The 'Infant' made its trial runs for the London and Brighton Steam-carriage Company in 1832, but nothing came of this enterprise.

Hancock built his 14th passenger-omnibus, the 'Enterprise', for the London and Paddington Steam Carriage Company, which Hancock personally ran between Paddington and the Bank for 16 days at 16 miles (25.7 km) an hour speed with full capacity, but the company cheated Hancock, dismantled the 'Enterprise', and assembled a steam-carriage of their own which was almost the same vehicle as that of Hancock's except having a different steering gear and complicated spring-wheels. This vehicle never steamed adequately, and the company failed.

Hancock designed an 18 passenger vehicle, the 'Era', for a Greenwich company, but again the line did not materialise. He rebuilt the body of the 'Era', named it 'Erin' and sent it to Dublin, Ireland, to attract the interest of entrepreneur James Anderson, but nothing came from these demonstrations and negotiations.

He redesigned the steam-omnibus for 22 passengers, similar to a previous steam-omnibus named the 'Autopsy', which had been built for the Brighton line. The new version of this omnibus Hancock named 'Automaton', which he considered justifiably his best steam-carriage. It was also the most powerful, having 12-inch (0.30 m) bore cylinders in its engine, and the most reliable steam-vehicle for its time, which developed a speed of 20 miles (32.2 km) an hour under full load. 'Automaton' made its last trip in 1840 with 32 passengers and a crew of 3, probably weighing altogether 4 tons. It is reported to have achieved a speed of 25 miles (40.2 km) an hour on its return trip.

Hancock's steam vehicles ran regular services in London and its vicinity at various times, apparently with profit. The 'Autopsy' ran between Finsbury and Pentonville for a number of weeks, and then together with the 'Era', from Moorgate to Paddington and back by way of the Bank. All together about 4,000 passengers were carried on this line.

In 1835, the 'Erin' ran demonstration trips to Marlborough, to Reading, and to Birmingham. The London to Birmingham Steam Coach Company was instituted but by that time, the exorbitant toll increases on the turnpikes and the competition from the railroads deflated the financial future of this company and it went out of existence.

In the middle of 1836 Hancock made the last great effort with his entire steam-carriage fleet to demonstrate that steam-vehicle transport on roads is viable and important. The 'Autopsy', the 'Automaton', the 'Era', and occasionally the venerable 'Infant', ran various routes for almost 5 months with considerable regularity: 525 trips were run from the City to Islington and back; 143 trips to Paddington and back, and 44 trips to Stratford and back. The city of London was crossed more than 200 times without any accidents despite the harassment by stagecoach drivers and other people hostile to road-steamers. On these trips 12,761 passengers were transported. Despite these successes the road tolls kept increasing beyond all reasonable bounds. For instance, the Liverpool-Prescott Turnpike charged only 4 shillings for a horse-drawn coach, but 48 shillings for a steam-powered coach!

James and Steam-Drag:

The last engineer who almost succeeded was William Henry James. He built his first steam-carriage in 1823, when he patented a high-pressure watertube boiler for steam-carriages. He constructed a steam-drag with 2 separate two-cylinder engines of very small dimensions by the 1825 standards, each one of the engines driving one rear wheel. It was fitted with a control mechanism coupled to the steering mechanism which automatically reduced steam to the inside wheel and increased the steam to the outside wheel on turning the vehicle.

In 1829, James built a large steam-coach for James Anderson weighing 3 tons and carrying 20 passengers. It incorporated James' boilers which worked under 250 psi (1724 kPa) steam pressure. Unfortunately, the standard boiler-tube connexions incessantly leaked and, finally, the steam-coach was dismantled and its engines used in a steam-drag fitted with 4 small boilers which had stronger tubes. Unfortunately, James committed the mistake of elevating the steam-pressure in his steam-drag to 300 psi (2069 kPa), which again led to leaks in the joints.

James designed a steam-drag in 1832, which had only one boiler working with 200 psi (1379 kPa) steam-pressure and its twin-cylinder engine mounted diagonally over it. This steam-vehicle was supplied with a 3-speed gear [used a century later in the 'Frazier-Nash' automobile], which allowed a smaller engine than required by a single-transmission, that always wastes power, but was designed to allow gear changes when the vehicle is in motion. This feature in his carriages reappeared again 70 years later. The motion of the crankshaft was transmitted by sprocket wheels and pitch-chains to the chain wheels, which were of different sizes to provide different gear ratios. The sprocket wheels on the countershaft were selectively clutched to the chain wheels by pedal-operated claw clutches with an interlock mechanism which released one clutch before another was engaged. By momentarily shutting off the steam and pressing the appropriate pedal, the gear was changed without an undue jerk at the chain. The driver, for the first time, was provided with a footbrake. The final drive was by means of side-chains, and for this reason the back axle could be suspended on springs. The final-drive sprockets on the countershaft were connected to the back-axle by friction-clutches controlled by the steering-gear in such a manner that the inside-wheel was disconnected when turning the vehicle. It was an ingenious substitute for differential gearing, and it was an improvement on the usual design of his contemporary steam-carriage builders. This vehicle had a condenser to conserve water, gaslighting and feedwater heater. Unfortunately, the entrepreneur James Anderson withdrew from the steam-carriage business, and this most practical design of a steam-drag was, therefore, never built.

The first steam-carriage with differential gear was built by Richard Roberts (1789-1864) in 1833, but it was an experimental vehicle which made only a few short trips before its boiler exploded. F. Hill designed a differential gear after the example of Roberts, but he did not use it in his 1840 steam-vehicle. It has been speculated that the differential gear invented by Hellenistic engineers may have been used in a Chinese vehicle known as the 'Yellow Emperor's South-Pointing Chariot'. The first modern differential gear had been designed by Joseph Williamson for his bifacial astronomical clock between 1719 and 1725, and it had been used in 1822 by James White (c.1760-1825) in his dynamometer, which was known to Roberts.

All the steam-carriages discussed *supra* used the horse-carriage steering in which the front axle is turned about a central pivot. The radial-axle steering of Georg Lankensperger (1779-1847), a German horse-drawn carriage manufacturer, in which each front wheel turns on its own radius had been used in the steam-carriage of Redmund built in 1832, and in another unsuccessful steam-carriage designed by Julius Griffith and built by Joseph Bramah (1748-1814). Unfortunately, the steam-carriage of Griffith never ran at all, because its patented boiler primed so badly in stationary tests that it could not sustain the motion of the carriage. This vehicle failed because of its useless boiler, but at the time, despite this shortcoming, it had been regarded as an engineering masterpiece. This type of steering mechanism had been used in windmill sail-carriages by a French millwright Du Quet already in 1714. One of the first successful road vehicles with another geometrically precise steering was the elliptical cam-steering designed by the French engineer Amédée Bollée (1844-1917) for his steam omnibus '*L'Obéissante*', built in 1873. The elliptical cam-steering unfortunately did not operate in a satisfactory manner since the stretching of the chains in this steering mechanism led to losses of kinematic precision in steering. Bollée also invented the independent front suspension and a differential gear for the rear axle.

Dietz and Steam-Tractor:

Meanwhile on the continent, a German engineer Jean Christian Dietz and his sons, Charles and Christian, developed a new line of steam-carriages: a 3-wheel steam-tractor which pulled a train of carriages between Brussels and Antwerp in Belgium. Charles Dietz established in 1834 a regular passenger service in Paris using a steam-tractor similar to that of his father. In 1839, Christian Dietz built another road-tractor which was really a road-going railway locomotive with 8 wheels of which 6 steered and 2 propelled. It was introduced into service in Bordeaux in 1841, but it aroused the hostility of local coachmen, and Dietz soon gave up his road-engines for lack of commercial success, after which the initiative for road-traction engines moved back to England. Some British inventors, such as Thomas Rickett of Birmingham, made attempts in the middle of the 19th century to build some private steam-carriages for rich people, but the infamous 'Red Flag' Act of 1865 as amended in 1878, put an end to the steam-carriage and road motor-vehicle development in Britain by requiring that self-powered vehicles travel no faster than 4 miles (6.4 km) an hour in the country, and two miles (3.2 km) an hour in towns. In either case a man carrying a red flag had to walk ahead of the motor vehicle. For this reason, the British motor-vehicle engineers built self-propelled road vehicles strictly for foreign markets.

The famous Scottish engineer, Thomas Telford (1757-1834), was a promoter of steam-powered road vehicles, because he noted after studying the Stockton-Darlington Railway for the Commission granting loans to the railway company, that all traffic on a railroad is necessarily the monopoly of the railway company. At that time it was generally accepted, that any route of communication, such as canals or turnpike roads, must be free for use by anyone on payment of the statutory toll. On account of the rigid schedules by which the trains needed to operate, this rule had to be rescinded. He realised that steam-power when applied to land transport by road vehicles are not subject to this restriction. He considered railways to be indirectly uneconomical because it required a costly special track for its exclusive use, and restricted movement. Telford saw railways only as an adjunct to a canal system for use where land conditions did not favour canal construction, or where water supply was inadequate. This conclusion is certainly valid today in United States: wherever the waterways have been extensively developed, the railroads cannot compete even for hauls over thousands of miles.

The steam-carriage builders advanced the boiler design by minimising the volume of steam which could escape on a burst. They developed the primitive forms of flashboilers, with the heating area exposed to fire being very large in relation to the overall dimensions of the boiler. Burstall and Hill invented the flashboiler in 1824, but it did not work very well. It was reinvented as an instantaneous steam-generator by the Frenchman Léon Serpollet (1858-1907) in 1894 for his steam-carriage in which cylinders were jacketed for efficiency, and forced-draught fans were used in preference to induced draught. Blow-off steam was discharged into the furnace to avoid nuisance and increase draught, and the éboilers had fusible plugs as well as safety valves. The 'dead-man's pedal' was fitted to steam-carriages which disconnected the valve-gear if the driver lifted his foot, and braking was applied by handbrake to the tyre, or by reversing the valve-gear. In either case the wheel did not lock and damage the roadway as much as the drags employed on horse-drawn vehicles. Overall Serpollet's steam-carriage was a highly competent piece of engineering.

Heroic Engineering in the Victorian Age
Age of Railroads :

The railroad age from 1830 to 1860 ushered in a period of heroic engineering dominated by such prominent railroad engineers as Robert Stephenson (1803-1859), son of George Stephenson,

Isambard Kingdom Brunel (1806-1859), son of Marc Isambard Brunel, and Joseph Locke (1805-1860), a former assistant engineer to George Stephenson.

The railroad era, beginning in 1830, brought with it a new wave of daring bridge building, tunnelling, and excavation. This era of bold engineering by outstanding individuals was led by the triumvirate of Victorian engineering: Robert Stephenson, Isambard Kingdom Brunel, and Joseph Locke.

Robert Stephenson built the London-Birmingham Railway, the first major railway, and a number of important bridges. In 1846, he built the High Level Bridge over the Tyne river, at Newcastle, which consisted of 6 bowstring girders of 125-foot (38.10 m) span each. In 1843, Stephenson introduced for the first time in bridge construction the steam-powered pile-driver, which delivered from 60 to 80 blows a minute with a 4-ton hammer. The steam pile-driver was designed and built by James Nasmyth (1808-1890) in 1845, and had been first used in building a 1½ mile (2.4 km) extension to Devonport Docks. The steam pile-driver rammed a 70-foot (21.34 m) long pile with 1½-foot (0.46 m) square cross-section into the ground in 4½ minutes, a task that took 12 hours with a gravity pile-driver.

Between 1846 and 1848, the Conway Tubular Bridge on the Chester-Holyhead Railway, which spanned a distance of 400 feet (121.92 m), was built. It is the first tubular plate-girder bridge made of wrought-iron ever built. Stephenson used the boiler plate rivetting machine of William Fairbairn (1789-1874), a manufacturer of wrought-iron beams and Lancaster boilers, on this job. The huge tubular plate-girder was fabricated on shore, floated into position, and then lifted into place with hydraulic jacks – the first such use of hydraulic jacks. Stephenson called the construction of the Conway Tubular Bridge 'a trial run' for his next bridge, his largest in Britain, the Britannia Tubular Bridge.

Between 1846 and 1850, the Britannia Tubular Bridge over the Menai Straits in Wales, located about 1 mile (1.61 km) southwest from the famous suspension bridge of Telford also spanning the Menai Straits, was built to carry Stephenson's railroad across the strait. The Britannia Tubular Bridge was a huge wrought-iron tubular girder, 28 feet (8.53 m) high and 14 feet (4.27 m) wide, which has two 459-foot (139.90 m) central spans, and two 230-foot (70.10 m) side spans. Stephenson hired two consulting engineers who were experts on iron structures to advise him on the design and construction of this bridge: William Fairbairn on the design and model testing of the tubular plate-girder, and Eaton Hodgkinson, whom Stephenson had invited to assist him in the stress analysis of this plate-girder bridge with cellular flanges. Hodgkinson, who was professor of engineering at the University College in London, had carried out tests on cast-iron I beams, and was familiar with the theoretical mechanics of the French engineers. Fairbairn, who had considerable experience with the design and testing of wrought-iron girders and plate-girders (having patented a solid plate-girder in 1846) in his boiler factory, tested a 75-foot (22.86 m) scale model of the tubular plate-girder to failure, and carried out other experiments on different cross-sectional shapes of the tubular girder and on the properties of wrought-iron. He was assisted in these tests by Hodgkinson who made the necessary theoretical calculations. Hodgkinson conducted an experimental investigation on the buckling of thin-walled tubes under compression in connexion with this bridge which contained the first experimental study of buckling of plates. The 4 spans of the Britannia Tubular Bridge were rivetted together whilst the girders were slightly tilted over their supports, a procedure of construction which created prestressing in the form of negative bending moments acting in the girders at both of their supports, and thus making the spans continuous, and, thereby, reducing the overall deflexion of the girder. The girders were provided with closely spaced vertical stiffeners rivetted to the plate to avoid lateral buckling of the plates.

Fairbairn and Edwin Clark, resident engineer of Stephenson, carried out the detailed design of the tubular plate-girder. Hodgkinson, at Stephenson's request, made theoretical calculations of the strength and deflexion of the Conway tubular plate-girder in 1846, and again in 1847. William Pole prepared alternative calculations for the Conway and the Britannia tubular plate-girder after the design had been completed.

In the construction of this bridge, Stephenson introduced a few novel mechanised methods of construction: a multiple rivet-hole punching machine controlled by Jacquard punched-card mechanism invented by Richard Roberts to overcome the effects of a crippling strike of construction workers, the rivetting machine of Fairbairn, and the hydraulic presses of Bramah used as jacks to lift the 3 million pound (1.36 million kg) tubular plate-girders by 6-inch (0.15 m) increments into place. Although this tubular plate-girder bridge was very expensive [it cost 5 times as much as Telford's suspension bridge over the same Menai Straits], it made a decisive contribution in its engineering to the plate-girder design because in the experiments, in the design of the plate-girder, and in structural theory the fundamental understanding of the way the plate-girder functions was involved. The thermal motion of the Britannia Bridge in the middle of the span was as much as 2½ inches (6.4 cm) both vertically and horizontally, and the deflexion due to self-weight about ¼ inch (0.6 cm). The self-weight of the plate-girder tube constituted at least ¾ of the total load on the bridge.

The structural design of the Britannia Tubular Bridge was criticised by a young Russian railroad engineer, Dimitrii I. Zhuravskii (1821-1891), who had solved the difficult theoretical problem of shear stresses in bending. Zhuravskii carried out simple cardboard model studies of the plate-girder, and showed that the diagonal stiffeners in plate-girders were much more effective than the vertical stiffeners. He also criticised the even spacing of rivets in the Britannia Bridge girder for Zhuravskii was able to calculate the correct uneven rivet spacings for the plate-girder with his shear stress formula which would have considerably reduced the 2,190,000 rivets used in this tubular plate-girder. He, moreover, recommended that model tests of the plate-girders be based on the working stress criterion rather than the ultimate strength criterion used by Fairbairn. Zhuravskii's criticism of Stephenson's plate-girder bridge was published in a French journal in 1860. In 1973, the Britannia Tubular Bridge was severely damaged by a fire accidentally set by boys seeking bird nests inside of the tubular bridge and, therefore, the Britannia Tubular Bridge for reasons of safety was reinforced underneath by new steel arch trusses.

In 1835, both George and Robert Stephenson were consulted on the first railway construction in Belgium, and Robert Stephenson later built railways in France, Italy, Sweden and Norway. The younger Stephenson constructed two tubular bridges for the Alexandria and Cairo Railways, one spanning the Damietta branch of the Nile at Benha, and the other the Karrineen Canal at Birket-el-Saba. Both bridges were opened for traffic in 1855.

Robert Stephenson's last, and the longest, tubular plate-girder bridge, The Victoria Tubular Bridge, at Montreal, in Canada, was partially prefabricated in Britain, and its erection which began in 1854 was finished after his death in 1859. The Victoria Tubular Bridge, which was opened by Prince of Wales, spanned the St. Lawrence river at Montreal, and its construction was directed by Alexander McKenzie Ross, who had been assistant engineer to Stephenson for the Britannia Tubular Bridge. The entire length of this tubular plate-girder bridge was 6,600 feet (2 km), and it consisted of 24 spans of 242 feet (73.76 m) and one central span of 300 feet (91.44 m). This bridge was the longest tubular plate-girder bridge ever built. The foundations and piers for this bridge were designed under Ross' direction by the Canadian engineer, Thomas Coltrin Keefer (1821-1915). The construction of the Victoria Tubular Bridge required a total work-force of 3,000 men. However, the tubular plate-girder bridge had some serious disadvantages: its initial cost was very high, the smoke from the locomotive was entrapped in the tube and had a corrosive effect on the plate-girder, and the noise made by the locomotive was objectionable. In the Victoria Tubular Bridge the problem of smoke and noise was somewhat alleviated, but not eliminated, by leaving the section of the tubular bridge open on top. For these reasons, the steel truss, which eliminated all these shortcomings, terminated the era of tubular plate-girder bridges. The tubular plate-girders of Stephenson in the Victoria Bridge were later replaced by steel trusses, but the original foundations and piers, popularly called 'Keefer's Shoes', are still there.

When Robert Stephenson died in 1859, the railroad engineers in Britain were public idols at the peak of their popularity. This high public esteem had been created largely through the remarkable engineering feats of Robert Stephenson and Isambard K. Brunel. The situation had been quite different when George Stephenson began building his railway. The opinion of a prominent landowner, and Member of Parliament, Col. Sibthorp, is a typical example of the contempt that a country gentleman had for the engineer: "I would rather meet a highwayman, or see a burglar on my premises than an engineer." The public's regard for engineers had changed dramatically since 1825, for a holiday was declared on the day of Robert Stephenson's funeral when he was buried like a royalty in the Westminster Abbey next to Telford as he himself had wished. Never again have engineers enjoyed such high public esteem.

Robert Stephenson, who was a methodical engineer, tended to be rather conservative and cautious in his engineering enterprises. For instance in 1847, George Stephenson studied the Isthmus of Suez for a possible canal, but owing to his aversion for taking risks, he deemed it unfeasible, and opposed the French-backed de Lesseps canal in the Parliament in 1857 as its Conservative member. In contrast to his father George, Robert Stephenson supported protectionist government policies.

Isambard Kingdom Brunel, Railroad and Naval Engineering:

The same could not be said about his illustrious contemporary, Isambard Kingdom Brunel, who was gifted, imaginative, high spirited, inventive, versatile, and ready to take great personal risks, both physical and financial. He had a romantic temperament of a Frenchman, a sense of grandeur, and a boundless confidence in his own abilities. At the same time he was one of the most exacting and demanding engineers of the Railroad Era. Brunel who was short in stature, stood tall in his professional intellect according to the opinion of his contemporaries. Therefore, he was affectionately called the 'Little Giant'. Brunel was always willing to invest his own money in his engineering projects, something Robert Stephenson, a risk-conscious man, never chose to do. Brunel did have several financial failures in his engineering enterprises, such as the atmospheric railroad which worked well in a small scaled-down model but did not work at all in full size, and his colossal steamship, The Great Eastern.

When Brunel was still a callow teenager, he assumed the direction of the driving of the first subaqueous tunnel in the world, The Thames Tube in London, from his father. This tunnel, which was designed by his father, Marc Isambard Brunel (1769-1849), was built from 1825 to 1843, the construction being interrupted for a period from 1828 to 1835 for financial reasons. The Thames Tunnel having been an object of admiration, became an object of public ridicule during the period of interruption in its construction, and was called by some wags 'the Great Bore'. The construction of this tunnel had been estimated by Marc Brunel to last only 3 years. This tunnel is 1,200 feet (365.76 m) long and is made of double brick vaulting. Young Brunel worked at first as the resident engineer on this tunnel construction, but he was put in charge of the entire project during the illness of his father. The elder Brunel had designed the first known excavation shield consisting of an assemblage of 12 separate sections each one of which could be advanced individually for the driving of the Thames Tube. The excavation shield was built by Henry Maudslay (1771-1831).

The first attempt to build this tunnel had been made by Richard Trevithick in 1808 who drove a drift [a drainage tunnel], the 3-foot (0.91 m) wide and 5-foot (1.52 m) high cross-section of which was maintained by timber shoring, about 1,000 feet (304.8 m) in length over 6 months, a Herculean achievement, before the tunnel caved in. Trevithick used no excavation shield in his failed effort to cut a tunnel under the Thames. After the failure of driving the drift for the tunnel, Trevithick proposed to build the tunnel by sinking a series of cofferdams along the line of the tunnel onto the river-bed, and then excavate a ditch within them, and bury a sectional pipeline of cast-iron. This method of open excavation for a tunnel was regarded as impossible at the time and, therefore, Trevithick's proposal was turned down, although a number of river-beds in United States were provided with tunnels in the next century by Trevithick's method of open-excavation.

After Trevithick's failure to drive a tunnel underneath the River Thames, many other proposals were offered to drive this subaqueous tunnel. Dr. Hutton and William Jessop were called to examine 54 different plans for the driving of this tunnel, but decided not to recommend the construction of the tunnel by any of the methods proposed. Three major cave-ins of the tunnel occurred during Brunel's cutting of the tunnel, mainly because a vital part of the tunnel roof remained unprotected during the advancing of any section of the excavation shield, and the young Brunel's heroic efforts in saving the lives of workmen caught in the flooded tunnel almost cost him his own life.

Whilst recuperating in Bristol from the tunnel disaster, he entered a competition for a bridge over the Avon river near Bristol, a competition which he won with his design of a suspension bridge, called the 'Clifton Suspension Bridge'. It spans 702 feet (213.97 m) and its construction, which was subjected to several delays, lasted from 1834 to 1864. It was finished after Brunel's death by his friends, and is still in service.

Isambard K. Brunel was one of the great pioneers of railroad. He built the Great Western Railway from London to Bristol in 5½ years, a record time. In the course of its construction, Brunel cut the famous Box Tunnel which was nearly 2 miles (3.22 km) long, a record length at that time, but the cost in human lives was high for 100 workmen perished in the cutting of this tunnel in a number of construction mishaps.

In the period from 1843 to 1851, Brunel built the well-known Chepstow Bridge, his first important bridge structure, crossing the Wye river. It was a Pratt truss type of bridge the top chord of which was a circular wrought-iron tube 9 feet (2.74 m) in diameter. The bridge had three spans, each 100 feet (30.48 m) long, making the total length of this wrought-iron bridge 300 feet (91.44 m). It was replaced by a steel bridge in the 1960s.

Brunel built his celebrated Saltash [or the 'Royal Albert'] Bridge near Plymouth from 1853 to 1859. It was a lenticular truss consisting of 2 arches with hollow elliptical cross-section, each spanning 455 feet (138.68 m). The major diameter of the elliptical cross-section was 16 ⅔ feet (5.08 m) and the minor diameter was 12⅓ feet (3.76 m). This bridge had a unique foundation for the central pier which was constructed using a large wrought-iron cylinder caisson floated in place and sunk 86½ feet (26.36 m) below the water level, so that excavations could be conducted under 40 psi (275.8 Pa) air-pressure in the caisson down to the rock-foundation. This bridge was much more economical than Stephenson's Britannia Tubular Bridge. The Saltash Bridge was replaced by a new steel bridge after the Second World War because the tubular arch made of wrought-iron had sustained some inelastic buckling deformations under the heavy wartime traffic which made it unsafe.

Brunel was invited to build a railroad in Italy from Genoa to Turin but he relinquished this project in frustration on account of many delays. However, he and his assistant engineer Benjamin Herschel Babbage did design and construct a short railroad north of Florence. Brunel was consultant to the East Bengal Railway but his early death prevented his carrying out this project. W.A. Purdon, as recommended by Brunel, became the chief engineer, and Brunel's assistant Bradford Leslie designed the bridges required for this railway. One of Brunel's assistant engineers, John Brunton, built the Scind Railway and other railroads in northern India.

Isambard K. Brunel also made fundamental and pioneering contributions to steam-powered navigation and to the structural design of steamships. In 1835 Brunel proposed to 'extend' the Great Western Railway by building the steamship The Great Western to ply between Bristol and New York. Brunel had recognised that the capacity of a ship increases as the cube of its dimensions whereas the resistance of the ship to its motion through water and, therefore, the power to propel it, only increases as the square of those dimensions. Therefore, Brunel maintained that to design a steamship for any given length of voyage reduces to a question of determining its right proportions. Subsequently, Brunel showed that larger size in ships is a definite advantage. This original idea, which apparently had not occurred to any other naval engineer, was the basis of design of all 3 great steamships engineered by Brunel. The Great Western Steamship Company was formed for the purpose of constructing such steamships, and the construction on the first ocean-going steamship designed by Brunel began in 1836. In 1838, Brunel launched the first true ocean-going steamship, The Great Western, which displaced 1,340 tons. It was constructed of oak by traditional methods, but great strength was built into it, especially in the longitudinal direction. It was the largest ocean-going steamship for its time, with an overall length of 236 feet (71.93 m) and a breadth of 59-feet 10-inches (18.23 m). It had two Murray's side-lever engines of 750-horsepower made by Maudslay which operated under 5 psi (34.5 Pa) steam-pressure.

In 1840, Brunel built The Great Britain, the first large iron steamship which displaced 3,400 tons and had a chain-drive turning a 15½-foot (4.72 m) screw-propeller. It was the first screw-propeller driven ocean-going steamship. This steamship achieved a speed of 12 ½ knots (7.6 km) an hour, and was powered by a novel 1,500-horsepower steam-engine working with 15 psi (103.4 Pa) steam-pressure, and using a chain-drive. It had a length of more than 320 feet (97.94 m). His design of The Great Britain anticipated the modern principles of ship design and construction for it had a strongly built and well-shaped iron hull without external keel, a double bottom, and a 'balanced' rudder.

From 1853 to 1859, Brunel built his last steamship, The Great Eastern, a gigantic vessel which displaced for its time an unbelievable 18,914 tons. This colossal vessel was 692 feet (210.92 m) long, 118 feet (35.97 m) wide, and carried 4,000 passengers and a crew of 100 sailors. The steam-engines for this giant ship built by Joshua Field (1787-1863) of Maudslay, Sons & Field developed 8,300 effective horsepower at the steam-pressure of 25 psi (172.4 Pa), the greatest concentration of power known for its time, the huge fuel supply of coal for its steam-engines was bestowed in the ship's hold. Brunel designed a novel hull structure for The Great Eastern consisting of double cellular walls, and 9 transverse and 2 longitudinal bulkheads, creating a ship's structure consisting of watertight compartments. Altogether 10,000 tons of iron, and 3½ million rivets driven by a punch-cards controlled rivetting machine, were used in its construction. This type of ship structure is still being used in hull design today. The ship was constructed in the shipyard of John Scott Russell, a former builder of steam-carriages. The Great Eastern was designed to be driven jointly by the screw-propellers and the paddle-wheels at the speed of 14 knots (8.51 km an hour), and it took this vessel only 11 days to cross the Atlantic whilst carrying 18,000 tons of coal and cargo, and 4,400 passengers and crew. Despite the unprecedented power concentration, its paddle-wheel steam-engines consisting of two cylinders of 6-feet 2-inches (1.88 m) bore and 14-foot (4.27 m) stroke, and the screw-propeller steam-engines consisting of 4 cylinders of 7-foot (2.13 m) bore and 4-foot (1.22 m) stroke failed to supply the indicated 10,000-horsepower that was required to drive the huge hull at the design speed of 14 knots (16 miles or 26 km an hour). Therefore despite the 4 steam-engines, The Great Eastern was still underpowered for its huge size.

Brunel, worn out by excessive work, was fatally stricken on the deck of his masterpiece of naval design, just when The Great Eastern was preparing for its maiden voyage. Unfortunately, this huge ship was not a commercial success. It was meant to carry cargo around Africa to Australia for local distribution, but soon the Suez Canal was opened which made this mammoth ship superfluous. It remained for half the century the largest ship afloat. It is still the largest iron ship ever built. American entrepreneur, Cyrus Field, used The Great Eastern to lay the trans-Atlantic telegraph cable from 1866 to 1874 because it was the only ship capable of carrying the heavy load of submarine cable weighing 5,000 tons. After only 2 trips of The Great Eastern across the Atlantic its white-metalled stern bearing of the screw-propeller shaft had developed about 2½ inch (6.4 cm) wear from the friction of the rotating propeller shaft. One of the technical problems with the screw propulsion was the problem of lubrication of the propeller shaft bearings which had to run submerged in sea water. Brass bearings of enormous length were first used in an effort to keep down the pressure per inch of the screw shaft's bearing surface to 60 lb (27 kg). The worn metal bearing of the screw shaft of The Great Eastern was replaced at Milford Haven by the water-lubricated *lignum vitae* bearing of the Greenwich shipbuilder John Penn (1805-1878). The *lignum vitae* bearing gave no further bearing problems during the life of The Great Eastern. Penn, a well-known Greenwich shipbuilder, had invented in 1856 his water-lubricated *lignum vitae* bearing which could withstand pressures up to 8,000 psi (55,160 kPa)without noticeable wear. This type of bearing has now been superseded by a thrust-block bearing fitted with pivoted pads which adjusts itself automatically according to load and speed. The thrust-block bearing was invented in 1942 by the Australian engineer, Anthony George Maldon Michell (1870-1959), whereas the conical thrust-bearing had been already devised by Leonardo da Vinci (1452-1519).

Another great contrast between Robert Stephenson and the younger Brunel was that Stephenson engaged experts to help him to do special engineering designs for him, whilst the versatile Brunel did all the engineering himself.

Locke, Brassey, Peto, and Modern Railroad Builders:

The 3rd outstanding engineer of the railroad era was Joseph Locke, a former assistant to George Stephenson and the most economical of the 3 great engineers. Locke made very accurate financial estimates of the cost of his engineering projects, and he believed that railroads should be built with the minimum initial cost. Some of his railroad lines cost only half as much as Stephenson's railroad lines. For this economic reason, Locke preferred steep gradients and unspectacular engineering on his railroads. For instance, on his Lancaster to Carlisle line built in 1846, the gradient was 1- to –74 for 4 miles (6.4 km) at Shep Fell, whereas Stephenson's maximum gradient was 1- to –330 on the London-Birmingham line. Locke collaborated with Thomas Brassey (1805-1870), originally a land surveyor and quarry manager who became the first modern type of contractor and who built railroads in Britain, all over Europe, in South America, in Asia, and in Canada. Brassey became the greatest railroad-builder of the Industrial Revolution, who constructed about 1,700 miles (2,736 km) of the early British railways and 8,000 miles (12,875 km) of railways in virtually all parts of the world. In his 35 years in business, Brassey carried out successfully 170 different contracts. Later Brassey formed an association with another prominent railroad contractor, Morton Peto (1809-1889), who built railroads in England, and overseas in Canada. Peto transported his workmen around the world in groups and provided them with benefit societies, books, teachers, and savings. Both of these contractors employed a large staff of men of science and commerce as consultants to whom difficult problems could be assigned, thus perpetuating the practice of great canal constructors such as Telford. It was Locke and Brassey who left the deepest imprint of railroad engineering on the map of Europe, although Robert Stephenson and Isambard K. Brunel also engineered railroads on the continental Europe. Locke and Brassey were responsible for the first important commercial railroad line in northern France. Locke's assistant engineer, Alfred Sanistrec Jee (1816-1858), was building the Santande Railroad when he was killed on the job. The great extent of railroad construction can be seen from the 6,620 miles (10,654 km) of railroad tracks which had been laid by 1850 in Britain alone.

Other International Activities of British Consulting Engineers:

Many British engineers took active part in consulting railroad companies in various countries. William Cubitt (1785-1861) not only advised in the building of the Paris-Lyon and Boulogne-Amiens railroads, but carried out harbour construction in Hamburg and built a new water supply in Berlin. Charles Blacker Vignoles (1793-1875) was consultant to several continental European railroads, as well as the builder of the multiple-span suspension bridge which carried a railroad across the Dnieper River in Russia. William Lindley (1808-1900) constructed a number of railroads in northern Germany, as well as building modern waterworks in Hamburg and in Frankfurt-am-Main in Germany, and a sewerage system in Warsaw, Poland. J.J. Berkley, on the recommendation of Robert Stephenson, Brunel and Cubitt was appointed in 1849 chief engineer of the railroad system in India, a system which had a length of 1,237 miles (1,991 km) when it opened in 1853. A.M. Rendel, son of the well-known harbour engineer James M. Rendel, served as a consultant to East India Railway and other Indian railroads, including his advice on the Landsowne Bridge over the Indus river and the Hardinge Bridge over the Ganges river. By 1859, 8 railroad companies began to construct 5,000 miles (8,047 km) of railroad track covering the subcontinent of India.

John Fowler (1817-1898) surveyed railroads in 1870's in Egypt, and advised British government on the conduct of their foreign affairs in Egypt for which he was knighted in 1885. He was also a consultant on the railroad projects in New South Wales, and on the Cockatoo Island docks in Sydney, Australia. Scottish engineer Sandford Fleming (1827-1914) emigrated to Canada in 1845, and built the eastward link of the Intercontinental Railway to Halifax in Nova Scotia. He surveyed the route of the Canadian Pacific Railway which connected British Columbia with the eastern provinces, a railroad completed in 1885. Edward Woods constructed railroads in Chile, Peru and Argentina. J. Brunlees built railroads in Brazil and Uruguay. In 1873, Robert Francis Fairlie (1831-1885) built railroads in Brazil.

Sir Charles Metcalfe (1853-1928), a railroad engineer and a born baronet, was engaged in large-scale railroad building projects, which included a section of the projected Cape-to-Cairo railroad line, and the building of the great bridge over the Victoria Falls in Africa.

Great Exhibition in London
Crystal Palace in Hyde Park:

In 1851, Prince Consort Albert of Great Britain organised against a vehement opposition the Great Exhibition of industrial products in London, England which were housed in an immense glass and wrought-iron exhibition hall that occupied 18 acres of land in Hyde Park in London. The building, called the 'Crystal Palace' in the magazine Punch, was a revolutionary prefabricated modern building designed in 10 days by Joseph Paxton (1803-1865), superintendent of Duke of Devonshire's gardens at Chatsworth, Derbyshire, who became a modern self-made architect. Paxton had designed and built two greenhouses for the Duke of Devonshire at Chatsworth, one of which served as a model for his design of Crystal Palace. This revolutionary modern prefabricated structure consisting of a lofty central arched nave, transepts and many aisles was almost entirely enclosed with 900,000 feet of glass panes fastened to a graceful weblike skeletal structure. The entire skeletal frame structure was erected by the engineering firm Fox, Henderson & Company of Smithwick in 17 weeks from 3,300 prefabricated wrought-iron columns [which both supported and drained the glass-covered skeletal structure] and 2,300 wrought-iron girders bolted together *in situ*, 300,000 large panes of glass held in position by mass-produced sash bars, and laminated wood. The erection was directed by Charles Fox (1810-1874) with the collaboration of William Cubitt (1785-1861), the builder of the Croydon atmospheric railway. The Crystal Palace displayed 13,000 industrial exhibits and was visited by more than 6 million people. The Great Exhibition of 1851 in Hyde Park, opened by Queen Victoria on May 1, had a pervasive influence on the general standard of industrial design. Paxton's revolutionary architectural and structural design of the Crystal Palace was widely imitated in building design both in Europe and in America. Joseph Paxton, Charles Fox and William Cubitt were knighted by Queen Victoria 5 months later for their efforts in the creation of the landmark prefabricated Crystal Palace. In 1854 the Crystal Palace was privately purchased, then dismantled and re-erected by Fox in a 200-acre park at Sydenham located at 8 minute distance from London as an exhibition and amusement center. The Crystal Palace at Sydenham not only housed a museum of sculptures, pictures and architecture, but its spacious halls were also available to the public for the staging of concerts and for such public events as the first automobile show. The Crystal Palace was finally purchased for Great Britain through a public subscription in 1913 but, unfortunately, the re-erected Crystal Palace was completely destroyed by fire in 1936, although it had previously survived damage by wind in 1861, by fire in 1866, and by water in 1880.

Victorian Engineering After the Great Exhibition in London
Tunnel Engineering :

Besides railroad engineering, the British engineers were also the leading experts in the driving of tunnels, irrigation engineering and municipal sanitation.

In tunnelling, Marc Isambard Brunel (1769-1849) invented an excavation shield in 1818 which brought about the modern era of tunnelling. The tunnelling shield had the purpose of holding the face of the heading of the tunnel as the miners advanced. This tunnelling shield consisted of a 12-foot (3.66 m) augur-blade encased in an iron cylinder. The augur-blade was to be turned manually by miners whilst the cylinder was to be pushed forwards by hydraulic jacks working against the brick lining of the tunnel. Brunel called the tunnelling shield his "ambulating cofferdam, travelling horizontally." This shield, which was never built, would not have worked since the power required to turn the auger would have been far beyond the physical strength of any team of miners. Brunel must have realised this, for in 1824 he designed a different tunnelling shield in the form of an 80-ton cast-iron structure consisting of twelve 3-foot (0.91 m) wide and 22½-foot (6.86 m) high cast-iron frames joined together to form a 38-foot (11.58 m) tunnelling shield which contained 36 working cells, 3-feet by 6-feet (0.91 m by 1.83 m), each one holding a miner. The shield made by Henry Maudslay (1771-1831) was divided vertically into 3 sections which could be moved ahead individually by means of large screw-jacks pushing against the tunnel lining. The earth was supported in front of the shield by 500 poling-boards. However, Brunel did not use compressed air method in building his tunnel.

After the stoppage of work on the Thames tunnel due to the break-in by the river, the compressed-air method of tunnelling was patented in 1830 by Thomas Cochrane (1775-1860), the tenth Earl of Dundonald. Cochrane was an adventurer, an admiral, a member of the Parliament, and a stock speculator. He was arrested for stock fraud, thrown into prison from whence he escaped. His honour was later restored, and upon his death Cochrane was buried in Westminister Abbey, the resting place of the most outstanding British engineers. Cochrane himself used compressed air in the sinking of air-tight caissons. The use of compressed air was not new, for excavation under naturally compressed air had been done for sometime with diving-bells. During the Renaissance, Leonardo da Vinci in his notebook showed the use of a diving-bell under naturally compressed air. In 1778 John Smeaton (1724-1792) had supplied air under pressure to his diving-bell by means of a force-pump when he made repairs to a bridge pier at Hexham. John Rennie (1761-1821) used the diving-bell in the same way for the construction of the Ramsgate Harbour in 1807. In 1813 he improved the design of the diving-bell and built with it a pier head in 17 feet (5.18 m) of low water. The basic principle of Smeaton in using compressed air in the diving-bell was to introduce air pressure in the diving-bell which just exceeds, or equals the water-pressure to keep the water out.

Cochrane's chief contribution in excavation with compressed air was his invention of the 'air lock', a small air-tight entrance chamber to the cylindrical caisson which had two doors each in the opposite end of the chamber, the inner door leading to the sealed-off section of the caisson, or tunnel, which contained the air-pressurised working area. The air pressure in the air-lock was gradually increased after the workmen entered it through the outside door, and when the air pressure in the air-lock reached the air pressure in the working area the inner door of the air-lock could be opened to gain ready access to the working area. Marc Isambard Brunel did not take advantage of Cochrane's compressed-air caisson method of excavating to finish the Thames Tunnel. It was first used in practice in 1834 by a French mining engineer, Triger, in the sinking of a small mine shaft.

Isambard K. Brunel did use a limited kind of Cochrane's compressed-air method to sink the central pier of his Chepstow bridge in 1851.

The British bridge engineer, Peter William Barlow (1809-1885), whilst constructing a suspension bridge across the Thames river at Lamberth and driving vertical cast-iron caissons into the clay of the river for the bridge foundations came to the idea that iron cylinders could also be driven horizontally and used for tunnelling under a river bed. Barlow patented his open-ended cylindrical tunnelling shield, in the form of a huge cartwheel, almost 8 feet (2.44 m) in diameter and about 3 feet (0.93 m) in length. The cylinder had a cast-iron rim as a cutting edge supported radially by 6 wrought-iron spokes and plates from a central hexagonal frame of a central opening, and the cylinder was slightly tapered towards the open end to reduce skin friction. Barlow invented a prefabricated segmental cast-iron lining for the tunnel which was an important advance in tunnel engineering. The outside diameter of the tunnel lining was slightly smaller than the inside diameter of the shield so that it could be assembled inside of the shield at the open end. Six screw-jacks acting against the tunnel lining were used to push the shield forwards.

When Barlow was placed in charge of driving the next tunnel through the clay under the Thames river in 1869, the memories of the catastrophic cave-ins of the tunnel of Brunel was still fresh in public mind, and no contractor was willing to bid on the tunnel. Therefore, Barlow's own firm had to undertake the construction of the tunnel. Barlow put his 24-year old assistant engineer, James Henry Greathead (1844-1896), who hailed from South Africa, in charge of driving the first subaqueous tunnel through the clay of the river bed with a shield. Greathead improved on Barlow's method of tunnelling by introducing his 'grouting pan', a method by which liquid grout is forced by air pressure through small holes in the lining which filled the annular cavity left behind by the shield when it is jacked forwards after a 1½-foot (0.46 m) ring of tunnel lining consisting of 3 segments had been erected. It was this vital improvement introduced by Greathead to Barlow's tunnelling method which eliminated the possibility of catastrophic cave-ins that had overtaken Brunel's tunnel.

The Tower Hill passenger tunnel under the Thames river was 1,350 feet (396.24 m) long and ran from Tower Hill to the intersection of Vine Street and Tooley Street. It was finished in less than 6 months without any difficulties or loss of life, and it cost only 16,000 pounds. The passengers were transported through the tunnel by cable-hauled omnibuses which were powered by steam-engines located at both terminals of the tunnel. The Tower Hill tunnel was the first tunnel driven by a modern tunnelling method and its construction was distinguished by the speed, simplicity and economy of its construction. This tunnel continued to be used until the Tower bridge was opened in 1894, after which the tunnel was closed for the public and revamped to accommodate water mains crossing under the Thames river.

At the time Greathead was driving the Tower tunnel under Thames, Alfred Ely Beach (1824-1896), editor of the journal,The Scientific American, and an inventor, drove an 8-foot (2.44 m) diameter tunnel lying 21 feet (6.40 m) under Broadway between Warren and Murray Streets in New York by means of a tunnelling shield made of boiler plates which was quite similar to that of Barlow. Beach, however, was the first engineer to advance the tunnelling shield by hydraulic rams. Beach wanted to demonstrate the proper method of tunnelling under city streets, as well as to promote pneumatic railway in subways as the only proper underground rapid transit system in which the passenger car was moved forwards by air pressure and pulled back by vacuum. This system was opened to public in 1871 and rapidly became a big tourist attraction in the city, but despite its popularity Beach's subway was not a commercial success.

About 1875, Greathead combined Cochrane's compressed-air method with his 11-feet 5-inches (3.47 m) diameter and 6 foot (1.83 m) long tunnelling shield, and incorporated Beach's method of advancing the shield by hydraulic rams. Having perfected the method of engineering of subaqueous tunnels, Greathead became the leading promoter of underground railway systems. Although the compressed-air method was first used with complete success in constructing a small tunnel in Antwerp in 1879, in a major tunnel construction it was first used in 1886 after the passage of the Subway Act in 1884, when Greathead was appointed to build a 3.4 miles (5.6 km) long underground railway tunnel between the monument in the City and Stockwell, south of Thames, constituting the City and South London Railway, at first called the City and Southwark Railway, which became the nucleus of London's great underground railway system. This tunnel consisted of two tubes of 10½ foot (3.20 m) diameter placed one above the other 5 feet (1.52 m) apart to reduce the cost. This first tube railway, which was completed in 4 years without a fatal accident, marks the beginning of the subaqueous tunnelling and underground railways. This railway originally was intended to use cable-hauling of cars by stationary engines but it was abandoned in favour of electric traction.

John Fowler, who was for a time an assistant to the railroad engineer and locomotive designer John Urpeth Rastrick, became in 1860 the chief engineer of the Metropolitan and the District Railways in London which is now called the Inner Circle. Benjamin Baker (1840-1907), who joined Fowler's staff in 1862, was appointed by Fowler in 1869 the Chief Assistant Engineer of the District Railway from Westminster to the City.

The railway lines comprising the Inner Circle have been called the world's first underground railway, which is inaccurate for only 2 short sections, a 1,263-foot (385 m) section at Campden Hill in the west and 2,184-foot (666 m) section at Clerkenwell in the east, are true tunnels. The rest of the line is either open cutting subsequently arched over, or open cutting between retaining walls. For this reason, this line was first called the 'Arcade Railway'. The first suggestion for a true tunnel railway was made by John Person, a London solicitor. The first such true underground tunnel railway, the City and South London line, was constructed by Greathead. Baker not only acted as consulting engineer of this 'tube'-line, but was also responsible for the Central London railway and all its later extensions. In 1892 five more underground railway lines in London were authorised. At the same time, Baker and Greathead were also consulting engineers for the driving of the Hudson river tunnel in New York. Greathead, worn out by the difficult underground work, died in 1895 whilst designing the Central London Railway. After Greathead's death, Baker and Galbraith were responsible for the engineering of the Baker Street & Waterloo tunnel, popularly called the 'Bakerloo Tunnel'.

Rock Drill

The availability of compressed air as a power supply that can be transmitted with relatively little loss over considerable distances provided a ready means to operate power machinery by one central power source in tunnel-cutting operations. Vacuum and atmospheric pressure had been first used by William Murdock in the Soho Foundry of Watt & Boulton in Birmingham to drive small machines by one central engine in a power transmission scheme first attempted by Denis Papin. Soon compressed air was used by many inventors to drive mechanical equipment to drill, to cut rock, to rivet, to lift, and to haul.

It is believed that the first rock-drilling machine was invented by Richard Trevithick in 1813, but no direct evidence of it has survived. In 1849, Jonathan Couch of Philadelphia invented the percussion rock drill which was powered by steam. The drill rod was hurled through a hollow reciprocating piston against the rock and after rebound caught by a gripper and hurled again by the forward moving piston. It was built by a Philadelphia mechanic, Joseph Fowle, who invented a percussion rock drill of his own which was also driven by steam, and in which the drill rod was attached to the piston and following its reciprocal motion was turned by a ratchet and pawl, and fed by a chain moved by an endless screw. He took out a patent for his rock drill in 1851 and used compressed air to operate it. All subsequent successful rock drills were essentially developments of the Fowle's percussion drill.

Railroad Tunnels of the Alps
Fréjus Tunnel:

By 1854, an English engineer Thomas Bartlett (1818-1864) had constructed a rock drill operated by steam-power. This rock drill was redesigned to operate on compressed air by a Sardinian commission investigating suitable rock-cutting methods for the projected Fréjus, or Mont Cenis Railroad Tunnel between Italy and France. It was soon replaced by a German rock drill, but since this drill was also not entirely satisfactory, the Italian engineer Germain Sommellier (1815-1871), chief engineer of the Fréjus Tunnel, devised an improved version of Bartlett's rock drill with an automatic feed which was driven now by compressed air and, moreover, was impacting as well as rotating. Sommellier was able to convince the Committee of the practicality of his rock drill. In 1861 Sommellier invented a 12-ton boring machine on rails incorporating a battery of 4 to 9 rock drills which was used in the drilling of rock in pilot headings of the Fréjus Tunnel. The tunnel pierces the Cottic Alps, between Mt. Tabor in the southwest and Mt. Cenis in the northeast. The two portals of the 8-mile (12.9 km) long Fréjus Tunnel, a record length for its time, were located at the Bardonnecchia hamlet in Italy, and at the Fourneaux village in France. Some years later features contained in the direct-action rock drill invented by Joseph Fowle of Philadelphia were incorporated into Sommellier's drills to improve the rock drills of the Fréjus Tunnel. Construction started on the Fréjus Tunnel in 1858 and it lasted until 1871. Two hydraulically operated compressor plants in which air was compressed by falling water, one at each portal, supplied the compressed air for the rock drills. The tunnel lies 1mile (1.6 km) below the highest point of the mountain range. This tunnel broke new ground in underground engineering, but the brilliant engineer, Sommellier, who was responsible for all of this, died from overwork just a few months before the project was brought to a successful conclusion. About 26 million cubic feet (736,242 m^3) of rock had been excavated by blasting 2,954,000 rounds of gunpowder requiring 580 tons of gunpowder.

Rock Drills:

The rock drill is a machine that is subjected to the most severe punishment among all the machine-tools, and it incorporates probably more inventions for its size than any other machine-tool in existence. Large number of patents on rock drills were taken out between 1850 and 1875, 110 in United States alone. Since Fowle lacked capital to produce his drill, he sold his patent in 1865 to Charles Burleigh, who after introducing some improvements in its design, manufactured the Burleigh rock drill, many of which were used in the driving of the Hoosac tunnel in the United States, because it was the first robust drill that was able to withstand the work on hard rock. Otherwise it was a rather clumsy machine which weighed 372 pounds (169 kg) and lacked some automatic features of its competitors.

In 1871 American inventor Simon Ingersoll bought the Fowle-Burleigh patents, and his company, which merged with that of Burleigh, produced the best contemporary rock drill in existence. It was compact, and the rotation of its chisel was produced by the rifle-bar and the feed-screw. In 1882, an American drill manufacturer, Henry C. Sergeant, borrowed the 'eclipse valve' from a pump made by A.S. Cameron for reversing the air supply valve at the end of piston stroke and simplified it by operating this valve by the travelling piston. In 1884, Sergeant's firm merged with that of Ingersoll and began to manufacture the famous Ingersoll Eclipse Drilling Machine. In these rock drills the chisel was impacting the rock and moving back and forth with the piston, a method which used up large volumes of compressed air.

It was more economical to keep the chisel, or the 'steel', stationary in the hole, and use the piston as a hammer hitting the end of the chisel, a method which had been used in pneumatic rivetting machines and chipping hammers. At the same time the action of a drill had to be introduced by turning the chisel through a small angle in order that new material in the bore hole could be struck by the hammering action of the chisel. English inventor George Low (1833-1894) took out a patent for a 'hammering rock drill' in 1865, which incorporated 2 cylinders, one housed within the other, and made to rotate slightly between the 300 and 500 blows delivered a minute. Henry Sergeant took out a patent on his 'hammering rock drill' in 1884. At this time a hammering rock drill delivering 8,000 blows a minute and weighing 16 pounds (7.3 kg) was set to work in the Manfeld Copper Mine in Germany, but it proved not to be a practical success.

About 1890, the American machinist, C.A. Shaw, from Denver, Colorado, invented the 'hammering rock drill' which had an 'airleg feed' of the chisel for overhead drilling. The 'hammering rock drills' were handicapped by the great difficulty of removing rock chips from the bottom of the drilled hole. For overhead drilling this difficulty did not exist, since stone-cuttings dropped out of the bore hole by gravity. Unfortunately, Shaw did not recognise the great merit of his invention, the 'hammering action' combined with the 'airleg-feed', and he failed to protect it with a patent.

The difficulty with the removal of the rock chips was solved by another mechanic in Denver, John George Leyner (1852-1920), who patented his 'air-blown rock drill' in 1897. Leyner's rock drill had a hollow chisel, a chisel with a tubular hole along its axis through which rock cuttings were blown out by compressed air. Since the blown rock chips proved dangerous to the drill operator, Leyner redesigned his rock drill which now forced water down the drill to flush out the stone dust as sludge. Leyner's 'hammering rock drill' was universally adopted after 1914, and subsequently his company also merged with the Ingersoll-Rand Company, which specialised in the manufacture of compressed-air tools.

St. Gotthard Tunnel:

After the successful completion of the Fréjus Tunnel in 1871, the second tunnel project in the Alps, the St. Gotthard Tunnel, was undertaken in 1872 to carry the Zurich-Milan railroad underneath the St. Gotthard Pass. A few years earlier this engineering project consisting of driving a 9¼-mile (14.89 km) long tunnel through solid rock would have been considered impossible, but the availability of new drilling machines, new improved air-compressors, and the new explosive, dynamite, made it look feasible. Seven construction firms submitted tenders in a keen competition which was won by Louis Favre (died 1879), a contractor from Genoa, with a bid so low that it practically allowed no contingencies. This tunnel which is 26 feet (7.92 m) wide and 24½ feet (7.47 m) high was driven simultaneously from both the Swiss Göschenen portal and the Italian Airolo portal. The St. Gotthart Tunnel was completed in 8 years despite the fact that it became a major debacle in tunnel engineering. Almost from the start not only the water bursting out of the rock, often at the rate of 3,000 gallons (11,356 liters) a minute with a force of a firehose, became a major problem, but also bad air for lack of adequate hydraulic turbine power at the portals, high temperatures reaching 122°F (50°C) due to increasing heat in the rock, rock dust, dynamite and oil lamp fumes, exhalations of men and animals all of which caused 'miner's anaemia' that incapacitated workmen in a few months and made them chronic invalids in a year if they survived. The nightmarelike working conditions had become almost lethal by 1879. 30 horses and mules died each month and at the end not only 310 men had perished including Favre himself, but 877 men had been permanently incapacitated. Favre, by working men ruthlessly around the clock, was able to maintain the daily rate of 13 feet (3.96 m) advance of the tunnel through the massif but at a hideous human cost. Favre not only lost $1,400,000 with this contract, but also his own life. The tunnel was finished by two of his assistants, Stockalper and Borsi in 1880, a year after his death although some additional work still remained to be done to finish the project. To avoid excessively steep gradients at the entrances to the tunnel additional loop-tunnels had to be driven to provide easy grades for locomotives entering the tunnel.

Simplon Tunnel:

The debacle of the St. Gotthart Tunnel delayed by 17 years the driving of the third, and the longest Alpine railroad tunnel, the Simplon Tunnel, under the Monte Leone between the towns Brig in Switzerland and Iselli in Italy. It has a length of 12½ miles (20 km), and is still the longest tunnel in the world. Even more severe physical difficulties were encountered in the construction of this tunnel. The temperatures which rose to 131°F (55°C) and massive flooding at the rate of 2,000,000 gallons (7,570,800 liters) a day as well as labour strikes and unrest retarded the advance of this tunnel. The Simplon Tunnel was driven from two parallel pilot headings which were 55.2-feet (16.82 m) apart and connected by oblique crossheadings. The second, smaller-bore tunnel was mainly used for air-conditioning the main tunnel. The chief engineer of the Simplon Tunnel, Alfred Brandt (1845-1898), like all the other great engineers of the 19th century, died on the job in his early fifties by having worked himself to death in maintaining a 24-hour daily schedule with just a few snatches of sleep over several months during the early critical stages of the project. Brandt invented a new rotary high-pressure rock drill which proved to be an efficient rock-drilling machine, but it was not used after the Simplon Tunnel because the American Leyner-type of compressed-air rock drill was considered more versatile, easily maintainable and economical to use. The driving of this tunnel lasted from 1898 to 1906. A parallel Simplon Tunnel was driven from 1912 to 1921. Since then a number of other shorter Alpine tunnels have been driven under more favourable working conditions.

Sanitary Engineering
French and British Sewerage Systems :

In 1842 a great conflagration destroyed a large part of the German city Hamburg. In rebuilding the city, a British engineer, William Lindley, who had been educated and was practicing in Germany, was engaged to design a comprehensive water supply and sewage system for this city, probably the first modern sewage system ever built.

In France, after several severe cholera epidemics, Baron George Eugene Haussmann (1809-1891), who was entrusted with the modernisation of Paris by the Emperor Napoleon III, began to widen the streets of Paris, for the improvement of traffic, the circulation of air and light, and riot control, and at the same time to provide it with what was known as the 'separate system', sewage and sewage disposal by 'sewage farming' where more than 10,000 acres (40.5 km^2) were irrigated in this manner. The major part of the planning and design for this sewage system was done by Marie Francois Eugene Belgrand (1810-1878) who was appointed an inspector general and director of the sewers and waters of Paris in 1867. Upon Belgrand's death in 1878, Jean Charles Adolphe Alphand (1817-1891) took over this responsibility. The sewers of Paris have a large tunnel-like cross-section.

In London, 3 men were mainly responsible for developing a common system of drainage for the storm water and domestic sewage. The London sewerage problem had to be solved by disposal in a tidal river which had an outwards ebb-movement of about 11 miles (17.7 km), and a matching tidal-movement, which required long outlets and pumping-plants to transport sewage 14 miles (22.5 km) below the city to the discharge point in the river, where large storage reservoirs were located which permitted the discharge of the sewage under favourable tidal conditions. The London sewers have a relatively small oval cross-section which facilitate flow under small discharge. The slopes of the sewers were designed to produce a self-cleaning velocity of the sewage flow, and skilfully designed overflows allowed the cleaner flow after heavy storms to discharge directly into the tidal river. This very successful sewerage system which, however, did not solve the bacterial problem, was a result of the labours of 3 men: Robert Rawlinson (1810-1898), who was a pioneer of public health and an expert on town drainage; William Haywood (1821-1894), a former architect who became the Sewer Commissioner of London in 1846 which position he occupied for nearly 50 years; and Joseph William Bazalgette (1819-1891), a former railroad engineer who became the chief engineer of the London Board of Works in 1855. The major part of the planning of the London Sewerage System was done by Haywood, and the design and construction of the Metropolitan drainage system between 1855 and 1864 was done under the direction of Bazalgette, who was also responsible for the building of the Thames Embankment between 1862 and 1874.

Irrigation Engineering
British Colonial Engineering:

British engineers undertook large colonial irrigation projects which established the modern irrigation engineering. In irrigation engineering, Colonel Arthur Thomas Cotton (1803-1899) of the British Imperial Army in India, designed and built a modern irrigation system in India beginning in 1846. The Godavari River irrigation project, lying north of Madras, incorporated 4 diversion dams, 2¼ miles (3.6 km) in length, to divert the water flow at the tapping point of the river. Another irrigation project in the delta region, was constructed by Cotton's assistant Captain C.A. Orr, which incorporated a 3,350-foot (102 m) long and 15-foot (4.57 m) high diversion dam constructed on porous sand foundation forming a wide apron through which water percolated so slowly that scour was avoided. Colonel Cotton was an eminent pioneer of modern irrigation engineering which led to notable advances in the design and construction of large dams for irrigation and water supply.

The modern Ganges Canal was built in northern India by Colonel Proby Thomas Cautley (1802-1871), who became director of canals of North-Western Provinces of India. Cautley's assistant engineers were Robert Smith (who was in charge of the Eastern Jumna Canal when Cautley arrived in India), Robert Napier, and Richard Strachey. This canal required large masonry aqueduct-structures to carry the canal over deep valleys. The Ganges Canal opened in 1878, as part of one of the largest irrigation systems in the world. The British engineers also engineered the Indus river for irrigation purposes making India the nation with the largest irrigated land-area in the world.

After Cautley retired to England, he continued to consult the British government on their hydraulic engineering projects in India. In this capacity, Cautley became embroiled in a bitter controversy with Cotton concerning the efficiency of Cautley's design for the Ganges Canal in India, a dispute that was unworthy of either engineer.

Colonel Colin Campbell Scott-Moncreiff (1836-1916) came from India to Egypt, where he built a large dam at Assiut on the Upper Nile River in 1884.

Another large dam was built 700 miles (1,126 km) up the Nile river in the first cataract at Aswan between 1898 and 1902. The preliminary plans for this dam had been made by William Willcocks, an assistant to Scott-Moncreiff, but Benjamin Baker, who also served on the International Commission reporting on the dam, later became the consulting engineer of the Aswan dam, and to him belongs the major credit for the final structure of this dam.

The British colonial engineers in India were responsible for the modern masonry dams which had record sizes for their time. Poona Dam, in the Murtha Canal project of 1869-1879, was over 1 mile (1.6 km) long with a maximum height of 107 feet (32.6 m). Nearby the Bhatgarh Dam was almost 1 mile long and 127 feet (38 m) high. The Tansa Dam providing water for Bombay was 1⅔ miles (2.7 km) long with a maximum height of 118 feet (36.0 m). The Periyar Dam in the Southern tip of India, reaching a record height of 173 feet (52 m) was built between 1887 and 1897.

From these examples it is apparent that the British colonial engineers were the first great dam builders of modern times.

CHAPTER XVIII

AGE OF METAL INDUSTRY: From Steel and Alloy Steels to Light Metals

STEEL and ALLOY STEELS

Traditional Steelmaking Processes

Industrial Revolution in Great Britain reached its zenith in 1851, at the time of the Great Industrial Exhibition in the Crystal Palace located in Hyde Park, London. Remarkable as it is, the Industrial Revolution in Great Britain brought a dramatic change to Western civilisation in manufacturing, transportation and construction that had employed materials such as wrought-iron, cast-iron, wood and brick. Machines in the factories, ships, locomotives, bridges and buildings were made without the use of steel as construction material. Yet the need for a stronger and safer construction material was becoming acute, since the manufacturers needed stronger machines and machine tools, and the military required more powerful and stronger firearms and cannons.

Steel as an alloy of iron and carbon was available for small articles such as cutting tools, knives and swords, because steel was very expensive to produce costing about 50 British pound sterling per ton at the time. Ever since antiquity steel had been made by the so-called 'Wootz' method which originated in India where the method of making good steel had been known since ancient times. In Roman times the main centers of iron production were in Austria and in India where ingots of high quality steel were produced and exported all over the civilised world.

In the Indian 'Wootz' method steel was produced on a very small scale by placing crushed iron ore in a closed clay crucible of about one pound [about half a kilogram] capacity, enveloped by sawdust and acacia leaves, sealed, and put into a small conical clay furnace. A charcoal fire was lit in the furnace and fanned by a crude form of bellows made of animal skins which produced in the crucible intense heat for several hours that brought about the necessary carbonisation of iron. After the crucible had cooled down it was broken and a small cake of iron, the so-called 'natural' steel, was retrieved. In Damascus the ingots of Indian steel were forged by the Syrian smiths into strips, and the strips of Wootz steel and wrought-iron were twisted together and then welded by hammering to produce the celebrated 'Damascus steel'. The sword blades produced by such a process were then heated to a 'red-hot' state, quenched in water, and finally reheated slightly to 'draw the temper' which resulted in steel blades of great hardness and flexibility. Habitually charcoal dust was hammered into the hot metal to produce the blades of the so-called 'Damascene swords' which were extremely flexible. The steel of the blade of the Damascene sword was of such high quality that the modern steelmaking methods have been quite unable to match it. The process of making Damascus steel was soon transmitted by the Muslims from Damascus to the city of Toledo in Spain, and subsequently the 'Damascus steel' produced in Toledo became known as the 'Toledo steel'.

In 1614 William Ellyot and Mathias Meysey patented the 'cementation' method of steelmaking. In this method wrought-iron cakes were hammered into bars, packed into stone boxes filled with charcoal, and placed into a 'muffle furnace', a furnace in which heat is generated externally. Then the charcoal was ignited and kept burning which created intense heat in the furnace for several days and even as long as a week, the period depending on the required degree of carbonisation absorbed by the wrought-iron from the charcoal to produce the so-called 'blister steel' which had a blistered surface caused by the expanding carbon monoxide gas that had penetrated the metal and removed impurities. The carbonisation of wrought-iron achieved by this method was superficial and nonuniform so that the physical properties of the 'blister steel' when worked up was unpredictable and frequently unsatisfactory. The prominent shortcoming of the 'blister steel' was its highly nonuniform hardness: the inside of the 'bloom' was much softer than the surface of the 'bloom' which made it virtually useless for the manufacture of cutlery, swords and springs. In order to remove this shortcoming of the 'blister steel', the cake of steel was cut up into short lengths, bound into faggots and welded together into a single bar of 'shear steel' in a forge from which were made good cutting tools such as shears and knives. The quality of the shear steel could be further enhanced by repeating the cementation process. Despite the elaborate and expensive method of production of the 'shear steel' which required highly skilled smiths, the quality of the steel was still uncertain, and its heterogeneity remained a prominent defect and a problem, particularly to people who made springs for clocks or tools. The cementation method in metallurgy had its origin also in antiquity.

The indifferent quality of steel springs supplied to Benjamin Huntsman (1704-1746), son of a Dutch immigrant who was a Doncaster clock- and instrument-maker, motivated him in 1743 to begin an extensive series of experiments which after many failures finally led him to the successful production of 'crucible steel' in 1746. He developed the 'crucible' method of steelmaking to produce steel of highly uniform quality required for his clock springs. In his crucible method Huntsman placed a charge of the 'blister steel' cut into small pieces together with steel scrap and a small quantity of charcoal into a clay crucible that was enveloped by coke in a crucible furnace. The furnace fired to a high heat remelted the 'blister steel' after 5 hours, which made the carbon redistribute more uniformly throughout the steel. The resulting more homogeneous Huntsman steel was very expensive because of the great quantities of fuel used, and the slow, elaborate, and exacting process of its production which frequently still led to uncertain results. On account of the large quantity of heat required in the crucible process, only small quantities of crucible steel could be produced, at most 60 pounds (27.2 kg) per crucible. When 'puddling furnaces' were used to produce the wrought-iron for steelmaking, it required the heaviest skilled labour in ironmaking. It is no wonder, then, that about 1851 the total production of steel in the world was only a few thousand tons a year. Immediately after 1746 Huntsman continued to practice his trade as a watchmaker and produced steel only in small quantities, but later he began to produce crucible steel on a commercial scale in his works at Sheffield.

The industrial Western civilisation being a technological civilisation depends mainly upon iron and its alloys such as steel. Since iron combines easily with other material elements it is normally found in impure form in iron ores such as iron carbonates, iron sulphides and iron oxides such as *haemitite*, the natural magnetic iron ore *magnetite*, and *limonite*. To recover a more purified form of iron from the iron oxide ore, the iron oxide ore must be reduced by smelting. The fundamental purpose of smelting is to reduce the iron oxide ore to free iron in the blast furnace. Many very complex chemical processes take place in the blast furnace, but the fundamental chemical change upon which iron smelting is based consists of heating iron oxide ore with carbon in the form of coke that changes the iron ore into metallic iron and carbon oxide gas, although many other associated chemical reactions take place in the furnace at the same time. Secondary chemical reactions with limestone flux convert earthy nonmetallic materials present in the iron ore, such as clay, sand etc., into slag, but the chemical processes which produce slag are in general quite complex. The smelting process consists of charging the blast furnace with the proper mixture of iron ore, coke [an almost pure form of carbon], and crushed limestone called 'flux' [a reducing agent for slag formation] in a furnace lined with refractory brick, and heating the charge with burning coke. The primary purpose of the air blast is to intensify the burning of coke and produce a very high temperature in the furnace. The air blown into the furnace through nozzles called ***'tuyères'*** combines with the burning coke to form the carbon monoxide gas [that burns with oxygen] which in turn reacts with iron ore producing oxide-free iron and carbon dioxide gas [a stable compound that does not support combustion]. In ancient times the air introduced into the furnace which fed the burning of coke and, thereby, raised the temperature in the furnace to the melting point of iron, was fanned into the furnace with bellows made of animal skins. Other impurities of the iron ore, such as sulphur and silica, are removed by the chemical combination of these impurities with limestone flux. The temperature generated by this chemical process is sufficient to melt the iron as it is formed in the blast furnace which is the hottest just above the level of ***tuyères*** where the temperature is about 3275° F (1800° C) which is nearly 575° F (300° C) higher than the melting point of iron. The silica in the ore is converted to calcium silicate slag by the action of lime liberated from the limestone flux by heat. Complex secondary chemical reactions with limestone flux convert earthy nonmetallic materials such as clay, sand etc. in the ore into slag. The semi-purified heavy molten iron produced by this process separates under gravity from the lighter slag and forms a pool in the bottom of the furnace called the 'hearth' made of refractory brick or carbon blocks, whereas the lighter impurities float to the top of the molten iron in the form of slag. The wall of the furnace hearth is pierced with a small hole that is plugged with a mixture of clay and coke dust during the smelting process, but the plug could be removed every few hours to tap the molten iron from the hearth. A similar plug higher up in the furnace wall could also be removed to tap the slag. The molten iron from the blast furnace is usually teemed to cool in channel-like moulds called 'pigs' having the outlines resembling a sow suckling her piglets which is why the solidified iron in the moulds is called 'pig-iron' that is about 91% pure iron. Pig-iron which contains about 4% carbon (that accounts for its low melting point), still has too many impurities such as carbon, manganese, sulphur, phosphorus, and silicon for immediate industrial use and, for this reason, it serves only as a raw material which must be further refined to obtain 2 other kinds of iron: the cast-iron and the wrought-iron.

Cast-iron, wrought-iron, and steel are all iron-carbon alloys which vary in their physical properties according to the amount of carbon they contain. The cast-iron is made directly from pig-iron by resmelting the pig-iron with coke in a smaller furnace. The burning coke again raises the temperature of the pig-iron charge and removes the residual impurities from pig-iron. The remaining cast-iron is tapped from the bottom of the furnace and poured into moulds. The cast-iron that has a high carbon content, between 2% and 4%, is brittle and when heated it does not soften but rather melts rapidly at about 2,250°F (1,232°C). When poured into a channel-like mould the cast-iron becomes very hard and brittle and, therefore, it is not malleable, that is, it cannot be hammered or forged into shapes. Since cast-iron does not distort when 'red hot', and corrodes only slightly in water, it is used for making stoves, fireplaces, water pipes, troughs, gutters, and vessels that are exposed to water.

A further smelting of cast-iron produces wrought-iron, an almost pure form of iron. The wrought-iron, or malleable iron, is an iron-carbon alloy which contains very little carbon, usually less than 0.1 of one percent, and like cast-iron it does not easily corrode because it contains slag inclusions. Wrought-iron is relatively soft, no harder than brass, but 'tough' in its resistance to impact fracture, and it can be forged, or 'wrought', into thin shapes. Wrought-iron becomes soft and doughlike when heated, but it does not melt until the temperature of about 2,800°F (1,538°C) is reached. Wrought-iron cannot be cast because it is not fluid enough in its molten form to fill the mould completely. Wrought-iron was in great demand for the building of almost every kind of machine and bridge until the Tay Bridge disaster in Scotland in 1879, after which the mass-production methods of steelmaking ended the dominance of wrought-iron in the Western civilisation. Wrought-iron is still used for the manufacture of such iron products as crane hooks, chains, and railroad draw-bars, because under large strains such wrought-iron products stretch and distort long before they suddenly break, thereby giving an early warning of impending failure of the product.

A large family of iron-carbon alloys called 'steels', that have a carbon content between 0.3% and 1.7 %, lies between the wrought-iron and the cast-iron. Steel as an iron-carbon alloy combines the malleability of wrought-iron with the hardness of cast-iron and, in addition, steel is stronger and more elastic than both wrought-iron and cast-iron. The 3 principal groups of steel alloys are the low-alloy structural steels, corrosion-resistant stainless steels, and high-speed tool steels. Each of these iron-carbon alloys is a different type of steel, made with the precise properties required for the particular application of the steel such as spring-steel for clockworks, shear-steel for cutting blades, and fatigue and high temperature resistant alloy steels, called 'superalloys', for gas-turbine blades and shafts which rotate with a high speed of 30,000 rpm, and are subjected to both high stress and high temperature. The only drawback of steel as an iron-carbon alloy in comparison with wrought-iron is that it rusts easily because it does not contain the slag inclusions like wrought-iron.

In 1709, about the time Abraham Darby (1676-1717) introduced coke as the fuel in iron smelting, the ironmakers began to remelt iron in special furnaces, called 'finery furnaces', where iron was exposed to the decarburising effect of the air-blast to improve, or refine, the purity of iron. The method of turning pig-iron into malleable wrought-iron was 'puddling', that is, by remelting it in finery furnaces and stirring it manually for days with a long iron pole, called 'rabble', until the molten iron lost enough carbon through oxidation to convert it into a plastic paste, an operation requiring great strength, endurance and skill.

Rise of Metallurgical Science

French naturalist, physicist, chemist and mathematician, René Antoine Ferchault de Réaumur (1683-1757), was the first person to distinguish between different types of iron by their internal structure as observed in a freshly fractured iron bar under magnification. He was the first physicist to assert that steel is nothing more than iron into which are diffused 'salts and sulphurs'. He experimented with making steel by fusing malleable iron with cast steel. Réaumur made an attempt to rationalise ironmaking, and to reduce the conversion of iron into steel to a scientific process. Réaumur was essentially interested in the cementation process in steelmaking, and he carried out a great number of experiments for the purpose of determining the best iron, the best packing material for carburation, the appropriate periods and temperatures of heating, and the most suitable furnace equipment. In 1722, Réaumur published a method of making iron castings that were malleable, tougher, and stronger in the collection of his memoirs, <The Art of Converting Wrought-iron into Steel and the Art of Softening Cast-iron> (***L'art de convertir le fer forgé en acier et l'art d'adoucir le fer fondu***), which he had formerly published as individual memoirs since 1711. Réaumur's tests of iron represent the first metallurgical tests for the quality control in iron production. He specified two metal tests: the bending test and the hardness test. According to Réaumur a sound strip of steel could be bent double around a prong without cracking, and hardness can be measured by the size of indentation of the steel specimen caused by a blow with a hammer.

The science of metallography, initiated by Réaumur, was developed further by the Swedish physicist, mathematician, chemist and mineralogist, Torbern Olof Bergman (1735-1784), who proved that the 'salts and sulphurs' of Réaumur were actually carbon, and that steel was the product of chemical interaction between carbon and iron. Bergman was the first scientist to analyse ferrous materials for carbon, and to indicate that steel has between 0.3% and 0.8% carbon, and cast-iron between 1% and 3.3% carbon. In 1786 the English chemist, Joseph Priestley (1733-1804), who with the Swedish chemist of German stock, Karl Wilhelm Scheele (1742-1786), is the independent codiscoverer of oxygen, came to the same conclusion as Bergman and the French chemist Guyton de Morveau (1737-1816) that the chemical combination of iron and carbon produces steel. Although about 1820, Karsten had emphasised that the difference between pig-iron, wrought-iron and steel depends upon the amount of carbon present in the metal, but not until 1831 was the German chemist Justus von Liebig (1803-1873) able to perfect a method for the exact determination of the amount of carbon present in steel. The most important element of steel, other than iron, is carbon. The higher the carbon content, up to about 0.85%, the greater the ultimate tensile and yield strength of steel. However, the ductility of steel decreases with the increase of carbon content.

In 1861 another important scientific advance in metallography was made by the English metallurgist Henry Clifton Sorby (1826-1908), who founded the science of microscopic metallography which consisted of a systematic microscopic examination of the microstructures of metals, and the correlation of such properties of metals and alloy metals as melting point, strength, hardness and electrical properties. The study of the microstructure of metals and the relationship between the composition, the structure, and the properties of metals constitutes the very foundation of modern science of metals.

Kelly-Bessemer-Mushet Steelmaking Process

In 1851, at the time of the Great Exhibition in London, mostly luxury articles were manufactured from steel, a metal which cost 4 or 5 times as much as wrought-iron or cast-iron. It is to the credit of 4 inventors that steelmaking became a highly economical industrial mass-production. Already in 1847, an American ironmaker, William Kelly (1811-1888), son of an Irish immigrant in Pittsburgh, had blown cold air through molten pig-iron to convert it into steel, or white cast-iron. Kelly called his process of steelmaking the 'pneumatic method'. In 1851 Kelly built his first 'pneumatic furnace' consisting of a cylindrical chamber with holes in the bottom for the air-blast, and enclosed it into a 4-foot (1.22 m) square brick structure. At first he had some trouble blowing air through the molten metal with his side-blown converter in his Suwanee furnace at the Cumberland river, but he improved his furnace as he went along. He built altogether 7 such furnaces, each an improvement upon the previous one. Kelly had never bothered to patent his 'pneumatic method' of steelmaking since he had trouble controlling the refining of the metal with his equipment, but he did file a patent in 1856, when he heard that an Englishman, Henry Bessemer, had applied for an American patent on a similar steelmaking process a few days earlier. The Patent Office decided in Kelly's favour and granted him the patent as the first inventor of the 'pneumatic process', but Kelly became bankrupt the same year and was unable to exploit his new pneumatic process of steelmaking.

Henry Bessemer (1813-1898) was the son of a Dutch mechanical engineer who had worked in the Paris Mint when the French Revolution forced him flee to England where he operated a type foundry in partnership with the printer Henry Caslon at Hitchen in Hertfordshire. Henry Bessemer worked in the foundry operating a lathe, modelling and casting type when he was 16 years of age. A year later his family moved to London where he earned his living reproducing works of art and casting them in a white metal alloy, and coating them with a thin layer of copper. He became an all-round professional inventor, who produced almost a continuous flow of inventions in virtually all existing branches of industry from the early 1830's to the late 1870's when he retired. He was at his inventive peak about 1851 when he filed for patents on new inventions on the average of one patent every two months. Altogether, Bessemer took out 117 British patents and many foreign patents. His first invention was a perforated die that cancelled the stamps according to date, for which he was knighted 46 years later by Queen Victoria. He invented a machine that made bronze powder and gold paint, an invention from which he made a small fortune. He developed new alloys for engraved rollers used in paper embossing and printing, invented a glass polisher, a machine for continuous rolling of glass sheets which was capable of working without skilled labour, a ventilator, a series of improvements of the machinery for the crushing of sugar cane which won him a gold metal struck by Prince Consort Albert, a method of making waste plumbago into pencil leads, a process for casting type under pressure, a typesetting machine, a process of making a kind of plush, a more efficient railway brake, and several other inventions which were not successful.

After Britain entered the Crimean War in 1854, Bessemer concentrated his attention to armaments and weapons which had not changed much since the Battle of Waterloo in 1815. He invented a new type of artillery shell which was made to spin upon leaving the barrel of a smooth bore cannon, but the existing cannon barrels with smooth bore proved to be too weak for the firing of the new shell. Bessemer, who was no metallurgist, set himself the task to produce an improved cast-iron with the characteristics of steel which could be cast in fluid state without losing its strength.

In 1854, James Nasmyth had taken out a patent to blow a hot jet of steam through a pool of molten metal to refine pig-iron which replaced the 'rabble' of the puddler in a puddling furnace. He introduced the blast just below the surface of the liquid metal to expose as much metal to the oxygen present in the steam-jet as possible. The very purpose of the traditional puddling was to expose as much molten metal as possible to the oxygen present in the puddling furnace by stirring the metal charge vigorously with hobbles and shovels, and so constantly exposing new metal to the oxygen present in the furnace. It never occurred to Nasmyth to use cold air in the blast, because it seemed absurd to blast cold air into the hot metal since it was generally thought that it tended to cool down the molten metal and harden it in its container. The idea that a cold air-blast intensifies combustion in the furnace which not only burns up impurities such as carbon, silicon and manganese present in the molten pig-iron but also produces great quantities of heat which keeps the metal bath in a molten state was contrary to the commonsense of the contemporary ironworkers, and was summarily rejected as being absurd by most ironmasters. It is precisely this aspect of seeing what everybody else sees, but thinking what nobody else thinks, that indicates the native ingenuity of Kelly when he invented in 1847 the method of mass-production of steel by the cold air-blast. Kelly was able to produce high quality steel in 1857 subject to Mushet's U.S. patent of steelmaking with *spiegeleisen*.

In 1855, Bessemer came to the same conclusion as Kelly, and also introduced an air-blast to refine pig-iron. Bessemer, at first, had experimented with a reverberatory furnace and had come close to the open-hearth method of refining pig-iron, but he failed to introduce heat regeneration which makes it successful. Bessemer promptly designed all the effective machinery for steel production by the pneumatic process. In his 1855 experiments, Bessemer, like Kelly before him, used a stationary furnace in the form of a vertical converter lined with a close-grained siliceous stone called 'gannister', which looked like a modern cement mixer that was fitted with conical air-blast nozzles, the so-called *tuyères*, made of fireclay in the bottom of his converter. He had seen a converter of this type being used in John Brown's Atlas Steel Works at Sheffield. In 1860, Bessemer patented his improved tilting converter mounted on trunnions in which he was able to blow 700 pounds (318 kg) of pig-iron supplied by the blast-furnace in about 30 minutes without the need of any external fuel for the converter by using a steam-driven blower to supply the air-blast. The tapping of the converter by tilting it and pouring its content into a ladle took about 15 minutes. The air blown through the molten metal caused the oxygen from the air-blast to combine with some of the iron, producing iron oxide, which then dissolved in the molten metal and reacted with silicon, manganese, and carbon which, in turn, became oxidised. The oxides of iron, manganese, and silicon then combined to form the slag whilst the carbon was removed as the combustible carbon monoxide gas which burnt in the air at the mouth of the converter with a spectacular flame producing carbon dioxide. Since the subsidence of the flame at the mouth of the converter indicated that the impurities have been oxidised, the blast of air is stopped and the proper amount of carbon and deoxidiser is added to the melt in the form of a high-carbon alloy as was done by Mushet.

The Kelly-Bessemer process converted high-carbon pig-iron first into steel and then into wrought-iron by burning out the carbon, as well as the silicon and manganese impurities with the air-blast. In practice this process actually produced soft wrought-iron rather than steel because the air-blast could not be turned off at the right moment to produce harder mild steel. The first 'steel' samples produced by Bessemer were not at all perfect for they were not only brittle, soft, and contained blowholes due to the presence of excess oxygen in occluded form in the metal, but also 'hot-short', that is, the 'steel' was brittle and cracked when worked under heat. Bessemer had been lucky in another way because he had used, by a mere chance, a Blaenavon pig-iron made from one of the few British iron ores which was very low in phosphorus, a predominant impurity of iron ores in continental Europe. Nasmyth had run into a similar difficulty with the presence of excess oxygen in occluded form in the metal in his steam-blast experiments, and given up the method. Bessemer demonstrated his steelmaking process to George Rennie (1791-1866), the President of the Mechanical Section of the British Association for the Advancement of Science at Cheltenham, who urged Bessemer to announce his process at the next meeting of this Association. After reading his paper at the next meeting, Bessemer sold in the first blush of excitement licenses for his steelmaking process to 5 independent firms who paid him a total of £27,000, but then the bubble of success burst because the ironmasters failed to produce 'steel' that was useful in practice by means of the Bessemer pneumatic process. In their anger the ironmasters denounced Bessemer, and some even accused him of fraud and charlatanism. As a result of this scandal, Bessemer had to return all the money paid to him for the licenses, and embark upon a 2-year research programme to find out why the ironmasters were not able to produce even the soft wrought-iron Bessemer himself had produced in his own experiments. The success of Bessemer's process depended on the accurate control of the low but critical carbon content of steel in his converter. Since the stopping of the air-blast at the right instant to achieve the low carbon content of steel was difficult and unreliable, a much better method was to burn first all the carbon out of the molten metal charge, and then to restore the correct amount of carbon by adding an alloy with the proper carbon content to the molten metal charge.

This important step in the production of mild steel was taken by the talented Scottish metallurgist, Robert Forrester Mushet (1811-1891), who owned a small ironworks near Caleford in the Forest of Dean and who had been experimenting since 1848 with an iron-manganese carbide, the so-called ***spiegeleisen*** [meaning in German, *mirror iron*], a Prussian iron ore containing from 15% to 30% manganese and from 4% to 7% carbon. William Reynolds of Colebrookdale had filed a patent already in 1799 for the use of ***spiegeleisen*** in ironmaking. Robert's father, David Mushet, who discovered the 'black-band ironstone' of Scotland in the 1830's that was mainly responsible for the rapid growth of a flourishing Scottish iron industry, had been associated with the ironmaster J.M. Heath, who demonstrated in 1839 that Huntsman's crucible steel was substantially improved if some manganese were added to the melt before the molten metal was poured into ingots. Soon after 1848, Robert Mushet recognised that the addition of manganese to wrought-iron not only greatly improved its forgeability but also had a supplementary deoxidising effect in curing the wrought-iron defect of being 'red-short'. In 1856 one of the unsuccessful Bessemer licensees, Thomas Brown of Ebbo Vale Ironworks, showed Mushet a sample of the brittle wrought-iron Brown had obtained by the Bessemer pneumatic process. Mushet readily recognised that the trouble in the Bessemer pneumatic process was the result of overheating the pig-iron with the air-blast which lasted too long thereby making the metal brittle and full of blowholes, but he also knew how to remove this shortcoming. Since Mushet was aware from his previous experience that manganese had a particular affinity to iron oxide, he remelted the sample and mixed it with an appropriate amount of pulverised *spiegeleisen*, and thus recovered the particular quality of wrought-iron that is now called 'mild steel'. Mushet rapidly worked out a method to produce mild steel with Bessemer converter and patented it in 1856. He added *spiegeleisen* to the molten metal after its complete decarburisation with the air-blast, which made the manganese in the *spiegeleisen* react with the surplus iron oxide occluded in the molten metal whilst the carbon in the *spiegeleisen* raised the carbon content of the iron to the exact percentage required for mild steel. Manganese in the molten metal became manganese oxide which passed into the slag formed on top of the melt during the blast.

Fox, Henderson & Co. of Smethwick, the engineering company which built the renowned Crystal Palace in 1851, constructed a small Bessemer steel plant for Mushet at his Coleford Works, where the first Bessemer-Mushet mild steel was produced. Only after Mushet's important improvement was the Bessemer pneumatic process capable of producing mild steel. Meanwhile, Bessemer had to experiment for 2 years with his pneumatic process until his metallurgical chemists were able to determine that his process only worked on iron ores which had a very low phosphorus content. Since the Bessemer pneumatic steelmaking process had lost credibility owing to its initial failure, Bessemer was unable to sell licences for his process and, therefore, he felt forced to establish in partnership with his brother-in-law Robert Longsdon in 1859 his own steelworks, Henry Bessemer & Co. in Sheffield, where large quantities of steel were produced at a price of £ 10 a ton compared to £65 or £70 a ton by earlier processes of steelmaking. When he sold his steelworks after 14 years of operation, it fetched 81 times the original investment, and when his patent of steelmaking expired Bessemer had earned in royalties from his patent £1,057,748. Bessemer was elected a Fellow of the Royal Society in 1879, and in the same year he reminded the British Prime Minister Lord Beaconsfield that the government was still using his invention of stamping deeds without compensation. The government acknowledged Bessemer's complaint by rewarding Bessemer with a knighthood rather than paying him for the invention.

The ironmasters, who were unimpressed, still regarded Bessemer steel to be 'treacherous' because steel undergoes internal structural changes when hot-worked and cooled in fabrication, something that was not understood at the time. Bessemer's process of steelmaking would have been rather worthless without Mushet's improvement, but through one of those peculiar circumstances, a trustee who was holding Mushet's patent let it lapse in 1859 by failing to pay the fee of the annual patent tax required in Britain. This misfortune of Mushet allowed Bessemer to use Mushet's crucial improvement in the production of mild steel by the Kelly-Bessemer pneumatic process without any payment of royalties. Bessemer, who was a hardheaded businessman, refused to pay royalties to Mushet because, legally, he was not obligated to do so. Whilst Bessemer became a very prosperous and wealthy inventor in England, Mushet spent his old age in a state of penury after the failure of his own metalworks. Finally in 1876, Bessemer acceded to the pleas of Mushet's daughter and approved paying Mushet an annuity of £300, as well as awarding him the Bessemer Medal of the Iron and Steel Institute for Mushet's outstanding contributions in steelmaking. However, Bessemer still had limited success with his original pneumatic method of steelmaking despite his use of low phosphorus pig-iron, because the steel he produced was 'hot-short', that is, it was brittle when hot and therefore could not be forged. Moreover, Bessemer was unable to establish a precise control over the duration of the air-blast to obtain the correct carbon content of steel, primarily an iron-carbon alloy containing anywhere from 0.1% to 1.5% carbon, directly from the blow. Only in 1885 did the conservative British government finally accede to use Bessemer steel in gunfounding and in shipbuilding.

It is not impossible that Bessemer learnt of Kelly's 'pneumatic method' of steelmaking from his brother-in-law William Allen, who might have briefly worked for Kelly before he returned from United States to assist Bessemer in the experiments to make steel by the pneumatic process. It is certain that Bessemer was aware of Nasmyth's patent which used steam-blast in ironmaking.

Much of Bessemer's wealth came from foreign licences, because most British ores had high phosphorus content, and good Bessemer-Mushet steel could only be produced from low phosphoric haematite ores of Monmouthshire, Cartmel and Cumberland. Bessemer, himself, imported non-phosphoric iron ore from Sweden. The huge Krupp Industries in Essen, Germany, acquired in 1862 the first foreign license to the Bessemer steelmaking process, and Krupp soon became the largest producer of steel in the world. Various ironworks in France, Austria and Belgium followed Krupp's example by acquiring Bessemer's licenses. Bessemer applied for his American patent 2 days before Kelly, whose patent was granted by the Commissioner of U.S. patents, although Bessemer received most of the royalties because of his converter patent. The Bessemer steelmaking process had its greatest success in United States of America where the Carnegie Company after acquiring a license from Bessemer began to roll steel beams in large quantities. In 1885, the first 12-storeys high skyscraper was constructed in Chicago, the lower 5 floors of which had an iron framework, but the remaining 7 floors above the iron framework were completed as a steel skeleton incorporating rolled steel beams made by Carnegie. Within a few years taller and taller skyscraper buildings began to rise in New York, all of which were constructed as skeletal structures of steel.

Bessemer's U.S. patent rights had been acquired by a group of ironmakers in Troy, New York, headed by a talented American engineer Alexander Lymon Holley (1832-1882), a graduate of Brown University and a former technical writer on the staff of the New York Times who as the agent of American businessmen bought the American rights of Bessemer steel process in 1863. Kelly's backers in Johnstown, Pennsylvania, who had bought Mushet's American patent rights, also had the patent rights of the 'pneumatic process' of Kelly. In 1866 both groups had to merge after a long and confused legal struggle since the Holley group could not use the 'pneumatic process' in United States, and the Kelly group could not use the tilting converter of Bessemer which made the steelmaking process practical in industry. Holley, as the consulting engineer of Cambria Steel, Bethlehem Steel and the Scranton Steel Works, was commissioned to design the first Bessemer steel plant at Troy (New York), and other Bessemer steel plants in Joliet (Illinois), in Pittsburgh (Pennsylvania), and in St. Louis (Missouri). Holley, who was the foremost steel plant engineer in the United States, not only improved the Bessemer converter but also raised the efficiency and output of the Bessemer steelmaking process by making the process continuous with the alternate use of 2 converters. Holley also insisted upon the use of chemical analysis and physical tests in the manufacture of steel products. In 1870, only Kelly's patent was renewed by the U.S. Patent Office.

In 1858 a Swedish metallurgist, Göran Fredrik Göransson (1819-1900) bought the license to use the Bessemer steelmaking process in Sweden, but he soon discovered that it did not work properly. He redesigned the Bessemer converter, and was able to make the first commercially successful steel with it by controlling the air-blast through his larger *tuyères* than had been suggested by Bessemer, thereby increasing the volume of air, reducing the pressure of the air-blast which slowed it down, and examining the change in the colour and the intensity of the flame issuing from the converter during the blow which indicated to him the moment when the carbon and manganese content of the charge corresponded to that of the mild steel. Soon Göransson made use of Professor Eggertz's method of carbon analysis, that measured and compared the colours of steel solutions in nitric acid. It was a rapid method which afforded the practical means for more accurate control of the blowing process in steelmaking. Göransson's improved Bessemer process made possible the production of all grades of steel by stopping the air-blast at the instant when any specified carbon content of the steel had been reached. The modern Bessemer converters use pure oxygen, rather than air, in their blast of about 15 minutes duration, and the quality of steel this rapid process produces is good enough for most everyday purposes.

The remarkably talented Mushet took out a number of other patents for iron alloyed with chromium, tungsten and titanium: chromium alloys gave stainless steel for cutlery, and tungsten-manganese alloys gave steel for high speed cutting tools. Presently the chromium-vanadium alloys give steel used for propeller shafts, chromium-molybdenum alloys give steel for aircraft frames, silicon-manganese alloys give steel for car springs, silicon alloys give steel for electric motors, and cobalt alloys give steel for permanent magnets. One of the greatest successes of Mushet was his invention of a special self-hardening alloy steel for tools which had immensely superior cutting power in comparison with the traditional cutting tools made of carbon steel.

Owing to his previous unfortunate experience with Bessemer, Mushet kept the production method of his high-speed tungsten-manganese alloy steel for cutting tools a carefully guarded secret by taking very elaborate, and extraordinary precautions which were so effective that his precise production process remains unknown till this very day.

Already in 1818, the ingenious self-taught English physicist Michael Faraday (1791-1867) had developed a high-grade alloy of rust-resistant steel, the first stainless steel, from alloying iron with platinum, rhodium and silver. In 1819 Faraday performed many experiments in the laboratory of the Royal Institution with alloy steels in which he had the able assistance of a Sheffield cutler James Stodard. The firm Green, Pickslay & Company in Sheffield did make cutlery from Faraday's alloy steels. In 1822 Faraday predicted in his article in the Philosophical Transactions that in the future alloy metals will be more important than the metals themselves, a prescient prediction which has been fulfilled. Faraday's investigation of alloy steels, and also his formula to produce extremely clear glass with high refractive index helped to put these products on a scientific basis, and to show the way to regularise the use of scientific methods in industry. Although Faraday had prepared many small samples of alloys, including the nickel alloy steel and the chromium alloy steel, it was Mushet who became the pioneer industrial producer of alloy steels. The nickel alloy steel was first produced in 1824/1825 by Johann Conrad Fischer at Schaffenhausen in Switzerland, but first marketed commercially by Schneider Steel Works at Le Creusot in France. The tungsten alloy steel was first made in 1855 by the Austrian metallurgical chemist Franz Köller, and commercially produced in a special steelworks at the Ems river in Austria. The chrome alloy steel was commercially produced in 1877 by Jacob Holtzer at Unieux in France. Moreover, Faraday's quantitative laws of electrolysis proved to be of major practical importance to electroplating, electrotyping and electrochemical industries. In 1843, Faraday discovered how to deposit by means of electrolysis a thin protective layer of rust-resistant nickel upon another metal.

In 1862, Mushet founded the Titanic Iron & Steel Company in the Forest Dean that was financially backed by members of the Pease family, who in 1825 had supported George Stephenson (1781-1848) in the building of the Stockton & Darlington Railway, the first railway using steam locomotion. Mushet carried out his experiments with manganese, tungsten, and chromium alloy steels in the crucible melting furnaces at his Titanic Iron & Steel Company in great secrecy. It appears that Mushet took pig-iron cast from an iron ore rich in manganese, pulverised it, and mixed it with finely powdered tungsten known as 'wolfram' ore. Tungsten, a hard and dense metal which is about two and a half times as heavy as iron, has the highest melting point, 6,098°F (3,370°C), of all the metals, yet it is both malleable and ductile. This ore mixture was melted in a crucible and poured into an ingot mould. Part of the ingot was forged into a bar of tool-steel under a tilt-hammer. This tungsten-manganese alloy steel was a major discovery in metallurgy, for it was immeasurably superior in its cutting ability to carbon steel and it proved to be self-hardening, a characteristic of metals which Mushet himself had never thought possible. Carbon-steel tools had to be hardened by heating and quenching in water, a process the success of which depended upon the skill and practical experience of the toolmaker. The tungsten-manganese alloy steel correctly hardened itself when left to cool in the open air after forging. The character and great potential of the tungsten-manganese alloy steel cutting tool was not completely grasped even by Mushet himself, because he thought that his cutting tools must be run dry. Many engineers considered the Mushet alloy steel as a means of cutting tough materials rather than as an improved means in ordinary metal-cutting operations. Mushet's tungsten-manganese alloy steel at first was used in machining tyres and axles of railroad wheels made of the manganese alloy steel.

Mushet's father, David, had already recognised in 1840 the high wear-resistance of iron-manganese alloy. Mushet discovered that manganese steel was also nonmagnetic, a property which later proved important. Although Mushet's Titanic Iron & Steel Company failed commercially, his alloy steel as a cutting tool material, which after 1872 also contained chromium besides tungsten and manganese, created the most important change in the machine shop practice since the time of Maudslay and Whitworth. After the financial collapse of Mushet's company, his special tool alloy steel was produced and marketed by Samuel Osborne and Co. of Sheffield, and its success was immediate and worldwide. Machine tools everywhere had to be redesigned to take full advantage of the far greater cutting power of the new cutting tools made of tungsten-manganese alloy steel of Mushet. In fact, the new alloy steel components of the new machines could not have been machined at all with the old cutting tools made of carbon steel.

The mighty Krupp Works in Essen developed the next more powerful generation of cutting tools, the tungsten-carbide alloy tools. Krupp initially produced the tungsten-carbide alloy dies for the German Osram Company who used the dies in the drawing of tungsten filaments for light bulbs, but in 1926 one of Krupp's workmen discovered its great value as a cutting tool material when he tried a piece of tungsten-carbide alloy in the lathe. Subsequently, Krupp developed the tungsten-carbide alloy cutting-tool and exhibited it under working conditions at the Leipzig Fair in 1928, where it created as great a sensation as Taylor's high-speed steel tools had created at the Paris Exhibition in 1900. The tungsten-carbide tool could cut the new tough alloys of steel and aluminium with great speed and efficiency. The machines using tungsten-carbide tools required from 3 to 4 times more power than the machines using high-speed steel, and they had to be 75% heavier than their predecessors. The tungsten-carbide alloy was manufactured under the trade names, 'Widia' and 'Wimet', in Europe, and 'Carboloy' in United States. The tungsten-carbide alloy was so hard that no existing abrasive wheel could cut it. In 1934, the Norton Company in United States produced a small diamond-bonded wheel which was capable of cutting the tungsten-carbide alloy, and the molybdenum-titanium-carbide or the tantalum-carbide alloy which followed it. The cobalt-chromium-tungsten alloy appeared on the market in 1917, and was subsequently improved. The high-speed steel cutting tool was able to cut metal 75 feet (22.8 m) a minute, whereas the tungsten-carbide alloy cutting tool could cut metal 400 feet (121.9 m) a minute.

Siemens-Martin Open-Hearth Steelmaking Process

The pneumatic process of steelmaking was not the only process being developed. Almost at the same time another steelmaking process was being developed in England, which enjoyed certain advantages over the pneumatic process and soon took over the lion's share of the steel market.

Karl Wilhelm Siemens (1823-1883) was the seventh of the 14 children of a cultured and knowledgeable German farmer, a believer in the application of scientific methods to agriculture, at Lenthe, near Hannover. Wilhelm was 16 years old when his parents died, and his elder brother Ernst Werner Siemens (1816-1892) guided his education as the new head of the family. Werner arranged for his younger brother Karl Wilhelm to attend the *Polytechnikum* (*Polytechnical Institute*) in Magdeburg where he studied science, mathematics, English and French for 3 years, and later to attend the *Universität Göttingen* (*University of Göttingen*) in Göttingen, Germany, where he graduated with a diploma in engineering. After his schooling, Karl Wilhelm Siemens worked in the manufacture of steam-engines in the Graf [Count] Stolberg's factory.

Ernst Werner Siemens, a lieutenant in the engineering corp of the artillery regiment of the Prussian army, studied mathematics and sciences at the *Ingenieur- und Artillerie-Schule* (*Engineer and Artillery School*) in Berlin from 1835 to 1838 where he was taught by such outstanding teachers as physicist and chemist Heinrich Gustav Magnus (1802-1870) and mathematician Martin Ohm (1792-1872), brother of Georg Simon Ohm (1789-1854). In 1840 Werner was commanded to

Magdeburg where he took care of the education of his younger brother Karl Wilhelm by teaching his brother mathematics in his spare time. Werner was sentenced to a 5 year term in prison by a military court for acting as a second in a duel which was forbidden and severely punished in Prussia. When in military prison, Werner pursued his experimental investigation of electrolysis, and he was able to continue his scientific and technical studies when he was pardoned after one month incarceration, and transferred to the artillery arsenal in Berlin for technological duty. In Berlin he had an opportunity to go into partnership with a prominent Berlin company interested in exploiting his new electroplating process. Since Werner Siemens could not financially afford to patent his new electroplating process in the numerous German principalities, he asked his 19-year old younger brother Karl Wilhelm, who had collaborated with Werner in the electroplating experiments in Berlin, to go to England and sell Werner's new electroplating process to an English firm, Elkingtons in Birmingham, a firm which pioneered in electroplating in England, but whose electroplating process was inferior to that of Siemens. Britain was one of the countries where an inventor could obtain long-term protection of his ideas for a relatively small fee. Wilhelm went to London in 1842, and after a successful negotiation he sold the British rights of Werner's electroplating process to the English firm for £1,600 of which Wilhelm received £800 for his part of the transaction, and then returned to Germany. After completing his training, he returned to England in 1844 where he hoped to sell his 'chronometric' governor of steam-engines. Karl Wilhelm Siemens, who was a scientific inventor, liked the unlimited industrial possibilities and the rapidly expanding economy of England which was hospitable to inventors, and he settled there permanently. In 1859, he became a naturalised British subject with the anglicised name, Charles William Siemens. After working on various electrical processes, printing and hydraulic apparatus, he turned his attention to the important problem of wastage of heat in industry, and to a method of using the heat in waste gases for the regeneration of heat in industrial processes. He had become interested in the 'regenerative' steam-engine when he worked for 2 years in Graf Stolberg's factory of steam-engines after his schooling. The principle of heat-regeneration had already been used and patented in 1816, by Dr. Robert Stirling (1790-1879), a Kilmarnock clergyman, in his closed-cycle hot-air engine.

  William Siemens intended to pass the steam in his regenerative steam-engine from the hot cylinder of the steam-engine through a 'metallic respirator' the purpose of which was to absorb most of the heat in the steam, and from the 'respirator' the steam was to enter the condenser in a semi-cooled state. The water in the condenser was then to be pumped back through the 'metallic respirator' where it was intended to absorb the heat conserved in the 'respirator' before returning to the boiler. William Siemens worked on this heat-regenerative system for steam-engines for 12 years but since success eluded him, he finally gave it up in 1859.

In the meantime, William Siemens invented a simple but very effective water-meter which he patented in 1851. His water-meter incorporated a drum that was rotated by the flow of consumed water through the meter which recorded the quantity of water in gallons on a dial. Since his water-meter was of great practical utility to the new central public water-supply system pumped through pipes to the consumers in London, its immediate commercial success brought him reasonable income for over 100,000 water-meters of his were sold in a relatively short time. In 1851 William joined his younger brother Friedrich Siemens (1826-1904) in the employ of the engineering company, Fox, Henderson & Co. in Smethwick, before he struck out on his own in 1852 as a consulting engineer in London. Charles William Siemens, who had a good scientific and engineering education from the University of Göttingen, knew how to use scientific ideas in technological applications. He invented a device that reproduced printing which remained a standard until photography was perfected. Encouraged by his substantially improved financial condition, he embarked on a study of various industrial manufacturing processes to which the principle of heat-regeneration could be profitably applied such as in the salt production where large quantities of fluid had to be evaporated, and in industrial manufacture of glass. William enlisted the assistance of his younger brother Friedrich Siemens, an intelligent and well-educated younger brother whom his elder brother Werner considered to be a born self-critical inventor.

Friedrich Siemens and his brother William had observed that relatively small quantities of heat were generated in existing furnaces using solid fuels such as coal and coke. By that time the reverberatory open-hearth furnaces were mostly heated with gaseous fuels by burners at one end of the furnace which directed flames across the hearth in such a manner that the charge in the hearth was heated both directly and indirectly by the reflection from the roof of the furnace. The burnt hot gases were exhausted into the chimney stack at the end of the furnace opposite to the burners, thereby allowing an appreciable quantity of heat present in the spent exhaust gases to go to waste. The two Siemens brothers invented a method of 'heat-regeneration' for industrial reverberatory furnaces that made the hot exhaust gases of the furnace preheat the incoming air for the combustion of the furnace. In 1856, Friedrich Siemens invented the 'regenerative furnace 'consisting of a heating chamber flanked by 2 hearths from which the hot combustion gases passed through ductwork into 2 regenerative chambers below the furnace which were built as honeycombs of loose refractory bricks that absorbed the heat from the hot waste gases of combustion. When the bricks were completely hot, the waste furnace gases were passed into the second regenerative chamber and the incoming blast air supply for the furnace was guided to pass through the hot regenerative chamber where the blast air was preheated before it entered the furnace. To make the heat-regenerative process continuous, the 2 regenerative chambers were used alternately, one heating the incoming blast air whilst the other absorbed heat from the hot waste furnace gases. This method reduced the temperature of furnace gases exhausted in the chimney to 200°F (93°C) thereby resulting in a substantial fuel economy. In this process the preheating of the blast air for combustion raised the temperature of the air to 1,470° F – 2,200° F (800° C – 1,200° C) when arriving at the furnace. Friedrich Siemens suggested to his brother that they use the stored heat of the furnace for the preheating of both the incoming air of the blast and the gaseous fuel of the furnace.

After several years of experimenting with the idea of heat-regeneration, William Siemens and his brother Friedrich substantially improved the performance, and the economy of their regenerative reverberatory furnace by gasifying the fuel in a separate gas-producer, and preheating regeneratively both the incoming blast air and the gaseous fuel necessary for the combustion in the furnace to the temperature of 2500°F (1371°C). This improved regenerative reverberatory furnace patented in 1861 was fitted with a separate gas-producer invented by William Siemens in 1851 that produced gas from low-grade fuels such as cheap coal, peat or even sawdust, which was used as the fuel of the regenerative furnace. The hot waste gases of the furnace left through 2 sets of regenerative chambers, one set at each end of the furnace, and the air and the gas flow in the regenerative chambers was reversible by means of a valve arrangement so that the hot exhaust gases were used to preheat regeneratively both the incoming air and the gaseous fuel in separate regenerative chambers. In 1861, the improved regenerative reverberatory furnace of Siemens was first used in glassmaking by the Chance Brothers in Birmingham, and it rapidly became an immense commercial success in glassmaking industry because the extremely high temperatures of the regenerative furnace conspicuously increased the glass production.

The heat produced by the open-hearth regenerative reverberatory furnace of Siemens was more than sufficient to reduce a charge of pig-iron and steel scrap to a completely molten state, something the puddling furnace was unable to do. This regenerative process brought with it a remarkable economy in the use of fuel in the furnace. In the Siemens' regenerative reverberatory furnace, the preheating of the incoming air was a great improvement upon the preheating system of James Beaumont Neilson (1792-1865) for it used heat which would have been wasted in Neilson's stove and, furthermore, the Siemens' regenerative reverberatory furnace used only between 20% and 30% of the fuel consumed by conventional furnaces. Moreover, the regenerative reverberatory furnace of Siemens was the most effective method of creating unprecedented high temperatures of combustion, as high as 2,900°F (1,600°C) in large regenerative reverberatory furnaces, high temperatures which were of cardinal importance to steelmaking industry.

As the blast-furnaces increased in size, even the new steam-driven or water-driven compressors could not produce adequate air-blast to create the high temperature required for the smelting of iron ore. Neilson, who was the manager of a Glasgow glassworks, thought to achieve such high temperatures by preheating the blast-air which was contrary to the contemporary belief of the ironfounders, and he began experimenting preheating at the Clyde Ironworks in 1828 using an auxiliary stove to preheat the incoming air to a temperature of 50°F (10°C), and discovered that the heated air indeed intensified the combustion in the blast-furnace. In 1838 he patented his invention of passing the air for the furnace blast through cast-iron pipes that were heated in the auxiliary cast-iron stove enclosed in brickwork with coal or coke fire burning on a grate, but after 1834 he improved the efficiency of the hot-blast process of ironmaking by heating the iron tubes in the stove with hot waste gases tapped from the blast-furnace to the temperature of 800°F (427°C), this being the limit that the cast-iron pipes through which the blast air passed in the stove were able to withstand without melting. In order to prevent the blast-furnace nozzles, called *tuyères*, from melting, Neilson enclosed the cast-iron nozzles in a watercooled coil. By such means, Neilson increased the output of the blast-furnace 3 or 4 times without a proportional increase in fuel consumption. Neilson's system of preheating the incoming air to blast-furnace made the smelting of low-quality Scottish black-band ironstone for the first time possible with raw Scottish coal which did not coke well, and that was of the greatest economic importance to the Scottish iron industry. Neilson's preheating stove is one of the earliest instance of a heat-transfer equipment incorporated into an industrial process to raise the temperature and save the heat of the process. Neilson's preheating stoves for blast air were improved in many minor ways until in 1857 Cowper created the modern high towers of refractory brick for preheating the air-blast to blast-furnaces that made possible the processing of huge quantities of pig-iron in large blast-furnaces.

The English inventor and associate of Friedrich Siemens, Edward Alfred Cowper (1819-1893), applied a similar regenerative system to ironmaking in blast-furnaces using hot air-blast to produce pig-iron. He created for this purpose the Neilson's preheating 'stoves' in the modern form of high towers of fireproof brick which are nearly as large as the blast-furnace itself. These regenerative brick towers contained a chamber and a labyrinth of refractory brick, which were first heated by burning from 14% to 30% of the hot blast-furnace gases containing between 20% and 25% combustible carbon monoxide tapped from the blast-furnace, and then were used to preheat the blast air on its way to the blast-furnace. The tall firebrick towers were worked in pairs, whilst one tower was being heated the other tower was transferring the heat from the hot brickwork to the blast air, thereby combining the inventions of Siemens and Neilson, which resulted in a further saving of coke. Cowper's invention not only raised the temperature of hot waste gases from the blast-furnace to 1300° F (704°C) but also reduced the temperature at the discharge chimney to a mere 200°F (93°C) – a great improvement upon the Neilson stove which discharged waste gases at the rather high temperature of 1,250°F (677°C), and only heated the blast air to 800°F (427°C). Cowper's method of preheating made it possible to attain very high temperatures in the large modern blast-furnace, and process massive quantities of iron ore. The modern blast-furnace usually has 3 large regenerative brick towers of Cowper, 2 of which are heated while the 3rd tower raises the temperature of the blast air as it passes to the blast-furnace.

Cowper was a versatile inventor. In 1868 he patented the so-called 'suspension wheel' with tensioned wire spokes, rubber tyres and anti-friction roller bearings, a wheel which made modern cycles possible.

A successful attempt was made by George Parry of Montmouthshire to cap the mouth of the old blast-furnace with a 'bell' consisting of a conical device, which could be lowered to draw off the gases, the heat of which was transferred to the air blast by means of the regenerative brick towers of Cowper acting as heat exchangers. Soon also the 'beehive' coke ovens to coke coal, as a replacement for open-coking, became a general practice in saving heat, but it still wasted some valuable byproducts of the coking process such as the tar. Finally by 1881 an installation was devised which conserved both the spent hot gases and the tar of the blast-furnace.

The Siemens brothers soon realised that their regenerative process had a much wider industrial potential than merely utilising waste heat, for it could supply combustion at a much higher temperature, something that was essential in steelmaking. In Siemens' open-hearth regenerative reverberatory furnace which was fired with producer gas generated by Siemens' gas-producer that was separate from the furnace, the temperatures of the combustion with the mixture of preheated coal gas and preheated air reached as high as 2,900°F (1,600°C) which readily melted not only the pig-iron, but also the scrap iron charge in the furnace. Iron oxide was added to the metal bath in the form of iron ore, or 'scale', which together with the oxygen in the furnace gases oxidised the impurities, and removed carbon as carbon monoxide whilst the flames of coal gas and hot air wafted over the surface of the molten metal. The silicon and the manganese in the iron ore were also changed into their oxides, and they reacted with the additional sand or limestone by becoming slag which was separately removed. At the end of this process which reduced the impurities to required levels, and after the melt had been tapped, adscititious ferro-silicon, and ferro-manganese with higher manganese to carbon ratio than *spiegeleisen* were added to the melt to lend the steel its correct composition. This process could take twenty-four hours in the Siemens open-hearth regenerative furnace, a long time when compared to the 'blow' of 30 minutes in a Bessemer converter, but it brought with it some very important advantages. The Bessemer converter could only accept molten pig-iron from the blast-furnace or from the remelting cupola, whereas the Siemens open-hearth reverberatory furnace was completely self-contained since it could accept pig-iron and metal either in the molten or in the solid state. Moreover, Siemens' open-hearth regenerative process permitted the steelmaking process to be far more accurately controlled than the Bessemer pneumatic process, an aspect which became increasingly important as the grades of alloy steels began to multiply, and the steelmaking process itself became more scientific. Moreover, Siemens' open-hearth regenerative reverberatory furnace produced steel of high quality and purity, and in much larger quantities [up to 500 tons of molten pig-iron] which made it a less expensive process. Siemens' regenerative reverberatory steelmaking furnace is called the 'open-hearth' regenerative furnace because the molten metal lies in a comparatively shallow pool in the bottom of the combustion chamber of the furnace, called the 'hearth'.

In a way, Siemens' steelmaking process was a logical development of the 'boiling process' of Joseph Hall, who improved upon Henry Cort's ironmaking process in 1838 by lining the furnace with iron oxide. As oxygen combined with iron ore to produce carbon monoxide it reduced the carbon content of iron ore that raised the melting point of iron which consequently changed from the liquid to the pasty state, and the slag was more readily separated from the pasty metal in the rolling forge. In the so-called 'boiling process', Hall was able to reduce the phosphorus and silicon content of the charge in the furnace and, thereby, obtain a much superior metal, since the oxides of silicon and phosphorus produced by his 'boiling process' passed into the slag.

The French metallurgist, and a graduate from the *l'école de Mines* (School of Mines), Piérre Émile Martin (1824-1915), an ironmaster at Sireuil in the Charente district in France, was one of the first steel manufacturers to obtain the license from William and Friedrich Siemens in 1863, when the 'open-hearth' steelmaking process was already in its last stages of development. William Siemens had an opportunity to build a full-size regenerative reverberatory furnace for Martin, and he went to Sireuil to supervise the construction of it. In his early experiments Siemens had discovered that in the refining of pig-iron in the open-hearth regenerative reverberatory furnace, the lining of the furnace burnt out in the extremely high heat long before the impurities such as carbon and silicon were removed. He overcame this shortcoming in his process by adding iron ore to the molten pig-iron charge because the oxygen present in the iron ore accelerated the removal of the impurities. Addition of scrap metal to the pig-iron is called 'diluting' the pig-iron, and it considerably reduced the time of refining the steel. Martin's method of making steel of a given carbon content was to melt a charge of cast-iron made from iron ore from the Island of Elba, wrought-iron, and steel scrap. Martin patented his process of steelmaking in 1865, and in 1878 claimed the success of the open-hearth steelmaking process as his own, but after a long litigation Siemens was declared the sole inventor, although the 'open-hearth' method of producing mild steel by using scrap iron is still called the 'Siemens-Martin Process'.

Siemens decided to promote the acceptance of his open-hearth process of steelmaking in Britain and, therefore, like Mushet before him, Siemens erected a small demonstration plant in a rented factory in Birmingham, which was known as the Sample Steel Works. Open-hearth steel production on a commercial scale began in his own plant, the Siemens Steel Company, that Siemens founded in 1870 at Landore in South Wales, which had 24 open-hearth regenerative furnaces, where by 1875 mild steel was produced in the quantity of 1,000 tons a week. Parts of Siemens' steelworks were still in production during the Second World War.

As a direct result of Kelly-Bessemer and Siemens-Martin steelmaking processes, steel prices fell 50 % between 1856 and 1870. The Kelly-Bessemer steelmaking process which was responsible for most of the steel made in 1875, became secondary to the open-hearth steelmaking process before the end of the 19th century, and currently little more than 10 % of the world's steel is made in the converter, 2% or 3% is made in the electric furnace, and a negligible amount is made in the crucible. By the end of the 19th century ⅔ of British annual steel production was made by the Siemens open-hearth process, that is, about 3,156,000 tons of steel.

In 1879, Charles William Siemens applied the principle of electric arc to steel production by inventing the electric-arc furnace for steelmaking in which the metal charge in the hearth is heated by the electric arc struck between the graphite electrodes suspended through the furnace roof and the metal charge in the hearth. The electric-arc furnace has a refractory lining, and since no fuel is burnt inside the furnace, the furnace atmosphere, the slag and the temperature of the steel can be controlled with great accuracy. The modern electric-arc furnaces, which use alternating current transformed to give 30,000 amperes current at 100 volts tension, are mostly used today for the processing of high-grade alloy steels

After the success in steelmaking, William Siemens turned his talents to electrical engineering like his elder brother Werner Siemens, who was the 19thcentury pioneer of heavy current electrotechnology. He collaborated with his brother Werner to produce a very efficient dynamo fitted with a new and remarkably effective shuttle-wound armature in 1856, and a highly sensitive and reliable polarised relay for the electric telegraph which was adopted worldwide.

In 1857, William Siemens and his brother Werner accepted the invitation of the British firm Newall & Co. to carry out the electrical testing of the submarine telegraph cable of the Telegraph Construction Maintenance Company which was laid from Cagliari in Sardinia to the Bona Island in Algeria. Siemens Brothers were actually responsible for the supervision of the entire cable-laying project that was successfully completed with the Siemens Brothers' cable paying-out gear. Siemens Brothers were also collaborating with the Newall & Co. in the successful laying of the submarine telegraph cable from Suez via Aden to Karachi, at the Bay of Indus River in India, from May, 1859 to January, 1860.

In 1858 William Siemens was made a partner and the British representative in London of the big German electrotechnical company **Siemens & Halske A.G.** in Berlin. In 1864, the London branch of **Siemens & Halske A.G.** at Charlton in Kent under the new name Siemens Brothers, Electrical Engineers, was made commercially independent from its branches in Berlin and in St. Petersburg, Russia, by contractual arrangements which regulated the relationship between the 3 branches of

Siemens Brothers Company because the parent company **Siemens & Halske A.G.** in Berlin was no longer able to oversee directly the business of all 3 branches. Werner Siemens had founded in 1847 his influential electrotechnological company **Siemens & Halske A.G.**, for the primary purpose of building telegraph lines and manufacturing telegraphic equipment, together with Johann Georg Halske (1814-1890), a highly skilled mechanic who built the first sound and reliable dial telegraph incorporating a self-acting current breaker which Werner Siemens had devised in 1846 as his first important telegraphic invention. Halske left the firm in 1867.

In 1847 Werner Siemens was the first to begin to insulate subterranean telegraphic cables with the mixture of gutta-percha and sulphur, and in 1848 he devised a screw-press machinery for the production of gutta-percha cover for copper wires of the telegraph cable which rapidly became the general practice in the manufacture of insulated copper wires for subterranean and submarine telegraph cables. Already in 1848, **Siemens & Halske** built the first long telegraph line from Berlin to Frankfurt at Main. A third brother Karl Siemens (1829-1906) was in charge of St. Petersburg branch, and William Siemens was in charge of the London branch of the Siemens Brothers Company. The 3 Siemens brothers, Werner, William and Karl, had considerable influence upon the electrotechnological developments in Germany, in England, and also in Russia. From 1868 to 1870, the Siemens Brothers Company built the 6,215-miles (10,000-km) long telegraph line from Europe to India.

Werner Siemens followed the precedent of Krupp by setting a part of the annual profits of the worldwide Siemens enterprise into an old-age benefit, sickness and accident insurance fund for his employees and workers. Werner, together with other Siemens brothers engaged in business, set up the **Siemens Stift** (*Siemens Charitable Foundation*) for the promotion of wellbeing of all the members in the Siemens family.

Werner Siemens' main rival in Germany was Emil Rathenau (1838-1915), who returned from United States with the German patent rights for Edison's electric light and created the German Edison Company in 1883 which became the **Allgemeine Elektrizitäts Gesellschaft** (*General Electricity Company*) known by its acronym **A.E.G.**. Because of the intense competition with the Rathenau's company the firm **Siemens & Halske A.G.**, which after the death of Werner in 1892 was directed by his surviving brother Karl, amalgamated in 1897 with another large electrotechnical company **Elektrizitäts Aktiengesellschaft Schuckert & Co.** (*Electricity's Stock Corporation Schuckert & Co.*), founded by the industrialist Johann Siegmund Schuckert (1846-1895) in 1893. The amalgamated company was named the **Siemens & Schuckert Werken G.m.b.H.** (*Siemens & Schuckert Works*). The 2 giant rival cartels soon led the world in electrotechnology, and by 1913 they produced 34% of the world's electrical products.

William Siemens was also a pioneer in the development of electric power and electric lighting. In 1881, Siemens caused a public sensation in England by illuminating the square outside the Bank of England and the Royal Exchange with electric lighting. The Siemens Brothers Company under the direction of William Siemens and with the collaboration of John Hopkinson (1849-1898) applied electric power and electric traction to railways, when they built in 1883 the Giants Causeway Electric Railway near Portrush in Northern Ireland for which the electric power of 240 volts tension was supplied by electric generators driven by hydraulic turbines, and picked up from the 3rd rail. In 1879, **Siemens & Halske AG** had built at the Berlin Industrial Exhibition the first electrically powered railway in small-scale with a circular track of 984-foot (300 m) length which was powered by 2 identical Siemens dynamos, one dynamo working as a generator of electric current but the other dynamo working in reverse as an electric traction motor of a small locomotive which drew its electric current from a 3rd rail. The dynamo producing from 4 to 5-horsepower gave 2-horsepower at the motor in the locomotive which hauled 3 or 4 cars carrying passengers with the speed from 15 to 20 miles (24 to 32 km) an hour. In 1881 **Siemens & Halske AG** built in Gross-Lichterfelde, a suburb of Berlin, the first electrically powered tramway 1½ miles (2.5 km) long, and in 1884 the firm built a tramway 3¾ miles (6 km) long from Frankfurt at Main to Offenbach. In 1890 the Siemens Brothers Company designed the electric locomotives used for the City-Stockwell railway, the first electric underground railway in London which was over 3 miles (5 km) long. In 1896 the Siemens Brothers Company built in Berlin the electrified narrow-gauge elevated railway, which the firm had proposed at first in 1880 as part of a much larger project. The first broad-gauge electric railroad locomotive was built in 1899 for the Burgdorf-Thun railway line in Switzerland by 2 Swiss companies: **Lokomotiv- und Maschinenfabrik** in Winterthur and **Brown, Boveri & Co.** in Baden.

The firm **Siemens & Halske AG** also manufactured an electrically powered hammer in 1879, and the first electrically operated elevator in 1880.

In 1880, Werner Siemens founded the **Elektrotechnischer Verein zu Berlin** (*Electrotechnical Society at Berlin*). In 1887 he supported financially the **Physikalisch-Technische Reichsanstalt** (*Physico-technical State Establishment*) at Charlottenburg, a suburb of Berlin, by donating land for its buildings worth half a million Reichsmarks. Both of these organisations supported by Werner Siemens, particularly the second one, had a decisive influence upon the progress of scientific technology and engineering in the 19th century.

William Siemens became an internationally recognised expert in the design, manufacture and laying of submarine telegraph cables insulated by the rubberlike substance, gutta-percha (tested and approved by Faraday as a dielectric), and first used in 1847 by Werner Siemens as the proper insulation of subterranean telegraph cables. In 1859/1860 William Siemens as an expert on telegraph cables gave evidence to the historic Commission of Inquiry set up by the British government because of the government's involvement in having issued financial guarantees to cable-laying companies which faced huge financial losses owing to various disastrous failures experienced during the laying of submarine telegraph cables, particularly the submarine cable from Egypt through the Red Sea to India, with losses amounting to more than a million pound sterling.

In 1874, William Siemens designed and built a special ship named <u>Faraday</u> for the laying of submarine telegraph cables in collaboration with William Froude (1810-1879), which was used in the laying of the 6th trans-Atlantic telegraph cable – the first submarine telegraph cable to link Great Britain directly with United States without going through Canada.

In 1858 Froude introduced his 2 dynamometers: his friction dynamometer for the recording of the power output of any power producer, and his belt dynamometer for the measurement of the load demand of power consumers. Both dynamometers were important to engine builders for the accurate measurement of the power output of their engines under variable conditions of load.

The Siemens Brothers Company grew with the boom in telegraph cable industry, and the professional reputation of William Siemens as an engineer and inventor was much enhanced by his contributions to the technology and engineering of the submarine telegraph cables on which topic he published both theoretical and practical articles. William Siemens as an inventor took out altogether 113 patents in England.

In 1883, William Siemens was knighted by Queen Victoria for his numerous technological inventions which had benefited the British industrial economy, just like his brother Werner Siemens was ennobled by Kaiser Friedrich III in 1888 for his technological and industrial accomplishments in Germany. William Siemens was elected a Fellow of The Royal Society in London already in 1862.

The Kelly-Bessemer and Siemens-Martin steelmaking processes were 'acid' processes because the converter and the open-hearth were lined with silica refractory bricks known in the iron trade as 'acid' to differentiate them from the 'basic' refractories. The 'acid' process was unable to remove phosphoric impurities from pig-iron which are harmful to steel and, therefore, both the Kelly-Bessemer and the Siemens-Martin steelmaking process had to use pig-iron with very low phosphoric content. This limitation of the 2 processes of steelmaking was a major drawback in steel production because the iron ore deposits in Britain, in the Continental Europe, particularly in Germany and in Belgium, and even in United States, had high phosphoric content and, therefore, remained unexploited. This prominent shortcoming in the 2 acid steelmaking processes was eradicated by a young but ingenious court clerk in London, who was an amateur chemist and metallurgist.

Thomas-Gilchrist Basic Steelmaking Process

The problem of steelmaking with phosphoric iron ore was solved by a young Welshman, Sidney Gilchrist Thomas (1850-1885), who was a junior clerk in the Metropolitan Magistrates' Court at Stepney in London. He was an idealistic and religious young man, who wanted to use his learning for philanthropic ends. Thomas, who had been schooled at Dulwich College in London, attended night school at Birbeck College in London, and in one of his courses the lecturer, Dr. Chaloner, referred to the prominent shortcoming in steelmaking and expressed his opinion that great financial rewards await the person who can make steel from the phosphorous iron ore. Thomas saw in this challenge a way to realise his intentions, and subsequently he spent all his free time and effort between 1871 and 1875 to solve this difficult problem, first theoretically, and then in practice.

The addition of lime as flux seemed like a straightforward method to eliminate phosophorus, and some previous investigator may have thought of it, but lime directly attacks the 'acid' silica lining of the converter and, therefore, made this approach unsound. Thomas soon came to the conclusion that the answer to his problem lay in the proper refractory lining of the converter, which had to provide conditions under which the reactions with lime could take place without destroying the lining of the converter. He came to the conclusion that the lining of the converter should be the 'basic' refractory brick of 'dolomite', a carbonite of lime and magnesium known as 'magnesian limestone', calcined or burnt to resist the extremely high temperature of molten steel and particularly the damaging action of lime on the lining of the furnace in very high heat. For the long series of practical tests of his process which were necessary, he had to enlist the help of his cousin, Percy Carlisle Gilchrist, who was a chemist at Blaenavon Ironworks in South Wales. The dolomite lining was alkaline and capable of removing phosphorus, and promoting the formation of phosphates in the basic slag. In the Thomas process limestone was added to the iron bath as flux which combined chemically with phosphorus, silicon, and sulphur, and thereby reduced these impurities in the pig-iron by forming a slag that contained large amounts of calcium phosphate and calcium silicate. This process of removal of phosphorus increased the quantity of slag and, therefore, the amount of heat required in the Thomas 'basic' steelmaking process had to be considerably greater than in the acid Kelly-Bessemer steelmaking process. To supply this great amount of heat, the pig-iron had to contain between 1.8 % and 2.0 % phosphorus, since this impurity, which is transformed by chemical reaction from iron to slag, generates enough heat in its chemical reaction to make the entire process feasible.

Thomas, who was an extremely careful investigator, finally had satisfied himself of the correctness of his theoretical and practical ideas in 1878, when he took out his patent and revealed his process to the world, a process of 'basic steelmaking' which was rapidly adopted all over the Western world. In continental Europe, the industrial chemists in Germany found a method to make chemical fertiliser out of the phosphoric slag, a waste product of the basic steelmaking process which has a high phosphate content. Thomas carried out all his own legal negotiations in the protection of his patent, in the course of which he became a recognised legal expert on patent rights. The licenses of his basic steelmaking process which Thomas sold in United States alone brought him a quarter of a million dollars. Unfortunately, many years of intense work on his steelmaking process had taken a heavy toll on the delicate health of this ingenious young man, and Thomas died of tuberculosis in 1885 at 34 years of age. Thomas himself was fully aware that he had solved the greatest industrial problem of England, for his basic steelmaking process could be easily adapted to the Kelly-Bessemer and the Siemens-Martin steelmaking process.

The 3 amateur metallurgists, Bessemer, Siemens and Thomas not only gave a new, but shortlived, revival to the British Industrial Revolution, but they also brought about the 'Age of Steel'. However, the use of steel in industry did not spread as rapidly as expected. Just like in the case of wrought-iron, the use of cast-iron as structural material came to a sudden end with a disaster in 1847, when the Dee Bridge of Robert Stephenson (1803-1859) which was made of trussed cast-iron girders collapsed under a train resulting in the loss of 5 lives. In 1879, the age of wrought-iron came to an abrupt end as construction material with the collapse of the celebrated wrought-iron bridge over the Firth of Tay in Scotland during a heavy gale whilst a mail train was crossing it. This multiple-span bridge carried a single-track railway of the main North British Line from Edinburgh to Aberdeen, and it crossed the 2 miles (3.22 km) wide open estuary, the longest bridge over water in the world at the time. To meet the navigational requirements, the 13 longest spans of the bridge, each spanned by a 245-foot (74.68 m) latticed girder of wrought-iron, stretched over more than ½ a mile (0.8 km) in the middle of the estuary. These latticed girders were set on high cast-iron columns standing on brick and concrete piers. This bridge had been hailed as a modern wonder of the world, and its engineer, Thomas Bouch (1822-1880), was knighted after the bridge was opened. On December 28, 1879, a strong gale blew the 13 central spans of the Tay Bridge together with the mail train from Edinburgh travelling on the bridge into the waters below, a catastrophe in which 78 lives were lost. This spectacular bridge disaster received worldwide attention, and it not only discredited wrought-iron as a sound structural material but also induced structural engineers to use ductile steel as its replacement. Before this structural catastrophe, many engineering authorities had considered 'steel' a risky construction material, although steel had been already used in boilermaking and shipbuilding because steel plates allowed much higher pressures than the wrought-iron plates, and since 1870, steel rails had been replacing wrought-iron rails in railroad tracks at a steadily accelerating rate because one steel rail would outlast 20 iron rails. After the Tay Bridge disaster, the use of steel in all kinds of construction spread rapidly, particularly in United States, where the Carnegie Company after 1885 began rolling steel beams which were used in skyscraper construction.

The first important use of steel in structures had taken place in United States before the Tay Bridge was built. The 3 tubular arches of the St. Louis Bridge, each spanning 502 feet (153.0 m), were made from nickel steel alloy. The bridge was constructed from 1869 to 1874 under the direction of its chief engineer and entrepreneur, James Buchanan Eads (1820-1887). The new Tay Bridge made of steel was built from 1882 to 1887 on new foundations alongside the old bridge by William A. Barlow (1812-1907). The next bridge made of steel which mostly influenced the engineering profession is the celebrated cantilever bridge over the Firth of Forth at Queensferry in Scotland, the engineering of which was the responsibility of John Fowler (1817-1898), Benjamin Baker (1840-1907) and his assistant Allan Stewart, who was responsible for the stress analysis and the detailed design of the bridge. It was built by the contractors, Tancred, Arrol & Co.. The Firth of Forth bridge has 2 long central spans of 1,710 feet (521.21 m) each, and 2 side spans of 675 feet (205.74 m) each. The approach viaducts consist of girders supported on masonry piles of which 15 girders have a span of 168 feet (51.21 m) each, and 5 girders have a span of 25 feet (7.62 m) each. The main compression members of this bridge were tubular struts made of curved plates which were fabricated according to the workshop and site practice of ship construction. Tension members were mostly latticed girders. It was a bridge without a comparable precedent in engineering, and its erection took place from 1883 to 1890. The structure of this cantilever bridge was made of 54,160 tons of steel members fastened with 6½ million rivets, and it had a painted surface area of 135 acres (54.6 ha). The completion of the Firth of Forth Bridge marks the end of the experimental period of cantilever bridges. It had the longest span of a bridge in the world until the construction of the longer cantilever bridge, which has a central span of 1,800 feet (548.64 m), built from 1900 to 1917 across the St. Lawrence river in Quebec, Canada. The consulting engineer for this giant bridge was Theodore Cooper (1839-1919), a former resident engineer of James Eads in the building of the St. Louis arch bridge, who for financial reasons could not reside at the construction site of the Quebec bridge. This bridge, which had a latticed girder structure, collapsed twice during its construction, killing altogether 86 workmen.

Precision Pyrometer

The problem of measuring high temperatures in the furnaces with precision was solved in 1886 by the French chemist Henri Louis Le Chatelier (1850-1936) when he invented his thermoelectric pyrometer which consisted of a thermocouple made of two wires, one of platinum, which melts at 3218°F (1770°C), and the other of platino-rhodium alloy, joined together at the ends. If one of its ends is heated, a tiny current begins to flow through the wire the strength of which is proportional to the temperature. He also invented an optical pyrometer which measured high temperatures by the nature of the light radiated from the hot objects. His pyrometers, which were essential for the exact measurement of high temperatures in the metal refining furnaces, made the control of temperature in such furnaces much more precise.

Le Chatelier was one of the first European chemists to discover the fundamental importance of the work of the brilliant American theoretical physicist Josiah Willard Gibbs (1839-1903) to chemistry, and he devoted his experimental researches to finding out all the implications of the ingenious Phase Rule of Gibbs in chemical processes.

Modern Steelmaking Processes

Linz-Donawitz Process

The Linz-Donawitz process of steelmaking known as the 'L-D process' was developed by the metallurgists of 2 Austrian steel plants located at Linz and at Donawitz. Austrian iron ore has a medium phosphorus content which makes neither the basic Thomas process nor the acid Kelly-Bessemer process economical. The Siemens-Martin open-hearth furnace is also not economical owing to the limited availability of scrap steel in Austria. Therefore, the Austrian metallurgist introduced a new converter method of making steel in which a jet of almost pure oxygen instead of air is blown with supersonic speed onto the surface of the molten iron inside the upright converter. The oxygen-blast process, which Bessemer himself had anticipated in his 1856 patent, is as rapid as the Kelly-Bessemer pneumatic process yet it produces high quality steel like the open-hearth process, since up to 25% scrap metal can be added to the charge. The only drawback of this process is its lack of analytic chemical control owing to the high speed of this particular steelmaking process. The original L-D process was restricted to iron ore with a phosphorus content of less than 0.3%. The French and the Luxembourg metallurgists discovered that by injecting lime with oxygen, iron ore with a phosphorus content up to 2.5% can be processed by this method. The modified process is called the L.D.A.C. process, and it could replace the open-hearth process.

Kaldo Process

The Kaldo process of steelmaking [***Kaldo*** is derived from ***Kalling*** at ***Domnarfvet***] was invented by Professor Bo Kalling at Domnarfvet in Sweden. It uses an inclined Bessemer-type converter which is rotated about its axis during the blast. The Kaldo top-blown converter has a capacity up to 125 tons and its rotational speed is as high as 30 rpm. The oxygen is blast against the surface of the molten iron, and the variation in the oxygen supply and the rotational speed during the 90-minute operating cycle affords a better control of the process than the L-D process. The Kaldo process can burn out the phosphorus content of any iron ore without reducing the carbon content, which makes the production of high carbon steels possible. The disadvantages of the Kaldo process of steelmaking is the rapid wear of the refractory lining, and the mechanical problems attending the rotation of the converter. For these reasons, a second converter is required which is being relined whilst the other converter is in use.

Rotor Converter Process

Rudolf Graef developed the Rotor Converter Process at Oberhausen in Germany. The converter is a long cylindrical drum which rotates every 2 minutes about its axis. The primary jet blows oxygen into the molten iron itself, whilst the secondary jet above the metal bath converts carbon monoxide into carbon dioxide by burning, thus providing considerable economy in heating. This converter, which can produce high-grade steel from pig-iron containing a maximum of 2% phosphorus, has a capacity up to 100 tons with a yearly production of 500,000 tons.

Grades of Steel in Modern Steelmaking

4 basic grades of steel are produced by the modern steelmaking processes:

1. *Low-carbon steels* contain 0.15% carbon and are sufficiently soft to be cold-pressed in the forge to make body-panels of cars.
2. *Mild steels* contain 0.25% carbon and are mainly used in construction.
3. *Medium-carbon steels* contain 0.5% carbon and can be toughened by heat-treatment to make axles and crankshafts.
4. *High-carbon steels* contain up to 1.4% carbon and are used in making springs, files, axes and hammers.

High-Temperature Resistant Metals

Superalloys

After the First World War many new products and machines required metals which would not only maintain their strength but also resist corrosion at high temperatures. Common metals, such as copper which is too soft, steel which corrodes easily, aluminium which melts, and magnesium which burns, do not meet these requirements. Only nickel and titanium among the most common metals do not have such drawbacks.

Pure nickel has properties which closely resemble those of iron, but since it is too expensive to substitute for iron, it is used in nickel alloys. The Swiss physicist Charles Edouard Guillaume (1861-1938), who joined the newly established International Bureau of Weights and Measures (**Bureau International des Poids et Mesures**) at Sevrès in 1883 and became its director in 1915, had the task to increase the precision of the standard measures. In his search to find an inexpensive substitute for the expensive platinum-iridium alloy from which to construct standards of length and mass, he discovered in 1896 an alloy of 64.4% iron and 35.6% nickel which remained almost invariant in length when subjected to temperature change by virtue of having a linear expansion coefficient $\alpha = 0.0000012$ per degree of Celsius. It was an outstanding contribution to the science of metrology and precision physics. He named this alloy 'Invar' because of its invariant property of length with respect to the temperature change. 'Invar' was commercially used in manufacturing measuring tapes and components of clockworks. In 1919 Guillaume discovered another important alloy of 60% iron, 38% nickel and 12% chromium, the elasticity of which remained almost invariant with respect to temperature changes. He named this alloy 'Elinvar' because of the property of its elasticity which remained invariant with respect to temperature changes.

He also discovered a 3rd alloy of 54% iron and 46% nickel which had an enormous technological importance since it expands with the increase of temperature at exactly the same rate as glass for which reason it is used as the lead-in wire in electric lamps because both the wire and the glass expand the same when heated thereby leaving the lamp air-tight. He named this alloy of iron and nickel 'Platinite'.

'Invar' and 'Elinvar' were of great importance in chronometry since 'Elinvar' hair-springs and 'Invar' balance wheels greatly reduced the temperature effects in precision chronometers. In 1920 Guillaume was awarded the Nobel Prize in physics for his discovery of 'Invar'.

It took more than half a century before the American physicist Charles Hard Townes (born 1915) conceived the Maser (**M***icrowave amplification by simulated emission of radiation*) which could be used as an 'atomic clock' for time-measurement, thereby finally making chronometry independent of the gross properties of matter.

Superalloys were developed not only to resist high-temperature corrosion but also persistent high stress which causes plastic 'creep' of metals. Such superalloys normally contain a high percentage of nickel and chromium, or of cobalt, nickel and chromium, but small additions of aluminium, titanium, molybdenum, tungsten, and carbon are also important.

The 80% nickel and 20% chrome alloy called 'Inconel' or 'Nimonic' has a high resistance to electric current and, therefore, gives off large quantities of heat when the electric current is passed through it which makes it a useful material for heating coils. Moreover, it loses little of its great strength at high temperatures, and does not corrode or scale at bright red-heat. The improved

'Nimonic' series of nickel-chrome alloys are not only capable of resisting high stresses at high temperatures but corrosion as well. For this reason, 'Nimonic alloys' are used in the construction of blades and other components of gas-turbines which have to withstand high stresses at high temperatures and resist corrosive effects of burnt fuel. For instance, the 'Nimonic alloy 115' can withstand a stress of 34,810 psi (240 N/mm^2) for 1,000 hours at 1,500°F (815°C) temperature.

Pure titanium, a much lighter metal, has great resistance to corrosion at high temperatures but unlike the nickel-chromium alloy it is prone to 'creep' and suffer fatigue when subjected to high stresses. Its lightness and ability to withstand high temperatures, however, makes it an ideal material for use as the heat-protective skin of satellites, compressors, and exhaust pipes of jet engines.

Alloy Steels

Steel, an iron-carbon alloy containing no substantial amounts of other metallic elements is called carbon steel. Steel to which other elements have been deliberately added in appreciable amounts to confer special properties to steel are called 'alloy steels'. The impurities in steel such as sulphur makes the steel brittle when hot, and therefore unsuitable for rolling or forging in a hot state, yet it makes steel easily machinable. Phosphorus impurity increases fluidity of steel when molten but makes finished steel more brittle. Silicon impurity decreases the incidence of blowholes from the release of dissolved gases occluded in finished ingots, and decreases the hardness of steel. Manganese impurity scavanges oxygen in the melt and removes deleterious effects of sulphur by combining with it to form manganese sulphide slag.

The properties of alloying elements when added to steel:

Nickel increases tensile strength, toughness to impact fracture, hardness and resistance to fatigue.
Chromium increases hardness, resists wear and corrosion.
Tungsten increases hardness and strength at high temperatures.
Molybdenum increases elasticity and strength at high temperatures.
Manganese makes steel nonmagnetic and lends it work-hardening properties.
Vanadium increases strength and toughness, and improves resistance to fatigue.
Cobalt increases hardness and retention of magnetism.

The 3 major groups of alloy steels are the low-alloy structural steels, corrosion resistant stainless steels, and high-speed tool steels. In low-alloy group the alloying elements used are nickel, chromium, molybdenum, vanadium, manganese, and also silicon. Correct alloying additions to steel can improve the strength, the toughness, the corrosion and wear resistance, and the hardness of alloy steels. Alloying, a method of adding other metals and elements to the base metal for the purpose of imparting particular properties to the base metal, is performed during the melting operation of the base metal.

The corrosion resistance of steel can be somewhat improved by the addition of about 0.25% copper alone or with 0.2% phosphorus. But for more severe corrosive conditions the addition of some combination of nickel, chromium and molybdenum up to about 10% may be necessary. The most common form of 'stainless steel' may contain 12% to 30% chromium, or 6% to 40% nickel with 15% to 30% chromium, 2% molybdenum and some other elements in smaller quantities.

In high-speed tool steels the principal carbide-forming elements are tungsten, molybdenum, and cobalt. Tungsten can be added alone or with the other 2 elements, but cobalt and molybdenum are added only in combination. The basic alloy steel to which these carbide-forming elements are added usually contains 0.7% to 1.5% carbon, about 4% chromium, and 1% to 1.5% vanadium.

Spring steel has 0.5% to 1% silicon, 0.7% to 0.9% manganese, and minor proportions of chromium and molybdenum in the range of 1% to increase the tensile strength and hardness of steel. Vanadium up to 1% is frequently added as an alloying element to improve hardenability of the spring steel. The manganese alloy steel, known as the 'Hadfield steel', is extremely tough and abrasion resistant. About 10% of United States steel production is alloy steels.

LIGHT METALS AND THEIR ALLOYS

Aluminium

The age of light metals began with aluminium. Aluminium ore, the bauxite, being an aluminium oxide is very stable and, therefore, cannot be easily smelted.

Although the Danish scientist Hans Christian Ørsted (1777-1851) was been able to produce a few particles of metallic aluminium in 1825, it was the German chemist Friedrich Wöhler (1800-1882) who in 1827, was able to reduce anhydrous aluminium chloride with potassium, a chemically active metal discovered by Humphry Davy (1778-1829) in his electrochemical experiments. In 1845, Wöhler succeeded in producing a small button of aluminium metal which was large enough for testing the metal for some of its physical properties. Wöhler published his results in 1854.

The first practically significant production of aluminium was achieved by the French metallurgist and geophysicist, Henri Étienne Sainte-Claire Deville (1818-1881) in 1854. Deville used basically the same method as Wöhler, except that he substituted sodium for potassium. This elaborate process first converted aluminium oxide (*bauxite*) to aluminium chloride, and then reduced the chloride with metallic sodium, a method simulated in the production of zirconium and titanium today. Using Deville's method, in which he later substituted a mixture of calcium fluoride and the mineral cryolite for the aluminium chloride, aluminium could be produced on a small industrial scale, particularly after the Emperor Napoleon III had supported Deville's work in the hope that aluminium could be used in military equipment such as helmets. The price of aluminium in the Deville process were further reduced after 1886 when Hamilton Young Castner (1858-1898) of New York had reduced the cost of sodium to ¼ of its earlier price by manufacturing it by means of the electrolysis of a brine consisting of common salt dissolved in fused caustic soda in a specially designed cell fitted with a mercury cathode. Castner used dynamos in his electrochemical manufacture of caustic soda from brine. The Castner cell was subjected to various modifications and a number of diaphragm cells were developed during the period from 1900 to 1925 in United States. Already in 1883, efforts had been made by R. Gratzel in Germany to use direct electrolysis of aluminium salts. However, to obtain metals from their oxides by heating metal oxides with sodium or potassium, as was done by Sainte-Claire Deville to produce aluminium, was an expensive process.

In 1879, Charles William Siemens had extended the electric-arc method to steelmaking by inventing the electric-arc furnace for steelmaking. In 1880, the French inventor Louis Clerc using his improved electric-arc furnace also developed the electric-arc method to fuse refractory metals. In 1886, two brothers, Eugene H. and Emile A. Cowles of United States developed a process in which an electric current was passed through a mixture of carbon and bauxite, and after reaching a very high temperature the aluminium oxide breaks down to form aluminium carbide. They introduced molten iron to alloy with aluminium carbide which produced iron-aluminium alloy, and aluminium. However, the Cowles process depended upon electric heating and not electrolysis. This process had a very short life, although it popularised the importance of the electric method in metallurgical practice, since in the next year a revolutionary method of producing aluminium on commercial scale was introduced in 1886 by the metallurgists, Paul Louis Toussaint Héroult (1863-1914) in France and Charles Martin Hall (1863-1914) in United States, and in 1887 by Martin Kiliani (1858-1895) in Germany.

Héroult and Hall had careers which were remarkably similar except for their worldly success. The two men were born in the same year and died in the same year. They invented the electrolytic method of reduction of aluminium at the same time, only Héroult received his patent in France some months before Hall received his in United States, but here the similarities ended. Hall made a fortune from his patent, whilst Héroult died in penury.

Hall was inspired to find an industrial method for the production of aluminium by Frank F. Jewett, a former doctoral student of Wöhler who was one of his professors at the Oberlin College where Hall was a student, very much like Sidney Gilchrist Thomas had been inspired to find a method of steel production from phosphoric iron ore by his teacher Dr. Chaloner. Dr. Jewett mentioned that anyone who can devise an inexpensive method to refine aluminium in large quantities will make a fortune. Hall based his research of this problem on the discovery of the British chemist Humphry Davy who had used electrolysis to deposit purified metals on the cathode plates by passing electric current through the molten bases as the electrolyte. Hall discovered by chance that the mineral cryolite with a chemical composition of sodium-aluminium-fluoride was the ideal basis for the electrolytic reduction of aluminium because it melted much more easily than alumina (aluminium oxide). If about 5% alumina (aluminium oxide) was dissolved in cryolite the mixture did melt when heated to 1,832°F (1,000°C) by electrolysis, and the electric current passing through it kept the mixture in the molten state and split up the aluminium oxide into aluminium and oxygen. In 1886, a year after graduation, Hall succeeded in producing pure aluminium globules by electrolysis in his aluminium reduction-pot from which he prepared several ingots. Hall, together with some businessmen, organised the Pittsburgh Reduction Company in 1888, which retained a monopoly of aluminium production in United States for over half a century. This company, which in 1907 was renamed the Aluminum Company of America [known by its acronym, **ALCOA**] was the first major industry to take advantage of the inexpensive hydroelectric power generated by the new hydroelectric power plant at Niagara Falls which became available in 1895.

Héroult, who developed a similar electrolytic process for the refining of aluminium, also invented an improved version of the electric-arc furnace for smelting which is still in general use today. He set up his aluminium reduction plant at Neuhausen in Switzerland where hydroelectric power was readily available at the Rhine waterfalls, but the use of aluminium in industry was much slower in Europe than in United States and, therefore, his enterprise was not a financial success.

The Héroult-Hall aluminium reduction process, which created a revolution in metallurgy, consists of electrolysis of alumina dissolved in molten cryolite (sodium aluminium fluoride), the low density of which lets the metal settle in a pool at the bottom of the electrolytic cell. This process depends upon refined carbon anodes which are consumed in large quantities, and the availability of inexpensive electric power. Aluminium oxide in the form of purified bauxite (hydrated alumina) served as the only suitable raw material. The electrolytic process of Hall and Héroult was the beginning of the era of light metals and their alloys.

In the igneous electrolytic process of Héroult and Hall, which consumed enormous amounts of electric current, very powerful, low-voltage current had to be used: from 8,000 to 50,000 amperes current at 5 to 6 volts tension. Production of one ton of aluminium required an enormous amount of electric power: 18,000 to 20,000 kilowatt-hours of electricity. In this process the electrolyte is a molten igneous mixture of purified bauxite and other compounds that is subjected to electrolysis between the carbon electrodes, in which process the electric current frees the metal and, also, keeps the metal bath in a molten state. This type of igneous electrolytic smelting method produces most of the aluminium and magnesium in commerce today.

A modern aluminium furnace of 100,000 amperes produces about 1,653 pounds (750 kg) of aluminium every 24 hours. It consists of a series of 150 to 200 'pots' or cells, called a 'pot-line', which is rated about 75 megawatts. Carbon paste is put in a mould where it is baked by the heat of the furnace as the electrolytic process proceeds. This is called the 'Soderburg anode'. The oxygen liberated at the anodes combines with the carbon to form carbon monoxide which burns at the top of the furnace resulting in carbon dioxide. The amount of carbon consumed in this process is enormous: half a ton of carbon anodes is burnt up for each ton of aluminium produced. The entire process of producing a ton of aluminium is an exercise of handling immense quantities of materials. In the electrolysis alone 15,000 kilowatt hours of electricity, 992 pounds (450 kg) of prebaked carbon anodes and 110 pounds (50 kg) of cryolite is required. Moreover, 4½ tons of bauxite is needed to produce 2 tons of alumina by the Bayer process which requires one ton of fuel oil and 353 pounds (160 kg) of caustic soda. The cost of electric power is from 15% to 30% of the total smelting cost of aluminium.

Despite the success of the Héroult and Hall process, the producers of aluminium had difficulties in selling their product until the relatively high-strength aluminium alloys had been developed between 1906 and 1909 by the German metallurgical engineer Alfred Wilm (1869-1937), who discovered the 'age-hardening' of aluminium alloys by adding 3.5% copper and 0.5% magnesium. The heat-treatment of this aluminium alloy did not harden it as Wilm had expected. When Wilm checked his results a few days after the heat-treatment, he was surprised to find the aluminium alloy not only harder, but also stronger. Wilm realised that leaving the aluminium alloy to 'rest' a few days after the heat-treatment had brought about certain microscopic changes in its structure which made it much tougher. Alloys which are aged at higher temperatures than the room temperature are called 'precipitation-hardened' alloys. ***Durener Metallwerke*** (*Durener Metal Works*) was the first producer of this age-hardened aluminium alloy which was commercially named 'duraluminium'. However, since duraluminium readily corrodes, it is coated with a thin layer of pure aluminium which quickly oxidises in the air to form a thin layer of aluminium oxide on the surface of the duraluminium that resists corrosion caused by the atmosphere or strong chemicals.

Magnesium

Magnesium is a lightweight metal, only ⅔ as heavy as aluminium, but it is a weak metal unless it is alloyed. It costs almost twice as much as aluminium, and more than 4 times as much as steel. However, where lightness rather than cost is of supreme importance, magnesium alloys are used.

Magnesium ores, such as magnesite (magnesium carbonate), dolomite (calcium-magnesium carbonate), and carnallite (potassium-magnesium chloride) are sparsely distributed throughout the

Earth's crust whereas the magnesium chloride exists in great abundance in the sea. About 6 tons of magnesium is contained in every cubic mile of sea water.

Magnesium is separated from its chloride by a method very similar to the Hall-Hérault electrolysis process used in producing aluminium. Pure magnesium looks like aluminium and it combines extremely rapidly with oxygen to form magnesium oxide. Once it is heated to its melting point at 1,171°F (633°C), it burns on its own with a blinding white flame the temperature of which is well over 1,832°F (1,000°C). For this reason, magnesium was first used in fireworks, in photographic flashlights, and in incendiary bombs.

Magnesium alloys are extremely strong although magnesium itself is a weak metal. Magnesium-aluminium-zinc alloys are tough, and magnesium-zirconium alloys tougher still. The magnesium-manganese alloys have a high resistance to corrosion. All magnesium alloys can be cast, rolled, extruded and pressed into any shape. Since magnesium is so easily inflammable, molten magnesium has to be kept away from the atmospheric oxygen by sprinkling the molten surface with a fine dust of sulphur that allows the sulphur which is exposed to the atmospheric oxygen to burn and form sulphur dioxide gas which shields the magnesium from air.

Pure magnesium is often used to protect steel pipelines from corrosion by burying a lump of magnesium connected by an electric cable to the steel pipe buried in soil, which makes the pipe and the lump of magnesium behave respectively like the cathode and the anode of a battery. The easily replaceable magnesium is gradually corroded away in the soil whereas the expensive steel pipe remains uncorroded.

Electrolytic processes have proven to be important in the production of metals. Copper is refined to a high degree of purity by means of electrolysis. In the case of light metals, aluminium is produced by electrolysis of molten salts, and magnesium is produced by electrolysis of concentrated brines from sea water.

Powdered-Metals and Sintering

A method of shaping metals, besides cutting or forging which use molten or solid metals respectively, is sintering which shapes items directly from powdered metal.

This method grew out of early efforts to produce larger lumps of platinum which occurs in nature only in the form of fine grains. Platinum is the heaviest of all metals, and has the highest resistance to corrosion than any other metal. All attempts to make the small grains of platinum into larger lumps failed in the 18th century since no furnace could reach the high melting temperature of platinum, 3,223°F (1,773°C). It was the English physician and scientist William Hyde Wollaston (1766-1828), the discoverer of the metals palladium and rhodium in 1803 and 1804, discovered in

1805 a method called 'sintering' to make large chunks of platinum by pressing powdered platinum metal into blocks, heating the blocks in a coke furnace, and then hammering the blocks into solids. Wollaston made a fortune out of this sintering process that he kept a closely guarded secret until his death in 1828, when his will allowed The Royal Society in London to disclose his method of sintering.

For a long time sintering, that is fusing powdered metals together under heat and pressure, was only used to shape those metals, the melting temperatures of which were beyond the capabilities of commercial furnaces. Even today, tungsten which has a melting temperature of 6,098°F (3,370°C) has to be sintered, because construction of a furnace that can produce such a high temperature would be far too expensive.

Sintering method is used today to produce tungsten-carbide drill bits made from powdered tungsten, carbon and cobalt. This mixture is pressed into a block and heated anywhere from 1,292°F (770°C) to 1,832°F (1,000°C) in an atmosphere of hydrogen since any presence of oxygen would oxidise the metal and burn up the carbon. In this state, the block is no harder than chalk and can be easily formed into its final shape such as a drill bit. Then it is heated again in hydrogen up to about 2,732°F (1,500°C) at which temperature the powder particles sinter and the tool becomes permanently hard. Only tools with diamond-tips are harder than the tungsten-carbide tools. Sintering is used to mix diamond powder into metals to produce extra hard grinding surfaces.

Sintering has recently been used to blend non-metals with metals for special purposes. Brushes in dynamos and electric motors undergo constant friction against the rotor and if made of pure copper the frictional heat would gradually melt the copper brush away. If brushes were made of graphite, a soft form of carbon, they would rapidly wear away. Sintering a mixture of graphite and copper powder produces a long-lasting brush in which graphite acts as a lubricant and copper grains give it the strength which prevents the graphite grains from wearing away quickly.

Since the sintered metal is porous like a sponge, its pores can be filled with another metal, or some lubricant. An extremely strong copper-iron solid can be made by dipping a very porous piece of sintered iron into molten copper which fills all the pores.

Self-lubricating bearings are made by such a sintering technique. A mixture of copper, tin and graphite is pressed into the desired shape and sintered. Then the sintered bearing is soaked in oil until all pores are filled with enough oil to last the lifetime of the machine.

Precious Metals

The following four metals, silver, gold, mercury and platinum, have one common property- they are expensive. Silver because it is quite scarce, gold because it is a standard for national currencies, mercury because it is an ore, cinnabar (mercury sulphide) which contains only 4% of mercury, and platinum because it is rather scarce and expensive to work.

Since silver is too soft a metal, it is alloyed with small amounts of copper to produce 'sterling silver' which contains 7 % of copper. Silver also resists organic acid and, therefore, it is used to line vats in organic chemical plants and drug manufacturing plants. Presently, the enormous rise in demand for this metal by the photographic industry has created a worldwide shortage of silver. Almost half of the world's production of silver is used to produce the light-sensitive emulsion in films and photographic printing paper.

Gold in modern technology has found a new usage along with silver: it makes a good and fairly light coating as a protection against harmful radiation from outer space. Thin films of gold and silver are applied to the delicate instruments in space probes to protect them against harmful radiation from space.

Gold and silver are mostly extracted from their ores by a process of amalgamation with mercury to form compounds with gold or silver the heating of which drives off the mercury in the form of vapour leaving a spongelike residue of pure gold or silver.

Mercury is also used to conduct electricity in mercury lamps known as 'sun-ray' lamps. In these lamps, mercury is heated until it vapourises and the passages of electric current between the atoms of the mercury vapour excites the electrons of these atoms which begin to emit a blue-green light together with the invisible stream of ultraviolet rays which produce sun-tan.

Since gold, silver and tin dissolve in mercury to form a soft alloy that hardens rapidly, they were mainly used in dentistry.

Platinum, the heaviest of all metals, has driven gold from its position as the world's most valuable metal for it costs twice as much as gold. Natural platinum was discovered in 1735 by the Spaniard Don Antonio de Ulloa in Peru. Rich deposits of platinum also exist in the Ural Mountains in Russia. Since platinum has a higher resistance to corrosion than any other metal, it is applied to line vessels which are used to process corrosive chemicals. Platinum, usually alloyed with iridium or ruthenium to impart platinum extra high hardness and corrosion resistance for certain applications, has durability, a low thermal expansion and low electrical resistivity, properties which make it useful in highly precise thermocouples, electrical instruments, and in electrical contacts in telephone circuits that repeatedly switch on and off. The high melting temperature of platinum, 3,151°F (1,733°C), which is well above the temperature of the electric spark in the contact, makes it last decades despite countlessly repeated switching. The platinum alloy of 90% platinum and 10% iridium was used to make national prototype 'Meters' and 'Kilograms' maintained by individual nations.

Metals in Electronics

The rapid development of the modern telecommunications and the computer industry led to the discovery of metals with new electrical properties: metals which conduct electric current at high temperatures, and metals which change their electrical properties under the influence of light rays. Among such metals, the most important are tungsten (also known as 'wolfram'), selenium, and germanium, all discovered a long time ago.

Tungsten, a heavy and hard metal which is both malleable and ductile, has a melting temperature of 6,098°F (3,370°C) that is much higher than that of iron. Since it glows at 'white-heat' without melting, it is suitable for making filaments of electric lamps. The electron ray in television sets which scans the screen is emitted from an electron gun the core of which is a fine filament of electrically heated tungsten.

Selenium was long known to be sensitive to light because its electrical resistance varies with the amount of light falling on it. Its resistance to electric current diminishes with the increase of light radiation, a property of selenium used in photocells operating doors. It was discovered later that selenium has another remarkable property: it generates a weak current when light shines on it. This property is called the photoelectric or photoemissive effect [in Greek, *phos* means light], which has been used in lightmeters.

Germanium was long known as a metal with a peculiar electrical property. At normal temperatures germanium is neither a good conductor nor a good nonconductor of electric current. Such metals are known as semiconductors. A peculiarity of semiconductors is that they allow electric current to pass more easily in one direction than in the opposite direction. Thus, if an alternating current is passed through a semiconductor such as germanium, then only the forwards current is let through, whereas the backwards current is blocked. It means that if the alternating current is fed into the germanium, only direct current emerges from it, an effect called 'rectification' of the alternating current. Therefore, germanium acts as a semiconductor 'rectifier'.

In 1948, William Bradford Shockley (1910-1989), John Bardeen (1908-1992) and Walter Houser Brattain (1902-1987) at the Bell Telephone Laboratories discovered that germanium could also be used to amplify weak high-frequency electric currents. By 1950, they had developed a semiconductor known as the 'transistor (*transfer resistor*)', which could be used to amplify as well as to rectify radio signals. Transistors, unlike electronic tubes, consume extremely small amounts of current, do not become hot, never really wear out, and can be made as small as a pinhead. This miniature electronic device created a revolution in electronic engineering, particularly in electronic computer engineering and miniaturisation of the telecommunication devices in satellites.

Alumino–Thermionic Method of Reduction

Sainte-Claire Deville had reported that in the process of reducing metallic oxides by heating, powdered aluminium could replace sodium or potassium. Since aluminium was now available it could be possible to develop the alumino-thermionic method of reduction in the production of other pure metals, which were difficult to fuse from their oxides. This process was investigated by the German chemist and metallurgist Hans Goldschmidt (1861-1923) of Krupp Works in Essen in the 1890's, and in 1898 he finally described his procedure by means of which it could be accomplished. The Goldschmidt alumino-thermionic process of reduction consisted of mixing aluminium powder with the oxides of other metals such as chromium, titanium, manganese and vanadium, which when ignited burn furiously and generate such enormous heat that melts the metal as well as the resulting slag of alumina. For instance, pure iron and chromium could be produced by this process. The alumino-thermionic method of reduction of Goldschmidt was used after 1898 for a few decades in producing new pure metals for alloy steels. One of the most important metals produced this way was tungsten, which has the highest melting point, about 6,098°F (3,370°C). Tungsten proved to be an important component of tool steels, since the chromium-tungsten alloy steel and the tungsten-carbide alloy steel provided the required hardness properties for industrial cutting tools. Since the oxide and aluminium mixture, called 'thermite', develops enormous heat when ignited, it could be used in welding, and for some purposes it is the best method of welding. Such other rare metals as vanadium, titanium, zirconium, tantalum, niobium [also known as columbium], molybdenum, and uranium, became commercially available only after 1900 when analytical and vacuum techniques were improved to the extent that high purity metals could be refined by such means.

The manufacture of waveguides and magnetron dischargers in electronics are made of copper with high thermal conductivity which presented a serious problem since it required machining copper into intricate shapes, and pure copper is a very difficult metal to machine. This problem was overcome by developing the copper-selenium and copper-tellurium alloys which contain about 0.3% of selenium and tellerium respectively. More recently, copper-sulphur alloys have been used for this purpose.

Electric Reduction Furnaces

Some metals have such high melting points that electrical methods of melting are the only feasible practical method. For instance, titanium is almost exclusively melted by the electric arc. Electric furnaces became important when temperatures up to 5450°F (c.3010°C) had to be reached, or when metal had to be melted in the presence of certain gases. In general, the electric furnaces are more important still for the highly accurate control of processes at high temperatures.

Electric-Arc Furnace

The efficiency of William Siemens' electric-arc furnace using graphite electrodes was improved in 1892 by the French metallurgist Henri Moissan (1852-1907). The modern electric-arc steel furnaces use alternating current transformed to give 30,000 amps at 100 volts. Each melt requires about 4 hours and can yield about 150 tons of steel. Since no fuel is used within the furnace, the slag, the temperature, and the furnace atmosphere can be controlled with great precision. Samples are taken and analysed to check the composition of steel. Electric-arc furnace is mainly used in the production of high-alloy steels, the bulk of stainless steels and cast-iron. William Siemens melted iron in crucibles with his electric-arc furnace at Paris Exhibition in 1879.

Induction Furnace

The induction furnace was introduced in 1887 by the young English electrical engineer of Italian stock, Sebastian Ziani de Ferranti (1864-1930), who was the foremost promoter of alternating current in Great Britain. In the induction furnace the chamber, or crucible, containing the metal charge is surrounded by a water-cooled coil carrying alternating electric current with the frequency that is usually less than 3,000 cps. Ferranti, who had an outstanding scientific ability in electrical science for a youngster of 17 years of age, joined the company of Siemens Brothers, Electrical Engineers, at Charlton in Kent, where he collaborated with William Siemens on research in electrical engineering. The alternating current with a frequency of about 1,000 cycles per second in the induction furnace gives 400 amperes at the tension from 2,000 to 3,000 volts. The rapidly fluctuating eddy currents induced in the metal charge in the crucible of the induction furnace by the high-frequency electromagnetic field create a temperature as high as 5,450°F (c.3,000°C). The metal in the induction furnace may be taken to be the core and the enclosing electric coil the secondary circuit of a transformer. The heating of the metal charge in the induction furnace is produced internally by induction in the absence of any direct contact with the electric current itself, or with gases or flames. Such electric furnaces make melting of various metals and alloys in the presence or absence of certain gases possible. Steel produced in induction furnaces has now replaced the crucible steel to a great extent. The induction furnace is used mostly for direct melting operations, and little if any refining is done in this furnace. Such electric furnaces have a capacity of one to five tons, and they are currently used for the production of small quantities of high-grade alloy steels. The induction furnaces are actually more important in the more precise control of metallurgical processes at high temperatures, and in the production of deoxidised and desulphurised steels whose properties are substantially improved by the removal of the last traces of oxygen and sulphur.

Electric Resistance Furnace

In the electric resistance furnace the electrodes are buried in the metal charge and the flow of electric current through the metal charge produces heat. The electric resistance furnace is mainly used in the production of silicon carbides.

Galvanising

Aluminium and other metals and special alloys are relatively resistant to corrosion particularly to atmospheric corrosion. The corrosion of wrought-iron, and particularly steel was a more serious problem.

A process for galvanising iron was invented by Hobson in 1805, which consisted of dipping sheets of wrought-iron into molten zinc. Soon steel sheets were galvanised by the same method but galvanised steel sheets lasted in corrosion only about one quarter as long as wrought-iron sheets. Corrugation of iron sheets which increased their strength was introduced in 1828 in England, and galvanised corrugated sheets have been used as roofing material ever since.

Tinning

A better method of protecting steel sheets from corrosion was to treat the sheets first with chemicals and then dipping them into molten tin. This method of producing a protective layer by dipping discovered by Sylvester and Hobson in 1805 was, and still is important to canning industry since tin is quite resistant to corrosion by the acids in food and by the atmosphere. Tin cans are also extensively used in petroleum and fatty-oils industries.

In antiquity, Romans had already tinned receptacles for the storage of food.

CHAPTER XIX

TELECOMMUNICATIONS: From Telegraph to Telephone

Electrostatic Telegraphy

The first important technological application of electricity that arose from the science of electricity was the electric telegraph. In 1753 the first proposal for a statical electric telegraph was made anonymously in an article signed 'C.M.' in the Scots Magazine. This anonymous author, who could have been the surgeon Charles Morrison of Renfrew, described a telegraph consisting of a series of wires between the stations, each wire representing one letter of the alphabet. At one station the ends of the wires can be set into contact with a friction machine generating static electricity with switches, whereas the far ends are supplied with metal balls suspended over grounded metal plates. Each plate had a paper with a letter of the alphabet written on it. When a particular wire was electrically charged at one end it attracted the paper below the metal ball at the other end, and thereby indicating which switch had been closed at the transmitting end. In 1767, Joseph Bozolus, a Jesuit professor of physics at the *Collegio Romano* in Rome, suggested a similar telegraph by merely observing sparks received at one end of the wires. The first experimental electrostatic telegraph was built in 1774 by Georges Louis Lesage (1724-1803), a French mathematician living in Geneva, Switzerland. Lesage's telegraph was similar to that suggested by the anonymous Scot in 1753. It had 24 wires attached to 24 electrometers consisting of suspended pith-balls each identified with a letter of an alphabet. He encased the wires in an underground tube of glazed earthenware accommodating perforated partitions with holes to separate the wires for the transmission of signals over some significant distances.

In 1787, a French inventor named Lomond demonstrated a telegraph similar to that of Lesage, but having the notable improvement of a single conductor instead of 26 conductors. The letters of the transmitted message were distinguished by a code which was not revealed. It is possible that synchronised clocks were used at both the transmitting and the receiving end of the line. In 1795, a Spaniard, Don Francisco Salvá (1751-1828) of Barcelona, proposed a statical electric telegraph similar to that of Lesage, but consisting of 22 pairs of wires, each pair of wires held by a man who called out the letter when he felt the electric shock. In 1798 he changed his telegraph to a single wire and detectors in the form of tin-foil plates. Salvá is supposed to have laid an experimental 26-mile (42 km) static telegraph line between Madrid and the summer residence of the Royal family of Spain in Aranquez, but there is no definite documentation to prove it.

An Englishman, Francis Ronalds (1788-1873), became interested in electricity through his acquaintance with DeLuc, a Swiss physicist who experimented with dry-pile batteries. In 1816 Ronalds developed a practical single-wire electrostatic telegraph incorporating two synchronised disks with letters of the alphabet on one disk and a sector cut out in the second disk, both of which were rotated by watch-mechanisms. A suspended electrometer consisting of two pith balls showed the charged state of the line at the instant a letter was exposed in the cut-out sector. Ronalds built an underground cable of 525-foot (160 m) length consisting of a 2 by 2 inch (5 by 5cm) trough coated with pitch, into which he placed lengths of thick glass tube containing the wire joined by soft wax. He operated his telegraph by frictional electricity. Although his single-wire static telegraph was quite original, the British Admiralty declined to adopt Ronalds' single-wire electric telegraph despite its successful demonstration.

Electrochemical Telegraphy

Frenchman Claude Chappe (1763-1805), who was in charge of the telegraphic services in France during the Revolution, built the first optical telegraph in 1794 between Paris and Lille, a distance of 144 miles (231.7 km). However, before he built his semaphoric telegraph line, Chappe had experimented in 1790 with an electric telegraph in which he used synchronised pendula, 2 clocks and a Leyden Jar to provide the electric discharge. He had serious troubles with the insulation of his equipment, and as a result of this difficulty, he abandoned his electric telegraph in favour of the optical telegraph the origin of which can be traced back to ancient Greeks.

In 1804, Francisco Salvá proposed an electrochemical telegraph in which hydrogen bubbles, that appear at the negative electrode in the electrolytic decomposition of water, serve as telegraphic indicator. The source of the electric current for this electrochemical telegraph was a Volta Pile.

In 1809, after having witnessed the success of Chappe's semaphoric telegraph in Napoleonic Wars, Bavarian Minister Monteglas requested Dr. Samuel Thomas Soemmering (1755-1830), a physician and a member of the Royal Academy of Sciences in Munich, to develop a telegraph. Soemmering, who was interested in electrochemistry, devised an apparatus similar to Salvá's proposal which he provided with an alarm to alert the receiving operator. Soemmering used 35 wires between the sender and the receiver which made his electrochemical telegraph quite expensive and, therefore, governments and officials to whom Soemmering had demonstrated his telegraph were not interested in adopting his system. He had been able to transmit messages over a 1,000-foot (305 m) cable.

A number of other inventors, such as Sharpe in 1816, Coxe in 1816, Edward Davy in 1838, Smith in 1843, and Alexander Bain (1810-1877) in 1846, devised electrochemical telegraphs but none of these instruments proved to be of lasting practical value.

German chemist Johann Salomo Christoph Schweigger (1779-1857) proposed several methods for the reduction of the number of conductors needed in the electrochemical telegraph. He was able to transmit messages using only two wires by employing two batteries of different power, changes in the duration of signals, changes of intervals, and a binary code. Ironically, Schweigger's 'current multiplier' invented in 1820, the first genuine galvanometer to measure the amount of electric current flowing in a circuit, laid a sound foundation for the practical telegraphy, the electromagnetic telegraphy.

In the United States, the pioneer in telegraphy was Harrison D. Dyar who, sometime between 1826 and 1828, worked out a system of electrochemical telegraph in which messages were recorded by sparks passing through a chemically treated paper and leaving dot-and-dash patterns which indicated the message. Dyar had to abandon his invention after a successful trial near a racetrack in Long Island because of his unprincipled business partners.

In 1838 the Englishman Edward Davy, and in 1846 the Scot Alexander Bain of Edinburgh revived the idea of an electrochemical telegraph. Davy never brought his invention to commercial application, but Bain was able to commercialise his invention. In Bain's telegraph the electrochemically produced marks on a strip of paper, which represented letters, had a high-speed transmission, and for this reason, it was widely used in Britain for some time. The imperfections of the apparatus of Bain finally led to the replacement of his electrochemical telegraph by other more reliable automatic telegraphic systems.

Electromagnetic Telegraphy

Paul Ludwig Schilling von Canstadt (1786-1837), a Baltic German native to Estonia, was engaged in the diplomatic service of the Embassy of Russian Empire in Munich, Germany. He was a friend of Soemmering, his physician in Munich. Owing to his personal interest in the means of communication, Schilling assisted Soemmering in experiments with the electrochemical telegraph. After meeting Johann S. Schweigger (1779-1857) in 1815 in Munich, and after Schweigger's discovery of the galvanometer known as the 'electric multiplier' in 1820, Schilling began his experiments with electromagnetic telegraphy about 1822 which he continued until his death in 1837. Schilling devised several different electromagnetic telegraphs: in some, he used a single electric multiplier and magnetic needle as indicator; in others, 5 or 6 electric multipliers and magnetic commutators, and a keyboard for signal transmission to the indicator needles which required 8 wires formed into a cable and a single return wire. The electric current for his electromagnetic telegraph was provided by a Volta Pile. Schilling used in his telegraph a binary code for different combinations of the movement of the needle as a single indicator, and an alarm device resembling that of Soemmering's telegraph. He anticipated the concept of damping in reducing the time required for the oscillation of the indicator needle by immersing the lower end of the suspended needle in mercury. Schilling gave many demonstrations of his electromagnetic telegraph on his extensive travels as a diplomat of the Russian Empire, and after the demonstration of his telegraphic equipment in 1835 in Heidelberg, Georg Wilhelm Muncke (1772-1847), professor of physics at the University of Heidelberg, prepared a demonstration model of Schilling's 3-needle telegraph for use in his lectures on physics. In 1836, Russian Tsar Nicholas I charged Schilling with the task of installing an 8-mile (12.9 km) submarine cable between the fortress in Kronstadt and the Imperial palace at St. Petersburg for the operation of Schilling's electromagentic telegraph, but Schilling died before he could carry out this project.

The first functioning electromagnetic telegraph was designed and built by the great German mathematician and physicist Carl Friedrich Gauss (1777-1855) and his colleague, the German physicist Wilhelm Eduard Weber (1804-1891). They built in 1834 an electromagnetic telegraph between the iron-free magnetic laboratory and the astronomical observatory at the University of Göttingen in Göttingen, Germany. Their telegraph consisted of a line of about 15,000-foot (4,572 m) length, over which electric impulses were transmitted in the circuit by magnetoelectric currents. The electric current was generated by a coil around a permanent magnet consisting of 7,000 turns of fine wire which could be moved up and down by the operator. This coil of fine wire could produce electric currents of high intensity but even faint electric impulses could be recorded by adding a small mirror and a scale to the receiving instrument, a device introduced by the German physicist Johann Christian Poggendorff (1796-1877) a few years earlier, and by reading the movement through an optical magnifier, Gauss and Weber were the first inventors to put the binary and the quintic code into practice in electromagnetic telegraphy. Gauss had seen Soemmering's electrochemical telegraph in Munich in the 1810's, and he subsequently met and corresponded with Schilling.

In 1836 Carl August Steinheil (1801-1870), professor at the Bavarian Academy of Sciences in Germany, completed an electromagnetic telegraph on Gauss' invitation, which was a simpler and more practical version of the needle telegraph of Gauss and Weber. The moving coil in Steinheil's telegraph was replaced by a Clarke-type of magnetoelectric generator, and the pivoted indicator needle in the multiplier could be used to strike bells or to ink dots on a recording tape. Various combination of tones, or dots were coded to various letters of the alphabet by means of which Steinheil was able to send six words per minute with his telegraph. In 1837, Steinheil built 3 telegraph lines: from the Bavarian Academy of Sciences to his home over a distance of 0.6 miles (0.9 km); to the shop of the Academy over a distance of 0.06 miles (0.1 km); and to the astronomical observatory over a distance of 3.1 miles (5 km). The telegraph wires of these lines were strung on poles.

In 1836 Steinheil constructed the first printing telegraph in which two pivoted magnetic needles with small stencils mounted at their ends made marks on a moving strip of paper. In 1838 Steinheil installed his electromagnetic telegraph line over a 5-mile (8 km) distance along the Nürnberg-Fürth railroad. Gauss had suggested to Steinheil to use the two rails of the railroad track as a telegraph circuit, but Steinheil used the Earth as the return circuit of his telegraph line, first used by William Watson (1715-1787). This telegraphic system was successful, and Steinheil installed another telegraph line along Munich-Augsburg Railroad. Unfortunately, the cost of installation of the Steinheil telegraph was high, and for this reason no new telegraph installations were ordered by the Government.

In 1835, William Fothergill Cooke (1806-1879), an English medical officer in the British Indian Army, during his leave of absence attended Professor Muncke's demonstration of Schilling's 3-needle electromagnetic telegraph at the University of Heidelberg. In 1837 Cooke, after having abandoned his medical profession, exhibited his 3-needle telegraph system, a development of Schilling's telegraph, in London. Since Cooke had inadequate scientific background in physics, he consulted Michael Faraday (1791-1867) who introduced him to Charles Wheatstone (1802-1875), who was a musical instrument maker by profession and an inventor interested in acoustics, because Wheatstone had also been working on a system of telegraphy at the same time. Wheatstone, who was professor of physics at the King's College in London, invented the concertina in 1829, the kaleidophon for combining harmonic motions, and the stereoscope. Subsequently, Cooke and Wheatstone pooled their resources and formed a partnership for the development of the electromagnetic telegraph in which Cooke assumed the responsibilities of the engineer and entrepreneur, and Wheatstone agreed to do the necessary scientific investigations. Only after the American physicist Joseph Henry (1797-1878) visited Wheatstone and informed him how an 'intensity electromagnet' and battery on a very long circuit had been used at Princeton to actuate a 'quantity electromagnet' and battery on a local circuit, were Wheatstone and Cooke able to overcome their difficulties in the transmission of telegraphic signals over long distances.

In 1837 Cooke and Wheatstone patented their 5-needle telegraph which incorporated a call-alarm based on a relay consisting of a coil and an anchor-magnet that opened and closed a local circuit. It was installed between Euston and Camden Town, covering a distance of about one mile (1.6 km). In 1838 they built a double-needle type telegraph working by a code between Paddington and West Drayton, covering a distance of about 13 miles (21 km). In 1836 Wheatstone and Cooke also invented a dial telegraph, similar to that of Ronalds, which they modified in 1840, and then again in 1858. In the final version of the step-by-step dial telegraph, known as the **ABC** instrument, the receiver dial was driven by the transmitting dial which made it a direct reading telegraph that was simple to operate. Wheatstone devised an automatic transmitter and a high-speed printer based on a polarised relay actuating an ink-wheel which permitted a transmission of 50 to 150 words a minute. The telegraphs of Wheatstone and Cooke were not original instruments in as much as they were developments of known discoveries and inventions of others, but they succeeded in practice because their telegraphs were founded on well-tried principles. Their needle telegraphs were essentially similar to that of Schilling. Notwithstanding this fact, Wheatstone cultivated a nasty habit of discrediting in public Cooke's technological contributions to their telegraph. This bickering interfered with the best interest of their commercial efforts and led to the dissolution of the partnership. Both famous railroad engineers, Isambard K. Brunel (1806-1859) and Robert Stephenson (1803-1859), advised their respective railroad companies to acquire the telegraphic equipment of Wheatstone and Cooke, since the simple but particular single-needle and double-needle instruments gave the railroad signalmen safe means for the regulation of rail traffic.

The most successful telegraph was invented by a distinguished American portrait painter and sculptor, Samuel Finley Breese Morse (1791-1872), a graduate of Yale College who was not only a man of high culture but also a technological tinkerer. Morse invented his telegraph about 1835, and unlike the telegraph of Wheatstone and Cooke it was quite original for it was literally a 'telegraph', a device for 'writing at a distance'. Morse, like Cooke, was neither a physicist nor a technician and, therefore, his first instrument was rather crude and not quite practical. It did work through 40 feet (12 m) of wire and it marked a paper strip by zig-zag lines because its armature moved horizontally. Morse sought the assistance of his colleague, Leonard Dunnell Gale (1800-1883), professor of chemistry at the new university of the City of New York, where Morse had been appointed professor of arts and design in 1835. It was Gale who informed Morse about Joseph Henry's improved intensity electromagnets which used many turns of insulated wire.

By 1837, Morse had invented his relay system consisting of a local circuit with a local power source which could repeat the signal from one circuit to the next. Morse assured Gale, whom he had made his partner, that if he can telegraph through 10 miles of wire, he can telegraph round the world. Gale set to work on Morse's electromagnet and battery to make them similar to electromagnets and batteries of Henry. By November of 1837, Gale could send signals through 10 miles (16 km) of wire. A visit in the late fall of 1837 by Alfred Vail (1807-1859), a former student of Morse and a graduate of New York University who was a mechanically talented young man and the son of the owner of Speedwell Ironworks in Morristown, New Jersey [which later became Baldwin Works], brought Morse a new and very capable collaborator and partner. According to agreement all inventions of the 3 men were to be assigned to Morse. Vail was impressed by Morse's relay concept and instrument, and he was willing to build the telegraphic instrument after Morse's design at his own expense, and pay the cost of patenting the invention. Vail, after joining the partnership, took an active part in designing the telegraphic equipment. In 1838, the new instrument, the armature of which now moved vertically making dots-and-dashes on the tape like in the Dyar's electrochemical telegraph 10 years earlier rather than making the horizontal zig-zag lines, could operate over 3 miles (4.8 km) of wire. It was apparently Vail who worked out the so-called 'Morse code' which was at first recorded by indentations on a paper strip, since only later an inked wheel was introduced for marking inked dots and dashes. The recording instrument of Morse clicked when it was recording and the telegraph-operators soon learnt to read the messages by sound without reading the tape. This disturbed Morse who was a purist, for telegraph to him was 'distant writing' which in his opinion was meant to be read, but despite all his efforts to the contrary he could not change this practical trend. Soon all messages were read by the clicking sound of the telegraphic equipment, and finally Vail designed a 'Morse sounder' for reading messages in 'Morse code' by ear. In 1844, Morse and his partners laid a telegraph line from Baltimore to Washington, a 40-mile (64.4 km) stretch financed by the Federal Government, which had the transmission speed of 30 letters a minute. The construction of telegraph lines continued at a steady pace, and already by 1861 telegraph lines had reached California.

A great advance of efficiency in the operation of the telegraph was achieved when the polarising principle was introduced into relays. The attractive force of the coil on the soft-iron armature was multiplied many times by placing a powerful permanent magnet near the armature so that the armature became magnetised by induction. It permitted double-current working of the relay circuit: the reversal of the current repelled the armature which not only made the controlling spring unnecessary but also increased appreciably the sensitivity of the relay. The most successful polarised electric relays was designed by the Siemens brothers, Werner and William, about 1858, a relay which was adopted worldwide.

The most important competitor to Morse's telegraph was the letter-printing telegraph. To meet this competition, Vail prepared a design for a 'printing Morse telegraph' in 1837, but it was never patented.

In 1846, an American inventor Royal Earl House (1814-1895) of Vermont constructed a printing telegraph the transmitter of which had a set of piano keys, one key for each letter. Each key produced a definite number of electrical impulses which at the receiving end turned a type-wheel to the appropriate letter that was stamped on a paper tape. House's telegraph which sent about 10 to 20 words a minute was used on short lines until it was replaced by a better printing telegraph patented in 1855 by David Edward Hughes (1831-1900), teacher of music in Kentucky who hailed from London, England. Hughes' instrument worked on the principle of synchronous movements of the corresponding components in the transmitter and the receiver. The movements of these components were maintained in synchrony by the setting of a vibrating spring so that it produced a musical note of a particular pitch. Although the speed of Hughes' printing telegraph was slower than that of House, it could operate on longer lines. The later modifications of Hughes' printing telegraph employed a piano keyboard like in the House's printing telegraph, and it was used in continental Europe until the turn of the century, just like Wheatstone's automatic printing telegraph was used in Great Britain. Hughes made a fortune with his printing telegraph.

In 1859, George M. Phelps of Troy, New York, combined certain features of the House and the Hughes instrument to produce the Phelps Combined Printing Telegraph, which initially sent 30 words a minute but improvements soon doubled this speed. It was a rather expensive and complicated machine which became successful on telegraph lines under heavy use.

Multiplex Telegraphy

The use of the telegraphy increased by leaps and bounds for the balance of the 19th century, and the urgent need arose to send a number of messages over the same line at the same time. The first such telegraph system was the duplex telegraph devised in 1872 by Joseph B. Stearns of Boston, Massachusetts, by adding a capacitance to the ordinary telegraphic system. Thomas Alva Edison designed the first practical quadruplex telegraphic system in 1874 by a similar approach. The disadvantage of such methods was the reduction in the speed of telegraphic transmission.

Another method of increasing the transmitting capacity of telegraph systems is 'multiplexing' the telegraph line by means of which a single line can be used to transmit messages between a number of stations. The multiplex system of telegraph was first proposed in 1852 by Moses Gerrish Farmer (1820-1893), a prolific American inventor. Farmer's method consisted of setting up commutators at each end of the line in such a manner that the motion of the brushes in the commutators was synchronised so as to connect corresponding stations at each end of the line.

Patrick B. Delany and Bernhard Meyer made the first attempts to devise a workable multiplex telegraph system, but it was J.M. Émile Baudot (1845-1903), a French engineer, who made the first practical multiplex system in 1878. The high speed of transmission of the Baudot multiplex system was mainly due to the Gauss-Weber 5 unit, or quintic code which Baudot used to replace the 2 unit, or binary code of Morse. The Baudot system was able to send 160 dispatches at the same time, and it rapidly spread in France in the 1880's, and was introduced in England in the 1890's. Subsequent improvements in the Baudot system, and its combination with other systems, ultimately led to the modern printing telegraph system.

Electrochemical Cells and Electric Storage Batteries

The operation of the early electromagnetic telegraph which functioned by relays required a constant source of light electric current. Since electrochemical batteries were suitable power sources of such light current, the telegraph brought on a rapid development of electrochemical cells as sources of continuous light electric current. Surprisingly, the first known electrochemical battery, which was unearthed in the house of a magician-physician near the ancient city of Babylon in 1936, is very old dating back to the first century A.D. when this region of Mesopotamia was under the domination of Parthians. Parthians were the ancient trouser-wearing expert horsemen of Iranian stock who became famous for their clever cavalry battle tactic, the so-called 'Parthian Shot', a cavalry ruse consisting of pretending to retreat at a gallop whilst shooting arrows backwards from horseback. The first electric storage battery consisted of a clay vase containing a copper cylinder, an iron stem covered with lead, and pitch. When acids or alkaline solutions were poured into the vase, it produced an electric current of ½ volt potential. This electric battery was a remarkable achievement, and apparently its light electric current was used to alleviate physical pain.

The modern development of the electrochemical cell had already begun with Alessandro Volta's invention of the so-called 'Volta Pile' before the electromagnetic telegraph existed. The original Volta Pile for the time being was the only available source of continuous electric current, but its life as a source of electric current was quite short. In order to extend the life of his battery, Volta invented his 'crown of cups' battery, consisting of a number of cups filled with a saline solution and containing in each cup dissimilar metals which were joined with different metals in the adjoining cups. This new source of electric current was, however, a much bulkier equipment than the Volta Pile.

In order to increase the power of such electric batteries, an Englishman, William Cruickshanks, devised his trough-battery in 1801, in which 60 pairs of 1½ inches (3.8 cm) square zinc and silver plates were cemented in a trough in such a manner that the zinc plates faced one way and the silver plates the other way. The cells formed by these metal partitions were filled with a diluted solution of ammonium chloride. William Pepys, in 1803, and Humphry Davy (1778-1829) in 1807 built more powerful trough-batteries with different solutions and different metal plates. English scientist William Hyde Wollaston (1766-1828) suggested the folding of the electrodes into U-shape in order to increase further the power of the trough-battery by replacing the customary wooden trough by a copper trough, which served as one of the electrodes, whereas electrodes of the other metal were inserted as partitions in the trough. This construction considerably increased the heating and sparking power of the battery.

Since the trough-battery, which lasted several weeks instead of just a few days as was the case with the Volta Pile, were difficult to clean once its power had declined, the 'plunge battery' was invented in which plates of one metal were connected together, and plunged into a trough containing the solution and the plates of the other metal.

Unfortunately, the chemical reactions of polarisation and the local action in the trough of the plunge-battery made the life of such electric batteries quite limited. In 1826, the French physicist Antoine Cesar Becquerel (1788-1878) explained the 'polarisation' phenomenon in the Voltaic cell as consisting of the hydrogen freed by the electrolyte which migrates to the copper serving as the positive pole called 'anode', where it creates an undesirable barrier to the flow of electric current that soon ceased to flow thus making the electric battery useless. The 'local action' in the electric battery was due to the impurity of the metal which created a local current that flowed through the electrolyte and the electrode thereby preventing the affected portion of the electrode from contributing to the output of the cell. As much as ¾ of the available energy of the zinc electrodes might be wasted in such local action. The local action problem was soon overcome, when in 1832 K.T. Kemp and William Sturgeon amalgamated the zinc plate with mercury, after the relation between the amalgamation and the local action had been explained by the physicist Auguste Arthur de la Rive (1801-1873) in 1830.

However, the polarisation was a much more difficult problem to overcome. The first attempt to overcome the polarisation in the battery was made by Antoine César Becquerel in 1829, but his solution was not practical. His son Edmond Alexandre Becquerel (1820-1891) was successful in building the first photo-Voltaic cell, also called the 'photoelectric cell' or the 'solar cell' in 1839, which directly transforms light radiation into electricity. The photoelectric cell, which is now made of the conductor and the semiconductor joined together, has assumed great importance in the modern communication technology in space. The first practical solution to polarisation was given in 1836 by the English chemist John Frederic Daniell (1790-1845), who had proven in 1831 that it is metal that travels to the 'anode' in electrolysis. Daniell adopted the 2-fluid principle advanced by Humphry Davy in 1801 which Davy had used to demonstrate that electricity was produced by chemical oxidation rather than by physical contact, an idea Becquerel had also used in his electric cell. Daniell based his design on Faraday's electrochemical theories, and developed the first self-depolarising electric cell. He immersed the zinc plate in dilute sulphuric acid and the copper electrode in a separate solution of copper sulphate. The two solutions were separated by a porous membrane, an ox gullet, which in later practice was replaced by an unglazed porcelain cylinder introduced by John Gassiot. Despite the fact that the 'Daniell cell' had a limited current and that its potential was a mere 1.1 volts, this voltage was so constant that it was used as a standard well into the 1870's. In Daniell cell the current from zinc to copper made the released hydrogen gas to combine at the separator with the copper sulphate to produce metallic copper which deposited onto the copper plate. The zinc plate in the form of cylinder stood in dilute sulphuric acid contained in an outer vessel which surrounded the porous porcelain container.

A number of investigators sought to improve the Daniell cell because of certain of its defects. In 1838, the English physicist William Robert Grove (1811-1896) devised a 2-fluid cell in which the zinc electrode was immersed in dilute sulfuric acid, and the platinum electrode in strong nitric acid contained in a porous ceramic pot serving as the membrane separator. In the original 'Grove cell', the hydrogen gas produced by the cell was allowed to form at the platinum electrode where it was instantly oxidised into water. The Grove cell gave a potential of 1.9 volts, almost double the potential of the Daniell cell, and owing to the Grove cell's low internal resistance it yielded electric currents as strong as 10 amperes. The disadvantages of the Grove cell were that it gave off very corrosive fumes, and was very expensive to fabricate.

Robert Wilhelm Bunsen (1811-1899), the famous German physical chemist, modified the Grove cell by replacing the platinum electrode with a charcoal rod, and the nitric acid with the fuming nitric acid. The 'Bunsen cell' had a slightly smaller voltage than the Daniell cell, but its current was more than twice the current of the Daniell cell, and its cost was also much less. Although the Bunsen cell also gave off strong fumes, it came into wide use.

The Daniell and the Grove primary cell avoided polarisation by employing 2-fluid solutions in the cell. However, other non-polarising cells were also invented which used single-fluid solutions. The first successful single-fluid cell was constructed in 1840 by an English physician Alfred Smee (1818-1877) in which a pair of amalgamated zinc plates were immersed in a dilute sulfuric acid with a platinum (later replaced by silver) plate covered with finely divided platinum. The 'Smee cell' had a simple construction, and despite the fact that its potential was only ½ a volt, its maintenance cost was low and it could be used in open circuits where the drainage of current was very low.

Other single-solution cells were devised by Bunsen in 1841 and by R. Warrington in 1842 which avoided polarisation by employing zinc and carbon electrodes in a bichromate and sulfuric acid solution. The typical form of the bichromate cell consisted of a glass bottle with a flat upper part and a bulbous bottom. An ebonite cap suspended two long flat carbon electrodes with a movable zinc plate supported by a metal rod placed between the carbon electrodes. In this cell the hydrogen gas is eliminated by a reduction of the chromic acid that visibly changed the colour of the solution. This type of chromic acid cell was invented by the Frenchman Grenet. The electrolyte used was bichromate of potash in sulphuric acid, or chromic acid. In 1842, the German physicist Johann Christian Poggendorff devised a chromic acid cell which yielded about 2 volts, and due to its internal resistance it gave a strong current over a short period of time.

In the 1850's, Cromwell Fleetwood Varley (1828-1883) in England, and in 1858, Heinrich Meidinger in Germany developed several simple forms of the Daniell cell which used 2 fluids of different densities by virtue of which their mixing was prevented when one fluid was laid on top of the other. These cells were popularly called the 'gravity cells'. The simplest form of the gravity cell was devised by the Frenchman A. Callaud, and a particular type of it, called the 'crowfoot cell', remained in use until quite recently.

Batteries of most of these cells were used to provide electric current to the early closed circuit telegraph and railroad signalling systems. The gravity cell was particularly popular in providing such light current.

After 1850, when the availability of the electric current became important in many industrial applications it became important to store electrical energy, and the development of a more rugged storage battery began in 1859 with Gaston Planté (1834-1889), a French physicist, when he compared the polarisation produced by solid films on electrodes of various metals. Planté discovered that lead electrodes produced a more intense and longer lasting secondary current than electrodes made from other metals, and he converted the disadvantage of polarisation into an engineering advantage by the creation of the first 'storage cell' as a secondary cell, by using solid electrodes. Planté discovered that when a current was passed through a dilute sulphuric acid between 2 immersed lead plates rolled into spirals, oxygen was freed at the 'anode', and combined with the lead to produce a brown deposit of lead peroxide (PbO_2) on the surface of the lead. The hydrogen bubbles released at the cathode rose to the surface of the electrolyte and escaped into the air. If such cells were disconnected from the power supply, and then connected to a new circuit it acted like a primary cell and discharged the stored charge which it had received from the power source. It was a definite improvement. In 1873 Clark perfected his standard zinc-mercury cell which further contributed to proper understanding of batteries. The 'Planté storage cell' which gave an almost constant potential of 1½ volts was revolutionised by another Frenchman, Camille A. Faure in 1881. He avoided the slow and expensive process of Planté by coating the anode with the paste of red lead oxide (Pb_3O_4) and sulfuric acid, and the cathode with litharge (PbO) thus increasing the capacity of the storage cell. When the cell was charged by electrolysis in dilute sulfuric acid, the paste on the anode was oxidised into lead peroxide (PbO_2) and the paste on the cathode became a spongy lead. The 'Faure storage cell' produced 2 volts and had a more stable operation than the 'Planté storage cell'. Improvements to the 'Faure storage cell' followed almost immediately, the most important of which was the grid-form of the lead anode with the red-lead paste pressed mechanically into the grid holes, the perforations of the cathode plate to permit the metallic lead to have the largest surface area in contact with the acid, and a more rugged construction. Faure's open gridwork for battery plates considerably reduced the weight of storage batteries. In 1888 Correns designed the present form of the battery grid.

Secondary cells employing acid electrolytes were difficult to handle, which motivated some inventors to resort to alkaline electrolytes. French inventors Felix de Lalande and Georges Chaperon developed in 1883 an alkaline form of 'Leclanché storage cell' which had iron or copper for one electrode and zinc for the other, and used copper oxide as a depolariser and caustic soda or potash solution for the electrolyte. It gave a potential of about 1 volt, but owing to the low internal resistance of this cell it produced a stronger current.

About 1908 Thomas Alva Edison (1847-1931) made another type of alkaline storage battery in which nickel-plated steel electrodes were coated with nickel peroxide and graphite to form the 'anode', and the finely divided iron and graphite formed the 'cathode'. These electrodes were immersed in caustic potash electrolyte. The 'Edison storage cell' produced a very strong current, even stronger than in the 'Lalande-Chaperon cell', which made it practical as a power source for electric traction, and it had from 1 to 3 volts potential. In 1914, Edison produced a 'basic nickel-iron and nickel-cadmium storage cell'. The efficiency of energy storage in alkaline battery is several times better than in the most common lead-acid battery.

The 'dry cell', a battery in which the electrolyte is in a state of humid paste, began with the 'Leclanché cell', which was invented in 1866 by the French railroad engineer Georges Leclanché (1839-1882). The Leclanché cell had a solid depolariser, a zinc anode and an amalgamated carbon-manganese dioxide for the 'cathode'. The electrolyte was the saline sal-ammoniac (ammonium chloride) solution. A porous ceramic pot separated the depolariser from the sal-ammoniac solution in the cell-battery, the most common form of the Leclanché cell. Leclanché cell produced about 1½ volt potential, but owing to its high internal resistance its current strength was not too high. Leclanché cell was employed extensively in the batteries for telegraph and telephone lines, and other signalling systems, but its use had to be intermittent because of its relatively low current strength. It was a significant contribution towards a commercial electric cell. Leclanché cell was improved much later by the French physicist Charles Féry (1865-1935), and is still in use today.

The real precursor of the modern 'dry cell' was the 'modified Leclanché cell' produced in 1887 by a German inventor C. Gassner who used a paste of zinc oxide, sal-ammoniac, plaster, and zinc chloride as the electrolyte, and the zinc electrode formed the container of his storage cell. The commercial production of such 'dry cells' began in 1890.

The standardisation of voltages became increasingly important in electrotechnology after 1850. In the beginning, the voltage produced by the Daniell cell was used as a reference voltage, but in 1873 Englishman J. Latimer Clarke devised a more stable cell, the potential of which was reproducible to an accuracy of one-tenth of one percent. By the turn of the century, the cell of the Anglo-American inventor Edward Weston (1850-1936) patented in 1893, which was even more accurate than the cell of Clarke, began rapidly to replace the Clarke cell, and soon supplanted the Clarke cell altogether as a reference voltage cell. Weston replaced the electrolyte of the Clarke cell by a saturated solution of cadmium sulphate, and the zinc negative electrode was replaced by cadmium almagam, the depolariser was mercurious sulphate as before.

Submarine Telegraph Cables

It was a natural development in telegraphy that after expanding the telegraphic network over lands and continents, there arose the need to connect telegraph lines which stopped at coastlines. In 1811 Soemmering and Schilling had laid a trial line across the Isar river near Munich for Soemmering's electrochemical telegraph. Colonel Pasley of the Royal British Engineers had laid a line across the Medway in 1838, and William O'Shaughnessy had laid a line across the Hooghly river in 1839. These lines were only temporary and they ultimately failed to work.

Serious considerations of laying submarine cables began only in 1840, when a committee of British House of Commons held hearings on the feasibility of laying a cable from Dover to France. Wheatstone suggested a cable consisting of 7 copper wires each lapped with hemp twine saturated with tar, and he soon carried out a successful trial in Swansea Bay. Charles West proposed a rubber insulated conductor, which he laid out in Portsmouth Harbour, and successfully transmitted telegraphic messages through it before a large group of spectators, but he failed to secure financial support. About the same time Morse and Ezra Cornell (1807-1874) carried out their trials with submarine cables in the New York Harbour.

In 1843 a Spaniard José d'Alameida had discovered a natural gum, called 'gutta-percha', in the Orient, and a Scottish surveyor with East India Company, Montgomery, brought a sample of it to England. After Faraday had tested gutta-percha as an insulator in 1847, Charles Wheatstone proposed gutta-percha as a suitable dielectric of submarine cables, and a trial length of it was laid in 1849 by C.V. Walker, an electrical engineer of South Eastern Railway Co., in the shallow waters at Dover. Later in 1849 a similar cable was laid under the Connecticut River at Middletown.

In 1847 the German electrotechnologist Werner Siemens (1816-1892) laid a subterrainean telegraph cable insulated by the mixture of gutta-percha and sulphur between Berlin and Grossbeeren in Germany. In 1848 he devised a screw press machine for the production of the insulating gutta-percha cover for the copper wires of the telegraph cable, a machine which was generally adopted for the manufacture of insulated wires for subterranean and submarine cables.

Two brothers, Jacob Brett (1808-1897) and John Watkins Brett (1805-1863) secured an exclusive concession for a 25-miles (32.2 km) long submarine cable from Dover to Calais. In 1849 the two Brett brothers founded the English Channel Submarine Telegraph Company for the laying of a single-wire submarine telegraph cable under the direction of C. J. Wollaston at a depth of 30 fathoms (55 m) in 1850. Unfortunately, it remained operational for less than a day because of the damage done to the cable by a French fishing boat. The Brett brothers then manufactured another 4-wire submarine cable reinforced with an armature of 4 strong iron wires over the gutta-percha dielectric with the assistance of Thomas Russell Crampton (1816-1888). The second submarine cable was laid in 1851 and remained in operation for a decade despite corrosion and severe distorsion of the dielectric. The success of this cable encouraged other submarine cable schemes.

Another submarine telegraph cable line was laid in 1853 from Dover to Ostend, a distance of 70 miles (113 km). The laying of the submarine cable between Scotland and Ireland where the maximum depth was about 180 fathoms (329 m) was much more difficult, and only the fourth attempt in 1853 succeeded. It was a 6-core submarine cable, similar to Crampton's cable of 1851, and it remained in service for 20 years.

A submarine cable was laid to The Netherlands in 1853 and from France to Algiers by way of Spezia, Corsica and Sardinia where depths of 300 fathoms (549 m) and 1,200 fathoms (2,195 m) were encountered. A submarine cable connecting Sweden and Denmark was laid in the same year. In 1857 a submarine telegraph cable was laid successfully by Newall & Co. in collaboration with Siemens Brothers from Cagliari in Sardinia to Bona Island in Algeria, after Werner Siemens had built a paying-out machinery for the cable in the project. Also the submarine telegraph cable of 3418-mile (5500 km) length from Suez via Aden to Karachi at the Bay of Indus River in India was successfully laid from May, 1859 to January, 1860 by Newall & Co. in collaboration with Siemens Brothers.

After this initial success in the submarine cable-laying it was natural to try to connect the European and the American continent by a submarine cable, particularly after a hydrographer, the

American Navy Captain Matthew Fontaine Maury (1806-1873), had discovered a submarine plateau between Ireland and Newfoundland, which offered a natural advantage for the laying of a trans-Atlantic submarine telegraph cable.

In 1854 an English engineer Frederick Newton Gisborne (1824-1892), who was constructing a telegraph line to Nova Scotia, entered into discussions with a retired American entrepreneur in wholesale paper business, Cyrus West Field (1819-1892), to secure a 50-year exclusive concession for the landing of a trans-Atlantic submarine cable in Trinity Bay of Newfoundland in Canada, and linking it up with the landlines just laid. John Watkins Brett joined the enterprise to lay a single 85-miles (136.8 km) long submarine telegraph cable across the Gulf of St. Lawrence to connect the Canadian with the American telegraph network.

In 1856 Field convinced John Watkins Brett and Charles Tilson Bright (1832-1888) to join him in the organisation of the Atlantic Telegraph Company for the purpose of laying a submarine telegraph cable from Valencia, in Ireland, to Trinity Bay of Newfoundland, in Canada. It was an unprecedented engineering project for the depth of the ocean floor varied from 1,700 fathoms (3,109 m) to 2,400 fathoms (4,389 m), and the length of the cable was about 2,327 miles (3,745 km). Physicist William Thomson, (1824-1907) [later Lord Kelvin], was appointed one of the directors of the company, and his contributions to the submarine cable technology were decisive for the final success of this enterprise. Apart from the great mechanical problems of laying the cable, there existed serious electrical problems of transmitting and receiving electrical signals over a telegraph line thousands of miles long without having any relays in the circuit. None of the commercial telegraph instruments were able to operate over such a long telegraph line that acts like an enormous Leyden Jar which smears out the faint definition of electric signals. William Thomson finally solved the problem of detection of the faint transoceanic telegraph cable signals. In 1858 he invented a new, and very sensitive telegraph receiver, the so-called 'speaking galvanometer', a needle galvanometer with a mirror, and it could be used on ships because it was not affected by the rolling motion of the sea. Cromwell Fleetwood Varley showed in 1862 how the addition of condensers at both ends of the telegraph cable sharpens the cable signals and compensates for the loss of signal definition owing to its passage through the enormously long transoceanic cable. In 1867, Thomson patented his so-called 'syphon recorder' which replaced the needle galvanometer with a mirror recorder that was able to provide a permanent record of the telegraphic messages. In 1871 Thomson produced his final improved galvanometer which was so highly sensitive that it could detect a current issued from a tiny chemical cell after it had been sent across the Atlantic and back again.

In the middle of 1857, the first attempt to lay the cable was made, but unfortunately sufficient time for the manufacture of the cable as well as for the modification of the two naval ships, Agamemnon of 3,200-ton displacement of the British Royal Navy and Niagara of 5,000-ton displacement of the U.S. Navy, for the proper laying of the transoceanic telegraph cable had not been allowed. The 2 ships were to begin laying-out the cable from both sides of the Atlantic, but the cable broke in a storm in August 7, 1857 after 335 miles (539 km) of cable had been laid owing to an accident in the paying-out machinery. Subsequently, the paying-out gear together with the 80-horsepower trunk engine designed by John Penn (1805-1878) driving the paying-out machinery was redesigned, and trials of it were carried out in the Bay of Biscay at 1,800 fathoms (3,292 m). The first cable was about 5/8 of an inch (1.59 cm) in diameter, and consisted of 7 copper strands in the conducting core which were packed in 3 thin gutta-percha layers that were protected by a tarred hemp sheath and reinforced by 18 strands of iron wire. In 1858 the second attempt to lay the cable was made by splicing the broken cable, but again the cable broke several times. Despite these mishaps the cable laying was completed by August 7 in the same year, but on September 1 the messages of the trans-Atlantic cable became weak and on October 20 the cable ceased to function at all. By that time about 700 telegraphic messages had been transmitted by the cable.

In the next 7 years a number of scientists and engineers worked on the cable to improve its insulation and mechanical strength, the receiving instruments and the strength of its signals, and the paying-out mechanism. Thomson's essential contribution to the electrical problems of the cable made the final success of the cable possible. This time Brunel's gigantic <u>The Great Eastern</u> of 22,500 tons displacement, the only ship in the world able to carry the entire cable made by the Telegraph Construction Maintenance Company weighing more than 5,000 tons, was rented in 1865 for the cable laying task, but after laying about 1,200 miles (1,931 km) of cable out of Valencia, Ireland, it snapped at 2,100 fathoms (3,840 m), and the work had to be abandoned for the rest of the year. In consequence, the Atlantic Telegraph Company floundered, and was succeeded upon its reorganisation by the Anglo-American Company. In 1866, a new, and this time, a successful attempt was made to lay the trans-Atlantic cable coiled in three 58-foot diameter water-filled tanks in the hold of <u>The Great Eastern</u> with Thomson and Varley aboard as scientific supervisors of the cable-laying. The old cable was also recovered and spliced thereby giving two operational trans-Atlantic cables. These 2 cables remained in use for a decade before they ceased to function, but by that time several other cables had been successfully laid. The new cable layed in 1865 was 1.10 inches (2.79 cm) in diameter and contained 3 times more copper in the conducting core bound by 9 layers of gutta-percha insulation. The outer shell of 10 substantial steel wires in pitch-treated hemp sheath protected the core of the cable. In 1866, the new cable was modified by omitting the hemp sheath because the use of steel wires galvanised with zinc made it superfluous.

The gigantic size of <u>The Great Eastern</u>, a ship 6 times as large as any other vessel afloat at the time, presented problems in the steering of the huge ship, problems which had never existed before. The vertical steering shaft to the rudder of <u>The Great Eastern</u> broke in the Atlantic storm because the helmsman lost control of the huge steering wheel on the bridge. In 1867, John McFarlane Gray (1832-1908) of Vauxhall Works in Liverpool invented the first successful steam-powered steering gear which was installed in The Great Eastern. The steam-powered steering gear made the huge ship perfectly controllable by means of a small steering wheel on the bridge. Superheating of steam was also installed in The Great Eastern to increase the power output of its steam engines. The white-metalled stern bearing of the great ship, which had suffered nearly 2½ inches of wear from 2 trips across the Atlantic, were replaced by John Penn's water-lubricated *lignum vitae* bearings that gave no further problems of wear in stern bearing.

In 1874, William Siemens of the Siemens Brothers company designed and built a special ship named Faraday for the laying of submarine telegraph cables in collaboration with William Froude (1810-1879). This ship was used in the laying of the 6th trans-Atlantic telegraph cable, the first submarine telegraph cable linking Great Britain directly with United States without going through Canada.

Telephone

In 1837 an American physician, Charles Grafton Page (1812-1868) of Salem, Massachusetts, discovered what he called the 'galvanic music'. Page observed that rapid changes in the electromagnetism of an iron bar produce musical notes of a pitch depending upon the frequency of the changes of magnetisation of the bar. This particular phenomenon was further studied in 1848 by a German physicist, Wilhelm Wertheim (1815-1861) in Paris, and these studies led to the examination of the inverse phenomenon of great practical value: the conversion of sound waves into electromagnetic vibrations, and then back into sound waves.

Since 1849, Charles Bourseul (1829-1912), a French telegraph clerk, had carried on studies of the inverse phenomenon, that of the telephone, for transmitting human voice by electricity, and published his investigations in 1854 in an article entitled '*Telephone électrique*'.

In 1860, a German high school teacher, Philipp Johann Reis (1834-1874), invented the first telephone and exhibited it more than 4 years in various cities in Germany. In 1863 Reis built an improved telephonic apparatus and gave a few demonstrations of it. The telephone of Reis became well-known as a result of these demonstrations, and his apparatus was reproduced in many countries [for instance, it was publically demonstrated in United States in 1869], and more than 50 articles were written about this particular simple experimental telephonic apparatus of Reis.

The telephonic instrument of Reis worked on the principle that an intermittent contact in a circuit can modulate the electric current flowing through the circuit. When in Reis' transmitter a diaphragm vibrated under the influence of human voice, a variation took place in the pressure of a metal point on a metal plate which, in turn, modulated the current in the circuit. The receiver of Reis' telephone consisted of a knitting needle inside an electromagnet which was attached to a resonator. The electromagnet gave the needle the rate of vibrations of the diaphragm in the transmitter, and the resonant box amplified the vibrations. Therefore, Reis' telephone was more like a rather crude, experimental toy than a practical telephone. Although Reis himself gave demonstrations in public by transmitting music by his telephone, he did inform F.J. Pisco in 1863, that he could also transmit words with his instrument.

An American inventor Antonio Meucci (1808-1889), an Italian immigrant, began to work on the telephone at the same time as Bourseul, but Meucci, unlike Bourseul, did succeed in building a successful experimental instrument. Already in 1852, Meucci was successful in transmitting speech electrically from one building to another in Havana. By 1860 Meucci had carried his development of the telephone so far that he was able to demonstrate his apparatus to the President of the New York District Telegraph Company. Meucci used vibrating metal membranes in his telephone, and he patented his instrument in 1871, an instrument which was quite similar to the instrument that Alexander Graham Bell later patented in 1876. Meucci was too poor to fight a legal battle for his right as the first inventor of a practical telephone but fortunately for him he did live to see Globe Telephone Company win a legal suit against Bell Telephone Company in which Bell's claim for originality in the invention of a practical telephone was refuted.

Since 1873, electrical engineer Elisha Gray (1835-1901), superintendent of Western Electric Manufacturing Company in Chicago, had been developing an instrument to change mechanical vibrations into electrical signals and back into vibrations. Gray also had been working on a transmitter of sound. One of his transmitters was based on the physical property that the friction between two frictional bodies changes with the application of voltage across the two bodies, and the other on the idea of Laborde that the mechanical vibrations produced in the armature of an electromagnet subject to an alternating current correspond to the frequencies of the current. After building his instruments, Gray gave a demonstration of his instruments to Joseph Henry in 1874, and also exhibited the same instruments in London. In 1875 Gray patented his instruments designed for his electric organ, harmonic multiple telegraph, and a printing telegraph. Gray also concocted a voice telegraph, and invented his so-called 'lover's telegraph' in which the transmitter was a cylinder at one end of which was a diaphragm with a light metal wire attachment, and the other end was immersed in acidulated water. The liquid and the wire were made part of the electric circuit with the battery and a receiver, and when the changing liquid to wire contact modulated the current produced by the battery, his receiver would duplicate this modulation in sound. In 1876 Gray applied for a patent on his harmonic telegraph, and at the same time he filed a caveat, an intention to patent his method of transmitting and reproducing speech, although he had not yet tested his transmitter experimentally. However, a few hours before Gray, a patent application for a telephone had been submitted by a Scottish inventor Bell.

Alexander Graham Bell (1847-1922) was a Scottish immigrant, a speech physiologist and a man of great refinement, who had been interested in developing his harmonic multiplex telegraph, which he finally patented in 1875, after demonstrating how 2 messages could be sent simultaneously over 200 miles (322 km) of telegraph line. Bell was well informed and had studied Helmholtz's famous treatise on sound reception, and he had seen the telephone of Reis during his visit to Joseph Henry. Bell set to work on the telephone with his Scottish assistant, Thomas August Watson (1854-1934), and from his attempts to send 3 messages over his multiplex telegraph which employed magnetised reeds, Bell discovered that a magnetised reed could actuate another one without a battery in the circuit. Watson built new instruments incorporating a magnetised reed attached to a membrane diaphragm, in which the reed itself acted as an armature to an electromagnet. In spite of the fact that Bell was able to transmit sounds and changes of pitch but unable to transmit articulated speech with his instrument, he submitted his patent application in which he claimed the basic method of telephony. His patent claims were so extensive that he had to defend against 600 patent suits, all of which he was able to refute in the court of law. In the middle of 1876, Bell constructed another type of transmitter which modulated the battery current and, therefore, was similar to the transmitter of Gray. Bell exhibited his multiplex telegraph and reed-membrane telephone at the Philadelphia Centennial Exposition in 1876 which was attended by William Thomson, but his telephone was weak and inefficient. Late in the same year Bell replaced the steel reed by a steel plate glued to the membrane, and Watson replaced the soft iron core of the electromagnet with a compound permanent magnet which finally made sustained conversation over a distance of several miles possible. Bell then replaced the membrane by an all-metal diaphragm fixed around its edge to produce his box telephone. After all his extensive efforts, Bell had been able to improve his receiver, but not his magnetotransmitter which was still too weak in his system of telephony, for it was necessary to shout into Bell's microphone to be heard at the receiver.

Rival companies of Bell had purchased the telephone patents of Gray and Amos Emerson Dolbear (1837-1910), professor of physics at Tufts College, who had invented an electrostatic telephone in 1881, and hired Thomas Edison to design a more powerful telephone transmitter. In 1877, Edison was successful in producing a microphone transmitter which was packed with carbon granules and supplied with strong current. The packing of the carbon granules was affected by the pressure exerted by the vibrating diaphragm set into motion by the speaker's voice. A variation in the packing of the carbon granules created a variation in the electric conductivity of the entire mass of packed carbon which modulated the current supplied to the microphone and transmitted a more powerful signal to the receiver than other competing microphones. Edison's telephonic transmitter was a rugged instrument and much better than Bell's magnetotransmitter. Shouting into the microphone required by Bell's transmitter was no longer necessary with Edison's transmitter. David Hughes had independently suggested a carbon pencil transmitter in 1878, but because his laboratory model of it was erratic, he did not patent it. Edison's microphone was so effective that it was still in use in the 1950's.

Bell bought the microphone patented in 1877 by a German-American inventor, and the chief inspector at the Bell Company, Emil Berliner (1851-1929), which they used mainly for their claim of priority since few commercial telephones were manufactured using Berliner's microphone that had metal electrodes. After15 years of litigation Edison retained the patent rights.

In 1878, Francis Blake (1850-1913) invented his telephonic transmitter in which a platinum bead attached to the membrane pushed against a carbon block, and it was promptly purchased by the Bell Company, but the practical use of this transmitter gave some difficulties until Emil Berliner showed how a harder carbon eliminated this trouble. An improved Blake's transmitter, patented in 1879 in United States, was extensively used by the Bell Company. An English clergyman Henry Hunnings replaced the single carbon in Blake's transmitter by coke granules in 1878, which made the telephone transmitter very efficient, and it was bought by the Bell Telephone Company in 1881, the year Bell himself retired from the firm. In 1880, the Bell Company was able to obtain full rights of the Western Union telephone patents by a financial arrangement and in 1887, the Supreme Court gave the Bell Telephone Company complete control of telephone business for the remaining duration of the Bell patents.

The modern telephone transmitter evolved from the works of Edison, Hunning and White. After 1890, there were no further improvements in the modulation or the reception of telephone signals for more than 50 years.

In 1899, the Danish engineer Valdemar Poulsen (1869-1942) invented the magnetic tape-recorder for message-taking in telephonic devices. In 1900, Poulsen's magnetic speech recorder, named 'Telegraphone' which recorded telephone messages on a moving magnetised steel tape or steel wire was successfully demonstrated at the Paris Exhibition. The steel tape slowly passed between the poles of an electromagnet connected in series with a telephone circuit, and received there from a series of transverse magnetisations corresponding to the sounds converted by a microphone into an electric current which flowed through a coil in the electromagnet. This magnetic track of speech was fixed on the tape, which through the reversal of the process, consisting of passing the magnetised tape between the poles of a magnet connected in series with a telephone receiver, was able to reproduce the speech.

In 1906, Poulsen was able to improve the design of the apparatus of his magnetic sound recorder by making it similar to a cylindrical phonographic recorder on which the electromagnetic head travelled longitudinally like a stylus recording or reading the magnetised steel wire wound spirally on a drum. The sound to be recorded was converted by a microphone into an electric current which flowed through the electromagnetic head that recorded the sound by changing the magnetisation of the wire on the drum. Reversing the process reproduced the sound. Demagnetisation removed the magnetic sound track. Poulsen also investigated besides wire other flexible materials covered with some kind of magnetisable powder in an anticipation of the modern recording tape which is a flexible plastic tape coated by a brownish iron oxide. Poulsen's tape-recorder had to wait for the application of electronic amplifiers and a flexible plastic tape coated with fine magnetic oxide particles, which was accomplished in Germany in the early 1940's, to make it into a really practical recording instrument. Poulsen also developed a wireless telephony in collaboration with the Danish physicist Peder Oluf Pedersen (1874-1941), professor of physics at the Technological University in Copenhagen.

CHAPTER XX

INTERNAL-COMBUSTION ENGINES: From Industrial Engine to Automobile and Aircraft Engines

Gas and Gasoline Engines

The steam-engine is an 'external-combustion' engine. The oldest known 'internal-combustion' engine is the cannon in which a controlled combustion of a quantity of gunpowder provides the power to propel the projectile. This type of heat-engine existed already in the Mediæval Age.

In 1666, the Dutch physicist and mathematician, Christiaan Huygens (1629-1695) suggested the use of a cannon-type heat-engine in which a cylinder open on top is fitted with a piston, and a small quantity of gunpowder is exploded under the piston to evacuate the space under the piston so that the atmospheric pressure can force the piston down in a power-stroke. In this scheme, the expanding gases from the explosion of gunpowder did no useful work in contrast to the cannon where they do. The experiments which Huygens carried out from 1673 to 1675 with his assistant Denis Papin (1647-1712) of Blois, and the later experiments of Papin in 1688, were not successful because the explosion was not able to evacuate the cylinder adequately. Moreover, its operation proved to be rather dangerous.

In 1791 an Englishman from Standsby, John Barber took out a patent, 'A Method of Raising Inflammable Air for the Purposes of Procuring Motion,' for an elementary exhaust-driven combustion turbine in which a mixture of inflammable gas and air was ignited and the hot expanded gases were used to drive a simple Branca-type impulse wheel.

A few years later an English inventor and varnishmaker Robert Street patented, and built the first 'reciprocating' internal-combustion engine which burnt a gaseous fuel. This engine operated on a cycle which began with the injection of a few drops of turpentine on a heated metal surface thereby vaporising it, then the piston drew air into the cylinder through a valve which mixed with the turpentine vapour, the mixture of air and turpentine vapour was ignited at the half-stroke of the piston that uncovered a flame. The ignited mixture of air and turpentine expanded after ignition and drove a heavy piston upwards in a working-stroke. Street had installed in his engine a cooling-jacket in the form of a water-tank surrounding the cylinder. He used his internal-combustion engine to operate a water-pump. The major shortcomings of the Street's engine were the lack of compression of the gaseous mixture before ignition which resulted in low power and poor efficiency, and the manual intervention required to raise the piston in order to draw more air into the cylinder because the engine could not run by itself.

In 1799, Philippe Lebon d'Humbersin (1767-1804), teacher of mechanical engineering at the **École des Ponts et Chaussées** in Paris and the inventor of gas illumination, modified a double-acting steam-engine in which the inflammable lighting gas mixed with air in the cylinder was electrically ignited, and the expansion resulting from the combustion of gases drove the piston in a power-stroke. The idea of igniting a fuel by electric spark had been proposed in 1777 by the Italian physicist Alessandro Volta (1745-1827). In Lebon's internal-combustion engine, the combustion of its fuel was not continuous like in the steam-engine, but intermittent, which made the method of ignition of the fuel one of the major problems of such engines.

Lebon, who was a well-educated engineer and chemist, designed a gas generator for the production of coal-gas which could be used both as an illuminating gas as well as a gaseous fuel for internal-combustion engines. Lebon began his attempts to produce illuminating gas from wood, oil and tar. In 1799 he patented his 'thermolampe', a miniature gas factory for domestic lighting consisting of a retort and a system of tubes conducting gas to 'fishtail burners'. Unfortunately, the candlepower of the 'thermolampe' was low, and besides it burnt with an unbearable odour.

The efficient and proper use of gas as an illuminant dates from 1886 when the Austrian chemist Karl Auer, Freiherr [Baron] von Welsbach (1858-1929), substituted an incandescent mantel, the so-called 'Welsbach mantle', for the 'fishtail burner' of gaslight. Welsbach discovered that a cylindrical fabric impregnated with thorium nitrate to which a small percentage of cerium nitrate was added produced a brilliant white glow in a gas flame, a discovery he patented in 1885. The 'Welsbach mantle' greatly improved gaslight which was only surpassed by Edison's electric incandescent light. However, Auer von Welsbach even improved upon Edison's electric light in 1898, for he was the first inventor to substitute metallic osmium filament for Edison's carbonised thread in the incandescent electric bulb. However, metallic osmium, a member of the platinum group, was too rare and expensive metal to be of practical value in the electrical manufacturing industry. But it prepared the way for the use of tungsten filament in electric light bulbs a decade later.

Auer von Welsbach made another discovery when he found that a metallic mixture of rare earth elements, consisting mainly of cerium, mixed with some iron was extremely pyrophoric, that is, when struck it gave off hot sparks. He was the first person to improve upon the neolithic invention of flint as a spark producer. Today the most prevalent use made of Welsbach's 'mixed metal' is as a substitute for flint in pocket lighters. In 1901, the versatile and ingenious Karl Auer was ennobled Freiherr [Baron] von Welsbach by Kaiser Franz Joseph I of Austro-Hungarian Empire.

In his design of the internal-combustion engine, Lebon examined several aspects of his engine such as the air-gas ratio, pumping the gas charge into the cylinder under pressure, spark ignition of the charge, and valves controlling the flow of gases. However, Lebon neglected the compression of the gas and air mixture in the cylinder before ignition to extract more thermal power from the combustion of fuel. In 1801 Lebon was granted a patent for spark ignition of gas engines. In 1804, Lebon was assassinated on **Champs Elyseés** in Paris, which stopped any further development of his engine, but fortunately Lebon's engine had aroused the interest of Niepce brothers in the internal-combustion engines.

In 1806, the French inventor Joseph Nicephore Niepce (1765-1833), the inventor of photography, and his brother Claude Niepce (1763-1828) built an atmospheric engine which worked by the internal-combustion principle. Niepce's internal-combustion engine was successful to the extent that it was able to propel small barges at twice the speed of the Saône river where the trials of the engine were conducted. Niepce's engine was actually a pumping-engine which drew in water from the bow and pumped it out at the stern of the barge thereby producing motion by reaction to the water expelled. The principle of boat propulsion by reaction had been first suggested by John Allen in 1724. The internal combustion principle of Niepce's pumping engine derived from Huygens' experimental internal-combustion engine which had not worked. Niepce used an upright cylinder of 12-inch (30.5 cm) bore which was open to the atmosphere on top and fitted with a piston. This atmospheric engine working by the internal-combustion principle in which small grains of solid inflammable fuel (lycopodium powder) were carried into the combustion chamber by a stream of air pumped by bellows, and then ignited so that the expanding hot gases from the explosion pushed the large piston to the top of the cylinder. In the returning power-stroke the atmospheric pressure pushed down the piston as well as exhausted the burnt gases. Niepce soon replaced the solid fuel with gasoline, a volatile liquid fuel. In this version of the engine, the bellows pumped the gasoline vapour into the combustion chamber with a stream of air resulting in an explosive gas mixture. Although the French engineer and General Lazare Carnot (1753-1823) reported favourably on the Niepce engine to the French Academy of Sciences in 1807, and recommended its use as a pumping-engine to replace Marly's pumping-engine, this engine was commercially not successful, and as the Niepce brothers had spent most of their family fortune on these experiments, they abandoned work on their engine after a few years of experimental success.

Isaac de Rivaz, a Swiss inventor, made considerable progress with the internal-combustion engine in its application to automotive use. In the period from 1802 to 1813 de Rivaz designed a vehicle with an upright cylinder open to the atmosphere on top and fitted with a piston. The portable gas supply carried in a leather bag was injected into the cylinder under the piston and fired by an electrical ignition system which made the piston move up towards the open end of the cylinder by the pressure of the burning gas. The downwards motion of the piston due to the atmospheric pressure and partial vacuum under the piston constituted the power-stroke of the engine, which was transmitted to the wheels of the vehicle by a series of chains, pulleys and ropes.

In 1809, English inventor George Cayley (1773-1857), who had created singlehandedly the incipient science of aerodynamics, gave a clear description of how a gas engine should work, after learning of the existence of the internal-combustion engine of Lebon. Cayley was interested in a small, compact engine as a power source for his heavier-than-air aircraft with fixed wings.

In 1817, an English parson, William Cecil, constructed an experimental model of an atmospheric gas-engine, which ran on a mixture of hydrogen gas and air, and carried out systematic experiments on its improvements. The model atmospheric gas-engine of Cecil worked rather well at 60 rpm, but he was unable to solve the problem of the huge consumption of hydrogen gas by the engine, about 17.8 cubic feet (0.5 m^3) an hour and, therefore, was ultimately forced to abandon it.

In 1823, an English naval engineer and inventor Samuel Brown (1776-1852) took out a patent on another type of atmospheric internal-combustion engine working by gas, which was somewhat like the atmospheric internal-combustion engine of the Niepce brothers, and he built several such practical water-cooled engines in which the water was circulated round the jacketed cylinder by a pump, and recooled in contact with the surrounding air. The water-cooling system of an internal-combustion engine was essentially Brown's invention. In 1826 Brown designed an ingenious carriage driven by a modified form of his patented gas-vacuum engine, and he demonstrated its ability to climb the steep Shooters Hill in London. He formed a company primarily for the exploitation of his gas-vacuum engine in the propulsion of canal boats, but it became apparent that Brown's engine was much less efficient, and used much more fuel than a steam-engine. His engine had a double-acting piston in the form of a float on water, and an external gas burner to ignite the gas mixture in the cylinders. When it became known that Brown's 2-cylinder engine for the carriage developed only 4-horsepower despite the 12-inch (0.30 m) bore of its cylinders and 24-inch (0.61 m) stroke, and that there was no practical way for the vehicle to carry the very large quantity of gas it used as fuel, the financial support for Brown's engine was no longer forthcoming from his backers, and as a consequence his company failed as a commercial enterprise.

At the time Brown was working on his gas-vacuum engine, Samuel Morley of Oxford, New Hampshire, was experimenting with his atmospheric internal-combustion engine in United States, the first internal-combustion engine in America. Morley, who had many years of experience with steam-engines, made a number of significant contributions to the development of internal-combustion engines. He introduced cam-activated poppet valves, a base-mounted crankshaft, and a novel idea for the fuel of the engine by vaporising the liquid fuel instead of carrying a large tank of ready-made gas, a concept which incorporated the earliest known carburettor. His carburettor had no moving parts but rather consisted of a box containing a labyrinth through which was passed the vapour of the liquid fuel produced by heating, and combined into an air-gas mixture by virtue of the turbulence created by the passage of the gases through the labyrinth. The power-stroke in Morley's engine was executed by the atmospheric pressure.

In 1833, an English inventor, Lemuel Wellman Wright built another internal-combustion engine that functioned by double-action, and was outfitted with a water-cooled cylinder in which the gas was ignited by a burning gas jet. It was the first internal-combustion engine in which the exploding gases drove the piston in a power-stroke. It was similar in design to the high-pressure steam-engines of the period, and an improvement upon the engine of Brown, but it was still not practical enough to go beyond the experimental stage. Both of these engines worked with the mixture of illumination gas and air as the fuel of the engine.

In 1838, an English mechanic, William Barnett introduced the essential method of compressing the fuel-air mixture in his engine before ignition to recover more of the available energy in the subsequent expansion of the burning gases. It was a 2-cylinder engine from which it was difficult to exhaust the burnt gases, and achieve the desired compression. Barnett suggested the use of volatile hydrocarbon as the liquid fuel for internal-combustion engines. This engine had an ignition system in which a steady flame lit an intermittent flame that ignited the precompressed fuel mixture in the cylinder at fixed intervals as part of the operating cycle of the engine. In 1843 another internal-combustion engine was patented in United States by Dr. Drake.

The gas engine of the Italian engineers Eugenio Niccolò Barsanti (1821-1864) and Carlo Felice Matteucci (1808-1887), patented in 1854, was the first engine which avoided the direct jolting action of detonation of the fuel on the driving force of the engine by incorporating a free-flying piston which was moved by the engine, but this engine was not a practical success.

In 1857, the English engineer John Cockerill (1790-1840) of Seraing, built an atmospheric gas engine, but again it proved not to be a practical success.

All these inventors, beginning with Huygens and Papin, intended to devise an engine which would burn the fuel inside the cylinder instead of burning the fuel outside the cylinder to heat an intermediary working substance which is also outside the cylinder, such as steam in the steam-engine, and, thus do away with the wasteful furnace, boiler, condenser, piping, valves and the great heat loss.

The first gas-engine patent was granted in Great Britain in 1794, and such engines were used decades later for pumping. In 1851, the Great Exhibition in Crystal Palace in Hyde Park, London, displayed only a single gas-engine invented by Dr. Drake in 1843 among the exhibits from United States.

The first commercially successful internal-combustion engine was the gas-engine of Jean Joseph Étienne Lenoir (1822-1900), a successful inventor and self-educated mechanic from Luxemburg who worked in Paris, and whose electroplating method for enamelling plates had made him financially independent. Lenoir's gas-engine employed a 2-stroke cycle modelled after the double-acting steam-engine and the gas-engine of Philippe Lebon. When it came on the market in 1860, it was the first engine of any kind besides the steam-engine able to work in industrial surroundings. It consisted of a single horizontal cylinder, a connecting-rod, a centrifugal governor, and a flywheel. In Lenoir's engine the illuminating gas [coal gas] took the place of steam, and there was a combustion chamber on both sides of the piston. As the piston moved towards one end of the cylinder a combustible mixture of air and coal-gas was sucked into the cylinder and ignited at the mid-point of the stroke by a spark delivered from a Ruhmkorff induction coil by means of something like a spark-plug and a battery made of Bunsen cells. Therefore, only half of each stroke of the piston was a power-stroke in Lenoir's engine. On the return-stroke the exhaust gases were expelled whilst a new charge of the explosive air-and-gas mixture was drawn into the other side of the piston. Lenoir used sliding valves in his gas-engine like in the steam-engine, and cooled the cylinder with water. Its action was very rough for each explosion was accompanied by a violent shock through the engine, a problem Lenoir tried to ameliorate by means of springs, and other devices. It was a massive engine which consumed 100 cubic feet (2.83 m^3) of expensive fuel to produce a power output of only slightly over one horsepower an hour at 100 rpm, that is, about 4% of the potential thermal energy in the coal-gas. The chief reason for such a low efficiency was that the air-gas mixture remained uncompressed so that at ignition the fuel charge was approximately under the atmospheric pressure, and, therefore, most of the potential energy in the fuel was not converted to work. Consequently the expansion of the burning gas remained incomplete thereby giving high exhaust-pressure and, as a result, Lenoir's gas-engine badly overheated and used up great quantities of cooling water, about 100 gallons (378.5 liters) for a ½-horsepower engine and 350 gallons (1,324.9 liters) for a 3-horsepower engine in an hour. Lenoir's gas-engine consumed about 100 cubic feet of expensive illuminating gas per horsepower-hour. Despite these shortcomings, Lenoir's gas-engine was the first internal-combustion engine which went successfully beyond the experimental stage, and was able to perform useful work in industry. Between 1860 and 1865 Lenoir was able to sell about 400 of his engines to factories and workshops which were too small to afford a large steam-engine. All Lenoir engines had small capacity, from ½ to 3 horsepower.

The principle of the induction coil had been discovered in 1831 by the English self-educated scientist Michael Faraday (1791-1867) in his self-induction experiments. In 1836, a French physicist Antoine Masson was able to produce high-tension currents by inducing rapid interruptions in normal battery current. In 1841, Masson and another French physicist Louis Breguet (1804-1884) built an induction coil which produced electrical discharge in rarefied gases. The induction coil was perfected in 1851 by the German instrument-maker Heinrich Daniel Ruhmkorff (1803-1877) which found general use in producing sparks in electrical equipment of various kinds, particularly in ignition systems of gas-engines.

In 1862 Lenoir also experimented with liquid fuels. He used the same 'carburation' method employed previously by Donovan, an English industrial chemist. Donovan constructed a device for the 'carburation' of coal-gas through evaporation by guiding it over the surface of a volatile liquid hydrocarbon. Lenoir used surface evaporation by guiding air over the volatile liquid fuel. These 'surface carburettors' were very large and worked well only with very volatile petroleum products.

In 1863, Lenoir installed one of his gasoline engines in a crude, wagon-like vehicle which ran from Paris to Joinville-le-Pont, a distance of 9.3 miles (15 km) in 2 hours, but even at that time it was not an impressive accomplishment.

After Lenoir's initial success in overcoming all practical difficulties in the production of a workable internal combustion gas-engine, other inventors attempted to improve his gas engine. In 1862, Pierre Hugon produced his gas-engine in which he tried to eliminate the overheating problem of the Lenoir gas-engine. He provided additional cooling to his gas-engine by injecting a fine spray of water into the cylinder. This modification did reduce the gas consumption, and did lower the temperature of the exhaust gases, but the main faults of the Lenoir gas-engine remained: the lack of compression of the air-fuel mixture, the incomplete expansion of the hot gases, the excessive heat loss through the wall of the cylinder, and the heavy use of lubricating oil.

Much later, in 1883 when the gas-engine of Otto had forced Lenoir's engines out of the market, Lenoir produced a 2-crank, 4-cylinder engine with enclosed crank chambers for marine works, and in the same year a 4-stroke gas-engine, but they were not commercially successful and could not save his business.

Lenoir was not the only successful gas-engine builder of the period. A German watchmaker, Christian Reithmann (1818-1909) in Munich, had been working since 1852 on a gas-engine independent of Lenoir. In 1858, he demonstrated a small practical double-acting experimental gas-engine, and in 1860, the same year as Lenoir marketed his gas-engine, Reithmann secured a local patent for his gas-engine in Bavaria. Reithmann used his small gas-engine to drive all the small machines in his workshop. He established an engine factory in 1867 but it was shortlived. Reithmann was dogged by bad luck: his gas-engine which was on exhibition at the Munich *Oktoberfest* broke its crankshaft, and his attempt to obtain a Prussian patent for his gas-engine was also unsuccessful. From 1872 to 1873, Reithmann reconstructed his 2-stroke gas-engine, and gave it a 4-stroke cycle. This flame-ignited, 4-stroke engine ran at 200 rpm and delivered ¾-horsepower. Reithmann's engine, which is preserved in the **Deutsches Museum** (*German Museum*) in Munich, unfortunately exerted no influence on the development of the internal-combustion engine, although the famous 4-stroke 'Silent Otto' gas-engine came 4 years after Reithmann's gas-engine which worked with a 4-stroke cycle.

In 1861, the inefficiency of the Lenoir gas-engines was critically examined and discussed by the Austrian mechanical engineer Gustav Johann Leopold Schmidt (1826-1883), professor at the *Technische Hochschule* (*Polytechnical Institute*) in Prague, who stressed the critical importance of the compression of the air-gas mixture before ignition, an idea first put forward by William Barnett in 1838, as an absolute necessity for the conversion of most of the thermal energy within the fuel into useful work.

In 1862, Adolphe Eugène Beau de Rochas (1815-1893), an eccentric French engineer who worked in technical services department of the Railroad du Midi, and after 1860 for a manufacturer of such physical equipment as telegraph apparatus, published a handwritten pamphlet on the improvement of high pressure locomotive design entitled <Researches on the Practical Conditions for the Greatest Utilisation of Heat, and, in General, of the Motive Force> (***Recherches sur les conditions pratiques de la plus grande utilisation de la chaleur, et, en géneral, de la force motrice***) in which he suggested compounding steam-engines with gas-engines. Beau de Rochas applied elementary ideas of thermodynamics and mathematics to the operation of heat-engines, and showed that Lenoir's gas-engine will never be able to develop much power because it violated the fundamental conditions of the most efficient use of elastic forces of gases :

(a) greatest possible volume of the cylinder with the minimum surface area,

(b) greatest possible speed of expansion,

(c) greatest possible expansion,

(d) greatest possible pressure at the beginning of expansion.

To accomplish all this, Beau de Rochas proposed an operating cycle of 4 strokes in the cylinder :

(1) induction of gas,

(2) compression of gas,

(3) ignition of gas,

(4) exhaust of burnt gas.

Beau de Rochas pointed out that the limit of compression of the gas-air mixture depends upon the temperature of ignition of the gas. Only one stroke out of the 4-stroke cycle of the engine as suggested by Beau de Rochas was a working-stroke, a concept that was quite foreign to contemporary steam-engine designers who believed that maximum efficiency of any engine always requires each stroke of the engine to be a working-stroke. It was a purely theoretical paper, and Beau de Rochas himself never did build an experimental engine, or develop his idea on the 4-stroke engine any further. The pamphlet of Beau de Rochas was rediscovered in 1883 by patent attorneys who used it to invalidate a part of Otto's patent of his revolutionary 4-stroke gas-engine, called the 'Silent Otto'.

In 1860, Nikolaus August Otto (1832-1891), a travelling salesman for a merchant in Germany, read a newspaper account of Lenoir's gas-engine in 1860 and set himself the task of designing a gas-engine with much higher efficiency than Lenoir's engine. Otto had even a more limited scientific and engineering background than Lenoir, and by virtue of making himself a competent engine designer he became even more autodidact as an engineer than Lenoir. After a period of experimenting with an 'atmospheric' gas-engine similar in design to that of Lenoir, Otto realised that the chief problem of the gas-engine lay mainly in the control of the richness of the gas-air mixture, the timing of the mixture, and the compression of the mixture for maximum power. In this way, Otto came to the idea of a 4-stroke gas-engine. Since Otto was initially unable to overcome the problem of shock brought about by the explosion of the gas-and-air mixture, and his experimental engine was destroyed by the violent explosions, he concentrated upon developing an efficient atmospheric engine, the principles of which were well-known at the time. Otto was successful in constructing such an engine, and he patented it in 1863. He was very fortunate in finding a valuable business partner in the German engineer Eugen Langen (1833-1895), a graduate of the **Technische Hochschule** (*Polytechnical Institute*) in Karlsruhe, the best engineering school in Germany at the time. Langen was as important to Otto in their business success as Boulton had been to Watt. In 1866, Otto and Langen patented the first entirely successful single-acting, internal-combustion engine in history in which the working-stroke was produced by the atmosphere. In the gas-engine of Otto and Langen the explosion of the gases, ignited by a flame, drove the heavy free-piston up a vertical cylinder whilst the gases cooled under the piston and formed a partial vacuum, upon which the atmospheric pressure and the weight of the heavy piston produced the downwards working-stroke. A clutch designed by Langen allowed the piston to rise freely, but engaged the output shaft by means of a rack to which the piston-rod was geared in the downwards working-stroke. Various links and levers operated slide-valves and a small slide which, for a brief instant, exposed a flame to the gas-air mixture. The Otto gas-engine operated with 80 to 90 explosions a minute in which the piston moved rapidly up and slowly down and, in so doing, it made a great racket; yet it provided an ingenious solution to the extraction of useful power from the explosion of the gases although the rack-and-pinion drive was a relatively poor design. The atmospheric engine of Otto and Lange had no shock problem like the engine of Lenoir because the piston was disengaged from the output shaft during the explosion, and it was twice as efficient as Lenoir's engine. Since the working-stroke in the Otto-Langen engine was executed by the atmosphere, it was called the 'Otto-Langen atmospheric engine'. This engine was relatively inexpensive, and became an immediate commercial success; about 5,000 Otto-Langen atmospheric engines were sold all over the world. However, since this engine worked by the atmospheric pressure which could not be increased, it put a limit on the power this engine could deliver; it was not practical to manufacture atmospheric engines with a larger capacity than 3-horsepower. Since it was a period of increasing industrial demand for more powerful engines, by 1876 the demand for Otto-Langen atmospheric engines had dropped dramatically.

In 1872, by the time they had built more than 700 engines, Otto and Langen reorganised their company into the famous Gas-Engine Factory Deutz, and hired Gottlieb Daimler (1834-1900), an experienced mechanic and theoretically educated engineer, as a production manager to organise the new factory. Daimler soon brought with him a young friend, Wilhelm Maybach (1847-1929), who became a very important automotive engineer. Otto, Daimler and Maybach formed a formidable team of experts in the field of internal-combustion engines, but there was one ingenious and eccentric Austrian engineer, Siegfried Marcus (1831-1898), who preceded the German triumvirate by building the first genuine road-vehicle with an internal-combustion engine which incorporated many modern features. Marcus was a specialist in electric machinery and his interest in internal-combustion engine and the automobile was only marginal. Marcus operated a workshop in Vienna for the construction of equipment and engines in electric and gas technology.

Already in 1864, Marcus had built his first handcart driven by a single-cylinder internal-combustion engine to test his 2-stroke internal-combustion engine which operated by benzene as fuel. This engine did not compress the fuel mixture before explosion and, therefore, was not much more efficient than Lenoir's gas-engine. This handcart had neither a transmission nor a clutch, and the rear wheels had to be lifted off the ground and spun to start the engine. However, the engine did run satisfactorily. In 1870, Marcus built an improved self-propelled road vehicle, but after a successful trial run in the streets of Vienna, the police prohibited any further road tests of his vehicle on account of the great noise made by its engine. In 1873, Marcus exhibited his automobile at the World Fair in Vienna. In 1875, Marcus resumed the development of the self-propelled carriage, and built a new improved automobile. This vehicle had a wooden chassis, wagonlike wheels with iron tyres, wooden brake-blocks, a steering-wheel and front-axle steering, a foot-pedal operated clutch and a differential gear. The 4-stroke internal-combustion engine of this simple wooden vehicle, however, was a sophisticated piece of equipment. Marcus' engine consisted of a single cylinder with a 96 cubic inch (1,573 cubic cm) capacity that developed about ¾-horsepower at 500 rpm, and it had a remarkably sophisticated carburettor which sprayed benzol into a mixing chamber out of a trough by means of a spinning brush whilst the strength of benzol-air mixture was controlled by a manually operated air valve. The carburettor, invented by Marcus in 1864, converted liquid petrol into inflammable gas. The intake-valve was of a sliding type whereas the poppet-valve was for the exhaust gases which were passed through the engine casing to warm the fuel-air mixture. The ignition of the engine was by a low-tension magneto, the first such device on any vehicle. Marcus' engine was water-cooled. He built 2 other similar self-propelled vehicles but they are not extant. Marcus lost interest in his self-propelled cars after his new gasoline engine proved to be a practical success. His vehicle is the first successful progenitor of automobiles powered by an internal-combustion engine operating with liquid fuel. Since Marcus did not choose to develop his automobile any further, it remained for Daimler, Maybach and Benz to produce the first practical automobiles. The automobile of Marcus is preserved in the **Technisches Museum** (Technical Museum) in Vienna, and it can still run at 5 miles (8 km) an hour.

Otto bought samples of all competitive engines in existence at the time to examine them carefully in his factory. All these engines acquired by Otto are extant in the **Deutsches Museum** (German Museum) in Munich, and constitute an excellent museum collection of gas-engines available during the period of Otto's activity.

In order to make Otto's engine more versatile and powerful, its dependence on gas and its manufacturers, a great inconvenience, had to be reduced if not completely avoided. Therefore, a search for other more efficient fuels than gas was instituted. In 1875, Professor Franz Reauleaux (1829-1905), the founder of modern machine design and a close friend of Langen, suggested to Langen to compress the gas-air mixture and to use volatile gasoline as fuel. Reauleaux came to this conviction after meeting Marcus in Vienna, and after examining the automobile engine of Marcus. Marcus had demonstrated that liquid fuel permitted installation of internal-combustion engines in vehicles of all sorts because a small volume of liquid fuel, such as benzol, produced great power and, therefore, required a small tank, whereas gaseous fluids required enormous tanks for the same output of power.

In his search for increased power for his engine, Otto returned to his first idea of a direct-acting internal-combustion engine with compression of the gas-air mixture before ignition. Daimler, Otto and Maybach finally created, in 1876, an internal-combustion engine working on a 4-stroke cycle using $2\frac{1}{2}$ – to –1 compression-ratio of the illuminating gas and air mixture. This engine had only one cylinder and, therefore, required a flywheel to carry the engine between power-strokes. Otto had to adopt the 4-stroke cycle in his new engine because it was the only way he was able to eliminate the shock effect the explosion of the fuel produced in the engine. The new engine of Otto, which developed 8-horsepower, was patented in 1877, and exhibited at the World Fair in Paris in 1878 where it won immediate worldwide fame. This 4-stroke engine designed by Otto and Daimler made very little noise in comparison with other existing engines and, therefore, it became popularly known as the 'Silent Otto'. Otto's engine was versatile for it could be adopted to operate with various fuels, and used in a wide variety of applications. It became an instant commercial success which compelled many of his competitors to produce a similar gas-engine. Since Otto's new engine was so superior in performance to all other existing engines, his competitors were, all of a sudden, facing bankruptcy unless they were able to use his 4-stroke cycle. Therefore, great efforts were made to invalidate Otto's patent. After a long search of literature for a prior discovery of the 'Silent Otto' type of engine, Beau de Rochas' theoretical paper of 1862 was found in which de Rochas had proposed a 4-stroke cycle for the internal-combustion engine. On the basis of this evidence the courts in Germany and in France cancelled Otto's patent, but not the English courts.

The ignition magneto, a small generator of low-tension electric current run by the turning of the flywheel of an internal-combustion engine, was invented in 1880 by a German technician Giesenberg. The magneto which provided the electric current to produce the electric spark in the spark-plug of an internal-combustion engine was perfected in 1883 by Nikolaus Otto, and by a French mechanic, Fernand Forest (1851-1914), who also developed a compact radial internal-combustion engine in 1888, which became a prominent type of aircraft engine.

By 1891, at the time of Otto's death, there were more than 30,000 engines with Otto cycle in operation, and by 1900 there were about 200,000 'Silent Otto'-type engines in existence providing power to machine shops, pumping stations, printing presses, all sorts of power machinery, and generating stations. Water-cooled Otto engines, ranging from 2 to 5 horsepower and fuelled by gasoline or kerosene, served on farms as the typical power source for pumping, cutting, sawing and other routine farm work requiring power until the advent of electrification of farms after the turn of the century. Beginning in 1890, large engines for industrial use, some over 100-horsepower, were built, but they were only economical where gas was inexpensive. The heavy stationary type of 'Silent Otto' engines burning low-grade gas fuel and ranging from 3 to 50 horsepower, were in general use until replaced by Diesel-engines and electric motors.

In 1875 Maybach built for Otto an experimental 8-horsepower engine which worked by benzol vapour. This engine made between 50 and 200 rpm, and weighed about 1,455 pounds (660 kg) per horsepower. In order to make this engine more powerful per its unit weight, its speed had to be increased dramatically but, unfortunately, owing to its flame-ignition the engine could no longer run reliably at 250 rpm because of the failure of passages in the slide-valve which had to carry a bit of burning gas from the external flame into the compressed charge. It soon became obvious that a fast-running engine required a radically different ignition system. Unlike the steam-engine, the internal-combustion engine was highly sensitive to the quality of the fuel, the fuel-air ratio, and the timing of ignition. Finally, Otto gave up the development of his gasoline engine because of patent difficulties, and also because of his opinion that gasoline was a dangerous fuel.

In 1881, the Deutz firm built 'Silent Otto' gas-engines of 10 and 12 horsepower and, in 1884, train-engines which produced an output from 30 to 60 horsepower to drive electric generators. In

1879 Otto and Daimler also built 2-stroke engines which had been included in Otto's original patent. The specific weight of Otto's engines at that time had reached about 495 lb (200 kg) per horsepower. In the same year, the 2-stroke cycle of operation, which had been used by Lenoir and Otto in their respective first atmospheric gas-engines, and the compressed charge were combined in a gas-engine designed by Dugald Clerk (1854-1932) of Scotland as an alternative to Otto's gas-engine with 4-stroke cycle. In Clerk's engine, the exhausting of the cylinder, a scavenging operation, was accomplished through ports in the wall of the cylinder exposed by the piston at the end of the expansion-stroke, and the induction of the new charge was done at the beginning of the compression-stroke. For this purpose a displacer is required to deliver the charge into the main cylinder at low pressure. In 1891, Clerk introduced the idea of supercharging by increasing the pressure exerted by the displacer. Joseph Day further simplified this engine in the same year, by using the bottom of the piston to compress the incoming charge in an enclosed crankcase, and by eliminating the cam-operated poppet-valves with an ingenious arrangement of intake and exhaust ports at the bottom of the cylinders. In principle, the 2-stroke cycle is more efficient than the 4-stroke cycle because every revolution of the 2-stroke engine has a power-stroke instead of every other stroke as is the case with the 4-stroke engine. Yet, the 2-stroke cylinder in actual operation is considerably less efficient than the 4-stroke engine because some fuel is lost with the exhaust gases in the scavenging operation. Since the 2-stroke engine has a high power-to-weight ratio it has become popular today as a small portable power-unit, such as an outboard motor and a power lawn-mower, in which light weight and simple design are more important than fuel economy.

    The Otto type of engine showed great promise as a power-source for self-propelled vehicles, but before this could be accomplished a number of improvements to the Otto engine had to be made. The original Otto engine burning illuminating gas was a heavy and slow engine. Its weight per horsepower was about 10 times that required for an engine propelling a vehicle. To develop an internal-combustion engine from the Otto engine which would be light and powerful enough to drive a vehicle under variable load, the fuel-air mixture of the engine had to be finely controlled, the weight-to-power ratio of the engine had to be reasonably small, the speed of the engine had to be about 5 to 10 times as fast as the speed of the standard Otto engine which required a new system of ignition and a new system of cooling, the problem of the frequent starting of the stopped vehicle had to be solved, and a mechanical gearing capable of variable transmission of motion which allows the wheels of the vehicle to turn at different speeds than the speed of the engine had to be devised. These basic problems of the internal-combustion engine were first adequately solved by 3 German engineers: Carl Benz, Gottlieb Daimler and his collaborator Wilhelm Maybach.

    Daimler himself was mainly interested in building a light, powerful and versatile internal-combustion engine burning gasoline which could be used as a stationary engine, or as an engine driving a vehicle. Since Otto and Langen were not in favour of developing such an engine, Daimler together with Maybach established in 1882 their own workshop in Bad Cannstatt to produce such a compact, high-speed engine. In 1883, Daimler patented his hot-tube ignition system consisting of a platinum tube which was screwed into the ignition chamber and was heated to incandescence by an external Bunsen burner. In 1884, Daimler patented a small experimental high-speed gasoline engine employing hot-tube ignition and a wick-carburettor to meter the fuel. Initially Daimler had planned to use the hot-tube only to start the engine when it was still cold, and to operate the ignition of the gasoline vapour and air mixture by compression only, but he discovered that for the relatively low compression ratio built into his engine the hot-tube ignition was required all the time. This was a simple and ingenious solution for the difficult ignition problem of a fast engine which made from 600 to 900 rpm. It was the first high-speed internal-combustion engine working by gasoline vapour. Its only weakness was that its external Bunsen burner presented a fire hazard, a risk that Daimler was willing to take. In 1885, Daimler patented a vertical single-cylinder engine with a crankcase and 2 flywheels, which remained a prototype for all subsequent Daimler engines. This engine was cooled by a fan. Its induction valve was operated by suction, and its exhaust valve by means of a mechanism controlled by a governor, which limited the speed of the engine. In 1885, Daimler also patented his surface carburettor which made it possible to run his engine on light petroleum products such as benzene. Daimler's internal-combustion engines developed from ½ to 25 horsepower and weighed 88 pounds (40 kg) per horsepower.

In 1885 Daimler fitted a safety bicycle with 2 outrigger-wheels and his ½-horsepower engine to make it into the first motorcycle. This motorcycle which attained the speed of 7½ miles (12 km) an hour, Daimler patented in the same year. It had wooden wheels with iron tyres, friction belts and a pulley transmission for 2 manually adjusted speeds. This small engine made 700 rpm. Since Daimler was interested in self-propelled vehicles only as a testing ground for his universal engine, he did not develop his primitive motorcycle any further.

In 1894, 2 German engineers, Heinrich Rupert Hildebrand (1855-1928) and Alois Wolfmüller (1864-1948), an aircraft constructor, built a genuine motorcycle with pneumatic tyres in Munich. Their motorcycle was powered by a horizontal twin-cylinder gasoline engine which drove the rear wheel by means of a crank and connecting-rod. The rear wheel in their first motorcycle was solid and acted as a flywheel, but in later models it was replaced by a regular spanned bicycle wheel. Hildebrand and Wolfmüller obtained a patent and established a factory for the production of their motorcycle which lasted from 1894 to 1900.

However, a really practical motorcycle with 3 wheels was first constructed in 1894 by 2 French engineers, Count Albert de Dion (1856–1946) and George Thadée Bouton, a former manufacturer of steam-driven vehicles. Their motorcycle engine running at 1,500 rpm was similar to the single-cylinder Daimler engine, but it had an electrical ignition with a battery and 'surface carburettor', an ignition system quite similar to that of Benz but improved by a cam distributor which gave much greater precision to the timing of spark. In 1896, another notable French engineer Léon Bollée (1870-1913), son of Amédée Bollée, also built a tricycle propelled by an air-cooled, single-cylinder engine.

An English mechanic, Edward Butler, had constructed a tricycle with a gasoline engine in 1884 which had a sophisticated float-feed spray carburettor with an automatic regulation of the strength of the mixture to fit the speed of the tricycle. Butler's carburettor had an air-cleaner which came into use only about 50 years later. Butler had to abandon his development of the self-propelled tricycle because he could not achieve commercial success with it, since the oppressive British turnpike law of 1865 was a great obstacle to the development of motorised road vehicles.

In 1886, Daimler installed his single-cylinder, 1½-horsepower engine in a 4-wheel horse-carriage, and achieved a speed of 11 miles (18 km) an hour in a trial run in Cannstadt near Stuttgart. Daimler also fitted a boat with his motor to prove that it can drive any moving vehicle, since his intention was to mass-produce a universal internal-combustion engine which can be used for any purpose. Maybach disagreed with this view of Daimler, because he believed that the engine and the vehicle carrying it should be designed from its very inception as a 'machine unit'.

In 1889, Daimler patented his renowned V-type twin-cylinder engine, the 2 cylinders of which enclosed a 15-degree angle between them and had their connecting-rods attached to a common crank. This engine, which used a hot-tube ignition system, found use as a stationary engine, as well as a power unit to drive motor cars, boats and even dirigibles. In 1890 Daimler finally founded the renowned Daimler Motor Company to produce his own engines commercially. In 1892, Maybach built the important float-feed jet carburettor for Daimler engines, which is the forerunner of modern carburettors. A float-controlled needle-valve governed the admission of gasoline to the float chamber. This carburettor did not have an air-filter nor an automatic mixture-regulator like the original float-feed jet-carburettor of Butler in 1884.

In 1891, Daimler built the first load-carrying car with a 4-horsepower motor, which had 2 forward speeds and one reverse speed, and the specific weight of 74 lb (30 kg) per horsepower. However, the great commercial success which Daimler had expected for this car did not materialise.

Before Daimler tested his motorcycle, another German engineer, Carl Benz (1844-1929), had road-tested his 3-wheel automobile in Mannheim, a city located about 60 miles from Cannstadt. Benz was a graduate of the **Technische Hochschule** (*Polytechnical Institute*) in Karlsruhe, and he had an extensive machine-shop and practical engineering experience. Since 1871 Benz had operated a factory for the manufacture of stationary 2-stroke gas-engines, but his main intention was always the manufacture of self-propelled vehicles. In 1886 Benz built a 3-wheeled automobile which was driven by his single-cylinder, 4-stroke, water-cooled gasoline engine, installed in the rear of the carriage. Benz's engine had a surface carburettor, a flywheel, a high-voltage electrical ignition system and a spark-plug, a spark-coil and a battery, and it delivered ¾-horsepower at 250 up to 300 rpm. This power was transmitted through a pair of bevel-gears from the crankshaft to the horizontal, spoked flywheel from which a friction belt turned a countershaft under the driver's seat incorporating a differential gear. The 2 link-chains from the countershaft drove the 2 rear wheels like in the modern bicycle. The friction belt could be moved from a fast pulley to a loose pulley to allow the engine to run when the vehicle was stationary. The carriage itself had wheels with taut wire spokes and solid rubber tyres, and a rack-and-pinion steering controlled by a tiller on the single front wheel. Although Benz had designed a differential gear, he had provided the car with only one speed. The main trouble with Benz's car was that its battery ran down so rapidly that Benz himself had to carry a spare battery with his car which ran at 7½ miles (12 km) an hour speed. Benz was not primarily interested in the speed of the car, but rather in its reliability and simplicity. In 1888, his wife and 2 teenaged sons made the first historic long-distance trip in an automobile when they drove from Mannheim to Pforzheim and back without his knowledge.

Benz built 2 more 3-wheel cars before he constructed his first 4-wheel car in 1890. It had an engine which developed 3½-horsepower at 500 rpm, and gave the car a top speed of 14 miles (22.5 km) an hour.

In 1892 and 1893, Benz developed his steering mechanism, which he patented. In his 3-wheel automobile he did not have to deal with the complex Lankensperger steering mechanism. Marcus and Daimler in their 4-wheel carriages were satisfied with the archaic front-axle steering which pivoted the front axle in the center and, thus, created lateral friction on the wheels on a turn. In turning a 4-wheel vehicle, the rolling axes of all 4 wheels should intersect at a point if frictional turning of the wheels is to be obviated. Usually this intersection point lies on the common axis of rotation of the rear wheels. Such an arrangement keeps all 4 wheels of a moving vehicle in an approximately pure rolling state. This principle of steering was discovered in 1714 for the first time by a French millwright Du Quet, who invented a mechanism of compensated gearing for pure rolling of windmill carriages. It was independently invented in 1818 in an improved form by a German carriage manufacturer Georg Lankensperger (1779-1847) of Munich for horse-drawn vehicles, and patented for him in England in 1832 by Rudolph Ackermann, a London printseller. Its first application to a self-propelled vehicle had been made already in 1832 by Redmund, who had used Lankensperger's steering mechanism in his steam-propelled carriage.

Benz improved the high-voltage electrical ignition, the efficiency of the engine, and the evaporation-type carburettor which he perfected in 1893 by adding a butterfly throttle valve to regulate the fluid-air mixture drawn in by the engine and, thereby, to control its turning speed and power, but otherwise little of his original automobile design was changed from 1890 to 1901. In 1898 he added an emergency low-gear to the 2-speed belt-and-pulley transmission of his car, so that it could climb any gradient of the road. **Benz A.G.** was the first firm to sell cars to customers, and the first Benz car was sold to Emil Rogers in Paris, and it was followed by other orders from France, England and United States. Already in 1896 a Benz car was used in Paris as a taxi. From 1890 to 1900 about 2,000 Benz cars were sold.

The first automobile exhibition was held in Paris in 1889 where both Benz and Daimler cars were on display. In 1894, the first international automobile race was run from Paris to Rouen and back, a distance of 78 miles (126 km), in which 102 vehicles competed: 38 gasoline cars, 39 steam cars, 5 electric cars, and 5 compressed-air cars. Only 15 cars finished the race, and the winner was a car built by the French engineers René Panhard (1941-1907) and Emile Levassor (1844-1897) using the Daimler patent, the most advanced motor car in the world at the time, which covered the distance with an average speed of 12 miles (19.3 km) an hour, an astonishing speed for the period. In 1896 another car race was run from Paris to Marseille and back, a distance of 1,063 miles (1,711 km), which was again won by 3 cars powered by Daimler engines. In 1900, two 16-horsepower Daimler-type of engines with specific weights of 10 lbs (4 kg) per horsepower were installed to drive the propellers of the dirigible of Count Ferdinand von Zeppelin (1838-1917). Later engines were manufactured by Maybach in his own company directed by his son, Karl Maybach (1879-1960).

The cars with Daimler engines may have been the fastest cars at the time but they were not the safest, for they were dangerous when overturned because of the presence of the external Bunsen burner. The Benz cars may have been slower than Daimler cars, but they were also safer, sturdier and more reliable. In the 1890's France was the leading manufacturer of automobiles, but many other cars were made at the time in England, in United States and in Germany. In England cars were being built by John H. Knight since 1895, Lord Austin since 1896, and Frederick W. Lanchester since 1896. Major Holden invented a unique 4-cylinder engine which he fitted to a motorcycle with pneumatic tyres. It had an electrical ignition and ran smoothly at 20 miles (32.2 km) an hour. In United States Henry Ford (1863-1947) built his first car in 1896, and his famous model T from 1908 to 1927 which was produced on a moving assembly-line leading to the method of modern assembly-line mass-production. By 1915 Ford had produced a million cars, and by 1927 had sold 16 million model T cars, which ran reliably for years without any need of expert maintenance. Ford's transmission in his automobile was a virtually automatic pedal-operated planetary transmission which, once the car had started, was easy to operate by a simple movement of a single pedal without any need of special skill.

In Italy the first automobile factory Fiat was established in 1899 in Turin, which produced 'Fiat' cars with 3½ and 8 horsepower engines. In Germany an automobile factory was set up in 1896 in Eiserach which produced the well-known 'Wartburger' car with only one cylinder. August Opel (1837-1895) changed his bicycle and motorcycle factory into an automobile factory, and became a prominent producer of German automobiles. In France, Armand Peugot (1848-1915) built his first car in 1896.

The practical problem of electrical ignition and workable spark-plug was finally solved by the German engineer Robert Bosch (1861-1942), who designed in 1902 an engine-driven, high-tension, magneto-generator ignition system which delivered a hot spark without battery and moving parts in the cylinder, and rapidly became the standard ignition system of automobile industry. The high-tension magneto-generator had been first devised by Frederick Simms in 1896, but it was Bosch who made it practical. Bosch, who had returned to Germany from United States in 1884, soon became the largest producer of automobile parts in the world. In 1898, the French manufacturer of automobiles, Louis Renault (1877-1944), introduced the drum-brake in his automobiles, which soon became the standard brake system for cars. Modern disk-brakes were first introduced in 1906.

In 1900 after the death of Gottlieb Daimler, Maybach finally had the opportunity to design an automobile and its engine as an organic whole. He called this organically designed car 'Mercedes', after the name of the daughter of the well-known sportsman, banker and Austrian Consul in Nice, Emil Jellinek, who was a close collaborator and an indefatigable promoter of such an integrated car. 'Mercedes' was the first true modern automobile which had a huge 4-cylinder engine cast in 2 blocks, first designed by Gottlieb's son Paul Daimler (1869-1945) as a 24-horsepower engine but redesigned and improved by Maybach to develop 35-horsepower at 1,000 rpm, a maximum speed of 45 miles (72 km) an hour, and specific weight of 16.8 lb (6.8 kg) per horsepower. It was provided with a float-controlled jet carburettor, magneto-electric ignition, inclined steering-wheel, and pneumatic tyres. Its specific weight was about half of that of any other contemporary automobile engine.

Pneumatic tyres had been first patented in 1845 by an English civil engineer, Robert William Thomson (1822-1873) from Adelphi, who 20 years later became a successful designer of steam-tractors. His pneumatic tyre consisted of an inner air-tube in the form of concentric air-sacs of rubberised calico enclosed in a bolted-on leather casing and provided with an air valve and an air pump, but few carriages were fitted with these tyres and his venture failed. In 1846 Thomson tried out solid bands of India rubber as tyres. In 1866 he used the solid rubber tyres on his 3-wheel steam tractor intended to draw omnibus trailers or freight wagons. Since the rubber was too soft and tended to 'creep', he protected the rubber tyre by steel-shoes fit loosely over the tyre. These armoured rubber tyres were expensive and troublesome, but despite this shortcoming many rubber-shod

Thomson tractors were made in the 1870's. Their use proved particularly suitable in India, Greece, Turkey and Egypt. A Scottish veterinarian surgeon, John Boyd Dunlop (1840-1921) in Ireland, constructed a pneumatic tyre for his son's tricycle which consisted of a caoutchuc tube inflated through a valve and protected by a rubber-impregnated outer casing wrapped around the tyre like in a mummy. He patented it in 1888. Punctures in Dunlop tyre could only be repaired by soaking the casing apart with benzol and then rebuilding it. Dunlop's tyre was called the 'mummy-tyre', or 'puddy-tyre', and openly ridiculed in public, but bicycle-racer William Huma found it faster than the rubber air-chamber types of tyres made by Charles MacIntosh in Glasgow since 1884, and the solid rubber tyres which had been long popular in England. Initially Dunlop produced his tyres with a bicycle-maker in the firm Edlin & Co., but after 1895 this company became Dunlop Pneumatic Tyre Company. Dunlop's tyre soon took over the entire tyre market from the MacIntosh tyre. Charles Kingston Welch from Coventry improved Dunlop's invention, which he patented in 1890. His tyre consisted of a caoutchuc tube covered with a casing of sailcloth soaked in caoutchuc and provided with a dished wheel rim through which ran a reinforcing wire. Welch also introduced the valve, and in 1893 the reinforcing web of cords in the casing. In 1892 the American inventors G.F. Stillman of New York and A.T. Brown of Buffalo had applied for a similar patent as Welch, which was acquired by the Dunlop Company. In France, André Michelin (1851-1914) and his brother Edouard Michelin (1859-1940) of Clermond-Ferrand learnt from a bicycle-racer about the pneumatic tyre in 1888. Michelins developed their own replaceable pneumatic tyre in collaboration with a French engineer Laroche with which the bicycle-racer Terront won the race from Paris to Brest in 1891. In 1895 Michelin successfully tested his pneumatic tyre for automobiles, after which pneumatic tyres became reliable tyres for self-propelled road vehicles.

Maybach's car 'Mercedes' won the important international car race in Nice, France in 1901, and its increasingly powerful annual models dominated car racing until 1914, which established its great international reputation.

In 1926, the companies of Daimler and Benz amalgamated, although the 2 men never met when Daimler was still alive, and the new company, still manufacturing automobiles today, is known as the Mercedes-Benz Company. Shortly before 1939, Mercedes-Benz built the first cars powered by Diesel-engines.

Maybach was responsible for most of the important developments in the automobile engine design such as its weight reduction, its variable gear transmission, its jet carburation, and its cooling by means of the honeycomb radiator, all of which gave the automobile engine its present form. Maybach employed a make-and-break device within the cylinder to produce the electric spark for ignition, rather than the high-tension electric spark system used in modern cars.

In 1909, Maybach and his son Karl established a factory as a subsidiary of the Zeppelin Dirigible Organisation for the production of aircraft engines, which was a new lucrative field for internal-combustion engines. The first dirigible engine designed by Maybach in his own company was a 6-cylinder engine of about 180-horsepower which made its first trip in a zeppelin in 1909.

In 1910, an American inventor, Charles Franklin Kettering (1876-1958), founder of the Dayton Engineering Laboratories Company, known by its acronym **DELCO**, improved the automobile ignition by introducing a high-tension distributor as well as the lighting system of the car. He also invented the 'self-starter', a battery-powered electric motor capable of turning the engine until it fired, which made hand-cranking unnecessary in the starting of the automobile engine, and was almost immediately adopted worldwide in the automobile industry.

By 1914 the automobile industry had become an important business, and there were altogether about 64,000 automobiles in existence.

Aircraft Engines

The talented English aircraft pioneer, George Cayley, a man of private means, had already outlined in 1799 the design principles of aircraft with stiff wings and body, and a stabilising tail unit with control surfaces. He performed extensive experiments with scaled-down and full-scale models. Cayley needed a prime mover, a compact engine to power his aircraft and, therefore, he prepared in 1807 a design for a compact heat-engine operating with air as a power source for his aircraft. He called it the 'caloric engine' and described in detail its mode of operation, although he did not attempt to build one himself. Cayley's caloric engine was an external-combustion engine, and the Scottish clergyman Robert Stirling (1790-1879), who followed Cayley's suggestion, patented his closed-cycle hot-air engine in 1827. It had 2 cylinders of 15.7-inch (40 cm) bore, a piston stroke of 47.2 inches (1.20 m), and produced 21-horsepower. It was manufactured industrially in 1844. In 1851 the Swedish inventor John Ericsson (1803-1889) patented his open-cycle hot-air engine which had 4 cylinders, and produced approximately 50-horsepower. Ericsson used it in 1858 to propel a ship bearing his name. Gottfried Wilhelm Leibniz (1642-1716) was the first person to suggest a hot-air engine in a letter to Denis Papin (1647-1713) in 1709. After Cayley had learnt about the internal-combustion engine of Philippe Lebon which operated with gas as fuel, he described in great detail how a gas-engine ought to work although he never attempted to build one himself.

A self-educated professor of physics at what is now Pittsburgh University, director of Allegheny Observatory, and later the secretary of Smithsonian Institution in Washington D.C., Samuel Pierpont Langley (1834-1906), after constructing a power-driven model aircraft with specially designed light engines which performed so successfully in flight tests in 1896, that he was encouraged to design a large-scale heavier-than-air experimental aircraft for which his assistant engineer Charles Matthews Manly (1876-1927) designed a special 5-cylinder radial internal-combustion engine. This excellent aircraft engine which had a circular form developed 52-horsepower at 1,000 rpm, and had a specific weight of only about 5 lb (2 kg) per horsepower. Manly designed his air-cooled engine, a type which became the standard aircraft engine, after examining the motorcycle engine of de Dion and Bouton. Unfortunately in its first flight test in 1903, Langley's aircraft instead of becoming airborne plummeted into the Potomac River during its mechanically assisted launching with Manly at the controls. Langley himself never attempted to test his experimental aircraft again because the failure of the first test brought him an enormous amount of undeserved public ridicule, although his restored aircraft was successfully flown in 1914 with a more powerful engine by Glenn Curtiss (1878-1930), an aviation pioneer.

A few weeks later, the Wright brothers, Wilbur (1867-1912) and Orville (1871-1948), achieved the first sustained flight in a heavier-than-air aircraft at Kitty Hawk. The Wright brothers had designed their aircraft using 2 important design principles : the warping of the wings and the vertical rudder, principles which had been discovered by the French artist Louis Mouillard (1834-1897) in 1890 from his observations of the flight of vultures. Mouillard had sold the rights for the use of his principles in aircraft design to the American civil engineer and aviation pioneer of French stock, Octave Chanute (1831-1910), who informed the Wright brothers of the 2 important principles of Mouillard and suggested the use of moveable control surfaces in their aircraft design to control the direction and the stability of the aircraft in flight. The Wright brothers proved the claims of Mouillard in their small windtunnel experiments, and they incorporated such moveable control surfaces into their aircraft design.

The Wright brothers needed a compact aircraft engine to power their pioneer airplane and they built for this purpose a 4-cylinder, water-cooled, horizontal engine which developed 30-horsepower as its maximum power, and 12-horsepower at 1,050 rpm as the cruising power. The specific weight of this simple but reliable aircraft engine was 15 lb. (6 kg) per horsepower, making it a relatively heavy aircraft engine.

One obvious way to increase the power of an Otto gasoline engine is to increase its compression ratio since it increases also the efficiency of the engine. However, this way of increasing the power and efficiency of the Otto engine was handicapped by a baffling problem of 'pre-ignition' consisting of the ignition of the fuel in the cylinder before the instant of maximum compression is reached in the cylinder. The 'pre-ignition' in gasoline engines is created by the pressure and temperature within the cylinder that is above the fuel's ignition point. In gasoline engines with high compression ratios the fuel may ignite spontaneously in the cylinder because of an overheated valve, spark-plug, glowing carbon deposit, or improper fuel. Since 'pre-ignition' creates irregular combustion of fuel before maximum compression is reached by the piston, the burning gases expand against the incoming piston thereby causing reduction of the power of the engine and 'engine knock', as well as overheating of the engine. The compression ratios for Otto gasoline engines in engineering practice vary from 4–to–1 to 10–to–1.

It took about a quarter of a century to find ways of avoiding 'engine knock', which entailed developing a fuel that could withstand high compression without premature ignition, and of finding new shapes for combustion chambers that would minimise the 'knock'.

Prosper L'Orange (1878-1939) in Germany and Harry Ralph Ricardo (1885-1974) in England examined hundreds of fuels, shapes of combustion chambers, and injection systems in the 1920's, and they were able to raise the power of the internal-combustion engine about 20% by appropriate modification of the combustion chamber. Then in 1921, the mechanical engineer at Dayton Engineering Company who was a self-educated chemical engineer, Thomas E. Midgley Jr. (1889-1944), discovered that a very small amount of tetraethyl lead added to the gasoline had a chemical property of suppressing the so-called 'engine-knock'. Apparently the compound of tetraethyl lead and gasoline in combustion gives up lead from its molecules when it strikes the hot cylinder surface and the released lead oxide on the cylinder walls of the engine helps to retard the pre-ignition of gasoline vapours in the cylinders which is the cause of the 'engine-knock'. The efficiency of the combustion, or the so-called 'octane rating' of the gasoline, is thereby increased by the properly timed combustion of the confined gases in the cylinders at peak compression which extracts maximum work out of the burning gases driving the pistons. This invention made high-speed internal-combustion engines possible.

In 1930, Midgeley also prepared Freon (*difluorodichloromethane*), a nonpoisonous refrigerant now universally used in refrigerators, freezers and air conditioners, as a replacement for the common refrigerants used in the 1920's such as ammonia, methyl chloride, and sulfur dioxide, all of which are poisonous. Midgeley, who was paralysed by polio in 1940, accidentally strangled himself with a harness operating over a pulley which he had made for the purpose of lifting himself out of bed.

In 1925, the American chemist Graham Edgar (1887-1955) established the so-called 'octane rating' or 'octane number', to represent the 'antiknock' properties or the combustion efficiency of a gasoline as a measure of compression that can be achieved without 'engine-knock'. The 'octane number' depends upon the percentage of the chemical additive to the fuel: the higher the octane number, the greater the 'antiknock' properties of the gasoline. This chemical additive to the gasoline which increased its octane rating also increased the power of the Otto gasoline engine about 20%.

Oil-Engines

The gas-engine proved valuable as the source of power, but it could not be used in localities where gas was not easily available. Therefore, many attempts were made to use other inflammable substances for fuel which were less expensive and safer than gasoline, such as illumination oil, or kerosene (paraffin), which was first drip-distilled by Abraham Gesner (1797-1864) in London. This type of hydrocarbon is not volatile until its temperature rises between 86°F and 122°F (between 30°C and 50°C), and therefore, it must be vaporised in the air before an inflammable mixture is obtained.

In 1873, an Austrian engineer, Julius Hoch of Vienna, patented a kerosene engine in which compressed air vaporised a jet of kerosene, but the incomplete combustion of the air-kerosene mixture, because the charge was not compressed, made it impractical. In the same year, an American mechanic, George B. Brayton (1830-1892) of Philadelphia, constructed a petroleum engine which had 2 cylinders, one for the compression of the air-vapour mixture, the other for the working-stroke. In the compression cylinder, air was compressed and forced through material soaked with petrol, thus making it combustible. This mixture of air and petrol vapour was then inducted into the water-cooled working cylinder and ignited by a flame which entered the cylinder through a diaphragm, the expanding gases resulting from the combustion of petrol as fuel produced the working-stroke, and the return stroke expelled the exhaust gases. Brayton-engine was double-acting, and had a 2-stroke cycle in which combustion occurred at constant pressure like in the later Diesel-engine. The engine itself provided a reservoir with compressed air for the purpose of starting the engine. However, the charge of the engine was not compressed and, therefore, the combustion was rather poor. It was mainly used in the United States until 1882. The Brayton-engine, which was quiet in its operation achieved its first real commercial success only in 1890, when it was converted to the 'Otto cycle' and made single-acting. The ideal thermodynamic cycle associated with this engine is often called the 'Brayton cycle', or the ideal air-standard gas-turbine cycle. Brayton-engine was used mainly in small workshops, and it was quite successful in propelling boats. Brayton's attempts to drive streetcars with his engine failed, since it was not powerful enough for that purpose. Brayton was probably the first to manufacture and market gas-turbines commercially in United States.

One of the first oil-engines to become commercially successful was built by the Priestman brothers of Hull in England. It had a single-cylinder with a 4-stroke cycle which developed about 11-horsepower using kerosene as fuel. The kerosene was injected by a jet of compressed gas into a vaporiser, a vessel heated by the exhaust gases, in which it immediately vaporised on the induction stroke of the piston. The kerosene vapour was then mixed with air and drawn into the cylinder, where the mixture was ignited after compression which produced the power-stroke. In 1886, it was patented by Dent and Priestman, and both horizontal and vertical types of oil-engines were produced by their company. A portable model mounted on wheels, which was entirely self-contained, was marketed in 1889 and won the silver medal of the Royal Agricultural Society.

The oil-engine of Hornsby-Stuart, designed by Herbert Akroyd Stuart (1864-1927) and manufactured by Ruston and Hornsby of Lincoln in 1890, was a horizontal, 4-stroke engine which incorporated a significant improvement of oil-engines for it dispensed with the external vaporiser and ignition system. It introduced a hot bulb at the top of the cylinder, which was initially heated by a blowtorch. When the engine was manually turned, air was sucked into the cylinder through a valve, whilst oil was pumped into the hot bulb, where it vaporised. On the compression stroke, air was forced into the hot bulb and mixed with oil vapour, and at the maximum stroke the mixture was ignited by the hot bulb, which was followed by the working-stroke when the hot gases expanded into the cylinder. After a few strokes, external heating of the hot bulb was discontinued because it received sufficient heat from the combustion of the fuel to maintain continuous operation of the engine. Since this oil-engine had a separate combustion chamber it was a forerunner of the Diesel-engine Akroyd Stuart had designed his simple and sturdy oil-engine for which the maintenance was easy and the fuel was cheap without any considerations of the science of thermodynamics. The oil-engines did not use high compression-ratios, but once started, several engines such as the Hornsby-Stuart oil-engine did run without the electric spark ignition.

Diesel-Engine

Rudolf Christian Carl Diesel (1858-1913), a German scientific engineer born in Paris, France, first studied in England and then attended the **Technishe Hochschule** (*Polytechnical Institute*) in Munich, Germany, where as a student in 1878, he heard a lecture by Carl von Linde (1842-1934), founder of the liquid-gas industry and an expert on refrigeration, in which Linde pointed out that in the steam-engines only 6% to 10% of the thermal energy of the fuel is converted into work, whereas with isothermal operating conditions in a heat engine according to the ideal Carnot cycle all the thermal energy of the fuel in the heat-engine would be converted into work. Whilst still a student Diesel decided to find out whether isothermal combustion in the operation of heat engines is possible. He graduated from the **Technische Hochschule** in Munich at the top of his class with the best scholastic record in the history of the school, and went into engineering practice, driven by the idea of finding a practical solution for the isothermal combustion process. He worked first for the **Gebrüder Sulzer Cie.** (*Sulzer Brothers Comp.*) in Winterthur, Switzerland, and then joined Linde's company that specialised in refrigeration, first as an engineer in Paris, and later as manager in Berlin. Diesel followed James Watt in the singleminded pursuit of an idea: "I must think of the machine night and day."

By 1893, Diesel had worked out his theoretical thermodynamic ideas in which his starting point was to produce a thermodynamically ideal prime mover the operation of which conforms as exactly as possible to the principle of the Carnot cycle: the heat to be converted into work has to be added at the maximum temperature of the cycle without any further rise in the temperature, and the combustion has to proceed at constant temperature. Diesel acquired a patent in December, 1892, and he published a booklet in which he explained the careful attention paid to thermodynamic principles in his design of his rational heat-engine: < Theory and Construction of a Rational Heat-Engine... > (*Theorie und Konstruktion eines rationellen Wärmemotors* ...) in 1893. Therefore, the design of the Diesel-engine as a prime mover was the first attempt to apply science to engineering design in the history of technology. Diesel soon discovered that the Carnot cycle deserves its reputation as being ' perfect' only in theory, and that in a practical heat-engine not the maximum temperature but rather the maximum compression is the deciding factor.

Diesel intended to achieve a much higher thermal efficiency in the internal-combustion engine, which in the beginning of the 4-stroke cycle, would compress pure air adiabatically in the cylinder in a compression-ratio of 250– to– 1 to achieve an extremely high initial temperature, then inject the fuel such as coal dust with an external air blast at a compression much higher than the compression in the cylinder, so that gradually the fuel would burn at constant temperature during the first part of the power-stroke. This isothermal combustion is followed by an adiabatic expansion to complete the power-stroke whilst the temperature of the exhaust drops to the atmospheric temperature. By such means, Diesel hoped to convert the total thermal energy of a fuel to useful mechanical work, and to come as close as possible to the ideal thermodynamic 'Carnot cycle'. His engine, according to Diesel's calculations, would have theoretically a 73% thermal efficiency because so little of its heat would be wasted and, therefore, it did not require a cooling system. Diesel counted among the additional advantages of his engine the fact that it required neither a carburettor nor an ignition system, and that it could use almost any fuel since all fuels would ignite in contact with an extremely hot compressed air having a temperature in excess of 800°F (427°C). Therefore, his engine could use any fuel, even coal dust, and offer a greater thermal efficiency. His intention was to use coal dust for the fuel in his engine which would eliminate the need for Germany to import expensive oil for such a purpose. Having this idea in mind, Diesel set to build his first experimental engine, which meant to translate his theoretical ideas into hard engineering practice, a task which almost the entire engineering profession considered impossible.

The German firm, the *Maschinenfabrik Augsburg-Nürnberg* (*Machine Factory Augsburg-Nuremberg*), known by the acronym **MAN**, was willing to provide the financial backing for Diesel to establish a laboratory to test his theory in practice by building his experimental engine. Diesel built his first experimental engine at Augsburg in 1893. This engine was of the piston-rod and crosshead type and its single cylinder had no water-jacket. The performance of this experimental engine fell far short of the ideal engine Diesel had in mind for he was nearly killed when this prototype engine exploded, but its thermal efficiency did encourage Diesel to continue his dangerous experiments with this type of an engine. From 1895 to 1897 the huge German firm *Friedrich Krupp* financed Diesel's new project to eliminate the defects from his first experimental engine design, during which time Diesel devoted all his efforts to the building of his new engine. It was the most discouraging period of his life for time and time again he met great disappointments, and practical engineering problems which seemed to be insuperable and doom his task to failure. Diesel thought to achieve isothermal combustion by proper rate of fuel injection, but this proved to be impossible in practice. One of the most difficult problems was to find a safe and reliable system of injection of minute but very accurate amounts of fuel into highly compressed air in the cylinders under extremely high pressure. After 6 years of incredibly strenuous effort, during which time he tried scores of injection and ignition devices, he was finally successful in producing a functioning experimental engine, which had a water-jacketed cylinder and a pump to supply compressed air to the cylinder since without it, a smokeless exhaust was impossible. During this time, Diesel carried out 6 series of tests. However, his new engine worked on a cycle quite different from the 'Carnot cycle', as have all subsequent Diesel-engines, and its compressed air system required a large amount of work to compress the air which used up a great deal of the power produced by the engine. Diesel's failure to achieve isothermal combustion demonstrated that the 'Carnot cycle' is a hypothetical cycle by means of which thermodynamic concepts, such as maximum motive power of an engine, can be analysed, yet no actual engine can be built to operate according to this cycle. In 1913, Diesel wrote a book entitled <The Creation of the Diesel Engine> (*Die Entstehung des Dieselmotors*) in which he described in great detail his tremendous struggle to build his engine.

In his experimental engine, Diesel was finally able to achieve a workable compression-ratio of 30- to- 1 which brought about an air pressure from 30 to 35 atmospheres and created temperatures of the compressed air high enough to satisfy the requirements of his combustion process. In the early experimental engine models Diesel injected coal dust as the fuel into the cylinders by compressed air, but he was profoundly disappointed in being forced to substitute crude oil for coal dust which proved to be quite unsatisfactory as the fuel of his engine. Despite all his exhaustive efforts, Diesel was not able to produce isothermal burning of the fuel, for the fuel in his engine burnt approximately at constant pressure. In the Otto cycle combustion occurs at constant volume, and not at constant pressure as in the Diesel cycle, so that the cycle of the operation of the Diesel-engine is quite different from that of the Otto gasoline engine. Since the Diesel-engine operated with much higher pressures than any existing engine of the period, it had to be considerably more robust in its construction than the Otto gasoline engine of comparable power output. The original Diesel-engine, designed to operate at low speeds, ran a little rough, but its thermal efficiency was 34.2%, whereas the thermal efficiency of a contemporary Otto gasoline engine was about 25%. Owing partly to its higher compression ratio, which implies a greater expansion of the burning gases resulting in cooler exhaust and therefore in a more complete use of the thermal energy of the fuel, the Diesel-engine is thermally a more efficient heat-engine than Otto's internal-combustion gasoline engine. However, as long as his engine had to have an air compressor and air tank, it could not be developed into the versatile engine Diesel had intended to create. In 1898, this engine of Diesel was exhibited in Munich, and it so impressed a number of prominent businessmen that they were prepared to finance its manufacture, although the originality of Diesel's engine was contested by Akroyd Stuart.

The second phase of the Diesel-engine development, its introduction into industrial practice, was almost as difficult, for Diesel had to overcome envy, illwill, tradition, and vested interests. His first, single-cylinder engine capable of producing from 20 to 25 horsepower was manufactured at Augsburg in 1898, and it was followed in 1899 by a 2-cylinder engine which produced from 60 to 67 horsepower at 180 rpm. In 1901, Diesel abandoned the external crosshead construction of his engine in favour of the trunk piston, an innovation introduced by the Diesel Motor Company of America in St. Louis, Missouri, founded in 1896, that was subsequently adopted by European Diesel companies. The Swiss company *Gebrüder Sulzer Cie.* (*Sulzer Brothers Comp.*) in Winterthur, began making Diesel-engines on licence. The **MAN** Company built a 250-horsepower 2-cylinder Diesel-engine in 1906. The first 4-stroke marine Diesel-engine was built in 1902 and 1903 by Adrien Bocket and Friedrich Dyckhoff. It was an unusual horizontal engine producing 20-horsepower which was fitted into a French canal boat. Since this engine could not be reversed it had to have a reversing gearbox. The marine Diesel-engine built by the *Gebrüder Sulzer Cie.* in Switzerland to propel a marine craft incorporated an electrical transmission. In 1907, Diesel himself worked on the designs of a Diesel locomotive and a Diesel-engine for ship propulsion. By 1908 about a 1,000 Diesel-engines were in use as stationary engines most of which developed from 50 to 100 horsepower. In 1909, Diesel constructed a 4-stroke reversible engine of 30-horsepower running at 600 rpm, which had an overhead camshaft and was meant to be tested as an engine for heavy trucks. For all his almost superhuman effort to make his invention into a commercially successful engine, Diesel personally earned over a million dollars.

The first important improvement of Diesel-engines came about with the development of the submarine engine, since submarines needed engines which saved in weight and space, yet these engines had to have 10 times the power of the contemporary Diesel-engines, and much smaller specific weight. Moreover, in submarines the use of Diesel fuel was much safer than the use of gasoline fuel in Otto internal-combustion engines. The development of the submarine Diesel-engine was so successful that by 1914 submarines were powered either by 1 or as many as 4 Diesel-engines, each of which developed 500-horsepower and had the specific weight of 62 lb (25 kg) per horsepower. By 1927, engineers had developed the direct (or solid) injection consisting of direct mechanical injection of fuel by fluid- pressure as a replacement of Diesel's compressed-air injection with nearly 1,000 psi (6,895 kPa) air-pressure, which allowed more precise injection metering, faster operation for high speed engines, internal pressures of about 30,000 psi (206,850 kPa), and an increase of the power of a single cylinder to 1,000-horsepower. By 1930 the typical marine Diesel-engine built by **MAN** Company was an 8-cylinder, double-acting, 2-stroke engine with direct fuel injection which developed 6,350-horsepower at 140 rpm.

First Diesel-engines used the 4-stroke cycle, but later Diesel-engines were built on the 2-stroke cycle, which had been first used by Dugald Clerk in his gas-engine in 1878. In the 2-stroke cycle every revolution of the crank is produced by a power-stroke, and the induction and exhaust take place simultaneously at the end of the power-stroke. The 2-stroke Diesel-engine eliminated the separate strokes for charge-induction and scavenging by using a blower to scavenge the exhaust gases out through the exhaust port when the piston is near the maximum volume position. Although the 2-stroke Diesel-engine requires theoretically only half the moving parts of the 4-stroke Diesel-engine, in practice the reduction in moving parts varies from 30% to 40%, its shorter crankshaft is stronger for a given power production, and its valving system is in general greatly simplified. The 2-stroke Diesel-engine developed in 1899 by a German engineer Hugo Güldner (1866-1926) increased the specific power of the Diesel-engine by nearly 20% for the same displacement. The 2-stroke Diesel-engines used throughout the modern industry are the most powerful internal-combustion piston-engines rating up to 50,000 horsepower. Since the 2-stroke Diesel-engines are easily reversible, the large 2-stroke marine Diesel-engines provide the propelling axle power in most naval vessels.

According to Diesel, Franco Tosi of Milan built a Diesel-engine and an equivalent power turbine with oil-fired boilers, which were shown at the Turin Exhibition in 1911. This Diesel-engine had an efficiency of 48%, and it consumed 2½ times less fuel than the turbine. In 1914, Dugald Clerk reported an average efficiency of 40% for the Diesel-engine, 23% efficiency for the Parsons turbine, and 3.8% efficiency for the Boulton & Watt beam steam-engine in comparison with the Diesel-engine for equal power delivery. From 1911 to 1914, the Swiss engineer Alfred J. Büchi (1879-1959) experimented with turbo-supercharging of Diesel-engines at the ***Gebrüder Sulzer Cie.*** in Winterthur, Switzerland, which improved the efficiency of Diesel-engines but there was too much resistance from the head of the Diesel division of this company and Büchi had to discontinue his experiments, although his method of supercharging later became a universal practice. Supercharging of the 2-stroke Diesel-engine by forcing more air into the combustion space in the cylinder to promote more complete combustion is possible by means of the engine-driven scavenger blowers.

Compression ratios of contemporary Diesel-engines vary from 12–to–1 to 22–to–1 compared to compression ratios of Otto-type gasoline internal-combustion engines which vary from 4– to–1 to 10–to–1. The 4-stroke Diesel-engine in comparison with the 2-stroke Diesel-engine has the advantage of slightly better efficiency, higher speeds, smoother engine performance, cooler engine, easier starting, and absence of scavenging apparatus. Smaller Diesel-engines can be started by an electric motor supplied by a storage battery which actuates the fuel injectors and the pistons. Larger Diesel-engines are normally started by compressed air forced into cylinders at a pressure of about 200 psi (1379 kPa). The high compression of air within the cylinders is hot enough to ignite the injected fuel oil which has an ignition temperature from about 600° F to 800° F (316° C to 427° C).

Contemporary Diesel-engines, which provide about 80% of the available power, are used in submarines, locomotives and trucks, and work as stationary engines in generating stations for electric power and in factories. The thermal efficiency of a contemporary Diesel-engine is about 40% whereas that of the Otto gasoline engine is about 30%. Of course, thermal efficiency is not the sole criterion for the application of heat-engines in industry.

Rudolf Diesel has left for posterity one of the unsolved mysteries of our time: he vanished without a trace on board of a ferry going from Harwich to Antwerp in 1913. He was returning from a business trip concerning the commercial production of his engine in England.

The high-speed Diesel-engine was made possible mainly by the experimental work of the German engineer, Prosper L'Orange and the English engineer, Harry Ralph Ricardo, who discovered the important role of turbulence in the mixing of the injected charge and the limits of combustion in the charge. Ricardo showed how by altering the design of the engine, the period of fuel injection could be greatly diminished by creating a violent turbulence of the charge in the cylinder. Ricardo preferred injecting the fuel into an auxiliary spherical combustion chamber which was connected to the cylinder by a D-port. Other engineers favoured direct injection of the fuel into the cylinder together with a hollow hemispherical piston crown to create the violently turbulent mixing. Ricardo's experiments led to notable improvements in the design of combustion chambers and valve ports.

Before 1930, oil-engines had an operating speed of 1,000 rpm, but by 1940, their operating speeds had been increased and now ranged from 2,000 to 3,000 rpm. Robert Bosch in Germany was successful in designing the first really reliable fuel pump which was able to inject precise, minute quantities of fuel 1,500 times a minute into the Diesel-engine running at 3,000 rpm, thereby overcoming the last major difficulty in the operation of the Diesel-engine. Diesel-engine as a marine engine rapidly gained predominance in world's mercantile ships. By 1939 almost ¼ of the mercantile ships in the world were powered by Diesel-engines.

CHAPTER XXI

TURBINE AS HEAT ENGINE: From Steam-Turbine to Turbojet and Rocket

Steam-Turbines

Philon of Byzantion, who flourished about 230 B.C., designed a spinning fountain in the form of an efflux-type hydraulic turbine the 4 radial tubes of which were bent into the shape of the *Swastika*, the venerable Indo-European symbol for wellbeing. The spinning hydraulic fountain functioned by the reaction principle of Archytas of Taras (c.428-c.347 B.C.), like the modern rotary lawn sprinkler.

Heron of Alexandria (*floruit* c.150-100 B.C.), designed a reactive steam-turbine, later called *aiolopyle* [the ball of the Greek God of Winds, *Aiolos*], which consisted of a hollow bronze sphere with radial arms bent into the shape of the Indo-European *Swastika*, in an apparent imitation of Philon's reactive hydraulic turbine. Steam in the hollow bronze sphere was supplied by a boiler through the hollow support tubes of the sphere.

In 1629, an Italian engineer, Giovanni Branca (died 1629), described the first known impulsive steam-turbine which was intended to operate a stamping mill. Branca was probably inspired in his design by the horizontal hydraulic ancient Greek mill known since the first century B.C.. Apparently, Branca's impulsive steam-turbine was never built.

In 1784, an Austrian engineer, Wolfgang von Kempelen (1734-1804), patented an impulse turbine which could be set into rotation by fire, air, water or steam. Kempelen was an unscrupulous international swindler who exhibited to the public for exorbitant admission fees a fraudulent chess-playing robot inside of which, concealed from the public, was an expert chess-playing midget. Kempelen extravagantly claimed that his turbine was superior to Watt's steam-engine. Watt found von Kempelen's impulse steam-turbine to be quite inefficient after examining its excessive steam consumption, high speed of rotation, and low output. Watt concluded from his study of the turbine that the spin of the turbine wheel had to be enormous in order to attain a satisfactory operating efficiency thereby rending the turbine wheel asunder by the action of its own centrifugal forces. Watt used Parent's waterwheel theory in his quantitative evaluation of Kempelen's turbine. By that time it was generally recognised that continuous rotary motion in prime movers is mechanically superior to reciprocal motion.

In 1791, an English inventor John Barber patented an engine in which the combustion of a mixture of inflammable gas and air produced a jet of expanding gases which drove a Branca-type impulse wheel. This engine was the first impulsive gas-turbine.

In 1799, William Murdock (1754-1839) patented a sound turbinelike device in which two engaged cogwheels were rotated by a jet of steam, but at the time the necessary manufacturing precision was not possible. This type of design is essentially used today in oil-pumps. About 1802, Oliver Evans (1755-1819) built an experimental *aiolopyle*-type reactive steam-turbine which operated with a steam pressure of 56 psi (3.86 kPa), made from 700 to 1,000 rpm, and was capable of producing more than the power of 2 men, that is, about one horsepower.

In 1802, a wealthy American inventor Colonel John Stevens (1749-1838) fitted one of his propeller-driven, flat-bottomed boats of 25-foot (7.62 m) length with a turbinelike steam-engine consisting of a cylindrical brass cylinder of 8-in (20.3 cm) bore and 4-in (10.2 cm) length in which was housed a two-bladed rotor mounted on the screw-shaft and driven by high-pressure steam. Stevens designed this turbinelike engine in an effort to avoid the inertia effect of the reciprocating piston in a conventional steam-engine. His screw-propelled boat was able to reach a speed of 4 miles (6.44 km) an hour.

In 1815, Richard Trevithick (1771-1833) patented and constructed an *aiolopyle*-type reactive steam-engine for marine propulsion, discussed above, by fitting two jets to the felloe of the wheel of a railroad engine which were supplied with steam from the boiler, but it was not efficient enough.

William Avery in 1831, Binstall in 1838, Pilbrow in 1843, Robert Wilson in 1848, and Harthman in 1858, had experimented and even patented steam-turbines, but none of these machines produced power in a thermally efficient manner. Since 1837, William Avery of Syracuse in United States, built about 50 Heron's *aiolopyle*-type steam-turbines, the rotor of which measured about 5 feet (1.52 m) across, and attained a linear speed at the periphery of the rotor about 880 feet (268.22 m) per second. Avery's steam-turbines, although mostly used for the driving of cotton gins and circular saws in lumber mills, also found application in locomotion. It was the only commercially produced steam-turbine of the period, but despite Avery's claim that the efficiency of his steam-turbine was comparable to that of the contemporary reciprocating steam-engines, its noisiness, hazardous operation and unreliability led to its abandonment. The overall efficiency of turbines requires very high speeds of rotation of the turbine rotor, a fact which the ingenious Swiss mathematician Leonhard Euler (1707-1783) had already established theoretically for his efflux hydraulic reaction turbine in 1753. The first experimental steam-turbine which proved to be an adequate practical success was constructed in 1846 by two English enginewrights, James Jamieson Cordes and Edward Locke from Newport, Monmouthshire. The action of this steam-turbine was based on the same principle as Branca's impulse-wheel, and it produced 32-horsepower. In the same year, Cordes and Locke built a prototype impulsive marine steam-turbine which had a rotor of 19-foot (5.79 m) diameter, and produced 150-horsepower at 300 rpm but, apparently, this impulsive steam-turbine was never fitted to a vessel. The idea of steam-turbine as a prime mover seems to have been quite popular at the time since more than 200 patents on steam-turbines had been issued by 1884 in Great Britain alone.

Since the steam-turbine is a thermomechanical prime mover, its overall efficiency is best achieved by a scientific design. For this reason, the successful designers of modern steam-turbines were all scientifically trained engineers. Since 1877, a scientifically educated Swedish engineer of French Protestant stock who graduated in 1866 from the **Kungliga Tekniska Högskola** (*Royal Institute of Technology*) in Stockholm with a mechanical engineering degree, and received his doctorate in 1872 from the University of Upsala with his thesis on the composition of chlorine and bromine, Carl Gustaf Patrik de Laval (1845-1913), had been trying to perfect the centrifugal cream-separator patented by the American electrical engineer, Elihu Thomson (1853-1937), which had to be rotated at very high speeds, from 6,000 to 10,000 rpm, to perform its function. De Laval decided to use the *aiolopyle*-type reaction turbine to drive this cream-separator, which was essentially a development of Avery's turbine. However, it is known that the English inventor Henry Bessemer (1813-1898) had exhibited in 1851 at the Crystal Palace Exposition in London a similar Heron-type of reaction turbine and a centrifugal separator for sugar syrup. De Laval tried to develop several versions of this reaction turbine which had an S-shape rotor, but came to the same conclusion as Watt had before him, that the efficiency of such a steam-turbine is not satisfactory unless its rotation attains speeds ranging from 30,000 to 40,000 rpm. He spent a great deal of time and effort to solve the difficult technical problems attending to such enormous speeds of rotation, and particularly the problem of the balance of the axle and the S-shape rotor. Even the slightest imbalance in the rotor and the axle created enormous centrifugal forces which were able to break the heaviest of steel axles. De Laval had finally mastered these technical problems by 1883, when he took out a patent on a reaction steam-turbine with an S-shape rotor of 5-foot (1.52 m) diameter. One type of de Laval's S-shape steam-turbine was designed to operate at 42,000 rpm, but it was still inefficient because of the poor nozzle shape. After this trying experience, de Laval decided to work on the action-type of steam-turbine first proposed by the Renaissance engineer Giovanni Branca.

Contemporaneously with de Laval, pioneering work was done on steam-turbines by the Irish physicist and engineer, Osborne Reynolds (1842-1912), a graduate of Cambridge University who was elected to the newly created chair of engineering at Owens College [now Manchester University] in 1868, after some years of civil engineering practice. Osborne, an exceptional scientific engineer, established his important similarity law of flow in hydrodynamics which also applied to aerodynamics, in two remarkable memoirs published in 1883 and 1884. Reynolds similarity law of flow of fluids made possible the determination of the transition from the laminar to the turbulent flow of fluids which can be easily found by means of the so-called 'Reynolds Number':

$$R = \frac{\rho v \ell}{\mu},$$

where ρ is the fluid density, μ the fluid viscosity, v the speed of flow, and ℓ a characteristic length. The similarity law of fluid flow of Reynolds together with the theory of dimensional analysis published in 1892 by John William Strutt, Lord Rayleigh (1842-1919), professor of physics at Cambridge University, made it possible for engineers to use experimental results obtained from inexpensive scale model tests in the design of large-scale engineering systems, such as turbines. In 1875 Reynolds took out a British patent on turbines and turbine-pumps operating with liquid or gaseous media, which dealt with both the multistage turbines and turbine-pumps with movable guide vanes and divergent diffuser passages the importance of which he scientifically recognised. The scientifically designed turbine-pump of Reynolds was considerably more efficient than the existing centrifugal pumps, particularly at high pumping heads. The first Reynolds hydraulic turbine-pump with fixed guide-vanes was built in 1887 for the engineering laboratories at the Owens College. The first commercial turbine-pump designed by Reynolds for the pumping of deep mines was manufactured by Mather & Platt, and his design was later used by many other companies the world over such as *Gebrüder Sulzer Cie.* (*Sulzer Brothers Comp.*) in Winterthur, Switzerland, a French company headed by Auguste Rateau, and also Laval Steam Turbine Company in Sweden. Reynolds' turbine-pump was quite similar in design to the hydraulic turbine of James Thomson (1822-1892) [brother of the Scottish physicist William Thomson, later Lord Kelvin], professor of engineering at the University of Glasgow, patented in 1850 which also operated with movable guide-vanes.

Soon after 1875, Reynolds constructed and operated a small experimental multistage axial steam-turbine which ran at 12,000 rpm, and in principle was essentially similar to the steam-turbine of Parsons, and even to that of Ljungström both of which appeared on the market many years later. It is to be regretted that Reynolds relinquished his work on the steam-turbine when he discovered that the clearance losses between the turbine blades and the casing were relatively high in his small-scale steam-turbine. He should have realised from his own remarkable researches that in large-size steam-turbines the steam leakage losses would have been proportionately less, and that his steam-turbine was able to expand the steam more completely than any piston steam-engine thereby realising more of the potential energy in the hot steam. In 1885, Reynolds gave a mathematical analysis of the convergent-divergent nozzle and its efficiency raising properties, so important in the design of steam-turbines, properties which were later independently rediscovered by de Laval and used in the latter's sandblasting equipment and steam-turbine.

Reynolds, moreover, carried out significant experiments on the problem of cavitation of ship's propellers rotating faster than 500 rpm which cause 'pitting' damage to the propeller material similar to the cavitation problem experienced in rapidly rotating hydraulic turbines with rotors shaped like propellers, on the similarity of model hulls, on river and tidal models, and even on the mechanical equivalent of heat. He also published a fundamental paper on the hydrodynamic theory of lubrication in 1886 which profoundly influenced the design of hydraulic turbines, ball bearings, and other machinery. Reynolds was obviously a scientific engineer of exceptional ability.

An English engineer and an honours graduate in 1877 from the Cambridge University in mathematics and applied mechanics, Charles Algernon Parsons (1854-1931), the youngest son of the Earl of Rosse, successfully developed an axial-flow steam-turbine which was really a turbogenerator because it had a suitable speed of rotation for the turning of his electric generators. This axial-flow turbine was based on the pressure-compounding principle, which reduced the steam-pressure to such an extent that lower blade speeds could be attained and, thereby, allowing the blade stresses to be kept within tolerable limits. In 1884 Parsons began the design of his predominantly reaction-type steam-turbine during his engineering apprenticeship at the Armstrong Works in Newcastle-upon-Tyne. Parsons' steam-turbine concept was influenced by the axial-flow hydraulic turbine of Henschel and Francis, and by the Laval steam-turbine. Parsons realised that if the total energy of the incoming steam is released in a single-stage, as was the case with the Laval turbine, almost prohibitively high rotational speeds would result, and when more than two stages are used by attaching more turbine wheels than two to the turbine shaft, huge fluid friction and turbulence losses would result. Therefore, he divided the entire range of steam-expansion of his turbine into 15 successive steps, or stages, each consisting of one turbine wheel with a ring of fixed blades attached to the inside of the turbine casing, and one turbine wheel with the ring of moving blades attached to the rotating shaft of the turbine. He let the steam flow between the fixed blades as guide vanes and meet the moving rotor blades at right angles. Parsons knew that this type of pressure-staging method had first been patented in 1848 for axial-flow and radial-flow steam-turbines by Robert Wilson of Greenock, Scotland. In the Parsons' turbine, the blades were shaped in a manner which allowed expansion of the steam during its passage between the blades, that resulted in a reduction of the steam-pressure and increase of speed of the steam. Parsons designed his multistage turbine in such a manner that the approximately 10% drop in steam-pressure across each stage of the total 15 stages was split to yield equal pressure drop across the fixed and the moving ring of blades. In Parsons' design steam expands also in passing through a succession of rotating blades. This arrangement accelerated the flow of steam in the moving ring of blades which produced a reaction in the turbine and, therefore, it became known as the reaction-type turbine. Since the turbine reaction is presently defined as the ratio of the pressure drop across the moving ring of blades to the total pressure drop over the stage, then the Parsons' turbine is really a 50 % reaction-type design. According to this criterion, the impulse turbine, which has expansion of steam not in the moving ring of blades but rather in the fixed ring of blades, represents a zero reaction-type turbine.

Parsons constructed in his turbine 15 wheels with rings of fixed blades and a rotor with 15 wheels with rings of rotating blades. The centrally admitted steam was split into two, half the steam flowed towards one end of the shaft and the other half towards the other end. Each half of the split steam when passing through a set of turbine wheels pushed against the turbine blades. The steam in the Parsons' turbine flowed symmetrically from the center towards both ends of the turbine in order to avoid the undesirable unbalanced end-thrusts which made a special thrust bearing of the turbine shaft unnecessary. The steam passing axially through each set of blades within its cylindrical casing under a maximum steam-pressure of 80 psi (552 kPa) set the rotor shaft into rotation at the high rate of 18,000 rpm, that produced the linear speed of 82,260 feet (25,075 m) a second of the blades and the centrifugal force acting on the rotor blades of nearly 13,000 times the force of gravity. This axial-flow turbine turned a 2-pole dynamo to which it was directly connected, and produced 75 amperes of direct-current at 100-volt tension. Since electric generators of the period were usually made to turn from 800 to 1,200 rpm, Parsons' design for the generator functioning at such a high rate of rotation was itself a remarkable achievement. A direct-drive screw-pump which circulated the lubricating oil to the main bearings of the turbine was a distinctly pioneering technique of forced lubrication which Parsons had already devised and patented for a revolutionary steam-engine of his design in 1877. The invention of forced lubrication of high-speed steam-engines is traditionally credited to Albert Charles Pain (1856-1929), a draughtsman at the Belliss & Marcom Works, a firm which manufactured high-speed steam-engines to drive electric generators as a successful rival to the high-speed steam-engines of Peter William Willans (1851-1892). Pain devised his forced lubrication system in 1889, and although the first steam-engine of 20-horsepower running at 625 rpm with forced lubrication was built by this firm in 1890, the designers of internal combustion engines were surprisingly slow to adopt forced lubrication to their engines. The steam-pressure varied over the blades in the Parsons' turbine because the blades made the turbine wheels into a series of diverging nozzles. To accommodate the increase in the volume of steam as it expands, the turbine blades increased in length in three increments. The leading edge of each blade was bevelled to facilitate the flow of steam. The steam consumption in a Parsons' turbine was 130 pounds (59 kg) per kilowatt-hour, whereas modern turbines consume about 8 pounds (3.63 kg) of steam per kilowatt-hour. The first Parsons' turbogenerator which had straight blades and supplied 7½ kilowatts at 18,000 rpm was built in 1884 for the lighting plant of a ship. It had a very sophisticated original servo-motor which governed the speed of the turbine directly by the output voltage of the dynamo. Parsons then developed a steam-turbine in which both the moving blades of the turbine wheel and the fixed rows of blades anchored in the casing were constructed in nozzle shape. In this type of steam-turbine the turning torque acting on the turbine wheel was primarily created by the reaction of steam issuing from the moving blades.

On account of the high rate of rotation of his turbine which was at least 10 times faster than any existing engine, Parsons had to face and solve a mechanical phenomenon known as the 'whipping' rotary motion of the shaft of his turbine. Each shaft has a characteristic critical speed of rotation at which even a very slight imbalance can send the shaft into a violent 'whipping' rotary motion. To prevent any physical damage to the shaft from such a phenomenon, Parsons had to devise a bearing for the shaft which allowed some lateral movement. He invented an ingenious bearing for such a spinning shaft in the form of an assembly of two sizes of washers compressed together by a spring. The vibration of the rotating shaft due to its lateral 'whipping' movement was promptly damped out by the large friction between the two sets of washers.

The Parsons steam-turbine was first installed in a public power station in Newcastle in 1888 where it supplied 75 kilowatts of electric power. It was built by Clark, Chapman & Company, a firm which Parsons joined as a partner in 1883.

In 1889, after almost 300 axial-flow marine turbogenerators had been produced, Parsons lost the patent of his axial-flow reactive steam-turbine, which represented an important technological breakthrough, in a dispute with his partners in the Clark Chapman & Company over the funding of a large-scale turbine development for central electric power stations, who retained his patent for an axial-flow steam-turbine and the dynamo with the drum armature when he left the company, and founded in 1890 his own company: C.A. Parsons & Co., Ltd., the Heaton Works of which at Newcastle-upon-Tyne soon became world famous. Therefore, Parsons had to continue his work with the compound radial-outflow type of reactive steam-turbine, which he had also patented but which was not as efficient as the axial-flow type. Parsons introduced a condenser to the radial-outflow turbine which reduced its steam consumption drastically and, thereby, made the radial-outflow steam-turbine more efficient than the contemporary reciprocating steam-engines of comparable power.

Parsons had great difficulties in selling his turbogenerators to the builders of central electric power stations who preferred the conventional reciprocating steam-engines despite the loud noise their operation created. Ultimately, Parsons had to found with his friends the Newcastle & District Electric Lighting Company in 1890 in order to build the first central electric power station using steam-turbines [which were not as noisy as the reciprocating steam-engines] to drive the dynamos. Even after this success Parsons had to assume certain financial risks in the electric power companies to be able to supply his steam-turbines for the driving of their electric generators. He built his first condensing 100-kW radial outflow steam-turbogenerator for the Forth Banks Station in 1891/1892, which produced single-phase alternating current with a frequency of 80 cycles a second at 2,000-volt tension. The rotational rate of the turbine was only 4,800 rpm, and its operating efficiency was very high. This turbine attracted worldwide attention among electrical engineers, and a number of his turbogenerators were built with ever-increasing capacity. In 1898, Parsons installed 2 turbogenerators of 1000-kW in the electric power station in Elberfeld, Germany, the largest turbogenerator in the world at the time, which consumed 20½ lbs (8.3 kg) of coal a kilowatt hour. He supplied 4 turbogenerators of 1,000-kW to Tyneside Power Stations, 2 to Fourth Banks Station, and 2 to Close Works. Parsons finally built a giant 25,000-kW turbogenerator for the electric power stations.

Meanwhile, de Laval had been working on an impulsive steam-turbine design, and in 1888, he produced his single-stage impulsive steam-turbine consisting of a single impulse wheel and a single convergent-divergent flow nozzle, which permitted an almost complete conversion of potential energy of pressure into kinetic energy of velocity by an efficient expansion of steam and an attainment of high impact speeds from 3,000 feet (900 m) to 4,000 feet (1200 m) a second. De Laval, who had an excellent theoretical background in thermodynamics and mechanics, derived the convergent-divergent nozzle design on the basis of experiments to convert potential pressure energy into kinetic energy, and to obtain the maximum steam velocity without creating turbulence in the rapid flow of steam. In 1870 he had done experiments on nozzle shapes for his sandblasting equipment to obtain maximum thrust from high-velocity nozzles, which proved useful to his steam-nozzle investigation. He took full cognisance of the point in the nozzle where the critical steam-pressure is reached at which point the relative rates of change of the velocity and the volume reverse and require a diverging nozzle. The ratio of the critical pressure to the initial pressure, called the 'critical pressure ratio', is about 0.57 for steam. The actual profile of the convergent-divergent nozzle de Laval calculated from thermodynamic theory to ensure a nearly adiabatic process. De Laval patented his convergent-divergent steam expansion nozzle in 1888. After his success in the nozzle design, he first developed his sandblasting machine using the convergent-divergent nozzle before he turned to the design of his impulsive, single-stage steam-turbine in which he was influenced by the Pelton Wheel, a successful hydraulic impulse turbine. In later designs, de Laval's single-stage steam-turbine had a set of peripheral convergent-divergent nozzles from which the steam jets impinged directly on the blades of the impulse wheel, giving the wheel extremely fast revolutions. In the de Laval single-stage steam-turbine, the steam exerted equal pressure over all the blade of the impulse wheel, since no further expansion of steam took place when steam traversed the turbine blades. The impulse wheels of different sizes of the Laval turbines varied from 3 inches (7.6 cm) to 30 inches (76 cm) in diameter which rotated with speeds ranging from 4,000 to 9,000 rpm. Owing to the extremely high rotational speeds, de Laval had to mount the impulse wheel on a flexible shaft which sought its own axis of rotation to avoid a violent vibration of the rotor at design speeds. The impulse wheel, together with its axle, was so proportioned that the inevitable imbalance present due to the practical construction of the assembly produced a brief, resonant vibration at a relatively low speed of rotation when the turbine was started or stopped. At resonance, the axle vibrated with an amplitude of about 0.4 inches (10 mm) or more, an amplitude which a resilient axle could withstand. Both ends of the flexible shaft were mounted in spherical bearings which permitted vibrationless running of the turbine near its design speed, after passing through a very brief period of severe vibration near the critical rotational speed in the lower speed ranges that gave the rotor and the axle momentarily a 'whipping' rotary motion during starting or stopping of the turbine.

Since the rotational speed of the Laval turbine was too high for the turning of electric generators, de Laval designed and perfected a method of manufacture of rather elaborate double helical speed-reduction [the so-called 'fish-bone'] gears for the turning of electric generators, which were much larger in size than the steam-turbine itself. In 1889 de Laval introduced a 2-stage velocity-compounded impulse turbine which gave higher efficiency and lower peripheral speed of the wheel. A number of Laval steam-turbines were produced and sold to drive electric generators, but the huge reduction gears required by the Laval turbine made it rather awkward and, therefore, other types of impulsive steam-turbines also using velocity-staging were devised, notably by Charles G. Curtis, to drive electric generators more economically. In 1892, de Laval constructed a small 15-horsepower experimental marine steam-turbine of the impulsive type, but it was never fitted to a vessel and tested because he had decided to produce instead turbines for generating electricity which ranged in output from 5 to 50 horsepower. The Laval steam-turbine was also applied to centrifugal pumps and blowers.

By 1897, de Laval was using steam at a very high temperature and a pressure as high as 3,000 psi (20,685 kPa), which is near the critical pressure 3,226 psi (22,243 kPa) at which a pound of steam has the same volume as a pound of water, and the latent heat of evaporation is zero. The Laval impulsive steam-turbine had a limited capacity and, therefore, it could not be used as a prime mover where great power was required. However, Laval turbines had an extremely high efficiency: a 6-horsepower reciprocating steam-engine consumed about 75% more steam than a comparable Laval steam-turbine. One of the reasons for the high efficiency of Laval steam-turbine was the highly efficient Laval steam boiler, which contained helical pipes of relatively small bore that were heated in the furnace at a high rate through forced combustion produced by automatically controlled electric fans.

De Laval was a versatile inventor: he patented a high-speed blower for the Bessemer steel converter, developed a method of smelting iron and zinc, a method of galvanising, a process of recovering nitrogen from air, a method of producing iron pipes by stamping, and many others. In 1873 de Laval installed a chemical factory for the production of sulfuric acid, and afterwards worked as an engineer in the **Kloster Bruck Eisenwerke** (*Kloster Bruck Iron Works*) in Germany where he developed methods of rolling iron, and experimented with centrifugal machinery. In 1876 he produced a work on high speed technology which brought him worldwide fame. In 1895, 39 enterprises exploited his patents and inventions which brought him a considerable fortune. He was a pioneer in contemporary Swedish industry, and finally became a deputy and senator in the **Riksdag**, the Swedish parliament, where he was an ardent supporter of protectionism.

In 1893, Parsons regained his axial-flow reaction steam-turbine patent by repurchasing his original patents from his former partners for £ 1,500, and he promptly designed an improved axial-flow turbine with an output of 1,000 kW, which was now produced by his own firm. Parsons continued to manufacture his own turbogenerators for electric power plants which increased in power from 7½ kW in 1884 to 200,000 kW in 1931, the year of his death.

Parsons applied his axial-flow turbine to marine propulsion, when in 1897, he outfitted a vessel of 44-ton displacement and 100-foot (30.48 m) length, named <u>Turbinia</u>, with 3 axial-flow steam-turbines : the high pressure and the intermediate pressure turbine at the sides, and the low-pressure turbine in the center which together supplied about 2,000 horsepower. Each turbine was coupled to a separate shaft to which were mounted 3 propellers. The steam from the boiler first passed through the high-pressure turbine on the starboard side, then through the intermediate-pressure turbine on the port side, and finally, through the low-pressure turbine in the middle of the ship, from which it was exhausted to the condenser. A separate turbine was coupled to the central shaft to propel the ship astern. Parsons' marine boiler was a double-ended watertube boiler operating at a pressure of 210 psi (1,448 kPa) which dropped to about 155 psi (1,069 kPa) at the inlet to the high-pressure turbine. <u>Turbinia</u> astonished the spectators in its unannounced appearance in the Naval Review held in celebration of the Diamond Jubilee of Queen Victoria at Spithead in 1897 where it attained for its time an unprecedented speed of 34½ knots (63.9 km an hour) and easily outran the destroyers sent to intercept <u>Turbinia</u>.

It was necessary to run the axial-flow turbines at much higher speeds than ordinary marine steam-engines to achieve their maximum efficiency. Since the high speed of marine turbines necessitated the introduction of speed-reduction gears to obtain appropriate speeds for the turning of the screw-propellers, Parsons outfitted a small twin-screw launch in 1897 with a 10-horsepower turbine which ran at 20,000 rpm, but he had to provide a reduction gear in the ratio of 14–to–1 in the form of single-helical spur-wheels to the two shafts which revolved at 1,400 rpm. Single-reduction gearing has mechanical efficiency of about 98½ %. Double-reduction gearing is used in merchant marine turbines where the screw-propellers turn at 100 rpm to avoid the damage done to propellers by cavitation at higher rates of rotation of the propellers.

Another important steam-turbine pioneer was Auguste Camille Edmond Rateau (1863-1930), a graduate from the *École Polytechnique* and the *École Supérior des Mines* who taught industrial electricity at the *École des Mines* in St. Etienne and analysis, mechanics and industrial electricity at the *École Supérior des Mines* in Paris. In 1894 Rateau began his work on steam-turbines in collaboration with the *Sauter, Harlé & Cie.* in Paris. After some experiments with a modified Laval steam-turbine which proved inefficient, he developed an original steam-turbine which was the multicellular, pressure-staged impulse turbine called the 'Rateau turbine'. Rateau founded his own company in 1903 for the purpose of producing his steam-turbine and other rotary machinery of his own design. Rateau's multicellular turbine consisted of a certain number of units in which each unit contained one fixed distributor and one moving wheel that revolved in a chamber where the pressure remained uniform, a unique arrangement which allowed the turbine wheel to rotate slower but without any loss in output, and made it possible to connect the electric generator directly to the axle of his turbine. There was no longitudinal thrust on the moving parts of the Rateau turbine and, moreover, it was possible to admit steam over a limited part of the circumference of the turbine. Rateau claimed that there was less steam leakage in his turbine than in the Parsons turbine and, therefore, his turbine required less precise and expensive workmanship. The Rateau steam-turbine became universally popular after it was exhibited in 1900 at the World Fair in Paris. At first Rateau was interested in designing centrifugal blowers used for the ventilation of mines, on which he had published a definitive memoir in 1892, but starting from 1902 he concentrated his efforts on the design of steam-turbines for central electric power stations. By 1902 over 700 axial-fans of Rateau were in use in Europe for the ventilation of ships, public buildings and theaters, and over 200 centrifugal blowers of Rateau were in use mostly for the ventilation of mines. He became rapidly the leading authority not only of steam-turbine and gas-turbine design, but also of rotary air compressors, turbopumps, and turbosuperchargers for piston-driven internal combustion aircraft engines as well as for Diesel-engines. After he published his outstanding multivolume work, <Treatise of Turbomachines > (*Traité des Turbo-machines*) in 1900, which was full of original contributions by Rateau, it was universally recognised as the masterpiece on all forms of turbomachinery. Rateau was not only a very sophisticated and extraordinarily competent theoretician in both pure and applied sciences, but also a highly skilled engineer and clever businessman. He was the only one among the 4 inventors of steam-turbines who was well-informed about all the previous work done in turbomachinery. Rateau was a perfect example of a scientific engineer: he designed his machine after forming an accurate scientific theory for it, then perfected that work by systematic experiments for which he created a new methodology and complex test apparatus. His experimentally measured quantities corresponded to his theoretical concepts and, therefore, they could be easily incorporated into the previously existing thermodynamic theory, which enabled him to refine and corroborate his theory. Rateau's multistage steam-turbines were manufactured under licence by such prominent engineering companies as *Maschinenfabrik Oerlikon* (*Machine Factory Oerlikon*) in Switzerland and the *Skoda Werke* (*Skoda Works*) in Czechoslovakia.

An American engineer Charles Gordon Curtis (1860-1953), who graduated with a civil engineering degree from the Columbia University in 1881 and a law degree from the New York Law School in 1883, founded in 1888 his own firm which produced electric traction motors for railways after a brief partnership in a firm manufacturing the first standard specification electric motors. In 1896, Curtis patented a velocity-staged steam-turbine which was similar but not identical to the 2-stage Laval steam-turbine. The main purpose of Curtis was to design a steam-turbine which did not employ excessive speeds and did not have such a great number of parts as required by the Parsons steam-turbine. To reduce the speed of the steam-turbine, Curtis used the principle of velocity-staging in his multistage impulse turbine by absorbing the steam in a series of concentric runners with blades which alternated with stationary guide vanes having opposite curvatures to the curvatures of the moving blades like in the Parsons steam-turbine. The guide vanes changed the direction of the steam jet but, at the same time, the steam expanded like in the Laval steam nozzle. In the Curtis multistage steam-turbine, the kinetic energy of the radially escaping steam was caught impulsively on successive rings of blades as the steam flowed radially outwards and, therefore, it was much more efficient than the standard Laval turbine. The limitation of the velocity-staged steam-turbines is that for more than 2 stages the friction and the turbulence losses tend to become excessive. The Curtis steam-turbine was mechanically quite different from the other steam-turbines designed by de Laval, Parsons and Rateau. However, Curtis was not the first to use the idea of velocity-compounding since de Laval in late 1880's had thought of the possibility of reducing the peripheral speed of his impulse wheel by introducing several rows of moving blades to his turbine which in succession absorb the kinetic energy of the steam flow, but he gave up experimenting with this type of steam-turbine for lack of real progress in its development. Curtis among all the other inventors of steam-turbines did not develop his design into a commercial product himself, but rather sold it for $1,500,000 to General Electric Company for development and execution, yet he retained his royalty rights in the use of his turbine in electric power applications. When the 500-kW electric power plant of General Electric was built in Schenectady the performance of its Curtis steam-turbine was quite impressive. The early Curtis steam-turbines had vertical axles, which required specially designed axle bearings and oil lubrication delivered under very high pressure, but later, when Curtis steam-turbines became larger and heavier, they were erected horizontally. Where great power was required, the Curtis impulse steam-turbine competed successfully with the Parsons reaction steam-turbine.

In the impulse turbine the rotor blades are moved in the same direction as the flowing steam, whereas in the reaction turbine the rotor blades are moved in one direction whilst the flowing steam impacting on the rotor blades is deflected in another direction.

Subsequently, many steam-turbines in which a few turbine wheels were impulsive and the rest were reactive were designed by using the advantages of both types of steam-turbines. For instance, the prominent engineering company in Germany, the **Maschinenfabrik Thyssen &Co.** (*Machine Factory Thyssen & Co.*), manufactured a steam-turbine which was a combination of the Parsons and the Curtis type of turbine, whereas the English Electric Company manufactured a combination of the Rateau and the Curtis type of turbine. Soon so many different combinations and developments of the steam-turbine types were marketed that the original turbine types for all practical purposes had lost their identity. Such turbines, designed by a Swiss engineer Hans Zoelly (1880-1950), were called in general the 'impulse-reaction steam-turbines'. The newer types of steam-turbines began to use progressively higher steam pressures, better boilers, and superheated steam.

Two Swedish inventors, Birger Ljungström (1872-1948) and his brother Fredrik (1875-1964), patented in 1906 a contra-rotating radial-outflow reactive turbine, the working principle of which had been abandoned by Parsons in 1894. In 1910 the 2 brothers were encouraged to develop such a turbine by Professor Aurel Stodola (1859-1942) of the **Eidgenössische Technische Hochschule Zürich** (*Swiss Federal Institute of Technology Zurich*), the leading expert on steam power and a promoter of turbines as prime movers, despite the fact that they found no financial backers for their turbine at the time. The principle of operation of this radial-outflow turbine was based on letting the steam flow radially from the axle to the periphery, whilst permitting the steam to expand and gain speed. This turbine had 2 turbine wheels which rotated in opposite directions. The steam expanded radially outwards through concentric rings of blades which were alternately attached to one or the other of the turbine wheels, and the blades were mounted in such a way that the blades of one ring served as 'guide vanes' for the next ring of blades, whilst absorbing the energy of the steam. The size and shape of the blades and their distance from one another varied with the distance from the turbine axle. In 1908 the Ljungström brothers founded their Steam Turbine Company which produced in 1910 the first large-scale prototype of this contra-rotating radial-outflow steam-turbine producing 500-horsepower at 3,000 rpm with a thermal efficiency of about 30 %. In 1913 their company was reorganised to become **Svenska Turbinfabriks Aktiebolaget Ljungström** (*Swedish Turbine Factory's Stock Company Ljungström*), popularly known by its acronym **STAL**, and subsequently their steam-turbines became known as **STAL**-turbines. In 1932, the Ljungström brothers built a 50,000-kW **STAL**-turbogenerator, a combination of the radial-flow and axial-flow steam-turbine, a type of steam-turbine they had already discussed in 1910 with Stodola, for the Swedish National Power Administration. In such large turbines, the blade stressing can become so severe that it requires axial flow stages to relieve such high stresses, which motivated the design of the compound turbine by the Ljungström brothers as an important variant of the axial-flow steam-turbine. In this compound steam-turbine, the steam at first expanded through a radial-flow turbine in the center of the axle until the volume of the steam became so large that its velocity almost reaches its maximum, then the steam jet was split and directed through the two symmetrically set axial-flow turbines where the steam underwent further expansion. The **STAL**-turbine was manufactured by Ljungström brothers to produce medium-range power.

The efficiency of steam-turbines improved rapidly with time: by 1884, a 7.5-kilowatt capacity steam-turbine of Parsons consumed 129 pounds (58.5 kg) of steam per kilowatt-hour; by 1900, a 1,250-kilowatt capacity steam-turbine consumed only 18.2 pounds (8.3 kg) of steam per kilowatt-hour; by 1923, a 50,000-kilowatt capacity steam-turbine consumed a mere 8.2 pounds (3.3 kg) of steam per kilowatt-hour. The result of all this improvement was that by 1900, steam-turbines and internal combustion engines were challenging the high-speed steam-engines as the best prime movers. Steam-turbines by that time were smaller and more compact than the high-speed steam-engines of equal power. The difference between the reaction-type steam-turbine and the impulse-type steam-turbine had been gradually reduced because the designers of impulse steam-turbines had introduced varying degrees of reaction into their turbine design.

Modern large steam-engines produce over 300,000-horsepower. Some modern turbo-generators produce hundreds of megawatts under tens of thousands of volts tension, with steam pressures over 2,000 psi (13,790 kPa). The steam-driven turbogenerators operate at fixed speeds to maintain a constant voltage output. In North America the turbogenerators run at 3600 rpm to accommodate the standard 120V alternating-current at a frequency of 60 cycles a second, whereas in Europe they run at 3000 rpm which conforms better to their standard systems of voltages.

The large quantities of steam required by modern steam-turbines are produced in modern boilers which have been improved as dramatically as the steam-turbines themselves. About 1925, boilers were operating with 650 psi (4,482 kPa) pressure at the temperature of 725°F (385°C) whereas recently, they operate with pressures around 2,000 psi (13,790 kPa) at the temperature of 1,000°F (538°C).

In modern steam-turbine design, the steam, after having done its work, is condensed and this condensate is reintroduced as feedwater to the boiler. The modern condensers have enormous cooling surfaces.

By 1950, about 80% of electric power in the world was produced by steam-turbines.

Gas-Turbines

The development of gas-turbines, which are relatively simple power plants consisting of an air-compressor, a combustion chamber, and a turbine wheel, lagged behind that of steam-turbines, mostly on account of the technological problems created by the very high gas temperatures. First gas-turbines were the mediæval 'smokejacks' which were installed in chimneys where the rising hot smoke turned a windmill-like turbine geared to a cooking spit.

The history of the gas-turbine as an internal combustion engine with continuous combustion begins with the patent of John Barber of Nuneaton in England, who had grasped rather well the principle of the gas-turbine, in which the combustion of a mixture of inflammable gas and air results in expanded gases that drive a Branca-type wheel. In 1847, the French engineer of Italian extraction, Claude Burdin (1788-1873), professor at the **École des Mines** (*School of Mines*), proposed to build a combustible gas-turbine, but he never carried it out. Although a great number of inventors after Burdin attempted to produce a practical gas-turbine, the first success came only between 1902 and 1911.

From 1881 until 1883, Parsons worked on gas-propelled torpedoes in collaboration with James Kilson of Leeds at the Armstrong Works. These gas-propelled torpedoes were driven by a gas-jet produced by the burning propellant which impinged upon the blades of a spinner that turned the screw propeller of the torpedo. Parsons built an experimental boat which had a propeller the shaft of which was fitted with a spinner consisting of a ring along the circumference of which were cut 44 spiral passages through which the burning gas jet was forced to produce the deflecting force. In 1884, Parsons prepared his patent specifications for the gas-turbine which included the modern gas-turbine cycle. However, he gave up his experiments on gas-turbines because suitable materials to resist the extremely high temperatures created by the burning gas were not available at the time.

The Norwegian engineer Ægidius Elling (1861-1949) received the patent of his first gas-turbine also in 1884, but it took him about 20 years of methodical and careful research to make his gas-turbine more efficient than 54%, the lowest practical limit of efficiency for the gas-turbine to be able to function under its own power since about ⅔ of the power generated by the gas-turbine is consumed by its compressor which forces air into the turbine's combustion chamber. Another major difficulty in the engineering of gas-turbines was the extremely high heat generated by the burning gases which tended to destroy the turbine's steel blades in a very short time. Since no heat-resistant materials existed at the time which could withstand such extremely high temperatures of the impinging burnt gases, the turbine blades had to be somehow protected against the intense heat. In 1903 Elling became the first person to succeed in building a working stationary gas-turbine, in which he cooled the extremely hot exhaust gases by passing them through a coiled tube submerged in water upon which the water began to boil since the hot exhaust gases gave off their extremely high heat to the water by convection. The steam created by the boiling water was then expelled together with the hot exhaust gases through the blades of the turbine wheel which increased the efficiency of his gas-turbine well over the lower limit of 54%. His ingenious solution eliminated the problem created by the lack of heat resistant materials. Unfortunately the industrial technology in Norway at the turn of the century was not adequately developed for the mass-production of Elling's gas-turbine.

In 1897, a Swedish engineering student, Nils Gustaf Dahlén (1869-1937), devised a simple principle for a working gas-turbine after being inspired by the lectures of Professor Aurel Stodola, the foremost promoter of steam- and gas-turbines, at the **Eidgenössische Technische Hochschule Zürich** (*Swiss Federal Institute of Technology Zurich*), in Switzerland. Dahlén and a fellow Swedish student Arthur Hultquist upon their return to Sweden attempted to build a light and compact gas-turbine to propel the contemporary flying machines. Their first gas-turbine achieved an efficiency of 50 % with 25,000 rpm turbine wheel speed, but the turbine wheel lasted only a very short time because of the extremely high heat generated by the turbine. In 1899, Dahlén and his collaborator tested their gas-turbine incorporating a redesigned turbine wheel but despite its 58 % efficiency, the turbine blades again failed in the high heat after a short time. Dahlén continued his gas-turbine development at de Laval's turbine company, but he finally had to abandon his gas-turbine experiments in 1902 for lack of suitable heat-resistant materials. He became chief engineer in a company specialising in gas accumulators, where he was responsible for a number of inventions for the automatic regulators used in lighthouses and light buoys for which he was awarded in 1912 the Nobel Prize in physics. Unfortunately, Dahlén had been blinded in an industrial accident caused by the explosion of acetylene gas.

An experimental gas-turbine which was partially successful was constructed by Sanford Alexander Moss (1872-1946), a former apprentice machinist and a doctoral student at Cornell University in Ithaca, New York, who in 1900 had written a Master of Science thesis on a gas-turbine at the University of California, but his gas-turbine was not quite capable of producing enough power to operate its own compressor which supplied compressed air for combustion. A gas-turbine has a double function: it drives a compressor to compress the incoming air, and provides output power. Moss' experimental gas-turbine equipment consisted of a converted Laval steam-turbine which he operated by burning gas in a pressurised chamber using an external reciprocating compressor, but it achieved an efficiency of only 3%. However, Moss' thesis made a beginning in the theoretical analysis of the gas-turbine.

When Moss returned to General Electric he was assigned to the steam-turbine department, where he quickly became a leading engineer. By 1917 General Electric had developed an extensive business in manufacturing centrifugal compressors for blast furnaces and for other applications of low-pressure compressed air, for the design of which Moss had organised a turbine research group. When the First World War broke out the research group of Moss was engaged in military work. Military aircrafts using more powerful piston engines during the war were able to fly faster and at higher altitudes than before the war which, unfortunately, led to a deterioration in the performance of the aircraft engines. At higher altitudes the air in its natural state becomes rarified and contains less oxygen which reduces the efficiency and power of the air-aspirating aircraft engine. This type of loss of power in the internal combustion engines could be virtually eliminated by using a booster which precompresses the air before it is driven into the engine where it is mixed with the fuel, compressed by the piston, and ignited, thereby boosting the engine power. The concept of boosting, or supercharging, was not new since piston-type air-compressors had been used before 1900 to produce proper air-fuel mixture, and to scavenge the cylinders of large gasoline and Diesel engines. In 1901 and 1902, Dugald Clerk (1854-1932), the British inventor of the 2-stroke piston engine, carried out one of the first thorough investigations of supercharging. In the period from 1909 to 1912, Hugo Junkers (1859-1935), the famous aircraft builder in Germany, and *Gebrüder Sulzer Cie. (Sulzer Brothers Comp.)* in Winterthur, Switzerland, had used different methods of supercharging Diesel-engines. A Swiss engineer, Alfred J. Büchi (1879-1959), who had worked on the early Diesel-engine development at the *Gebrüder Sulzer Cie.*, conducted in 1911 extensive experiments on the supercharging of Diesel-engines using a gas-turbine. He utilised the exhaust gases of the piston engine which were otherwise wasted, to drive a gas-turbine connected to a compressor which in turn drove compressed air back into the engine thereby boosting its power. Unfortunately, this type of turbosupercharging was soon abandoned by the *Gebrüder Sulzer Cie.* mostly because of the opposition of the head of the Diesel division of this firm who preferred conventional 2-stroke engines since the *Gebrüder Sulzer Cie.* had achieved their business success by building this type of ship and stationary engines. In 1916, Büchi's turbosupercharger concept was taken up by the French engineer Auguste Rateau, who designed a supercharger for the French Air Force consisting of a single-stage turbocompressor operating at very high rotational speeds which was capable of producing sufficient compression in the air to compensate for the altitude effect. The gas-turbine of the compressor driven by the exhaust gases of the aircraft engine did automatically increase its speed of rotation when the atmospheric back-pressure diminished owing to the climbing of the aircraft. Rateau tested his design in early 1917, and his turbocharger which weighed 56.8 pounds (23 kg), and produced 50-horsepower at 30,000 rpm, was installed in some French military aircrafts which participated in the war. William Durant, professor at the Cornell University and at that time the head of the newly established National Advisory Committee for Aeronautics (**NACA**), asked Moss as well as Rateau to develop and test a booster for the military aircraft of United States. The supercharger of Moss was crucially different from that of Rateau, since it provided air-cooling for the turbine whereas the supercharger of Rateau, as was the usual practice in steam-turbine design, did not. In the middle of 1918 tests were carried out on both superchargers, and the supercharger of Moss which gave superior results won the contract. The Liberty piston engine boosted by the Moss turbosupercharger and operating at 14,109-foot (4,300 m) altitude produced a 6-horsepower increase in the power of the aircraft engine operating at the sea level. Unfortunately, one month later the contract was cancelled because of the Armistice.

In his turbosupercharger design Moss had narrowly missed to develop a turbojet-engine. His turbosupercharger compressed engine gases and cold air between the centrifugal compressor and the turbine, and then fed the mixture of gas and air to the aspirating engine. If Moss would have installed a combustion chamber between the compressor and the turbine and heated the fuel-air mixture, he would have had in effect a turbojet-engine. Nevertheless, the efforts required to develop the turbosupercharger gave American engineers valuable early experience in devising special alloys for gas-turbine blades, and in the basic form of gas-turbines.

Between 1903 and 1906, the French engineers René Armengaud and Charles Lemale built their first functioning gas-turbine which consisted of a rebuilt old two-stage Laval impulse turbine driving a multistaged centrifugal compressor designed by Rateau. The multicellular rotary compressor with 25 impellers in series arranged in three casings on the same shaft supplied highly compressed air (from 3 to 6 atmospheres) to a combustion chamber in which fuel was burnt continuously as in a furnace and the resulting combustion gases were then discharged and expanded through the turbine wheel back to the atmospheric pressure. In some modern gas-turbines the heated combustion gases are sent back to a heat-exchanger to preheat the air leaving the compressor and entering the combustion chamber. Such a heat regeneration makes the design of the gas-turbine much more complicated, but it increases markedly its efficiency. Since the very high temperatures of the combustion gases required cooling of the turbine blades and disks with water, Rateau and Armengaud designed an internal water-cooling system for the multicellular turbocompressor which delivered a compression ratio of 3-to-1 at 4,000 rpm, and gave an efficiency from 65% to 70 % whilst using 328-horsepower. Another method of controlling high temperatures in a gas-turbine is to supply excess air from the compressor to reduce the temperature of combustion to the design temperature. Unfortunately, this also lowers the gas-turbine's efficiency because more air must be compressed than required, and as a result, the compressor absorbs more than half of the power generated by the turbine. This particular gas-turbine was fuelled with paraffin (kerosene) which was injected together with water into the single pear-shaped and carborundum-lined combustion chamber and then vaporised, thereby cooling the exhaust fumes and protecting the blades of the 2-stage Curtis-type turbine through which the combustion chamber was exhausted. This gas-turbine definitely proved that the principle of its design was correct, but its thermal efficiency was very low, only 3% since the complex cooling arrangement robbed the gas-turbine of its possible high efficiency. It ran at 5,000 rpm and developed 300-horsepower instead of the designed 500-horsepower. In 1908, Armengaud and Lemale founded the **Société Anonyme de Turbomoteurs** (*Limited Company of Turbomotors*) in Paris, and in 1908, produced a larger gas-turbine but it also had a very low efficiency. **Brown-Boveri & Cie.** in Baden, Switzerland, had manufactured the turbocompressor for the Armengaud and Lemale gas-turbine, and owing to its success they remained in the turbocompressor business.

In 1903, the German engineer, Hans G. Holzwarth (1877-1953), patented his explosion-turbine, the operation of which was based on a new principle, requiring preliminary compression, a combustion chamber with guided admission and exhaust valves, and a spark plug. Between 1905 and 1908, Holzwarth built the first really practical constant-volume gas-turbine in Hannover. His second explosion-turbine with several combustion chambers was built in 1910, and it became the first gas-turbine to be used in industry. Holzwarth designed at the Mannheim subsidiary of the **Swiss Brown-Boveri Cie.** a gas-turbine to produce 1,000-horsepower, but his 1913 prototype yielded a mere 200-horsepower. Only after the Swiss **Brown-Boveri Cie.** began to cooperate with Holzwarth in the development of his gas-turbine did gas-turbines become competitive with steam-turbines as practical sources of power. **Brown-Boveri Cie.** continued the development of Holzwarth's gas-turbine under the direction of Adolf Meyer (1880-1965) who was in charge of the thermal division of this firm. In 1930, **Brown-Boveri Cie.** built a 2,000-horsepower gas-turbine which operated with the blast furnace gas and was based on an advanced version of the Holzwarth design. Although the Holzwarth gas-turbine had achieved a limited success in special applications, yet its mechanical complexity, excessive weight and relative inefficiency finally caused **Brown-Boveri Cie.** to abandon its development as a primer mover.

In 1929, the **Bofors Aktiebolaget** (*Bofors Stock Comp.*) in Sweden decided to develop an aircraft which is propelled by a gas-turbine. A team under the direction of the engineer Alf Lysholm (1893-1973), professor of steam-technology at the **Kungliga Tekniska Högskola** (*Royal Institute of Technology*) in Stockholm and chief engineer of **STAL** company, began the development of a gas-turbine engine for the propulsion of aircraft. He had done considerable amount of design work on the contra-rotating radial-efflux types of steam-turbines of Ljungström [**STAL**-turbines], before he undertook to design a great number of different types of gas-turbines for aircraft. The first experimental gas-turbine engine of Lysholm built by the engineering company **Bofors Aktiebolaget** was tested from 1933 to 1935, which revealed severe problems with the multistage centrifugal compressor of the turbine. Lysholm, after trying a positive displacement compressor that proved inadequate, designed a special compressor with helical blades which unfortunately suffered thermal distorsions. After four years of intensive development, Lysholm had an experimental model of a turboprop engine ready for demonstration, and in 1934, 2 distinct versions of the turboprop engine in which the gas-turbine drove a propeller, were offered by him. One of these engines produced 1,200-horsepower. In 1935 **Bofors Aktiebolaget** tried to interest the Royal Swedish Air Force in the

Lysholm's turboprop engine, but did not succeed since the Swedish defence budget had been severely cut and, as a consequence, the turboprop project was abandoned by **Bofors**.

In 1940, Jakob Ackeret (1898-1981), professor of aerodynamics at the **Eidgenössische Technische Hochschule Zürich** (*Swiss Federal Institute of Technology Zurich*), invented the aerodynamic closed-cycle gas-turbine in which a constant quantity of gas is periodically compressed and heated in a continuous process of recycling the same mass of gas. This aerodynamic gas-turbine was developed into a gas-turbine engine that was mass-produced by the **Escher-Wyss Cie.** in Zurich, Switzerland.

The industrial type of gas-turbines gradually earned their place among the prime movers and found their use in driving the centrifugal compressors on gas pipelines, in maintaining pressure in oil wells, in compressing gas for blast furnaces, and in ship propulsion.

Turbojet-Engines

After the First World War, interest in using gas-turbines to power the propeller of an aircraft had arisen, and in 1920 Dr. W.J. Stern of the Royal Air Ministry Laboratory in South Kensington, was ordered to examine the possible use of a gas-turbine to drive the propeller of an aircraft. His thinking unfortunately was dominated by the industrial turbine practice and by assuming the worst industrial case to be the rule, Stern came to the conclusion that the weight of a 1,000-horsepower engine consisting of 7 single-stage compressors and a 2-stage gas-turbine would be 6,000 lb (2721.6 kg) which was far too heavy for any aeroplane and, therefore, his final judgement of such an aircraft engine was unfavourable. Stern otherwise recognised the advantages a gas-turbine can offer over the piston engine of an aircraft, but he was too much influenced by the conventional technology and lacking the imagination to foresee the future possibilities, he ventured to say that no inventor of a revolutionary design will be lucky enough to render the internal combustion turbine practical for an aircraft engine. Stern proved to be a poor oracle of the future of the gas turbine as aircraft engine: the turbojet-engine came into existence through the revolutionary design of 2 young amateurs who were not industrial gas-turbine engineers.

Both the reaction principle and the jet-propulsion were ancient Greek discoveries made by the ancient Greek mathematician and engineer Archytas of Taras (c.428-c.360 B.C.) about 400 B.C., who built a toy-dove which was propelled through the air by the reaction to a jet of hot compressed air expelled from the toy-bird. The principle of reaction of Archytas of Taras was later used by the Greek military engineer Philon of Byzantion about 240 B.C. in a spinning hydraulic fountain functioning like a modern rotative lawn-sprinkler, and by Heron of Alexandria, the head of the school of mechanics in Alexandria about 130 B.C., in his steam-turbine which functioned by the same reaction principle as Philon's spinning hydraulic fountain. In these ancient Greek machines the jet expelled from a machine exerted by reaction a counter-force against the machine itself, and not against the atmosphere.

In modern times a French inventor René Lorin proposed in 1908 a type of air-aspirating jet-engine in which the burnt-gas exhaust of the reciprocating piston engine provided the propelling jet. In 1913 Lorin patented a ram jet-engine which consisted of an inlet compression nozzle, a combustion chamber and an outlet-expansion nozzle but it provided no starting thrust and achieved satisfactory efficiency only in supersonic flight. Despite its remarkable simplicity, Lorin's ram jet-engine found no practical application at the time since this type of jet-engine must first be brought to a high velocity before it begins to function continuously by itself which, however, was not possible at the time. Dr. E.J. Harris of England patented in 1917 a piston-powered engine driving a fan which compressed the air to which fuel was added and the mixture ignited thereby producing a jet of burnt gas, but it gave little advantage over the piston-powered engine.

In 1921 a Frenchman Maxime Guillaume obtained a patent on an axial-flow turbojet-engine consisting of a multistage axial-flow compressor and a multistage turbine. This engine is quite similar to the modern turbojet-engine, but it is doubtful that it could have worked in practice. The patents of Lorin and Guillaume found no practical applications at the time, and were soon forgotten.

It now seems quite evident that traditional mechanical engineers and engine builders cleaved too closely to technological traditions of the industrial turbine design and, therefore, were only able to recognise the axle-power of the gas-turbine. They regarded the burnt exhaust gases which are expelled from the turbine as waste product. The concept of the turbojet-engine as a source of power required a revolutionary concept in the design of gas-turbines as reaction engines, and it was conceived by 3 men who were amateurs rather than professionals in power engineering: Frank Whittle (1907-1996) in England, Hans Joachim Pabst von Ohain (1911-1998) in Germany, and Herbert Wagner (1900-1982) in Germany. These 3 men were responsible for the so-called 'turbojet revolution' because right from the inception their intention was to create a turbojet-engine for the propulsion of high-performance aircrafts flying at high altitudes. The concept of the turbojet reaction-engine expelling a stream of air mixed with a small quantity of burnt fuel at high speed required revolutionary thinking in gas-turbine design which the conventional turbine designers were incapable of imagining. For this very reason the pioneers of the turbojet revolution were scientific amateurs such as aerodynamicists, physicists, and airframe engineers rather than industrial engine designers.

By the time Whittle, von Ohain and Wagner undertook to build their versions of the turbojet-engine, the aerodynamic theory of subsonic flow had been already published in 1918 by Ludwig Prandtl (1875-1953), professor of applied mechanics at the **Universität Göttingen** (*University of Göttingen*) in Göttingen, Germany. The first application of aerodynamics to the design of turbomachinery was made by the German engineering physicist, Walther Bauersfeld (1879-1959), who in 1922 applied Prandtl's aerodynamic theory to the design of the blades of turbocompressors, turbines and turbomachinery in general, which made a scientifically precise design of blades in the form of aerofoils possible. The aerofoil-shaped turbine blade was able to transfer as much energy as possible between the blade and the airflow which raised the efficiency of the turbomachines.

From 1918 to 1926, the Prandtl aerofoil theory was applied to propellers and windmills by Albert Johann Betz (1885-1968), a colleague of Prandtl at the University of Göttingen, and in 1927 Betz in collaboration with Walter Encke (1898-1982) applied this theory to axial-flow turbines. After Betz, the aerofoil theory was used in 1926 by the English engineer Alan Arnold Griffith (1893-1963), and Jakob Ackeret, a former student of Prandtl from Switzerland.

Griffith, who had received his bachelors degree in 1914, his masters degree in 1917, and his doctorate in 1921 in mechanical engineering from the University of Liverpool, was the second man in 1926 to use Prandtl's aerofoil theory in the design of an axial-flow compressor, and a hypothetical turboprop engine. Griffith had already demonstrated his scientific versatility when in 1917 he extended Prandtl's work in 1903 on the use of soap film analogy in structural analysis by applying it to torsion analysis and, when in 1920, he published an original classical work on the theory of rupture. Griffith had proposed in his report a turboprop engine in which a compressor and a gas-turbine mounted within the compressor drive a propeller by means of gearing, an arrangement which unfortunately lead to insuperable practical difficulties with the combustion chamber. He calculated that his turboprop engine would weigh only about one-fifth of what Stern had calculated in 1920. However, Griffith as an engineer was technologically much too convention-bound for he utterly failed to realise that the reaction to the exhaust jet expelled by the gas-turbine was all that was required for the efficient propulsion of aircraft. Despite his scientific versatility, Griffith was 'hidebound' in conventional aerodynamics since to him the exhaust jet of the gas-turbine was nothing but 'useless waste product', in a sharp contrast with the ingenuity and exceptional insight of the ancient Greek mathematician and mechanical engineer Archytas of Taras (428-c.347 B.C.), who not only discovered the reaction principle but also applied it effectively in his mechanical design more than 2 thousand years ago. Griffith was promoted to the post of the principal scientific officer at the Royal Air Ministry Laboratory in South Kensington in recognition for his work on the turboprop aircraft engine.

The Swiss aerodynamicist and mechanical engineer Jakob Ackeret was the next person to develop the aerofoil theory for turbocompressors, and the first to develop the supersonic aerofoil theory for infinite-span aircraft wings which later became important in the design of supersonic jet aircraft.

Whittle, a young but very ingenious English pilot officer at the Royal Air Force College, proposed a revolutionary turbojet aircraft engine which dispensed with the piston, crankshaft and propeller system in aircraft engines altogether. When Whittle tried to interest the Royal Air Ministry in his revolutionary aircraft engine, he was sent to discuss it with Griffith who was superintendent of the Royal Air Ministry's Laboratory at the time. Griffith, for reasons of his own, made every effort to discourage Whittle by asserting that Whittle's assumptions about the compressor efficiencies were too optimistic and that Whittle's calculations for his engine were faulty. Dismissing the negative criticism of Griffith, Whittle patented his turbojet proposal in January of 1930, and published his patent in 1932. By that time, Whittle had singlehandedly made correct basic deductions from aerodynamics about the speed and the altitude performance of the aircraft as well as the possible efficiencies of the gas-turbine components that could, and did, make the turbojet-engine successful. Griffith on the other hand stubbornly adhered to the prevalent technological tradition and conventional turbine principles such as the shaft power, and completely ignored the potential of the reaction to the burnt gas-jet expelled as the exhaust of the turbine to propel the aircraft.

Unfortunately, Whittle's turbojet-engine was neglected by the British aircraft industry, particularly after Griffith, probably out of sheer professional envy, had denounced Whittle's turbojet design in his unfair evaluation of it in 1936 for the committee of the Royal Air Ministry. Subsequently, a number of large commercial firms turned down Whittle's request for financial support for the building of a prototype of his turbojet-engine which was required for bench tests. Fortunately Whittle did receive modest initial fund of £ 2,000 supplied by a firm of merchant bankers to found his own fledgling company, the Power Jets Ltd., in March of 1936, through the kind assistance of a few of his former colleagues in the Royal Air Force. Since the Power Jets Company could not afford a large plant required for the testing of each turbine component separately, Whittle was forced to build the complete engine and test it as a whole. The first gas-turbine built by Whittle used one large combustion chamber, but his engine ran into performance difficulties with an inadequate centrifugal-flow compressor and the faulty fuel control system which leaked unburnt fuel into the combustion chamber before the engine was ignited and thereby built up a pool of fuel between tests carried out in 1937 which caught fire when the engine was reignited and made it run out of control. He redesigned the combustion system which now consisted of 10 reverse-flow combustion chambers in the form of 'cans' as well as the centrifugal-flow compressor which now incorporated new blades having an extra twist to compensate for differential radial velocity and pressure along the diameter of the blade based on the so-called 'free-vortex' blade design discovered by Whittle in 1935. After extensive use, the gas-turbine of Whittle's rebuilt jet-engine disintegrated during its testing in June of 1938.

The same engine was rebuilt with a new turbine, and 10 reverse-flow 'can' combustion chambers. This version of the gas-turbine engine was being tested from October of 1938 to February of 1941, when it was finally destroyed by a turbine failure. Since by that time the performance of Whittle's turbojet-engine had convinced even the most stubborn skeptics that it will work in practice, the Royal Air Ministry was finally willing to finance the third reconstruction of the Whittle engine. After witnessing a successful demonstration of Whittle's test engine in June of 1939, the Power Jets Company finally did receive a contract from the Royal Air Ministry to build a flight test engine which can be fitted into an experimental aircraft. In May of 1941, after a number of setbacks in the building of his experimental turbojet-engine, Whittle was finally able to fit the British experimental aircraft Gloster E28/39 with his redesigned turbojet-engine making 16,000 rpm and producing a thrust of 877 pounds (3900 N), which made the first test flight that lasted 17 minutes. It was the second jet plane in the world to fly successfully. It made only one more test flight for the press in December of 1941 before it was retired and placed in the Science Museum. Unfortunately, little work was subsequently done on the Whittle turbojet-engine because of the heavy hand of the bureaucracy in the British military forces as well as in the government. Not until 1944 was the aircraft Gloster Meteor F9/40 fitted with 2 Whittle's turbojet-engines producing 1,700 pounds (7,562 N) of thrust and maximum aircraft speed of 410 miles (660 km) an hour [in late 1945, Whittle's engine was upgraded to 3,500 pounds (15,569 N) of thrust which gave a speed of 550 miles (885 km) an hour] and allowed to enter the air war, but by that time it was too late for the British turbojet aircraft to have any direct influence on the outcome of the war. These jet-powered planes were mostly used to chase the 'German Flying Bomb V-1' which were propelled by pulsating jet-engines to a maximum speed of 435 miles (700 km) an hour. Scarce heat-resisting and strong metal alloys were readily available in Great Britain. Nimonic-80 nickel alloy containing 80% nickel and 20% chromium was used for the blades of the turbine, and Nimonic-75 nickel alloy containing 75% nickel and 20% chromium was used for the combustion chambers of the Whittle turbojet-engine manufactured by the Rolls-Royce Welland Company for the Gloster Meteor F9/40 aircraft. This turbojet engine had to be overhauled after each 100-hours of operation.

As stated above, Whittle singlehandedly deduced the correct assumptions about the efficiencies of the gas-turbine components, the aircraft speeds and the altitude performances which made the jet-powered aircraft a technological success and, thereby, created the aeronautical revolution. Whittle himself was professionally competent to design the radial-flow compressor and the turbine for the turbojet-engine himself, but in the design of the combustion system he lacked practical experience and, therefore, he needed outside help. In the development of the combustion system for his turbojet-engine he was assisted by the firm Laidlaw, Drew & Company. Whittle used Diesel oil as fuel in his experimental gas-turbine, but when the turbine became hot the Diesel oil began to vaporise and burn prematurely in its combustion chamber which remained a difficult problem to overcome for a long time. Whittle finally used a U-shaped combustion system in his first turbojet-engine which alleviated the problem of premature combustion of Diesel oil.

In 1933, almost at the same time as Whittle, another turbojet pioneer, a young and talented German physicist Hans Joachim Pabst von Ohain was working on his doctoral dissertation at the ***Institut für Physik der Universität Göttingen*** (*Institute for Physics of University of Göttingen*) as the assistant of the prominent physicist Professor Robert Wichard Pohl (1884-1976). Ohain, who had attended some lectures on aerodynamics at the University of Göttingen, became interested in aeronautical engineering which motivated him to study in his spare time the state of development of aircraft. He soon came to the conclusion that the speed of aircrafts with streamlined airframes should have been much greater than they actually were at the time. He, moreover, noted that the progress in increasing aircraft speeds had appreciably slowed down by the early 1930's. His studies of the piston-and-propeller aircraft engines led him to conclude that the power of such aircraft engines could not be so significantly increased that they are able to propel streamlined airframes to much higher speeds and, therefore, other possible sources of power had to be found. Ohain was unaware of Whittle's patent on the turbojet-engine, but he was well informed about the work done on axial-flow compressors by Albert Betz and Walter Encke at the ***Aerodynamische Versuchsanstalt Göttingen*** (*Aerodynamic Experiment Establishment in Göttingen*). On the basis of the current aircraft speed records, the results of the research of Betz and Encke on axial-flow compressors, and his own theoretical investigations, Ohain came to the conclusion that the turbojet was the only possible aircraft engine capable of producing real progress in high-speed aviation.

In 1934 during his doctoral studies in Göttingen, von Ohain performed theoretical calculations on the basis of which he designed a turbojet-engine with a centrifugal-flow compressor giving a pressure-ratio of 3–to–1, and a radial-flow turbine working at a temperature between 650°C and 760°C (1202° and 1400°F). The thermal efficiency of this turbojet-engine was about 20% and the jet efficiency about 60% in comparison with a total efficiency of 24% for a piston-type aircraft engine.

Ohain developed a simple form of construction for his engine and applied for a patent in 1935 which was granted in 1937. At the end of 1934 von Ohain designed a demonstration model of his turbojet-engine which was built at his own expense in the ***Reparatur-Werkstatt Bartels & Becker*** (*Repair Workshop Bartels & Becker*) in Göttingen where his Ford car was usually serviced. The master mechanic of this workshop, Max Hahn, whom von Ohain knew personally, was willing to construct the demonstration model, and Hahn right away suggested some simplifications to the construction of the engine. In the beginning of 1935, Professor Pohl allowed von Ohain to use the equipment of the Physics Institute to measure pressures and temperatures in his demonstration engine which was built from sheet metal and had a diameter of 4.00 feet (1.22 m). From these tests it became apparent that the combustion took place mainly in and behind the turbine rotor. Although the demonstration engine did not run completely under its own power, it clearly indicated the soundness of von Ohain's ideas concerning his turbojet-engine. The experiments also made it clear that lengthy and expensive combustion investigations were necessary for his turbojet-engine and, therefore, Professor Pohl came to his rescue by writing a letter in 1936 to the famous German aircraft manufacturer Ernst Heinkel (1888-1958) in which he gave information about von Ohain's turbojet-engine experiments. Heinkel, a known taker of great risks and a man fascinated by speed, hired the 24 year old von Ohain as well as Max Hahn for the development of a turbojet-engine which was carried out in secret at the ***Ernst Heinkel Flugzeugwerke GmbH*** (*Ernst Heinkel Aircraft Works, Comp. with Limited Liability*) in Marienehe at Rostock. They were helped by a small team of draughtsmen and Wilhelm Gundermann, an experienced engineer in different areas of machine design, such as hydraulic turbines and torsion converters, who had studied turbomachinery at the ***Technische Hochschule*** in Charlottenburg, a suburb of Berlin.

Von Ohain prudently chose hydrogen as the fuel for his prototype turbojet-engine, since it burns easily and being a gas at normal pressure it presented no problems of vaporisation as was experienced by Whittle with Diesel oil. Although the prototype turbojet engine incorporating 1.97-foot (0.60 m) diameter compressor and turbine wheels was ready by the end of 1936, it was bench-tested in March of 1937 when the engine running under its own power produced a thrust of 288 pounds (1280 N) at 10,000 rpm, and met all other design expectations. This prototype turbojet-engine of von Ohain finally convinced even the most pessimistic critics that the turbojet-engine indeed had a bright future.

Ohain had chosen the radial-flow turbine for his engine because even in the absence of proper milling machines, it was still possible to build a radial-flow turbine in which there exists a proper reaction between its centrifugal-flow compressor and radial-flow turbine of almost equal diameter since both do about the same amount of work and, moreover, both are subjected to approximately the same pressure. Therefore, the radial-flow turbine was less risky in its development and offered less difficulties in determining the matching of turbomachine elements.

After the success of the prototype turbojet-engine of von Ohain all efforts were devoted to the building of a turbojet-engine for flight tests, which can give a thrust of 1,102 pounds (4,900 N) and be ready for installation into an aircraft by early 1939. At this time during a thorough study of existing patents Whittle's patent for the turbojet of 1930 was discovered, a patent which included an axial-flow gas-turbine. But von Ohain and Gundermann did not dare to consider an axial-flow compressor since centrifugal-flow compressors were much easier and faster to build, and only in this way could they meet Heinkel's wish to be the first aircraft manufacturer to fly a turbojet-powered aircraft.

In designing his flight engine von Ohain kept it as much as possible similar to the prototype engine: a centrifugal-flow compressor which finally had 16 blades preceded by an axial-flow intake-fan, which finally had 8 blades, a single combustion chamber, and a radial-flow single-stage influx turbine. He was troubled by the combustion system but unlike Whittle he found no outside expert help to build an effective combustion chamber. It took Max Hahn a year of testing to solve the combustion problem with a reverse-flow combustion chamber, which Hahn also patented. Since the first flight engine tested in 1938 did not produce the designed thrust of 1,776 pounds (7,900 N) or more, primarily because the centrifugal-flow compressor and the combustion chambers were insufficient, the turbojet-engine was redesigned. The rebuilt engine was installed in He178 aircraft which also incorporated a conventional aircraft engine. This aircraft flew 6 minutes with one of the foremost test pilots, Erich Warsitz (1906-1983), at the controls on August 27 of 1939, 4 days before the outbreak of the Second World War, by the turbojet-power alone. It was the first flight of a jet-powered aeroplane. The turbojet-engine of von Ohain which produced 1,100 pounds (4,900N) of thrust at 13,000 rpm, had a specific fuel consumption of 1.6, an outside diameter of 3.94 feet (1.20 m), and weighed 795 pounds (360 kg), but the aircraft He178 reached a maximum speed of only 250 miles (402 km) an hour instead of the designed speed of 500 miles (804.7 km) an hour.

The experimental aeroplane He 178 had been designed by Heinkel's leading aircraft engineers to incorporate a single turbojet-engine which was provided with a long duct passing through the entire length of the aeroplane and ending in a propulsion nozzle, a design feature which led to a 15% loss in the effective thrust. Ohain had suggested to Heinkel to use 2 turbojet-engines attached to the wings of the aircraft instead of the single engine with the disadvantageous long thrust pipe which lost 15 % of the thrust, but Heinkel retained the present arrangement probably because he did not want to delay the historical first flight of a jet-powered aircraft. Warsitz made more than a dozen additional test flights with He 178, although such flights were forbidden by the government after September 1, 1939.

Based on this experience, von Ohain developed in 1941 a new and more powerful turbojet-engine which was designed to propel Heinkel's first aircraft powered by 2 turbojet-engines, the prototype service fighter aircraft He 280, which was already in the design stage before the flight test of He178. The new engine had a smaller diameter, an axial-flow intake-blower followed by a centrifugal-flow compressor straight-through the annual combustion chamber, and a radial-flow influx-turbine. This engine produced 1,596 pounds (7,100 N) of thrust at 13,500 rpm, weighed 851 pounds (386 kg), had a diameter of 2.54 feet (0.75 m) and a length of 5.50 feet (1.68 m). Although this turbojet engine did not meet all the design expectations and had certain shortcomings in its method of lubrication, combustion and cooling, it was flight tested on March 30, 1941 in the first genuine jet-fighter He 280 which attained the speed of 373 miles (600 km) an hour. Heinkel's jet-aircraft He 280 flew more than a month before Whittle's experimental jet-powered aircraft Gloster E28/39 made its first flight test on May 15,1941.

It is important to recognise that already a year before the first flight of He 178, von Ohain and Heinkel had agreed to include the axial-flow turbojet-engines in Heinkel's aircraft engine development programme. In fact, von Ohain and Heinkel had planned to use an axial-flow turbojet-engine in the fighter aircraft He 280. On the suggestion of von Ohain, Heinkel acquired the services of Walter Encke as a consultant on axial-flow compressors, but this collaboration was interrupted in October of 1939, when Heinkel acquired a group of 18 former collaborators of Herbert Wagner from the *Junkers Motorenwerke* in Magdeburg including their leading engineer Max Adolf Müller (1901-1962), who had been developing the axial-flow turbojet-engine of Wagner. Near the end of 1939, Heinkel employed over 120 engineers for the development of turbojet-engines, but at the time Heinkel had neither experienced workshop personnel in engine-building nor sufficient technical equipment for testing engines or their components. In April,1941, the *Reichsluftfahrt-Ministerium* (*State Aeronautics Ministerium*) authorised Heinkel to buy the *Hirth Flugmotorenwerke* (*Hirth Flight Motor Works*) in Stuttgart-Zuffenhausen with a subsidiary branch in Berlin-Waltersdorf for 6 million Reichsmark despite the strong objection by Willy Messerschmitt (1898-1978), Heinkel's most important rival. The *Hirth Flugmotorenwerke* had a staff of experienced practical engineers, master-mechanics, machine-tools, and testing facilities which were needed in order that Heinkel could compete with other large engine-building firms. Many of *Hirth*'s leading engineers, such as Dr. Max Bentele (born 1909), who was an expert in vibration problems and in charge of machine components and other accessories, were also competent scientists.

Since the *Hirth Flugmotorenwerke* had been building superchargers and aeroplanes mainly for sporting purposes, its facilities had to be extended and rebuilt to make them suitable for the building of turbojet-engines, which caused a delay in the development of various turbojet-engine projects in Heinkel company.

It is necessary now to discuss the contribution of the third pioneer in the turbojet revolution, the Austrian engineer Herbert Wagner who is responsible for the creation of the axial-flow turbojet-engine. Among the pioneers of the turbojet-engine Wagner was by far the most versatile. In 1922 he earned his *Dipl. ingenieur* (*diploma in engineering*) in naval engineering and marine engine design from the *Technische Hochschule* in Charlottenburg, a suburb of Berlin, in 3 years after studying 1 year at the *Technische Hochschule* in Graz, Austria. It was a remarkable scholarly performance since this is normally a course of 5 years and, moreover, he worked at the same time as a draughtsman and constructor in a firm building small engines in Berlin. In 1924 he received his doctorate from the **Technische Hochschule** in Charlottenburg with the thesis on the formation of dynamic lift on aircraft wings. His solution was based on a function which describes the time-dependent growth of lift that is still the basis of the instationary wing theory. This function is now called the 'Wagner Function'. He also wrote a seminal work on the impact and sliding processes on the surface of fluids which permits the application of the method and results of the wing theory to the motion of bodies on the surfaces of fluids, a work that Prandtl considered one of the best in hydrodynamics since Hermann von Helmholtz (1821-1894) wrote his famous paper on the vortex motion in perfect fluids in 1858.

From 1923 to 1924 Wagner was the chief assistant of the professor occupying the chair of marine steam-turbines and propellers. From 1924 to 1927, he became a leading structural engineer at the **Rohrbach Metallflugzeugbau** (*Rohrbach Metal Aircraft Manufacture*) in Berlin where he created his famous airframe structure, now called the 'Wagner Web', a framework consisting of closely spaced transverse ribs supporting a sheet-metal skin without diagonal ribs in which the shear stresses are absorbed by diagonal tension fields in the sheet-metal skin. Wagner's new construction method of aircraft bodies created a revolution in the design of aircraft structures which rapidly became a general airframe design practice. At this time he also wrote papers on the hydrodynamics of seaplanes, concentrating particularly on the landing and takeoff methods of the seaplanes. These works attracted the attention of aircraft engineers. In 1927 Wagner was appointed professor of aircraft technology at the **Technische Hochschule** in Danzig, a free city. There he founded the **Institut für Flugtechnik** (*Institute for Flight Technology*), and carried out important theoretical and experimental studies on the statics and strength of aircraft structures involving buckling of struts with open cross-section. In Danzig he promoted the growth of an academic pilot group, and became a pilot himself. In 1930 Wagner was appointed professor of Aircraft Design at the **Technische Hochschule** in Charlottenburg, where he also founded the **Institut für Flugtechnik** which he devoted to experimental investigations. Wagner, who by now was the foremost aeronautical engineer in the world, concluded from his fundamental studies that the streamlined airframes he was designing were capable of speeds at high altitude flights which no conventional piston-driven aircraft engine could provide, and for this reason such aircrafts in his opinion require a completely different type of aircraft engine. In 1934 Wagner, independently of the **Technische Hochschule** and the **Institut**, made a detailed design of a combined propeller and a jet-engine as an aircraft power plant, in which an axial-flow gas-turbine incorporating a combustion chamber drove a propeller by means of reduction gearing and also expelled a jet of exhaust-gases through a nozzle. He designed an aircraft engine in which the propeller provided about ⅔ and the jet about ⅓ of the propulsion. He patented his turboprop-turbojet engine in the beginning of 1935, and at the same time founded an open company consisting of himself and a draughtsman.

In 1935 on the suggestion of the **Reichsluftfahrt-Ministerium** (*State Aeronautics Ministry*), Wagner was given leave of absence from the **Technische Hochschule** in Charlottenburg to work for the **Junkers Flugzeug- und Motorenwerken A.G.** (*Junkers Aircraft- and Motor Works, Stock Corp.*) in Dessau to bring new creative energy to that important aircraft company. At first Wagner took charge of the department of special aircraft development in the company, but in 1937 he was put in charge of all the aircraft development at **Junkers**. Wagner, who was designing an aircraft for high-altitude flight in Dessau, suggested that **Junkers**, a firm which had many branches, also develop an appropriate aircraft engine to propel it. He sold his patent of turboprop and turbojet-engine to the **Junkers** which made its facilities in Magdeburg available for the development of his aircraft engine. Wagner acquired his former assistant and chief-engineer from the **Institut**, Max Adolf Müller (1901-1962), as the leader of a small group of engineers who were also transferred from the **Institut** to **Junkers**. Wagner was responsible for the actual development of his aircraft engine and he made regular visits to Magdeburg to direct this project. Already by 1936 Wagner was quite convinced that the turbojet-engine alone is capable of supplying the power necessary for the high-altitude flight of a streamlined aircraft. Wagner prepared a preliminary design for the first axial-flow turbojet-engine which consisted of a 14-stage axial-flow compressor with a pressure-ratio of 3-to-1, an annular combustion chamber, and a two-stage gas-turbine, and put Müller in charge of the further development of his turbojet-engine. Müller was assisted by Rudolf Friedrich (born 1909), an expert on compressors, and Hans Stabernack, an expert on turbines, both former close collaborators of Wagner, who designed a new kind of high-efficiency compressor in which the pressure increase is equal between the rotor and the stator blading, thus allowing the greatest possible increase in pressure for the given velocity of the flow relative to the blades. Wagner left **Junkers Motorenbau** in Dessau near the end of 1938, when Müller took over the direction of the development of the first axial-flow turbojet-engine in the final design of which his own design contributions were substantial.

The Wagner-Müller axial-flow turbojet-engine was a technically advanced, small and light aircraft engine but it was very difficult to construct. In early 1939, after Wagner had left the **Junkers Motorenbau** in Dessau, the axial-flow turbojet engine was ready for bench-testing using propane as fuel and a blower to create the ramming effect of the incoming air. It turned out that the axial-flow gas-turbine had a component matching problem with the very efficient but highly loaded axial-flow compressor which processed far too much air for the gas-turbine to cope with.

At this time a young German engineer Helmut Schelp (born 1912), although not being a pioneer creator of turbojet-engines he nevertheless had an important influence on the development of such aircraft engines in Germany as a government **Flugbaumeister** (Aircraft Master Builder). Schelp, after earning his Master of Science degree in 1936 at the Stevens Institute of Technology, in Hoboken, New Jersey, United States, returned to Germany in 1937 to join the **Reichsluftfahrt-**

Ministerium (*State Aeronautics Ministerium*) where he was assigned to participate in the programme which trained the so-called ***Flugbaumeister*** (*Aircraft Master Builder*), the aeronautical engineer who was broadly trained in all practical aspects of aviation as well as in aeronautical engineering. Schelp had to spend 6 months in the ***Daimler-Benz A.G.***(*Daimler-Benz Stock Corp.*), and an additional 6 months at the ***Deutsche Versuchsanstalt für Luftfahrt*** (*German Experimental Establishment for Aeronautics*) located in Berlin-Adlershof to obtain this experience. At the Berlin-Adlershof research center Schelp was assigned as part of his study programme the task of determining the limits of aircraft performance, and specifying the aircraft engine which would double the speed of the contemporary aircraft.

At the Zurich International Flying Meeting in 1937, a German aircraft BF109 designed by Willy Messerschmitt, one of the best aircrafts in the world, had given an exceptionally fine flying performance despite the fact that it was powered by quite unexceptional aircraft engine. In 1939, Me 209R, an aircraft particularly designed for speed by Messerschmitt, set a new world speed record of 755 km (469 miles) an hour which was not broken for 30 years. The remarkable aeronautical performances of BF109 and Me 209R had raised a fundamental aerodynamic question: What is the limiting speed of a streamlined aircraft? Schelp determined from his analysis that the limits of aircraft speed were set by the compressibility of air which from existing windtunnel tests would be just over 600 miles (966 km) an hour, that is, at 0.82 Mach, when the airframe speed begins to produce obtrusive compressibility in the air and the propeller as an airscrew loses most of its propulsive ability. The problem, as Schelp saw it, was how to bridge the gap between the compressibility limit of the air and the airframe speed, and to determine the speed at which the propeller becomes totally ineffective. In 1933, Adolf Busemann (1901-1986) from the ***Aerodynamische Versuchsanstalt Göttingen*** (*Aerodynamic Experiment Establishment in Göttingen*), who is regarded the founder of supersonic aerodynamics, had shown that thin airfoils delay and reduce the drag sharply near the speed of sound, that is, at 'Mach 1', and had suggested in 1935 that the supersonic flight of aircraft is possible only with swept-back wings. Later, in 1939, Albert Betz proved that swept-back wings delay and reduce sonic drag. Schelp, who had studied all the existing works on gas-turbines and on various reaction- propulsion systems, recognised that flight near Mach 1, that is near the speed of sound, requires some form of reaction-propulsion and his systematic analysis of all possible types of reaction-engines finally convinced him that a Wagner-Müller type of turbojet-engine with an axial-flow compressor gives the best solution on account of its lower drag. Thus, by 1938, Schelp had come independently to the same conclusion as had Whittle, von Ohain, and Wagner before him that turbojet-engine is the best type of aircraft engine for high-performance aircraft, a conclusion which he had deduced from fundamental aerodynamic principles. Schelp and his superior Hans A. Mauch, head of the ***Abteilung Sondertriebwerken***

(*Section of Special Propulsion Systems*) in the **Reichsluftfahrt-Ministerium** (*State Aeronautics Ministerium*) who was interested in **Strahltriebwerken** (*jet-propulsion*) and already sponsored the research of Paul Schmidt (born 1898) on the pulse-jet which was later used to propel the 'German Flying Bomb, V-1', thought that the time had come for the turbojet-engine development, but they believed that only experienced aircraft engine manufacturers could develop and mass-produce a practical turbojet-engine because they already had the facilities and the experienced work force. Schelp and Mauch visited all German aircraft engine builders and engaged each one of them in some kind of reaction-propulsion work. They had been favourably impressed by the Wagner-Müller axial-flow turbojet-engine and considered the axial-flow turbojet to be the aircraft engine of the future.

The airframe development had not been neglected by the **Reichsluftfahrt-Ministerium** (*State Aeronautics Ministerium*), and a young engineer Hans Martin Antz (born 1909) who had participated in the same course as Schelp, was the leader of the Airframe Group of the Ministerium. Schelp and Antz studied together the problem of speed of modern aircraft in great detail, and Schelp calculated the estimates of the thrust-per-unit engine weight and thrust-per-unit diameter for the turbojet-engine. Schelp and Antz concluded that 2 axial-flow turbojet-engines mounted in individual pods would probably be the best arrangement for a fighter-aircraft, whereupon Antz contracted Messerschmitt to build a turbojet-powered interceptor aircraft on the basis of Schelp's calculations. Messerschmitt determined that his interceptor aircraft, which became the famous Me 262, needed 2 turbojet-engines each producing 1,500 pounds (6,672 N) of thrust. Therefore, it is evident that the turbojet revolution had already occurred in Germany and in England before any aircraft powered by the turbojet-engine had actually flown. The only serious mistake made by Schelp was to cancel the development work on the Wagner-Müller axial-flow turbojet-engine which had been designed by Müller for Heinkel, an engine which initially had the basic problem of mismatching the characteristics of the axial-flow compressor and the gas-turbine: the first axial-flow compressor had compressed too much air for the gas-turbine to cope with. By the end of 1942 this basic problem had been solved and the axial-flow turbojet-engine of Wagner-Müller was able to produce about 1,808 pounds (8,040 N) of thrust at 13,000 rpm for an engine weighing only 838 pounds (380 kg). It had a 5-stage compressor designed by Rudolf Friedrich which produced a pressure-ratio of 3.2 and had an efficiency of 87 % despite its high speed, a two-stage axial-flow turbine the new blades of which were designed by Hans Stabernack, 9 individual combustion chambers, specific fuel consumption of 1.19, outside diameter of 1.90 feet (0.58 m), and length of 8.91 feet (2.72 m). It had the best thrust-to-weight ratio (21.16 N/kg) of any German turbojet-engine built during the war, and this engine could have been in mass-production by 1944. Its cancellation by Schelp was a major mistake in the German aircraft programme. However, another very optimistic turbojet-engine designed by von Ohain and contracted by Schelp was under development for a Heinkel aircraft. It was a very large turbojet-engine about 8.20 feet (2.5 m) long, with a diameter of 2.49 feet (0.76 m), and weight of 2,094 pounds (950 kg). It was designed to produce a thrust of 2,866 pounds (12,745 N) or more at the speed of 11,000 rpm, and to have a specific fuel consumption of 1.30. This turbojet-engine incorporated an axial-flow intake-blower, a combination of axial-flow and centrifugal-flow compressor in the form of a diagonal compressor as suggested by Schelp (which was extremely difficult and time-consuming to produce and, therefore, it had to be made by **Firma Voith** in Hildenheim), it was followed by 3 axial-flow stages, a large straight-through annual combustion chamber, and a 2-stage axial-flow gas-turbine with air-cooled hollow blades, followed by a moveable needle in the nozzle, but it had not reached the production stage by the end of the war. Therefore, the cancellation of the Wagner-Müller turbojet project at Heinkel Aircraft Company left the aircraft frame He 280, an extremely fast Heinkel aircraft frame which proved markedly superior to the outstanding fighter-plane powered by conventional engines Focke-Wulf 190 in a mock battle, without a turbojet-engine. It had obtained a speed of 317 miles (510 km) an hour at sea level and 578 miles (930 km) an hour at 19,685 feet (6,000 m) altitude when it was fitted with 2 centrifugal-flow turbojet-engines of von Ohain.

Heinkel, whose company had built the first rocket-powered aircraft He 112, the first turbojet-engine, the first turbojet aircraft He 178, and the first fighter aircraft He 280 with 2 turbojet-engines, was left without a production turbojet-engine at the end of the war. The reason for this failure probably was Heinkel's unfortunate tendency to work on too many problems at the same time, which dissipated the efforts of his workforce. For instance, Müller was not only in charge of developing the Wagner-Müller axial-flow turbojet-engine, but also a turboprop-engine, a double-stream jet-engine in which two turbines drove two jets, and a jet-engine driven by a piston engine, all these projects were seriously handicapped by the shortage of competent specialists. Furthermore, the entire operation of the **Heinkel-Hirth Motorenwerke** left an impression of being more like a research institute than a factory for the mass-production of aircraft-engines, a situation which created considerable friction between the practical engineers of the former **Hirth Flugmotorenwerke** in Stuttgart-Zuffenhausen with a **Zweigwerk** (*Branch Works*) in Berlin-Waltersdorf, and the scientific engineers of **Ernst Heinkel Flugzeugwerke.** If Heinkel would have concentrated all his staff on the development of the Wagner-Müller axial-flow turbojet-engine, in all likelihood he would have had this engine in production before the end of the war. But Heinkel was too ambitious and, therefore, missed his opportunity to develop the first production turbojet-engine. However, the experience of Heinkel engineers with the development of their different turbojet-engines was of great value to other engine-builders such as **Junkers Flugzeug- und Motorenwerke, A.G., Daimler-Benz A.G.,** and **Bayerischen Motoren Werke, GmbH,** firms that had been contracted by Schelp to study and build axial-flow turbojet-engines.

Schelp had requested the *Junkers Motorenwerke* (**JUMO**) to take over the development of the Wagner-Müller axial-flow turbojet-engine from the *Junkers Flugzeug* soon after Wagner had left it to return to teaching. Müller did not like this decision, and he also left *Junkers Motorenwerke* to join the *Heinkel Flugzeug Werke* in Marienehe at Rostock together with his team. The Austrian engineer Dr. Anselm Franz (1900-1994), head of the supercharger development at *Junkers Motorenwerke* (**JUMO**), was put in charge of this project but after Franz had examined the Wagner-Müller turbojet-engine he rejected it as being far too advanced in design for a rapid development of a production model, after which Heinkel himself took over the development of the Wagner-Müller axial-flow turbojet-engine which has been discussed above. Franz and his team of young engineers then designed an axial-flow turbojet-engine, the famous *Junkers Jumo 004*, a design based on the Wagner-Müller axial-flow turbojet-engine as a model, subject to the guiding principle that every design decision should minimise the risk of failure and lead rapidly to a practical operating turbojet-engine. It had a 5-stage axial-flow compressor designed by Rudolf Friedrich which produced a pressure-ratio of 3.2 and had an efficiency of 87% despite the high speed, a single-stage axial-flow gas-turbine having new blades designed by Hans Stabernack, 9 individual combustion chambers, specific fuel consumption of 1.19, outside diameter of 1.90 feet (0.58 m), and length of 8.91 feet (2.72 m). It was particularly important to determine the limiting heat-resistance of each engine component used. Franz employed a number of firms who specialised in certain particular details of his turbojet-engine design to solve his problems. For instance, during the endurance test of the *Jumo 004* engine in June, 1943, the engine suddenly ran into a serious problem with the fracture of both the turbine blades and the compressor blades. On the request of the State Flight Ministry, Dr. Max Bentele was borrowed from the *Heinkel-Hirth Motorenwerke* to solve this difficult problem for Franz. Bentele discovered after some months of intense investigation that the fracture of blades was caused by the resonance existing between the 6 combustion chambers and the turbine wheel, and he corrected it by changing the proper frequencies of the turbine blades and lowering the revolutions of the gas-turbine from 9,000 rpm to 8,700 rpm. He solved the breakage problem of the compressor blades by a similar method. It was a remarkable engineering achievement of Bentele to solve the serious blade-fracture problem of the *Jumo 004* engine, a problem which had puzzled Franz's engineers. The *Jumo 004* had an 8-stage axial-flow compressor designed by Walter Encke, a single-stage axial-flow gas-turbine designed in cooperation with *Allgemeine Elektrizitätsgesellschaft* (**A.E.G.**) (*General Electricity Company*), a firm experienced in the production of electrical equipment and in the steam-turbine manufacture. The 6 can-type combustion chambers were employed to simplify the development of this turbojet-engine, the axial-flow compressor of which produced a pressure-rise in the ratio of 3.1-to-1 at an efficiency of 78 %. It produced 1,980 pounds (8,807 N) of thrust at 8,700 rpm, weighed 1,590 pounds (721 kg), and had a specific fuel consumption of 1.4. This axial-flow turbojet-engine, which was 12.67 feet (3.86 m) long and 31.5 inches (0.8 m) in diameter, was a remarkable technical achievement in its minimal use of strategically scarce materials. It used hollow, internally air-cooled turbine blades made of folded and welded sheets of the Krupp alloy Tinidur (nickel 30%, chromium 15%, and titanium 1.7%), which was later replaced in order to conserve scarce metal and improve weldability by the alloy Cromadur (chromium 13%, manganese 18%, vanadium 0.7%, but no nickel). Franz found it necessary to make the turbine blades hollow so that they could be internally air-cooled in order to reduce their temperature to more tolerable levels. The hollow air-cooled turbine blades had been invented by the German engineer Christian Lorenzen who was a pioneer in improving the performance of automobile engines as well as the propulsion methods of aircraft. After the First World War, Lorenzen invented the variable-pitch aircraft propeller and investigated methods of supercharging internal combustion engines after the pioneering effort of the Swiss engineer Alfred J. Büchi (1879-1959) in 1911. In 1928, Lorenzen built a turbocharger for Mercedes-Benz automobile engine but the turbine blades warped owing to overheating. He then invented the hollow turbine blades which were internally cooled by circulating air. Although the internal losses in Lorenzen's air-cooled turbine proved to be too high and, therefore, his turbine was impractical, his concept of internally cooled hollow blades proved valuable to the design of the turbojet-engine *Jumo 004*. The combustion chambers and other hot parts of this jet engine, which were made of mild steel anodised with aluminium as a shield against intense heat, restricted the service life of the *Jumo 004* turbojet-engine to 25 hours. This jet-engine required no nickel at all and used less than 5 pounds (2.27 kg) of chromium. It was the most economical, technically skilful, and materially feasible pioneering engine design in history. German turbojet-engine designers, unlike the British engineers, were handicapped by the scarcity of suitable structural materials for their jet-engines. In contrast, the acute problem in Britain, where exotic materials suitable for jet-engines were readily available, was lack of sufficient funding for the turbojet-engine development. Franz introduced another innovation in his *Jumo 004* turbojet-engine known as the 'afterburner' : certain amount of gas in the exhaust of the jet which had not burnt up in combustion can be mixed with fuel in the exhaust and ignited, thereby producing an additional thrust. Whittle came independently to the same solution, and he also introduced afterburners in his turbojet-engine. *Jumo 004* was the world's first production jet-engine used in the Second World War, and more than 6,000 such engines had been produced before the end of the war. Messerschmitt fighter plane, Me 262, was first provided with two *Jumo 004* turbojet-engines in July of 1942 for its first pure turbojet flight, but the production of this aircraft was forbidden by Hitler until August of 1943. This aircraft was aerodynamically superior to all Anglo-American aircrafts in the war, even as late as in 1947. Its flight endurance was 1½ hours, and its turbojet-engines had to be overhauled after each 25 hours of operation. Later in the war the Me 262 aircraft was fitted with small 'Walter rockets' to assist takeoff and shorten its run to about 2,000 feet (610 m). Me 262 with 2 *Jumo 004* axial-flow turbojet-engines had a maximum speed of 528 miles (850 km) an hour, whereas the Gloster Meteor F9/40 with 2 radial-flow turbojet-engines had a maximum speed of only 478 miles (770 km) an hour. The radial-flow turbojet-engines of Whittle had to be overhauled after each 100 hours of operation, and not after each 25 hours of operation like the *Jumo 004* axial-flow turbojet-engines, mainly because of the superior heat-resistant materials used in the construction of Whittle's turbojet-engines.

Franz also fitted the *Jumo 004* turbojet-engines to the German Arado airplanes, Ar 234 and Ar 264, which took part in the Second World War as the first fast jet-propelled bomber aircrafts.

In 1946, after the end of hostilities in Europe, Franz found employment in United States where he was responsible for the development of the turbofan production engine based upon modular construction, which was guaranteed for 4,000 hours of service. Turbofan-engine consumes 20 % less fuel than the turbojet-engine, makes less noise at the exhaust nozzle and produces less pollution. Later Franz was directing the design and production of a gas-turbine with a heat-exchanger of 1,120 kilowatt rating which served as the power unit of the U.S. XM-1 Main Battle Tank.

In Britain the axial-flow turbojet-engine was also developed from the turboprop-engine concept of Griffith. Hayne Constant (1904-1968), another Cambridge graduate who was an assistant to Griffith at the Royal Aircraft Establishment in charge of the supercharger development, set himself the task in 1936 to design a small axial-flow compressor together with some gas-turbine to drive a propeller. However, by 1939, Constant had reluctantly become convinced that Whittle's turbojet concept was superior to the Griffith's turboprop concept for the optimal functioning of the gas-turbine aircraft engine. He and his team were successful in building with the help of Metropolitan-Vickers Electrical Company, a firm experienced in steam-turbine manufacture, the first British axial-flow turbojet-engine by November of 1941, which had its first successful flight test in 1943. Constant together with Griffith played a sinister role behind the back of Whittle by trying to downgrade his achievements and frustrate Whittle's intention to make his small company, the Power Jets Ltd., a commercial producer of his turbojet-engine. The conniving of the 2 men of science, the government bureaucrats, and the commercial manufacturers succeeded in forcing Whittle out of the commercial field, crippling his small company, and profiteering from this commercial plunder. Whittle, who had suffered several nervous breakdowns during his long ordeal in creating his turbojet-engine and struggling against all the obstacles that were set up to block his progress, was never too far from a complete mental breakdown, but he finally succeeded at least in his engineering. It is another prominent case of shabby treatment of an ingenious man of native talent by the British government bureaucrats, manufacturers, and envious fellow professionals. Compared to Whittle, who was rather despicably treated by his own country after he had served his country with great distinction by producing the first British turbojet-engine, Wagner, von Ohain, Bentele and von Braun were treated with professional respect by their former enemies who evidently recognised the great value of their unique talents, and their professional services were eagerly sought and handsomely rewarded. Schelp became the chief engineer of *AiResearch* in United States from 1952 to 1967 where he was responsible for a number of new developments of gas-turbines, particularly of axle-powered gas-turbines.

Wagner had already in 1940 established a department of remote control and guidance at the *Firma Henschel*, and after the Second World War he was one of the first German engineers who was brought to United States to work on gliding bombs, remote control and guidance, and the stability of control systems. Wagner had his own consulting firm in California, and even after he returned in 1960's to the *Technische Universität Berlin* (formerly *Technische Hochschule* in Charlottenburg), he always spent half a year in California as a consulting engineer.

In 1941 Whittle's ideas on the turbojet-engine were imported to United States where Bell Aircraft and General Electric Motors used the theory of the turbojet-engine to improve the supercharging of propeller-driven aircraft during the last war. Supercharged conventional engines allowed the fighter planes to perform well enough without having to rush a working turbojet aircraft through the production. The first turbojet aircraft produced in United States had turbojet-engines which were direct copies of the Whittle turbojet-engine since the emphasis of American research during the war was on the turboprop aircraft engines rather than on the turbojet aircraft engines. One of the improvements introduced by the American engineers to the Whittle turbojet-engine was better heat-resistant materials for the turbine blades. The development of turbojet aircrafts as the general means of air travel took place only after the Second World War. Further development of Whittle's gas-turbine was carried out by the British Thomson-Houston Company Ltd., and the Rolls-Royce Company, and this gas-turbine became initially the most influential turbojet-engine in aircraft industry.

When Whittle patented his bypass-engine operating on the dual-flow principle in 1936, he was well on his way to the so-called turbofan aircraft engine which came into the service of airlines only in the 1960's. After the Second World War economic and ecological considerations in the operation of turbojet-engines became paramount. Great efforts were made to improve the turbojet-engine by reducing its fuel consumption, increasing its thrust and thrust-to-weight ratio, and reducing the noise and smoke of its exhaust. The main improvement in the turbojet-engine was the introduction of the dual-flow principle which made it into a turbofan-engine anticipated in principle already by Elling, who used a separate high-velocity jet of steam to augment the jet of expelled gases, and by Whittle. A turbofan-engine, or bypass-engine is a form of a turbojet-engine which incorporates certain qualities of the turboprop-engine. In this engine a large multistage axial-flow air-compressor passes only a part of the compressed air through the combustion system and the gas-turbine which supplies power to the compressor, the rest of the compressed air passes the combustion chambers and the turbine by, and is expelled as an air-stream together with the exhaust which increases the mass-flow and reduces the exhaust velocity, thereby raising the propulsive efficiency at moderate speeds. Furthermore, a system of 2-spool or 3-spool arrangement was incorporated into the turbofan-engine which allowed different parts of the compressor as well as the propulsive fan to be driven by separate gas-turbines thereby enabling the turbofan-engine to operate more efficiently at different speeds.

The ram-jet of Lorin is basically a streamlined tube with no moving parts into which air is rammed at speeds over 300 miles (438 km) an hour so that it is compressed within a specially designed air-intake, mixed with fuel and ignited in the combustion chamber as part of the tube of the engine. The convergent-divergent duct converts part of the velocity of the oncoming air relative to the aircraft into pressure within the duct, where fuel is added in a constant combustion process, by retarding the velocity of the stream passing through it. The hot gas resulting from combustion expands and is expelled out of the rear end of the engine as a jet which provides the thrust. The ram-jet is the simplest aircraft jet-engine, an engine with no moving parts at all, but since this engine is least efficient when its velocity is less than several thousand miles (3,220 km) an hour it is primarily used in guided missiles.

The pulse-jet is a ram-jet which has a series of spring-loaded inlet valves which open inwards to take in air and slam shut when the compressed mixture of fuel and air is ignited in the combustion chamber which makes the exhaust leave through the tail pipe and cause the pressure in the combustion chamber to decrease when more air is drawn into the combustion chamber and ignited. The combustion in the pulse jet-engine is intermittent thereby making a characteristic buzz sound. A detonation pipe was invented in 1906 by a Russian engineer Karavodin which worked independently in tests. In 1909 an Italian engineer Marconnet proposed to use a pipe issuing a detonating gas-jet for the propulsion of airplanes. This principle was rediscovered in 1930 by a German engineer Paul Schmidt (born 1898), when he was able to obtain a patent on a jet-reaction pipe which worked by a pulsating combustion process, although he could not explain this process theoretically. The intermittently working pipe-jet was theoretically explained by means of gas dynamics by Adolf Busemann. He demonstrated that the efficiency of the pulsating jet-engine in the speed range of 310 miles (500 km) to 435 miles (700 km) an hour is higher than for the steady jet-engine. The weight of this jet-engine is low, but the fuel consumption is high. The pulsating jet-engine was independently discovered by a German engineer Günther Dietrich who worked for **Motorenfirma Argus** (*Motor Firm Argus*) in Berlin where they were able to produce a pulse-jet thrust of 812 pounds (3,610 N) at 435 miles (700 km) per hour speed after having access to the researches and investigations of Paul Schmidt. It led to the development of the 'Flying Bomb V-1' during the Second World War. A piloted 'Flying Bomb V-1' was also developed and tested by Hanna Reitsch, the leading female test pilot in Germany, but was never used in the war. Schmidt's original intention in devising the pulse-jet had been to develop a stratojet aircraft engine that could start under its own power. Although the pulse-jet is mechanically simple, lightweight and inexpensive, it is relatively inefficient.

Gas-turbines in general came into their own only after the Second World War. The great problem with gas-turbines is that they operate continuously at very high temperatures between 1,500°F (816°C) and 2,000°F (1,093°C) and it is difficult to find materials that can withstand such continuous high heat, particularly materials for the disk and blades of the gas-turbine. For instance, Whittle's turbojet-engine after extended use developed creep in its turbine blades made of austenitic steel, the hardest steel at that time. Later during the Second World War, Nimonic steel which contains large amounts of nickel and molybdenum were used in gas-turbine blades. By 1950 the steel alloys produced for gas-turbine blades had become so hard that they could not be easily forged into blades and a method of high precision casting of steel had to be adopted. Many fuels burn with temperatures between 3,000°F (1,649°C) and 4,000°F (2,204°C), which makes it necessary to provide some means of cooling to protect the gas-turbine rotor. The 2 types of high heat-resistant materials currently used in gas-turbines are cemented hard-carbides and ceramics. The cemented hard-carbides are produced by powder metallurgy from carbides of tungsten, molybdenum, titanium, zirconium and tantalum bound together with a cobalt matrix. The ceramic materials used are fused quartz, porcelain and aluminium sintered with iron. Some cemented hard-carbides can withstand temperatures almost up to 2,000°F (1,093°C). Moreover, gas-turbines are heavy users of fuel, which is usually some type of petroleum derivative.

The first turbine-driven car was the *Jet 1* built by the Rover Company in the 1950's, but it never went into production. Chrysler Company built their first experimental car powered by gas-turbine in 1950, but it also never went into production. The first jet-powered transport plane was the British de Havilland *Comet* which was put into service in 1952. In the 1970's, trucks and buses with gas-turbine motors were built by the British Leyland Company, General Motors, Mercedes-Benz and Ford. Gas-turbines are also used to drive centrifugal compressors on gas pipelines to supply pressure in oil wells, to compress gas in refineries, to provide hot compressed gas for blast-furnaces, and to propel ships. The first warship powered with a gas-turbine was launched in 1947, and the first merchant ship in 1956.

Mechanically, gas-turbines which have many times fewer parts than other prime movers are very simple engines weighing a fraction of the weight of other comparable prime movers.

Rocketry

The history of rocketry began with the ancient Greek mathematician and engineer Archytas of Taras (428-c.347 B.C.), who discovered the reaction principle and the jet-propulsion, built a wooden dove which was propelled through the air by the reaction to a jet of compressed hot air expelled from the toy-bird. More than a thousand years later, the Chinese experimented with gunpowder and their reaction-driven fire-arrows, **huo pien chien**, in A.D.1045. In the 13th century A.D. the Mongol hordes used rockets in their sieges of Baghdad and other cities. Muslims used Chinese rockets against the Crusaders who brought this knowledge to Europe. In 1379, the mediæval Italian pioneer in rocketry, Munatori, used for the first time the term, **rocchetta**, meaning rocket in Italian. Konrad Kyeser (1366-1405), Giacomo Fontana (c.1393-c.1415) and Jean Froissart (1338-c.1411), all described various types of rockets in their books on war machinery. The French military forces used rockets in their defence of Orléans in 1429, and in the siege of Pont Andemer in 1449, and they used rockets against Bordeaux in 1452 and Gant in 1453. In the 17th century the French were the world leading developers of rockets. The British became interested in rocketry after their military forces were subjected to rocket attacks in India during two battles in 1792 and 1799. Colonel William Congreve (1772-1828) at the Royal Woolwich Arsenal began to develop an improved type of rocket for use against the French army. He developed new rockets which had ranges from 12,000 to 15,000 feet (3,658 to 4,572 m), and the British military used it in their attacks against Boulogne in 1805, against Copenhagen in 1807 where 25,000 rockets were fired, and in the War against United States in 1812. Rockets were also used in the American Civil War. During the First World War, military rockets were fired from French and British war planes as well as from the ground, but their accuracy and range was not competitive with the existing artillery.

A rocket is a jet-engine only in so far as it produces thrust by the expansion of hot gases, or even liquids, expelled backwards. Rockets carry their fuels, liquid propellants or solid propellants, and the oxidising agent and require no contribution to their operation from the medium in which they travel.

The interest of several men was drawn to space travel and rockets by the book < From the Earth to the Moon > of the famous French science fiction writer Jules Verne (1828-1905). The first man to develop a sound scientific astronautical theory was a self-taught Russian schoolteacher, Konstantin E. Tsiolkovskii (1857-1935). Tsiolkovskii published a large number of works on rocketry, space travel, astronomy, dynamics, physics and philosophy. In 1883, he discovered the basic idea of the rocket device for space travel which he subsequently worked out in a great number of papers. However, he was a pure theoretician who never constructed a single rocket-motor, but rather concentrated on working out the basic theory of rocket dynamics. In 1903, he gave a design of a rocket-ship for space travel which was powered by a very modern propellant combination, liquid hydrogen and liquid oxygen, confined to separate compartments which upon mixing produces an explosive mixture of gases expelled from the rocket as a jet of burnt gases and the reaction to the jet propels the rocket in space. In 1914, he gave a more detailed design for such a rocket spaceship.

He made great efforts in the analysis of the components of rocket engines, and evaluated the power potential of various combinations of propellants, such as liquid hydrogen, liquid oxygen, kerosene, alcohol, and methane in consultation with the famous Russian chemist, Dimitrii I. Medeleev (1837-1907). He proposed the so-called multistage rocket, a device which had been used over a couple of centuries by firework-makers, to escape the gravitational field of the Earth. In 1935, he published his book on lunar travel < On the Moon > (*Grezy o zemle i nebe*) in which he also gave a new modified design of the rocket spaceship. Since Tsiolkovskii did not perform any experiments to demonstrate the practical feasibility of his theories of rocketry, he cannot be considered to be the founder of modern rocketry, the honour of which belongs to an American physicist and engineer.

The actual founder of modern rocketry is Robert Hutchings Goddard (1882-1945), who virtually singlehandedly developed and tested the modern rocket technology. Goddard graduated with an engineering degree from Worcester Polytechnic Institute in 1908, and earned the doctorate in physics degree from Clark University in Worcester, Massachusetts. In 1919, he was appointed professor of physics at Clark University. He began his research on liquid-propellant engines already in 1909 and proved that liquid hydrogen and liquid oxygen are the best fuel combination for propelling rockets. His theoretical and experimental research resulted in more than two hundred basic patents in rocket technology on such matters of rocketry as combustion chambers, propellant feed-systems, nozzles, multistage rockets, automatic launch sequential system, sophisticated automatic guidance system incorporating a gyroscope on gimbals to move vanes in the exhaust jet to steer the rocket, gyroscopic stabilisation systems, first gimbal-mounted rocket engine, and gas-generators driving turbines. Goddard developed the prototype bazooka rocket in 1918, and designed, built and tested the early high-altitude rockets. In 1926, he launched successfully the first liquid-fuel rocket, and developed the first smokeless solid-fuel rocket. In 1929, the popular aviator Charles Augustus Lindbergh (1902-1974), who visited Goddard, was able to convince the Guggenheim Foundation to grant Goddard $50,000 for rocket research. The Carnegie Institute gave Goddard a small grant for the founding of rocket testing facilities. In 1930, Goddard established a rocket testing facility in the desert of New Mexico and carried out the most remarkable technological development by a single man in the history of technology. He performed an incredible number of rocket firings using mostly liquid oxygen and gasoline as rocket fuel. His rockets finally reached the altitude of 9,000 feet (2.74 km) and achieved speeds in excess of the speed of sound. In 1931 Goddard patented his design for a rocket-propelled aircraft. Early in his scientific career, Goddard did research on vacuum tubes and invented the vacuum tube oscillator capable of generating undamped high-frequency radio-waves, which was of basic importance in radio technology. He patented it in 1912, and he was the first to do so.

The United States Army Air Corps and Navy had shown no interest in Goddard's important work on rockets until 1942, when the Navy put him in charge of the Navy project to develop a liquid-fuel booster-rocket to assist Navy planes in their lift-off from the deck of aircraft carrier. A more significant assignment of Goddard for the Navy was his successful development of variable-thrust rocket motors after hundreds of proving-stand tests.

In 1945, when Goddard died, his work was unappreciated and he himself remained virtually unknown in his own country, whereas his book on rocketry < A Method of Reaching Extreme Altitudes >, which was first published in 1919, had become a classic and the most highly regarded textbook on the technology of rockets in Europe particularly by German engineers. Working alone in complete isolation because he refused any offers of collaboration, Goddard had perfected most of the basic technology that is now a fundamental part of modern rocketry. Finally, in 1959, the United States government awarded Mrs. Goddard, who had assisted her husband in his research all throughout the many years, one million dollars as the settlement of the lawsuit instituted by Mrs. Goddard and the Guggenheim Foundation against the government for the government's use of the more than 200 patents of Goddard in rocket technology.

The third important rocket pioneer was a German mathematics teacher, Hermann Julius Oberth (1894-1989), who learnt about Goddard's book of 1919 and requested a copy of it from its author in 1922. Oberth published his own book < Rockets into the Planetary Space > (***Die Racketen zu den Planetenräumen***) in 1923, a book of less than 100 pages which represented his rejected doctoral dissertation and contained a thorough discussion of virtually every aspect of rocket travel in space including its undue effects upon the human body. He studied many propellant combinations, particularly alcohol and hydrogen, and gave a detailed design of a rocket to explore the upper atmosphere. Since his book became very popular, he published its 2nd , and much expanded edition, < The Road to Space Travel > (***Wege zur Raumschiffahrt***) in 1929, which was 4 times as long as the 1st edition and contained many novel thoughts on the problems of space flight. This book inspired many scientists and engineers to begin work on rocket technology. Oberth, unlike Goddard, publicised rocket flight and his own work about it as much as possible. In 1929, he became a scientific advisor to the famous German film director, Fritz Lang (1890-1976), who was making his celebrated film < Woman on the Moon > (***Frau im Mond***), and was given funds by **UFA**, the foremost German film company, to build a rocket and launch it as a publicity stunt for the film. Oberth, unlike Goddard, was more of a theoretician than a practical engineer, and the rocket he designed and built was not capable of operation unless many of its basic components were completely redesigned. The small test motor of the rocket designed by Oberth was successful in the static test done in 1930 at Berlin, but by then **UFA**'s funds had run out and he had to return to his teaching. After the Second World War, in 1958 his erstwhile assistant, Wernher von Braun (1912-1974), brought Oberth to work with him in the Army Ballistic Missile Agency in United States.

The first rocket-propelled aircraft worthy of its name was a glider which was powered by two slow-burning solid fuel rockets designed by the German engineer Friedrich Wilhelm Sander (1886-1938). Sander was a manufacturer of solid-fuel rockets for the rescue at sea. It made its short maiden flight in 1928 which lasted a little more than sixty seconds during which time the airplane travelled ¾ of a mile (1.2 km). In the same year the German automobile manufacturer, Fritz von Opel (1899-1971) of ***Opel Automobilewerke*** (*Opel Automobile Works*), piloted his glider plane powered with 16 Sander rockets, each of which provided a thrust of 60 pounds (267 N).

Rocket propulsion was also applied to automobiles. The automobile magnate Fritz von Opel designed a streamlined racing car in 1928 which was fitted with 2 short wings to hold the car down to the track when 24 Sander rockets which propelled the vehicle were being fired. Max Valier (1895-1930), a space flight enthusiast, was at first von Opel's collaborator, but later became von Opel's competitor in establishing speed records for rocket-propelled cars. In 1930, Valier was killed in his last attempt to break the speed record of land-driven vehicles with his Valier-Heylandt car powered by rockets using liquid fuels.

The first significant air-worthy rocket-propelled aircraft was He176 built by Heinkel and designed by Heinrich Hertel, which made its first rocket-propelled flight in June of 1939. He 176 was the first plane fitted with the ejection-seat for the pilot which was operated by compressed air. This aircraft which was fitted with the hydrogen-peroxide rocket engine of Hellmuth Walter (1900-1980) exerting 1,300 pounds (5,783 N) of thrust could have been capable of exceeding the speed of sound with a more powerful rocket engine burning alcohol-liquid oxygen fuel, but the He176 project was cancelled in the fall of 1939, after the beginning of the Second World War. An earlier rocket engine burning alcohol-liquid oxygen fuel designed by Wernher von Braun was incorporated into the fighter aircraft He112 also built by Heinkel, and successfully tested in flight by the test pilot Erich Warsitz in 1937.

The rocket-powered aircraft project was revived in the design of a squat, tailless, interceptor aircraft with swept-back wings, the Me163A, designed by Alexander Martin Lippisch (1894-1976), and built by the aircraft manufacturer Willy Messerschmitt. Already in 1921, Lippisch had advocated delta-shaped wings for modern high-speed aircraft since such wings make small side-proportions possible. In April of 1941 this rocket-propelled aircraft was successfully flight tested, and a modified design of it, the Me163B, was fitted with a more powerful rocket engine designed by

Walter that burnt a fuel consisting of hydrogen peroxide-hydrazine-hydrate/methyl alcohol. Messerschmitt built about sixty Me-163B rocket-powered interceptor fighter aircrafts which saw limited action in aerial defence of German cities in the Second World War.

Walter, at first a builder of steam-engines, proposed in 1933 to drive submarines with gas-turbines using hydrogen superoxide as fuel, but it did not work. However as he soon discovered, this fuel did work in rockets. In 1935, Walter designed a monopropellant rocket engine which used a single liquid fuel, hydrogen peroxide of 80 % to 85 % concentration, instead of the customary two liquid fuels. In 1937, Walter fitted his rocket which gave 220 pounds (979 N) of thrust as a starting engine to Heinkel's **Kadett** aircraft. It was the first rocket-assisted takeoff of an aircraft in history. In 1936 he built a ram-jet which he was able to make work, and it was the first time a ram-jet had ever functioned on its own. In 1938 he mixed some surrounding air into the gas-jet and, thereby, increased the thrust by 150 %. During the Second World War, Walter designed a rocket launcher for the German 'Flying Bomb V-1' in order to start the normal operation of its pulse jet-engine. He developed rockets for various practical purposes with great success that is unmatched anywhere else in the world. He also invented the so-called 'snorkel' and various gas-turbine engines for submarines. After the war he worked on the improvement of the submarine engines for the Royal Navy in Great Britain, and finally became vice-president of the Worthington Corporation in New Jersey, United States, where he built power stations and introduced many new inventions into oil technology.

Since the rocket-powered aircraft has certain obvious shortcomings such as a relatively short period of rocket-action, the turbojet aircraft eclipsed the importance of rocket-powered aircraft in aviation, particularly in commercial aviation. On the other hand, rocket-propulsion of aircraft in the post-war period has completely dominated the high-speed aerial research and space exploration.

It is quite evident from the preceding discussion that only Goddard deserves to be called the 'Founder of Modern Rocketry', since he was competent both in the theoretical and technological aspects of rocket technology in which he was an unchallenged pioneer inventor as well as the theoretical and technological expert.

CHAPTER XXII

ELECTRIC POWER: From Generator to Induction Motor

Electric Generators

The scientific principle of the electric generator had been discovered almost at the same time by Michael Faraday (1791-1867) and Joseph Henry (1797-1878) in 1831, but the first technological application of it came through the influence of André Marie Ampère (1775-1836).

In 1831, Hippolyte Pixii (1800-1835), an instrument-maker in Paris who had built instruments for Ampère, designed an experimental electric generator consisting of a permanent horseshoe magnet which was manually turned by means of a hand-crank and gearing underneath 2 stationary coils wound on soft-iron cores. The 2 iron bobbins had 4,000 turns of wire of total length of 3,000 feet (914.4 m), and the magnet exerted an attractive force of 220 pounds (979 N). This particular generator produced an alternating current. At the suggestion of Ampère in 1832, Pixii converted his machine into a direct but fluctuating current generator by providing another electric generator he had built with a mechanical commutator in the form of a rotating switch as suggested by Ampère. The fluctuation of the direct current was a source of trouble for larger machines because the fluctuation of current generated heat in electric machines.

Pixii's magnetoelectric generator had been anticipated about 1823 by William Sturgeon (1783-1850), an English manufacturer and a teacher at the Academy of the East India Company, who devised a magnetoelectric experimental machine with a shuttle-type of coil, which revolved between the poles of the magnet, and incorporated an all-metallic commutator. Already in 1825, Sturgeon had used more efficient electromagnets excited by batteries as alternatives to natural magnets, a concept which ultimately led to the principle of self-excitation in electric generators.

In 1833, a Scottish inventor, Reverend William Ritchie (1790-1837) of Edinburgh, professor of physics at the London University, built a magnetoelectric generator in which the rotating armature, consisting of 4 coils at right angles to one another and set between 2 disks, passed between the stationary horseshoe magnet. Since a single-loop armature produced a fluctuating current, Ritchie attempted to reduce these pulsations by increasing the loops in the armature to 4. His commutator switch was mounted on the axis of the armature. Since the magnetoelectric generators had permanent magnets for the creation of the magnetic field, they were usually called 'magnetogenerators' or, briefly, 'magnetos'.

An American instrument-maker of German stock, Joseph Saxton (1799-1873) of Washington, devised in 1833 a magnetogenerator which had a rotating, horizontal armature of 3 coils and a compound horseshoe magnet. It was a modification of Pixii's magnetogenerator. An English instrument-maker, Edward M. Clarke, after having repaired some of Saxton's machines in England, designed an improved version of Saxton's generator by using only a pair of coils but rotating them besides the poles rather than in front of the poles as in Saxton's generator. Clarke sought to increase the generated current and reduce the vibration of the machine during its operation.

In 1838, an American inventor Charles Grafton Page (1812-1868) of Washington, increased the intensity of the magnetic field in Clarke's magneto to augment the output current, by installing a second compound magnet parallel with the compound magnet in the Clarke magneto, and rotating the armature coils between the 2 compound magnets. Since 1838, the magnetos of Page were made and sold commercially by Daniel Davies, whereas magnetos of Saxton and Clarke were popular in laboratory demonstrations and medical experiments.

A German instrument-maker, Emil Stöhrer (1813-1890), further improved the magnetoelectric generator in his attempt to reduce their vibration. Stöhrer introduced the multipolar concept to the magneto design by devising a 6-pole machine consisting of 3 horseshoe magnets, the poles of which formed a circle, and over which were mounted 6 coils and a commutator. Although this machine was still manually driven, Stöhrer's magnetoelectric generator design represented a definite progress towards a more reliable practical machine.

In 1839 the invention of electrotyping in which copper is deposited by electrolysis on a mould of an engraving soon led to electroplating in general, a process which required much more current than electric batteries could provide. One of the earliest substantial commercial applications of magnetoelectric generator was made to electroplating by Professor of chemistry, John Stephen Woolrich in Birmingham, England. In 1841, Woolrich showed in a patent how the Saxton magnetoelectric generator can be modified for the electroplating process. In 1844 he designed a more ambitious magnetoelectric generator that was similar to that of Ritchie by using a disk armature of 8 uniformly spaced coils rotating between the poles of 4 magnets spaced at right angle. He increased the strength of the magnetic field by stacking several flat horseshoe magnets together, and improved the design of the commutator to produce more steady electric current required for electroplating. His magnetoelectric generator was driven by a steam-engine rather than by manpower. Woolrich's machine, which he sold to the Prime Plating Company, was $5\frac{1}{3}$ feet (1.62 m) high, 6 feet (1.82 m) wide, and 2 feet (0.61 m) deep. Subsequently, he designed several magnetoelectric generators for other plating companies. In 1851, William Millward of Birmingham, patented a magnetoelectric generator which was quite similar to that of Woolrich.

In 1849, the Belgian physicist, Florise Nollet (1794-1853), professor at the *École Militaire* (*Military School*) in Brussels, was the first person to propose the construction of large powerful generators of electricity able to supply adequate electric power for the illumination of lighthouses. He took out a patent in 1850 for an improved Stöhrer's magnetoelectric generator which was driven by steam-power. In 1852, Nollet filed another patent in which a modified William Millward magnetoelectric generator was presented. Nollet proposed to create the beacon of light in lighthouses with the Drummond 'limelight' by making a block of lime incandescent by heating it in the intensely hot oxyhydrogen flame. He proposed to generate the oxygen and hydrogen necessary for the oxyhydrogen flame by the electrolytic decomposition of water for which purpose Nollet designed a magnetoelectric generator incorporating a multiple-disk armature with a number of coils which rotated past the poles of horseshoe magnets, a design basically similar to the magnetoelectric generator of Woolrich, to provide large quantities of electric current for the electrolysis of water. Nollet died before his magnetoelectric generator was built.

Auguste Berlioz and Joseph van Malderen, a coworker of Nollet, bought Nollet's patent in 1855, and applied an improved version of it for the production of electric current in the illumination of French lighthouses which used Victor Serrin's arclight manufactured by their company, *Société l'Alliance*. Many lighthouses in France were illuminated by arclights using the electric current supplied by the magnetoelectric generators of the *Société l'Alliance* which were driven by steam-engines, generators which after 1870 were gradually replaced by Gramme's generators.

In England, Frederick Hale Holmes had already suggested in 1853 that magnetoelectric generator could be used to produce electric current to feed arclights for illumination purposes, a claim which the French expert in electricity, Alexandre Edmond Becquerel (1820-1891), considered ridiculous at the time. In 1856, Holmes patented a multiple-disk armature consisting of many Woolrich generators mounted on a single frame. Every other disk was displaced through a small angle to reduce fluctuations in the total induced direct current. The current was collected by rollers touching the commutator-bars set about 60 degrees apart, their position being controlled by the speed of rotation. Unfortunately, in practical application this machine was not very successful. Holmes' magnetoelectric generator was a precursor of several practical but inefficient generators. The next improvement in generator design consisted of replacing permanent magnets with far more efficient electromagnets.

Already in 1845, Charles Wheatstone (1802-1875), an English physicist and telegraph engineer, had employed electromagnets in a generator to supply electric current for his telegraph, but his arrangement required also fuel cells for its operation. In 1863 Henry Wilde (1833-1919), a manufacturer of electrical instruments in England, patented the replacement of natural magnets by electromechanical magnets in generators, and improved upon Wheatstone's generating system by replacing the fuel cells with a magnetoelectric generator the current of which was used to excite the electromagnetic field-coils of another generator. In 1866, William Ladd, an English instrument-maker, simplified Wilde's dual generating system by combining its 2 separate magnetic fields into one by placing permanent bar-magnets parallel and above each other with wire wrapped round the permanent magnets. An armature was made to rotate between each pair of poles at the end of magnets, so that one armature provided current for the coils on the permanent magnets and thus boosted the magnetic field of the second armature, whilst the output current was taken from the other armature. Both the Wilde's dual generator, and the Ladd's combined magneto-electromagnetic generators were used to supply the electric power to keep an arclight burning at the Paris Exhibition in 1867. These hybrid magnetogenerators were quite successful, but by that time a new development of electric generators had taken place which soon made the magneto-type generators of Wilde and Ladd obsolete. The new electric generator utilised the self-excitation concept first suggested by the English electrical engineer Jacob Brett (1808-1897) in 1848, which made the generator self-contained.

In 1851, a Danish civil engineer, Søren Hjorth (1801-1870), designed a self-excited electric generator in which permanent magnets provided the initial excitation, but after setting the generator into motion, part of the current produced was used to excite the electromagnets of the generator. Hjorth patented his generator in 1854 in Britain, but he never exploited it commercially. It was discovered 10 years later that the electromagnets contained enough residual magnetism in their soft-iron cores to provide the initial magnetic field necessary to start the generation of electric power from the electric generator.

The first improvement in the form of armature was produced in 1856 by Ernst Werner Siemens (1816-1892), a German electrical engineer from the famous Siemens family, when he substituted a weaver's shuttlelike armature for the disk-type armature which had an **H**-type of cross-section with the wire wound longitudinally in the cavity of the armature. This shuttle-type anchor armature of Siemens, the so-called 'double-T armature',was rotated about the longitudinal centroidal axis of the **H**-type of cross-section between the concave cavities of the poles. The shuttle armature was considerably more efficient than the disk armature, for it was more compact and rigid, had smaller gaps between magnetic poles and the armature, cut more dense magnetic flux, required less power to turn, and admitted higher speeds of rotation. For all these reasons, the electric generators with the shuttle armature dominated the industry of electric power for a decade. However, the increase of electrical efficiency in generators with the shuttle armature led for the first time to the heating problem in the generator's armature, which was mainly due to the increased speeds of rotation of the armature.

The next improvement upon the armature was the introduction of ring-armature. In 1860, an Italian student of physics at the University of Pisa, Antonio Pacinotti (1841-1912), who later became professor of physics at the same university, developed an electric motor with a ring-armature in which a single, continuous closed coil, tapped at intervals and placed between spokes, replaced the separate wire loops. The ring-armature had been invented in 1842 by the Dutch engineer P. Elias in his electric motor. The core of Pacinotti's armature looked like a horizontal toothed wheel with iron spokes rotating on a vertical spindle. A horizontal circular magnet, supported by 2 upright electromagnets, enclosed the ring armature with 16 coils connected in series, each juncture of which was connected to one of the 16 commutator bars.

Pacinotti discovered that his electric motor could function as an electric generator if operated in reverse. Pacinotti's generator with ring-wound armature incorporated great improvements in the armature design and in the magnetic circuit. He wound his coils on an iron ring that revolved in the plane of the lines of magnetic force between 2 electromagnets so that more of each coil was able to cut the lines of magnetic force at right angles whilst the iron ring improved the magnetic circuit by compacting the magnetic flux thereby letting more lines of magnetic force pass through the iron ring. Pacinotti's generator gave remarkably steady direct-current in comparison with other generators available in 1867/1868 which produced fluctuating direct-current that generated considerable heat in the generators. He published his results in 1863 but it received little if any professional attention. However, Pacinotti's generator did not use self-excitation principle. The principle of self-excitation was independently discovered by the Danish physicist Wilhelm Sinsteden (1803-1891) and the Hungarian physicist Anyos Jedlik (1800-1895) in 1861, and by the American electrical inventor Moses Garrish Farmer (1820-1893) in 1865. Unfortunately, the work of Hjorth, Sinsteden, Jedlik and Farmer did not lead to any further development of the electric generators. It appears that Farmer was the first to use the idea of shunt-connexion in the electric generator in which only part of the current generated is taken off by a separate field winding. The shunt-connexion in the electric generator represents a feedback principle of operation of a machine first used by Archimedes of Syracuse (c.287-212 B.C.).

By 1866, the self-excitation principle was sufficiently well established so that self-excited electric generators were proposed independently by 3 men almost at the same time: Samuel Alfred Varley (1830-1921), Charles Wheatstone (1802-1875) and Werner Ernst Siemens (1816-1892). In fact, Wheatstone and Charles William Siemens, Werner's younger brother, announced it at the same meeting in 1867. Varley had patented his self-excited generator in December, 1866. Werner Siemens named the self-excited generator, the 'dynamo electric machine', which was later abbreviated to 'dynamo'.

The principle of self-excitation recognises the fact that electromagnets contain enough residual magnetism in their soft-iron core to provide the initial magnetic field to start the generation of electric current, which is then fed back into the coils of the electromagnet to increase the magnetic field strength. The self-excitation in dynamos is produced by two methods of wiring: the shunt and the series wiring. In the shunt-wiring only part of the current generated by the armature of the dynamo is shunted, or returned to the coils of the electromagnet to excite the magnetic field. For instance, Wheatstone used shunt-connexion in his dynamo. In the series-wiring the total current generated by the armature is returned to the coils of the electromagnet to excite the magnetic field. Siemens used the series-connexion of the field coils in his dynamo.

The man who made the dynamo a real commercial success was a Belgian carpenter and model-maker, Zénobe Théophile Gramme (1826-1901), who had worked together with van Malderen in the workshop of the *Société l'Alliance*, and had left it to become an instrument-maker for Heinrich Daniel Ruhmkorff (1803-1877) and Gustave Froment (1815-1865). Afterwards, when working on his own, Gramme patented in 1867 a Woolrich type of multidisk magnetoelectric generator. In 1870 he rediscovered the ring-winding, and established with Hippolyte Fontaine (1833-1910) his own company. In 1872 Gramme marketed his first commercial dynamo which was based on Ladd's electric generator that had upright electromagnets closed at the top and at the bottom, several Pacinotti-type ring-wound armatures, and crescent-shaped pole-pieces to shunt the magnetic field from one electromagnet to the other, and by such means pass through several ring-armatures which were rotating about an axis at right angles to the electromagnets between the pole-pieces. One of the armatures supplied the current to the electromagnets of the other armatures and silver-plated copper wire brushes collected the current induced in the armatures. In principle, Gramme's dynamo was essentially the same as Pacinotti's generator, but the modified version of it became the standard electric generator for electroplating, and for arclighting in lighthouses and factories. The Gramme dynamos were manufactured in 2 types: the dynamo for high-current electroplating was of low resistance with 2 coarse-wire armatures and 4 electromagnets which produced 150 ampere direct current at 2 volts with about 40 % efficiency; the dynamo as an electric power source for high-voltage arclight illumination was of high-resistance with 3 fine-wire armatures and 6 electromagnets which produced 850 carcel-units of light at an efficiency of about 50%. Gramme's dynamo for arclighting produced about 4-times as much power as the 6-disk magnetoelectric generator of the *Société l'Alliance,* and the voltage equalled that of 105 Bunsen cells. However, the high-voltage Gramme dynamo tended to overheat in operation, and for this, and other reasons, Gramme improved the efficiency of both dynamos by reducing their size in a new design relying on self-excitation concept in which an increase of speed of rotation was achieved without excessive heating of armatures. These machines in turn were further improved, with the result that a new era in electric power technology began with the Gramme dynamos, which proved to be reliable sources of relatively inexpensive electric power.

Scientists such as Heinrich Friedrich Emil Lenz (1804-1865) in 1838, John Frederick William Herschel (1792-1871), Charles Babbage (1792-1871), and others, had maintained that both the electric motor and the electric generator work on the same scientific principle. In practice, Pacinotti had shown in 1863 that an electric motor becomes an electric generator when run in reverse, and the American electrical engineer Edward Weston (1850-1936) demonstrated the same in 1878. Gramme's partner, Hippolyte Fontaine, discovered through an accident that one dynamo can drive another dynamo as an electric motor. A workman at the Vienna Exhibition in 1873 accidentally connected the wires of the second dynamo to the leads of the first dynamo which was running, and discovered to his amazement that the second dynamo became a motor after the current from the first running dynamo was sent through the windings of the other dynamo. Fontaine promptly made it part of Gramme's exhibit in which the transmission of electric power was also demonstrated: a centrifugal pump, which supplied a 6-foot high waterfall was run by a Gramme dynamo operating in reverse as a motor that obtained its electric power with 50% efficiency from a Gramme dynamo located ¾ of a mile (1.21 km) away. A similar exhibit of Gramme was shown at the Philadelphia Centennial Exposition in 1876. By 1876 Gramme's dynamos were able to achieve an overall efficiency of 63% in this type of arrangement.

Gramme's dynamo with ring-armature did not become excessively hot when operated under normal conditions but it did have certain shortcomings among which the most important one was that essentially only the outer circumference of the coils produced electric current because the inside armature wires were shielded by outside wires, and, therefore, not all of the coil of the armature, which had a greater resistance than necessary, was electrically productive thereby reducing the efficiency of Gramme's dynamo.

About 1880, a Swiss engineer, Emil Bürgin (1848-1933) of Basel, improved Gramme's ring-wound armature, in which a series of 4 or more rectangular cores of iron were mounted on a common spindle. Bürgin's improvement made it possible to construct Gramme-type dynamos more efficiently, and the heating of the coils was reduced since the armature coils had a smaller diameter and its windings were composed of a series of sections. The English electrical engineer, Rookes Evelyn Bell Crompton (1845-1940), improved the Bürgin dynamo in Britain in the early 1880's by making the armature core hexagonal and increasing their number to 10, and by letting each hexagonal core rotate through 60-degrees relative to the preceding core.

In 1881, a Swedish electrical engineer, Jonas Wenström (1855-1893), who made many ingenious inventions in different fields, realised the great importance of an effective magnetic circuit in the armature, and was the first to countersink armature conductors in the slots of the armature core, a method of construction which has become a universal practice. His dynamo was the most efficient dynamo in existence at the time. Wenström invented an electric iron-ore separator and an electric-arc furnace as well as one of the first geared electric motors incorporating a planet-gear with 1–to–5.7 reduction ratio. This ingenious young man was able to design his electric machines according to the laws of the magnetic flux by relying on his remarkable intuition long before the theory of these laws existed. Wenström was working on the 3-phase system of electric power transmission when he died at 38 years of age.

The ring-form of armatures which were not very efficient were gradually replaced by the modern drum-armature invented by Friedrich von Hefner-Alteneck (1845-1904), chief engineer of the *Siemens & Halske Fabrik* (*Siemens & Halske Factory*) in 1872. He invented an armature which incorporated a method of symmetrical coiling that minimised the unproductive end-turns which did not cut the magnetic field. His armature was better than Gramme's ring-wound armature, yet it still retained the advantage of the Gramme ring-wound armature in commutation. It had 16 coils the terminals of which were reversed twice in each revolution to produce 'direct current', and the coils were interconnected at the commutator bars to form a single closed-circuit coil. The original drum made of wood was overwound with iron wire to form the armature core. In the original design, which was based on the Ladd machine, the heating of the armature was a problem but by 1876 this difficulty had been removed through a modified design. The armature core was now a hollow wrought-iron cylinder with 12 longitudinal armature coils laid on its surface and held in place by pegs and binding wire. The field magnet was formed of 14 horizontal wrought-iron bars, 7 on top and 7 below the armature, connected with iron spacers at their ends, and curved in the center to leave a space for the cylindrical armature. The 4 field-coils were so wound around the bars as to produce an upright electromagnetic field in which the drum-armature rotated. It produced an output of 20-ampere current at 50-volt potential. Soon the horizontal design was replaced by the vertical design of field coils, and the armature windings were countersunk into longitudinal slots in the drum following Wenström's innovation. In 1879 the English electrical engineer John Hopkinson (1849-1898) found the Siemens generator to have an efficiency between 88 % and 90 %, since Hopkinson had developed a graphical analysis for the study of magnetic behaviour of dynamos. In 1886, Hopkinson showed how to predict the performance of magnetic circuits by means of a general graphical method of analysis, which eliminated the trial-and-error method of dynamo design.

In the 1880's, efforts were made to improve the form of the field-magnets which ultimately gave a more compact form to the dynamo. In 1882, compound winding of the field-coils was introduced which consisted of the series-winding added to the customary shunt-coils to compensate for voltage drop on the load. The firm *Siemens & Halske A.G.* in Germany became a leading manufacturer of dynamos, and by 1890 they manufactured large dynamos weighing 1½ tons, which supplied 1,500-amperes of current with a drum-armature 2 feet (0.61 m) in diameter and 3 feet (0.91 m) in length. At the Paris Exhibition in 1900, Siemens exhibited a dynamo which produced 1,570-kilowatts of power running at 200 rpm.

In 1882, the French engineer François Daniels Marcel Deprez (1843-1918) found from his theoretical studies of the combination of the self-excited and the separately-excited generators that a constant voltage output could be achieved with a current supply from an outside source.

Alternators

Frederick H. Holmes had suggested in 1869 to use the 'alternating current dynamo', the so-called 'alternator', to supply electric power for arclights. Later that year, he did construct a pair of alternators for the South Foreland lighthouse.

The first experimental alternator using a permanent magnet was built by Pixii in 1832, but the first commercial alternator which was not dependent on permanent magnets was constructed by Dr. Henry Wilde in 1867. Wilde's rotating-field alternator consisted of armature-coils on iron bobbins fixed to two disks as in Holmes' magnetoelectric generators.

When a Russian engineer Paul Jablochkoff (1847-1914) developed the first effective arclighting system based on the so-called 'Jablochkoff candle', which worked with alternating current of high voltage, no suitable alternating current generators were available since the early direct-current dynamos of Siemens and Gramme could not be redesigned to generate alternating current. The later models of magnetoelectric generators of Holmes, and of the *Société l'Alliance* could easily be redesigned for alternating current, and they dominated the self-excited single-phase alternator market to produce alternating current for Jablochkoff's arclighting system. In 1878, A. DeMeritens of the *Société l'Alliance* invented a modified magnetoelectric generator which had a distributed winding instead of coils and gave a much more uniform current than previous magnetoelectric generators. DeMeriten's generators gave an output of 4½ kilowatts at 830 rpm, and were smaller than those of Holmes. In 1878, Gramme built a special alternator to supply power to Jablochkoff's arclighting system, which was a self-excited, single-phase alternating current generator. The magnetoelectric generators gradually lost the commercial market because they were less efficient than Gramme's dynamos yet cost almost twice as much. In 1880, Gramme designed a machine which served both as the alternator and a 4-pole dynamo in a single machine.

Soon Siemens produced his own alternators in 3 different sizes for the Jablochkoff's arclighting system, that resembled in construction the Woolrich magnetoelectric generators but in which the permanent magnets were replaced by more efficient electromagnets and the disk armature, now called a 'rotor', rotating between 2 stationary rings of electromagnets, now called 'stators'. Siemens also manufactured a self-excited alternator in 1882.

Thomas Alva Edison (1847-1931), in his Pearl Street Power Station in New York, tried to increase the power of field-magnets with his assistant Francis Robbins Upton (1852-1921), a graduate of Princeton University who devised a fundamental concept of dynamo design which recognised that the product of ampere-turns was the fundamental criterion of a magnetic circuit. Edison was able to locate the places of his power losses by inserting thermometers into the cavities located in various parts of the iron core of his field-magnets. With this knowledge, Edison was able to construct his renowned second dynamo, called the 'Long-Legged Ann', which achieved a high efficiency of 82%. His famous first dynamo, called 'Jumbo' was built in 1879 and consisted of 2 long magnets as high as a man, yoked together at the top and having pole pieces at the bottom between which the armature was rotated by a 150-horsepower Porter & Allen steam-engine. John Hopkinson, a consultant of the Edison company in England, redesigned Edison's Jumbo dynamo and showed that shorter, and more compact field-magnets were more effective than the long magnets of Edison. Hopkinson developed the 'Manchester' type of bipolar dynamos with side-coils and pole-pieces at top and at bottom, but by that time, multipolar dynamos were becoming more popular.

In 1882, James Edward Henry Gordon (1852-1893), a brilliant young Cambridge engineer, designed and built some of the largest alternators of his time. His alternators were of the rotating-field type, which had a diameter of 10 feet (3.05 m), and weighed 22 tons. Since the heating problem of the stator-coils was critical in his alternator, the core had to be laminated to overcome the tendency of the machine to burn out the coils of the stator.

In 1886, William M. Mordey (1856-1938), an American engineer, designed an alternator which was built by the Anglo-American Brush Corporation. The rotating magnetic-field was supplied by an iron rotor with 2 sets of 9 iron claw-like core ribs enclosing a single electromagnetic field-coil, which was concentric with the shaft of the alternator and supplied from a separate exciter. The 9 pairs of steel pole pieces were rotating with the field-coil, and the current was generated in a fixed ring of flat coils inside of a cast-iron frame positioned between the pairs of pole-pieces providing the longitudinal multipolar magnetic field which was cutting the coils of the fixed armature acting as the stator. There were advantages in keeping the coils in which the voltage was generated fixed, and such generators became popular until the advent of turbine-driven high-speed alternators. This alternator had the disadvantage that the coils of its stator when working at full load became distorted by being unable to expand outwards under high heating. It produced undulatory unidirectional current.

In 1881, William Thomson (1824-1907), later known as Lord Kelvin, suggested a modification to the method of rotor-winding, which the young English electrical engineer of Italian stock, Sebastian Ziani de Ferranti (1864-1930), made into a continuous zig-zag winding of insulated copper-ribbon in the rotor. This single-phase alternator of Thomson-Ferranti resembled that of Wilde produced in 1878, but Ferranti's windings were free to expand under heating which obviated the problem encountered by other alternators of the period. In 1886, Ferranti became the chief engineer of the Grosvenor Gallery Power Station in London for which he supplied his own alternators.

Ferranti's alternators, which were similar to those of Mordey, provided 2,400-volt service current. The transformers used at this power station were also designed by Ferranti.

In 1890, the Swiss engineer of English extraction, Charles Eugene Lancelot Brown (1863-1924), chief engineer of the **Maschinenfabrik Oerlikon** (*Machine Factory Oerlikon*) in Zurich, designed a compact 4-pole alternator similar to the alternator of Mordey, which produced an undulating alternating current. Although its shortcomings were magnetic leakage and irregularity in its magnetic fields, it led to the development of modern multipolar cylindrical dynamos. The 4-pole Oerlikon dynamos of Brown were nearly 8 feet (2.44 m) tall and produced 170-kilowatts of power at 500 rpm.

Electric Motors

The first electric motor was invented in 1748 by the American statesman and scientist, Benjamin Franklin (1706-1790). His electric motor functioned by the electrostatic principle and he suggested its use for the turning of a roasting spit or of a small orrery among other possible practical applications. Franklin was the first scientist-inventor to refer explicitly to the conversion of electric power to mechanical power by means of his electric motor.

In 1822, the brilliant English physicist and chemist Michael Faraday carried out a simple dual experiment in which he was able to rotate a wire about a magnet as well as a magnet about a fixed wire. In the same year, an English mathematician and physicist, Peter Barlow (1776-1862), constructed a simple apparatus in which a toothed wheel was made to rotate between a permanent horseshoe magnet when each of its teeth came into contact with a small pool of mercury located between the poles of the electromagnet which completed the circuit with an electric cell. Although these experiments demonstrated the principle of the electric motor, the apparatuses used were mere experimental toys.

In 1830, a few months before Faraday announced his famous experiments of electric induction, Salvatore Dal Negro (1768-1839), professor of physics at the University of Padua, was able to demonstrate an apparatus which produced rotary motion from electric current supplied by a Volta battery. Dal Negro kept a permanent magnet pendulum in continuous oscillation by an electromagnet the polarity of which was changed by a commutator switch. To the free-end of the pendulum was mounted a bar with a pawl which engaged a cogwheel attached to the spindle of a flywheel. When the pendulum rose in its swing, the pawl engaged the cogwheel and set the flywheel in motion. On the return stroke the pawl slid over the rotating cogs by means of a ratchet device which disengaged it from the cogwheel. Dal Negro was able to lift 0.11 pounds (5 grams) over 2 inches (5 cm) in one second with this electric motor. A similar instrument was built in 1834 by Giuseppe Domenico Botto (1791-1865), professor of physics at the University of Turin, Italy. In

1831, American physicist Joseph Henry built a laboratory apparatus which did represent a model of an electromagnetic motor that rocked an armature in the form of an electromagnet back and forth. This electromagnetic motor of Henry was supplied with a commutator, batteries, and a permanent magnet to provide the magnetic field.

A number of electrical experimenters built small-scale models of electric motors of various types. William Sturgeon built an electric motor in 1832 to run models of machinery which he exhibited in the next year. In 1833, the Scottish Reverend William Ritchie described a device by means of which he was able to set a magnet into rotation to raise a weight of several ounces over a pulley. In 1834, Dr. T. Edmundson of Baltimore, built a model magnetic paddle-wheel motor. In the 1820's and the 1830's, most electrical experimenters used chemical cells such as Hare's calorimeter, for their electric power source.

One of the first electric motors which was not a small experimental model was constructed in 1834 by Moritz Hermann Jacobi (1801-1875), who was professor of architecture at the University of Dorpat in Estonia and brother of the famous German mathematician Carl Gustav Jacob Jacobi (1804-1851). It could lift a 10 to 12 pound (4.5 to 5.4 kg) weight at a speed of 1 foot (0.30 m) a second. The motor consisted of 2 wood frames on each of which was mounted 4 horseshoe-shaped electromagnets with their pole-ends facing one another over a gap of 2 inches (5 cm). In the space between the magnets a wooden spider-frame with 6 arms carrying horizontal straight permanent magnets rotated about a horizontal axis. The battery current was delivered to the electromagnets through commutators which were so arranged that the currents reversed to maintain the attraction and repulsion between the permanent magnets and electromagnets always in the same direction. Jacobi's battery was made of zinc and platinum plates. In 1838, Jacobi installed his electric motor in a 28-foot (8.53 m) long boat of 7½-foot (2.29 m) width and 2¾-foot (0.84 m) draught. In the second trial of his paddle-wheeled boat on the Neva River in St. Petersburg, Russia, it reached a speed of 2.6 miles (4.18 km) an hour, The electric power to the motor was supplied by a battery of 64 Grove cells. Despite this initial success, Jacobi realised that a battery as a source of electric power was uneconomical and, therefore, he did no further experiments with his electric motor.

The first satisfactory electric motors which performed industrial work were built by the American blacksmith, Thomas Davenport (1802-1851) of Brandon, Vermont. His first electric motor, which had a 7-inch (0.18 m) flywheel, a speed of 30 rpm, and a magnetic field provided by permanent magnets, was built in 1834. Davenport was inspired to construct his first motor after seeing Joseph Henry's simple experimental electric motor. All his subsequent electric motors had electromagnetic fields. In 1837, Davenport built an electric motor with a 6-pound (2.7 kg) rotor of about ½-foot (0.15 m) diameter which had a speed of 1,000 rpm, and it was able to raise 200-pound

(90.7 kg) weight 1 foot (0.30 m) a minute when supplied by a 3-cell battery. He operated a drill and turned a lathe by using the electric motor in his workshop.

Davenport built a miniature 'electric train', a shunt-wound electric motor which ran on a circular rail-track 4 feet (1.22 m) in diameter. In 1838, one of Davenport's electric motors drove a small electric train of several carriages having a total weight of 80 pounds (36.3 kg) at a speed of 3 miles (4.83 km) an hour. In 1840, Davenport was able to build an electric motor which delivered up to 2-horsepower. All these motors were rotary motors.

In 1838, Davenport began work on a reciprocating electric motor based on the design of Joseph Henry. He was able to use what he called a 'sucking coil', in which one coil is sucked into another one as the basis of reciprocal action of his motor. Although Davenport built over one hundred electric motors, he was unable to secure financial backing for commercial exploitation of his electric motors mostly because an inexpensive source of electric power for his motors was not available at the time.

In 1835, Francis Watkins, an English inventor, built electric motors to operate mechanical models. In 1839, a Scottish inventor Robert Davidson of Edinburgh constructed an electric motor capable of turning a lathe and drive a small carriage. In the 1840's, a number of inventors produced rather ingenious electric motors. In 1842, Davidson built a carriage weighing about 6 tons which was driven by his electric motor a distance of about 1½ miles (2.41 km) with a speed of 4 miles (6.4 km) an hour between Edinburgh and Glasgow. In 1840, Uriah Clarke devised a reciprocating electric engine, and he used it to drive a model electric locomotive of 100 pounds (45.4 kg). Thomas Clarke proposed another reciprocating electric motor in the same year, but he regarded it as impractical. Even James Joule (1818-1889) made one electric motor in 1839, and another one in 1842.

A Dutch inventor, P. Elias of Haarlem, devised a very ingenious electric motor in 1840 which was the first electric machine making use of ring-wound armature, an armature which became important in the electric generator design more than 20 years later.

A well-known French instrument-maker Gustave Froment built many electric motors beginning in 1844, for his entire workshop was run by his electric motors. In 1857 he was awarded the Volta Prize by Emperor Napoleon III for this accomplishment. In 1849, the Danish civil engineer, Søren Hjorth, built a reciprocating electric motor incorporating a flywheel.

In the United States, a number of inventors were engaged in building electric motors. In 1847, an American dentist, Dr. G.Q. Colton, who gave public demonstrations on the wonders of science, installed a reciprocating electric motor, 14 inches (0.36 m) long and 5 inches (0.13 m) wide in his model locomotive which was supplied with electric power through the track by 4 Grove cells. Colton gave exhibitions of this locomotive throughout the northern and western United States.

The American dynamo pioneer Moses Gerrish Farmer, electrical superintendent of U. S. Naval Station at Newport, Rhode Island, constructed an electric motor in 1846 which in 1847 drove a model electric train in a public exhibition. Farmer's model locomotive was about 4 feet (1.22 m) in length. In 1850, Thomas Hall, an assistant of instrument-maker Daniel Davies in Boston, exhibited a miniature electric railway in Boston, and T.C. Avery patented an electric motor in 1851.

One of the most prolific American inventors was Charles Grafton Page (1812-1868), who later became a Patent Office examiner in Washington, D.C. In 1838, he constructed an electric motor to drive a drill in Davies' workshop. In 1851 he built a reciprocating motor delivering one horsepower, which a little later he improved to deliver 4 horsepower. In 1851, Page installed 2 of his motors in a 10-ton locomotive which ran at a speed of 10 miles (16.1 km) an hour. A few months later, his motors were able to deliver from 8 to 20 estimated horsepower, and the locomotive made a 39-minute journey from Washington to Bladensburg at a top speed of 19 miles (30.6 km) an hour. Each of Page's electric motors was powered by a battery of 50 Grove cells. This journey was so rough that the diaphragms of the Grove cells and the insulation in the motor were damaged. However by that time, Page had exhausted the funds provided by the United States Government in 1849 for this project, and he discontinued work on his locomotive.

In the reciprocating electric engines, the basic constructions consisted of the armature which was pulled into a solenoid, like in the Page's motor, or an armature hinged at one end was pulled down by an electromagnet, like in Clarke's motor. Linear motion in such an electric motor was changed into rotational motion by means of linkages. Another type of electric motor was the paddle-wheel type in which an armature was kept in constant rotational motion by a commutator switching on a magnetic field to promote the motion of the armature. The electric motors of Ritchie, Jacobi, Davenport, Davidson and Froment were of this type. No new basic types of electric motors were introduced until the appearance of the alternating current induction motor of Tesla. These electric motors did not succeed commercially mostly because of the high cost of electric power relative to that of steam-power. The electric power for these electric motors was supplied by batteries, the materials of which were very expensive, yet their storage of power was very limited.

In 1880, Thomas Alva Edison designed a tiny electric motor to operate his electric pen for the production of copying stencils. This motor had a size of 1 inch by 1½ inches (2.5 by 3.8 cm), and it had a 2-pole electromagnet and a flywheel armature carrying a short steel bar magnet. The motor ran at 4,000 rpm, had a small commutator to reverse the coil of the electromagnet when the poles of the magnet passed its core, and made a long needle vibrate in the pen-holder. A 2-cell battery supplied the current. This tiny electric motor of Edison has the distinction of having been the first commercially successful electric motor, for it was sold in large quantities, more than 60,000 pens, all over the world.

Werner Siemens exhibited his new electric elevator operated by his electric motor at the Paris Electrical Exhibition in 1881, which served as the model for later developments of lifts.

Electric Streetcar

The first application of electric motor to traction where electric power was provided by a dynamo instead of storage batteries was made by the German engineer Ernst Werner Siemens in an electric street-railway line in Gross Lichtenstein, a suburb in Berlin, Germany, after he had demonstrated his electric streetcar on a 1,640-foot (800 m) long track at the Berlin Exhibition in 1879. In his streetcars, Siemens used dynamos run in reverse as electric motors. The electric current was at first supplied to the streetcar through a 3rd central rail, but beginning in 1881, the current was supplied to the streetcar by an overhead cable.

In 1883, William Siemens constructed a 6-mile (9.67 km) long electric railway in a tourist town Portrush on the northern coast of Ireland which climbed 203 feet (61.87 m). The electric current at 240-volt tension, produced by two 52-horsepower hydraulic turbines driven by a 26-foot (7.92 m) head of water, was transmitted through a 3rd central rail to the electric train. In the next 5 years a number of street railways were built in Germany, in France and in England.

In the United States, Edison built and operated 2 experimental electric railways at Menlo Park in 1880 and 1882 respectively. Edison's first experimental locomotive was a crude 4-wheel flat-car with a dynamo operating in reverse as an electric motor, the electric current for which was produced by 2 Edison-type dynamos and returned through the rails and wheels of the locomotive. This primitive locomotive hauled 2 cars at 20 miles (32.2 km) an hour on a ⅓-mile (0.54 km) long experimental track. The second set of 2 experimental locomotives of Edison were more sophisticated, and they ran on a 2½-miles (4.0 km) long track with 40 miles (64.4 km) an hour speed. Edison and Stephen D. Field, nephew of the retired New York wholesale paper merchant Cyrus Field, produced a 3-ton locomotive in 1883 for a 3-rail elevated railway which was exhibited at the Chicago Exposition for Railway Appliances. This locomotive was powered by a standard Edison generator built by Edward Weston which worked in reverse as an electric motor, and was supplied with electric current from the 3rd center rail. During the 2 weeks of exhibition these locomotives ran altogether a distance of 446 miles (708.1 km). Since Edison did not consider his electric locomotive to be suitable for street railways, it did not lead to any further commercial development. However, it did create a global interest in electric traction.

American engineer J.C. Henry, the English engineer Leo Daft (1843-1922), the Belgian engineer Charles Joseph van DePoele (1846-1892), the American engineers Edward M. Bentley, Walter H. King, and most of all Frank Julian Sprague (1857-1934), a graduate of Naval Academy, developed the American streetcar and electric railway system after 1883. Henry installed a street railway in Kansas City in 1884; Daft electrified the railroad between Baltimore and Hampden in 1885; and van DePoele exhibited his small electric train at the Toronto Exposition in 1885. In 1884, Bentley and Knight ran their first electrically powered streetcar in Cleveland for public use. The problem of propelling the cars consisted of the design of an electric motor, the method of its mounting on cars, and the manner of transmitting power from the armature of the electric motor to the axle of the car. Bentley and Knight used one 30-horsepower, 500-volt constant current dynamo of Brush as the source of electric power, and another Brush series-wound dynamo of smaller capacity run in reverse as the electric motor. The electric motor was suspended by springs underneath the body of the car and connected to the front end and rear axles by belts of coiled wires. The electric power was buried in a wooden conduit in the middle of the track with a continuous slot on top through which a 'plough' made contact with the powerline and carried the current to the motor. They had trouble with the fracture of springs, failure of the wire belts, and overloading of the dynamo under a fixed current. Bentley and Knight were able to maintain one of their 2 streetcars in service through 1885 by using spur-gearing, constant voltage and varying current, but still they were not able to make a practical success of their street railway. In 1884, van DePoele, who had been a success in woodcarving business and operated his electric manufacturing company in Chicago, built an electric motor-car which pulled a train of cars in 1884 from the streetcar terminal in Toronto to the Toronto Annual Exhibition over a mile-long (over 1.61 km) railway. Van DePoele's method of supplying electric current to his electric motor consisted of a pivoted boom mounted on the roof of his motor-car with a contact wheel on its upper end pressed against the underside of the overhead electric powerline by means of a powerful spring at the lower end of the boom. A flexible cable connected to the wheel conducted electric current down to the electric motor. This was a novel method of supplying the streetcar with electric current and his trolley-pole method is still in universal use. When van DePoele applied for a patent for his streetcar in 1885 he came into conflict with Sprague, who laid claim to the invention of the under-running trolley in 1882 whilst he, as an officer of the United States Navy, was on a year's leave of absence. Owing to the technicality that Sprague had his residency on a U.S. Navy vessel, the patent was finally issued to van DePoele.

Sprague, together with Francis R. Upton, both college-trained engineers, were the theoretical assistants of Edison. In the United States, Sprague, Upton and Elihu Thomson (1853-1937) demonstrated what a theoretically educated engineer can accomplish in electrotechnology. Sprague resigned from the Navy and joined Edison in Menlo Park, but after discovering that Edison was more interested in developing electric illumination than electric power, Sprague organised in 1884 his own company: the Sprague Electric Railway and Motor Company. By the time Sprague entered the field, several electric exhibition lines for streetcar had been established, but the basic problem of establishing electric traction on a general commercial basis still remained. Only in 1888 was this problem first successfully met in an electric tramway in Richmond, Virginia. Sprague solved the basic problems of electric traction and designed his electric direct-current traction-motor which set the general design standard for traction-motors. His streetcars were supplied with electric power from an overhead conductor by a wheel mounted on a pole which could pivot on its support and was free to move up-and-down. The electric motors were geared to the wheel-axle remote from the armature and connected flexibly with strong springs at the other end to the frame of the bogie of the car. In his original design, Sprague used a horizontal bipolar magnet hinged to the axle at the yoke-end with the motor-pinion engaging a large wheel on an intermediate shaft that passed through the motor between the field-coils and engaged the car-axle with the second pinion.

The starting of the electric motor, the ability to withstand heavy power-surges, and the control of speed of electric motor of the streetcar were the most difficult problems in the design of the electric traction-motor. To start the motor, it was required to limit the current by setting resistances in series, and then to remove them section by section whilst the speed is increased. Two series-wound motors were used by Sprague, one on each axle, which were first coupled in series to divide the voltage, and then in parallel to obtain maximum power by means of a controlling switch in the form of a cylindrical drum. Sprague connected his motors to electric conductors by universally swivelling trolleys running under the current-carrying conductor which were maintained in contact with the conductor by strong springs. Sprague improvised a controller for his cars but he was not able to overcome one basic problem: when the controller of the electric power was moved a notch, it was equivalent to opening and closing a switch, and with that large amount of current a large electric arc was formed at the contacts which not only wasted power but also burnt the contacts of the controller to destruction. Elihu Thomson solved this electric arcing problem in 1892, too late for Sprague's Richmond tramway, by inventing his magnetic controller which consisted of placing a magnet near the contacts of the controller so that the magnetic field could 'blow-out' the spark. Thomson's method has been used in solving the controller's problem ever since, particularly in opening larger circuits in power stations. Only in 1895 was Sprague able to develop a multiple-control system for the operation of several electric traction-motors distributed along a long train, which permitted each motor to draw their current individually and directly from the track, and still be subject to the master-control in front of the train. This invention rapidly became the standard practice, and it made electric trains versatile and competitive with steam locomotives. Sprague, a daring engineer, was able to meet the incredibly difficult engineering problems posed by the streetcar in his Richmond tramway, but the impossible time limit defeated him. On account of his surpassing the time limit, he collected only $90,000 for this engineering project, whereas his costs were $160,000. However, his company became well established after the Richmond project, which was the first completely electric street railway system that made the horse-drawn railway systems obsolete. One electric traction-motor replaced 2 horses, and in the winter 4 horses. In 1888, 130 electric streetcars were in service in United States and by 1892, the number of streetcars in service had increased to more than 8,000. In 1890, 16% of street-railways in United States had been electrified, and by 1893 this percentage had become nearly 4 times as large. In 1902, when 22,576 miles (36,332 km) of electric streetcar tracks had been laid and the investment in it had reached over 2 billion dollars, the percentage of the electrified street-railways had become nearly 97%. Thomson-Houston Company also entered the electric traction field in 1888 and built street-railway lines which were very similar to Sprague electric railway lines. Even Westinghouse, a passionate advocate of alternating current, became interested in electric traction in 1889, and hired several engineers who formerly had been employed by Sprague to work on the problem of electric traction. In 1904, Westinghouse, who had continued experimenting with alternating-current motors, built the first street-railway which was powered by alternating-current.

Sprague had fulfilled 110 contracts to build electric tramways when in 1889 Edison finally became interested in electric traction and bought out Sprague's company in 1890. Sprague was also the most prominent maker of large direct-current motors, some as large as 15-horsepower. In 1887, more than 250 Sprague's direct-current motors were in industrial use in United States. From 1890 on, Sprague turned his talents to electric elevator design and construction. When this company was also bought out by Edison in 1902, he turned his engineering skills to the design of automatic train controls.

Induction Motor

The next advance in electric motors came with the alternating current. The function of the induction motor depended on two fundamental electric phenomena:
1. The rotation effect of the Arago disk.
2. The revolving magnetic field produced by the alternating electric current.

In 1824, Dominique Jean François Arago (1786-1853) had discovered that a magnetic needle freely suspended above a rotating copper disk spun with the revolving disk. In 1825, Charles Babbage, who later became well-known for his mechanical calculator, and his friend, the astronomer John Frederick William Herschel, rotated a copper disk by turning a horseshoe magnet underneath it, thereby demonstrating the inverse 'Arago effect' which they explained incorrectly on the basis of magnetic attraction by assuming that the motion of the magnet relative to the disk produced temporary magnetic poles in the copper disk that made the disk to spin owing to the attraction between the poles and the rotating magnet.

The experiments of Michael Faraday, carried out in 1831, discredited this explanation. Faraday supplied his own theory for the 'Arago effect' which assumed that the spinning disk of

Arago which cuts the lines of magnetic force creates temporary eddy currents along small closed paths in the disk. In his opinion these eddy currents bring about temporary magnetic fields which interact with the magnetic field of the needle and, thus, make it spin with the disk. This phenomenon of Arago which produces electrically one motion from another was not adequate for the design of electric motors since such motors require the conversion of applied electric power into mechanical rotation. What was lacking was the electrical means of producing a revolving magnetic field which had been supplied by a mechanically rotated magnet in the inverse Arago experiment of Babbage and Herschel. A method to produce a rotating magnetic field from electric current was first demonstrated by a British physicist Walter Baily in 1879. He set in a circle 4 upright electromagnets the coils of which were connected by a commutator-switch so that by manually turning the commutator the electromagnets were excited in a sequence thereby creating the rotating magnetic field which turned a horizontal copper disk suspended above the electromagnets around with the magnetic field. In 1883, the French engineer Marcel Deprez showed theoretically that Baily's rotating magnetic field can be produced without the aid of a commutator by exciting the 4 electromagnets with 2 out-of-phase alternating currents. However, Deprez himself only used this idea to develop a geomagnetic compass rather than an electric induction motor as an electromechanical power device. A number of engineers, Nikola Tesla in United States, Galileo Ferraris (1847-1897) in Italy, Friedrich Haselwander (1859-1932) in Germany, Jonas Wenström in Sweden, and Charles Bradley in United States, who were unaware of the work on the rotating magnetic field by Baily and by Deprez, did develop independently this device and claimed to have invented the constant speed induction motor. Haselwander was the first to build a polyphase generator, but by the time his induction motor was ready in the spring of 1888, Tesla had already submitted his patent application a few months earlier.

In 1885, Ferraris who was professor of technical physics at the University of Turin, in Italy, devised a demonstration apparatus for the 'Arago disk' which actually represented a simple model for a two-phase induction motor. Ferraris had 4 electromagnets laid out in a cross so that between their 4 projecting poles rotated a hollow copper cylinder when two-phase current was sent through the coils of the opposite pairs of electromagnets so that each pair of coils was supplied with an out-of-step alternating current. Ferraris followed Faraday's ideas in his formulation of the first theory of induction motor by assuming that the rotating magnetic field in the primary coils of the electromagnets induced eddy currents in the secondary copper cylinder which through interaction with the rotating magnetic field made the cylinder to rotate. Based on this theory, and the mechanical definitions of energy losses and output power, he concluded that the induction motor would produce maximum power when the speed of rotation of the cylinder is half of that of the magnetic field, thereby giving the maximum efficiency of 50 %. The experimental test of his model induction motor indicated the maximum efficiency of the motor occurred at a considerably reduced speed. He concluded that this kind of motor had no practical utility beyond its use as an alternating-current meter.

The ingenious American engineer from Croatia, Nikola Tesla (1856-1943), had already conceived the alternating current electric motor as a two-phase induction motor in 1882 and a three-phase induction motor in 1884, but he built the first standard induction motor model in 1887 when he applied for his patent in November of the same year. He conceived the polyphase induction motor at the same time. Tesla began his researches on the alternating-current electric motor by first producing the rotating magnetic field with out-of-phase alternating current and by trying out different types of armature. His 'reluctance' motor had an iron disk armature which was capable of rotation when its sides were cut away. His 'hysteresis' motor worked by magnetisation produced in the steel armature by the rotating magnetic field lagging behind the inducing field owing to the high resistance of steel to its magnetisation, a phenomenon known as 'magnetic inertia' or 'hysteresis'. In the induction motor, the field windings produce a magnetic field the effective direction of which rotates around the axis of the machine owing to the out-of-phase alternating current supplied to the machine. Electric currents are induced in the rotor by the surrounding field windings, which magnetise it, and the rotor is pulled around by virtue of the interaction of magnetic fields, both attractive and repulsive, between the stator and the rotor at almost the same speed as the rotation of the magnetic field. The speed of rotation of the rotor is determined by the arrangement of the windings and by the frequency of the electric power supply. Since these 2 quantities are fixed for a given motor, the speed of rotation of the induction motor is approximately fixed. Tesla, in contrast to Ferraris, was convinced from the very beginning that induction motors are of great industrial value. The electromagnets of Tesla's induction motor had inwardly-projecting poles between which rotated an iron-drum. The primary coils were wound around the pole pieces and the secondary coils consisted of closed-circuited copper windings around the iron-drum rotor. The pairs of primary coils were supplied by 2 alternating currents 90 degrees out-of-phase in which one current passing through one opposite pair of electromagnets achieved its maximum value whilst at the same instant the other current passing through the other opposite pair of electromagnets had its minimum value. Although Tesla's 'two-phase induction motor' functioned on the same principle as that of Ferraris, it was more efficient, about 60 %, and more powerful than the motor of Ferraris. Tesla used closed-loop windings on his laminated rotor so that the rotating electromagnetic fields of the coils of the stator induced magnetic poles in the ferromagnetic rotor, thereby producing a stronger interaction with the rotating magnetic field which was the real reason for the increase in the power and efficiency of Tesla's induction motor. His 3rd induction motor was essentially different from the induction motor of Ferraris.

In the induction motors built by Ferraris, Tesla, Haselwander and Bradley the 2 pairs of stator-coils representing the electromagnets carried alternating currents out-of-phase with each other which created a rotating magnetic field. This magnetic field, the strength of which varied with time, created by induction such eddy currents in the rotor-windings that the rotating magnetic field dragged the rotor along with it.

Tesla also invented the 'synchronous motor', the magnetic field of which was like that in the induction motor, only that its rotor was either a permanent magnet or an iron core provided with coils carrying direct current. The sets of coils on both the rotor and the stator were energised directly from the same generator. Since the rotor in such a synchronous motor follows the rotating magnetic field without any slip, synchronous electric motors run only at fixed speeds, whereas the speed of induction motors can be regulated within certain limits. The disadvantage of these electric motors was that special means had to be found to run them up to their operating speeds, and even when running satisfactorily, a temporary overload could cause the motor to come to a stop. It turned out that the induction motor ran most efficiently at constant speed, but its main annoying shortcoming was that it exerted a low starting torque, 2 operating properties that limited its industrial usefulness. At the time when Tesla worked on his induction motor the industrial electric current in United States was a single-phase current having a relatively high frequency of 133 cycles a second, and induction motors do not function well with high frequency current. Tesla, who sold his patent to George Westinghouse (1846-1914), tried to obviate the phase problem by inventing a few types of single-phase induction motors, but by the time he left Westinghouse in late 1889 only a 1/6th horsepower single-phase fan motor had been manufactured. Tesla used projecting pole pieces in his induction motor which left great air gaps between the pole pieces and thereby gave rise to magnetic leakage in which the lines of force fail to connect the primary and secondary windings. In 1890, Tesla's former collaborator, Charles F. Scott (1864-1944) redesigned Tesla's two-phase induction motor by giving a better distribution of the secondary winding over the rotor's surface, a revision which increased the power of the motor. The improved types of induction motor were exhibited by Westinghouse at the Pittsburgh Exposition in 1890 but only a few such motors were ever sold, and since George Westinghouse at the time was in financial difficulty he relinquished work on the development of the practical induction motor later in the same year.

Tesla did fundamental research on methods to transmit power without wires using the upper atmosphere, but he did not succeed in this effort. Tesla also invented a bladeless turbine functioning by the boundary-layer drag, an idea he had conceived when still a teenager. It could be used as a turbine, or a pump. Such a turbine used in reverse as a pump only functions at extremely high speeds which could be achieved only in the 1950's. A pump functioning on Tesla's boundary-layer drag is able to pump sludgelike material which can be 82% solid at a rate of 100 gallons (379 liters) a minute. In 1920's Tesla also conceived a vertical take-off aircraft, a type of aircraft which was first produced after the Second World War.

Further development of the induction motor was taken up by Charles Eugene Lancelot Brown of the *Maschinenfabrik Oerlikon* (*Machine Factory Oerlikon*) in Zurich, Switzerland, a talented son of Charles Brown (1827-1905), and by Michael von Doliwo-Dobrowolski (1861-1933) of the *Allgemeine Elektrizitätsgesellschaft* (*General Electricity Company*), popularly known by its acronym **A.E.G.**, in Berlin. Brown was essentially a self-educated electrical engineer, whereas Doliwo-Dobrowolski was a graduate of the *Technische Hochschule* (*Technical Institute*) in Darmstadt, a leading European polytechnical school in electrotechnology. At the time Brown was director of the electrotechnical department of the *Maschinenfabrik Oerlikon*, whereas Doliwo-Dobrowolski was the chief engineer of **A.E.G.**.

In 1891, the 2 firms, in collaboration with Oskar von Miller (1855-1934) who was in charge of the project, were responsible for the design and construction of the hydroelectric generators near Lauffen at Neckar in Germany, as well as the 175-kilometer long transmission line, the longest in the world at the time, carrying a three-phase alternating current at 15,000 volts, later increased to 25,000 volts, over 3 wires to the Electrotechnical Exhibition Hall in Frankfurt at Main. The 2 dynamos at Lauffen and the three-phase alternating current transformers invented by Doliwo-Dobrowolski in 1888 were designed by Brown and built by the *Maschinenfabrik Oerlikon*, whereas the three-phase induction motor was designed by Doliwo-Dobrowolski and built by **A.E.G.**. Doliwo-Dobrowolski had proposed the use of the three-phase alternating-current for the efficient transmission of electric power in preference over the two-phase alternating-current which had been suggested by Tesla and used in the electric power transmission from Niagara Falls in 1896. The two-phase alternating-current power transmission was doomed after the success of the Lauffen project.

In 1888 Doliwo-Dobrowolski, after learning of the work of Tesla and Ferraris on induction motors, took up the study of induction motors because he did not believe that Ferraris' theory which asserted that the maximum efficiency of such motors is limited to 50% was correct. He had considerable experience with direct-current motors, and on the basis of his knowledge and what he had learnt from Tesla and Ferraris, he built an experimental three-phase induction motor in late 1888 with a squirrel-cage rotor consisting of uninsulated copper bars set in slots of the iron core, and connected at their ends to metal rings. Next year he built an induction motor of 1/10th horsepower based upon his experimental model which had 80% efficiency and a high starting torque. He increased the power of his next motor to 5-horsepower, but learnt that this motor had increased magnetic leakage and reduced secondary resistance in comparison with the less powerful motor. Doliwo-Dobrowolski carried out his research of the induction motor empirically by a cut-and-try method which led him to the conclusion that in the design of induction motors it is important to obtain the least possible magnetic leakage, and to provide a higher secondary resistance for starting than for normal operating speed of the motor. He achieved the first aim by an improved primary winding, and the second by introducing a rheostat by means of which he was able to increase secondary resistance at starting the motor which produced a much larger starting torque.

On account of the collaboration of **A.E.G.** and the *Maschinenfabrik Oerlikon* on the project of the electric power station Lauffen at Neckar, the two firms decided, in 1890, to pool their research results on polyphase electric power. Brown became acquainted with Doliwo-Dobrowolski's induction motor, and immediately produced by the trial-and-error method an improved induction motor between 1 and 2 horsepower in which he countersank the primary winding of copper wires into slots around the inner surface of the stator, which reduced magnetic leakage, and manufactured the squirrel-cage secondary from copper bars insulated from the iron core of the rotor which provided more conductive paths for eddy currents. Later in the same year Doliwo-Dobrowolski designed a comparable induction motor between 2 and 3 horsepower but with the primary as the rotor and the secondary as the stator, a design which gave the motor 80% efficiency. He used a slip-ring with a lever-operated rheostat to increase the secondary resistance at the start of the induction motor to give it a larger starting torque. This design of Doliwo-Dobrowolski was so successful that the 100-horsepower version of it was used at the Frankfurt Electrotechnical Exhibition in 1891 to pump an artificial waterfall of 10-meter height. Brown also exhibited his induction motor of 20-horsepower at the Frankfurt Exhibition, a motor which he had scaled up from his 2 horsepower model. Brown's great improvement of the original Doliwo-Dobrowolski's design consisted of the reduction of air gap between the rotor and the stator, which decreased the magnetic leakage, and the insulation of the copper bars in the rotor which improved the passage of eddy currents in the secondary. In 1892, **A.E.G.** installed Doliwo-Dobrowolski's large induction motors to turn the generators in the three-phase electric power plant at Lauffen which illuminated the city around Heilbronn. The generators of the three-phase current, and the transformers at Lauffen power plant were designed by Charles E.L. Brown, and manufactured by the *Maschinenfabrik Oerlikon*. The transmission line for electric power from Lauffen to Frankfurt at Main had an efficiency between 68% and 75% which proved that the electric power transmission by means of three-phase alternating current was economical and efficient. The Lauffen electric power project also established the three-phase induction motor as an eminently useful and efficient electric motor in industrial application, since Tesla's two-phase induction motor seemed to offer no real competition to the three-phase induction motors of Brown and Doliwo-Dobrowolski. When electric motors of less than 10-horsepower were required, their construction was based upon Brown's squirrel-cage design because they were reliable, sturdy and inexpensive. Their only disadvantage was a relatively low starting torque. Electric motors larger than 10-horsepower were based upon Doliwo-Dobrowolski's design having a rotating primary, a stationery secondary, a slip-ring, and a manually operated external rheostat.

Tesla's induction motor worked on a higher frequency which, in general, reduced the starting and operating torques, because the torque exerted by the induction motor is inversely proportional to its speed, some of which Scott learnt from his experimental tests of Tesla's motor. It was learnt later that the induction motors can function properly when its circuits are properly proportioned, as theoretical and more comprehensive experimental research later indicated.

It became apparent that a sound engineering theory for the theoretical design of the induction motor was required after Doliwo-Dobrowolski found that the theory of the induction motor of Ferraris was faulty since this motor, unlike the direct current motor, did not obey the Ohm's law, as had been assumed by Ferraris, with respect to the resistance of the secondary or to the efficiency. In 1891, Louis Duncan (1867-1916), professor of electrical technology at the Johns Hopkins University, used Maxwell's theory to demonstrate theoretically why the induction motor had an efficiency greater than 50%, and why an increased secondary resistance increased the starting torque. A little later 2 French engineers, Maurice Hutin and Maurice Leblanc (1857-1923), published an extensive theoretical analysis of the induction motor which included the speed-torque relation, and the formulae for voltage and currents present in the motor, as well as power output and torque curves which offered considerable improvements over Ferraris' results, because they had considered in their analysis self-induction of the secondary which Ferraris had neglected. They also proved theoretically that resistance in the secondary should be high at starting and low at operating speed, a fact which had been experimentally discovered in 1891 by Doliwo-Dobrowolski. They showed that at the normal operating speed, the difference between the rotation of the magnetic field and the rotation of the rotor of the induction motor was small so that the torque became inversely proportional to the secondary resistance, whereas at the start this difference was large which made the torque directly proportional to the secondary resistance. They applied Maxwell's circuit theory in their analysis which assumed a perfect air-core machine, rather than an iron-core machine which exhibits magnetic leakage that substantially reduces the starting torque. It was known from experience with generators and direct current motors that their performance varies from the theoretical predictions owing to 3 phenomena associated with iron-core hysteresis (representing the resistance of iron to changes in its magnetisation), eddy-currents, and variable permeability (representing medium's ability to conduct magnetic lines of force). Hysteresis as well as eddy-currents produced heating in iron-cores, effects which reduced efficiency of the induction motor through increased energy losses. Permeability, a constant in air-core machines, varied with magnetisation in an iron-core. For this reason Maxwell's theory based upon constant permeability was inaccurate for iron-core machines, even if hysteresis, eddy-currents, and magnetic leakage were considered.

In order to incorporate variable permeability into the theory of the induction motor, engineering scientists made use of the theoretical method of dynamo design established by the English engineers Gisbert Kapp (1852-1922) and John Hopkinson in the years 1885 and 1886, a method they had based on the magnetic circuit theory of the American physicist Henry Augustus Rowland (1848-1901) that avoided Maxwell's induction coefficients in favour of the so-called 'magnetic resistance' measured by the ratio of the length of the magnetic circuit to the product of its area and permeability. The change in permeability with magnetisation was obtained from experimental graphs of flux-density versus current. Using the magnetic resistance of the dynamo, the resultant flux could be calculated which, in turn, could be used to find the generated voltage of the dynamo. The Swiss engineer Engelbert Arnold (1856-1911) of the **Maschinenfabrik Oerlikon** (*Machine Factory Oerlikon*) and the French engineer André Eugène Blondel (1863-1938) were the first to use the resultant-flux method in the analysis of the induction motor. Both engineers amalgamated the Maxwellian and the magnetic circuit theory in their methods. Blondel, in contrast to Arnold, expressed magnetic leakage in terms of the resultant-flux rather than in terms of induction coefficients used by Arnold, but he did not include the induction coefficients in the important speed-torque relation.

In 1894, Kapp and Charles P. Steinmetz simplified and improved this approach of Arnold and Blondel by presenting a pure resultant-flux method for the analysis of induction motors. Kapp, who was a cofounder of the resultant-flux method for dynamo design in 1885, applied it in 1887 to the transformer theory where he introduced a graphical method as a substitute for Hopkinson's differential equations. Steinmetz improved Kapp's transformer theory in 1891 by using a slightly different graphical method. Besides his original research in 1893 on hysteresis, Steinmetz established a method of analysing alternating current electric circuits based upon the algebra of complex-numbers, rather than on mathematically more difficult differential equations, a method which was more accurate than the graphical analysis. Both Kapp and Steinmetz obtained the pure resultant-flux method by applying their transformer theories to the induction motor as had been done by some other engineers before them, but who had used Maxwell's coefficients for ironless induction coils. Kapp and Steinmetz avoided this inaccuracy by using 2 new parameters from their transformer theories: the so-called 'leakage inductance', and the 'primary inductance', which represented better magnetic leakage, hysteresis and eddy currents than earlier theories and were defined in analogy with Maxwell's induction coefficients. Kapp and Steinmetz were able to calculate by means of their theory the performance of the induction motor more accurately than had been done by means of the earlier theories. In 1899, Steinmetz introduced an 'equivalent circuit' for the induction motor which made his theory more useful for practical design because engineers were able to use alternating current circuit analysis to determine the intricate electromagnetic interactions in the induction motor, a method that was much easier than his complex-number algebraic formulas. His 'equivalent circuit' method was subsequently applied to other electric devices such as generators, transformers, transmission lines, and motors which permitted a reduction of the intricate electromagnetic interactions to a single circuit on one diagram thus making the analysis of the entire, interconnected electromagnetic system feasible by means of the alternating current circuit theory.

The 'circle diagram' method, a competitor to 'equivalent circuit' method of Steinmetz, was developed by the German engineers Alexander Heyland (1869-1943) in 1894 and Bernhard Behrend (1875-1932) in 1896. This method offered engineers a graphical procedure for the design of induction motors. Heyland was still an undergraduate student at the **Technische Hochschule** (*Polytechnical Institute*) in Hanover when he invented his ingenious circle-diagram method. This diagram reduced the voltage, current, phase difference, output, and efficiency of induction motors to a circle the diameter of which is given by the magnetic leakage factor of the motor, all the pertinent relations such as the current and voltage relation and others being given by various vectors in this circle. In 1897 Heyland invented his single-phase induction motor which was distinguished by the absence of complicated starting apparatus whilst its starting torque was suitable for every type of work. In the application of the method of the circle diagram the designer needed to know only 3 parameters, 2 of which were obtained from experiments and one, the 'Behrend leakage coefficient', from calculations. This method was in use well into the 1950's. The engineering design equations for the induction motors were usually kept a commercial secret by such firms as General Electric.

In 1891, Walter Langdon-Davies (1867-1924), an English engineer who worked as a consultant to a power company in Vancouver, Canada, built an experimental single-phase induction motor in which the ring-stator with 6-pole pieces was wound in 6 sections forming 3 groups of 2 coils each, and one group had resistance in series and the others were connected directly to the mains. This design created a lag in the current of one coil relative to that in the other, and thus resulted in a rotating magnetic field. The rotor being a smaller ring was wound with heavy copper wire in 6 sections, each one of which was short-circuited. Langdon-Davies marketed his improved commercial induction motor in 1897.

The induction motor turned at nearly constant speed regardless of the load, and it was closed, rugged, and produced no sparking. The only disadvantage of the induction motor was that it ran at the speed determined by the frequency of the alternating-current supply and the arrangement of the winding, and this property of it was exceedingly difficult to change since the winding could not be easily altered. In contrast, the direct-current motor could easily be changed to different speeds by merely changing the rheostatic control. For this reason, electric traction motors are mainly direct-current motors. In 1891 an alternating current series-wound commutator motor was patented by Ernest Wilson (1863-1932), professor of electrical engineering at King's College in London, and a motor of shunt-type was patented by Johannes Görges (1859-1946) in Germany. The synchronous electric motor was patented in Germany and in United States in 1887, and appeared on the market in 1891, although the direct-current motors dominated the market until 1900 and even some years beyond. The greatest shortcomings of the direct-current motors were their commutators and their open, unprotected windings. The first completely protected heavy-duty direct-current motors were developed for tramways and railways. The greatest improvement in the performance of direct-current motors was the result of the introduction of a supplementary magnetic pole, the so-called 'interpole', in 1906. The copper wire-gauze commutator brushes were replaced by carbon commutators in 1889, which drastically reduced the commutator wear.

Small electric motors which are in use today are the so-called 'universal motors' as the most common electric motors, and the 'capacitor-start motors'. The universal motors are essentially direct-current motors in which the field iron is laminated in order that the polarity of both the field and the armature can be reversed magnetically at the frequency of the power supply. Such motors run both on direct current or on 60 cycle alternating current, and they have a high starting torque of a direct-current motor in contrast to a low starting torque of an induction motor and, moreover, they can run faster than induction motors by having a range of speed from 4,000 rpm to 10,000 rpm. These motors can be either series-wound or shunt-wound, but usually they are series-wound. 'Universal motors' are customarily used in domestic appliances and portable tools.

Small induction motors are dependent on their starting arrangement, since a single-phase induction motor is basically not self-starting. The so-called 'shaded-pole' motors are used where very low power, simplicity, and reliability are paramount. In such motors, part of the pole is surrounded by a short-circuited loop of copper strip, and the current induced in that loop retards the rise and fall of magnetic field within the loop, resulting in a 2-pole magnetic field with a rotating part that pulls the armature around. Hence, it is a two-phase induction motor, the efficiency of which is very low, that is used in fans and record-players where the efficiency of the motor can be very low.

Another small induction motor is the 'capacitor-start' motor, the field winding of which consists of two parts: one part of the field is connected directly to the power supply, the other part is connected through a capacitor which causes a phase displacement so that the motor effectively runs on a two-phase power supply. It is mainly used in washing machines.

The range of the power of electric motors is immense: varying from several thousand horsepower to a tiny fraction of a horsepower.

CHAPTER XXIII

ELECTRIC ILLUMINATION: From Arclighting to Transmission of Electric Power

Arclighting

The rapid development of electric generators was promoted by the gradual replacement of gaslighting with electric illumination in public places and factories.

In 1808, Humphry Davy (1778-1829) had demonstrated that a brilliant electric arclight can be produced with a 4-inch (0.1 m) long electric arc struck between two carbon rods. However, Davy's carbon arclight had a short life because of the limited power of his huge Volta cell battery consisting of 2,000 zinc and copper elements. Although Davy could not develop this type of electric lighting any further for the aforementioned technical reasons, it did serve as an inspiration for the search of electric illumination by later inventors.

An English astronomer and inventor, Warren de LaRue (1815-1889), produced in 1820 the first known incandescent lamp which had a platinum filament. De LaRue, and the English physicist William Robert Grove (1811-1896) realised that the life of the metallic filament in air is very brief owing to the oxidation of the metal. Therefore, both carried out experiments consisting of placing the platinum filament in a glass bulb and evacuating the air in the bulb as well as they could. The life of such platinum filament lamps is quite short in the imperfect vacuum because of the close proximity of the glowing temperature and the melting temperature of the platinum. After Grove, and the famous German chemist and 'element hunter', Wilhelm Robert Bunsen (1811-1899), had invented their electric cells as the source of electric power, experimenting on electric lighting began in earnest.

In 1842 a French mechanic Joseph Deleuil made the first attempt to illuminate a street in Paris with electric arclight. In 1844 he lit the ***Place de la Concorde*** with an electric-arc searchlight the wood-charcoal electrodes of which were enclosed in a glass globe evacuated of air to retard the burning of the electrodes. In 1849 the German architect and physicist Moritz Hermann von Jacobi (1801-1874) performed similar lighting experiments with arclights in Dorpat, Estonia, and later in St. Petersburg in Russia, but complete success eluded both inventors.

In 1844 when carbon purification had become possible, Jean Bernard Léon Foucault (1819-1868) suggested the use of harder retort carbon for electrodes in arclights but it proved to be unsuitable. Foucault also invented an automatic arcing-gap regulator in which a tiny variation of current in the carbon rods ran a clockwork mechanism which maintained the proper arcing-gap between the arcing carbon electrodes. The English engineer W. Edward Staite (1809-1854) gave a satisfactory demonstration of an arc-lamp in 1846. He had designed an arc-lamp supplied with the first automatic feed-mechanism to conserve the proper arcing-gap in the lamp, since one of the carbon-bar electrodes, the 'cathode', was consumed much more rapidly than the other carbon-bar electrode, the 'anode'. William Mathew Flinders Petrie (1821-1908) considerably improved the arc-lamp mechanism and collaborated with Staite in demonstrating it in 1848, but they failed to obtain financial backing although their arclight did last for 5 hours. Staite and Petrie were essentially defeated by the limitation of their power source, the Daniell electric cell.

In France, Jules Duboscq, and Victor Serrin continued experimenting with arclighting. In 1855, Duboscq modified Foucault's automatic regulator of the arcing-gap, that was based on the arc-regulator of Staite, after which the arc-lamp became sufficiently reliable for such practical applications as illumination of opera houses, laboratories, and lecture theaters. In 1857, Victor Serrin devised a regulator for arclights incorporating some of the best features of Duboscq arclight. Serrin's arclight regulator after some improvements in 1859 dominated electric lighting everywhere until the late 1870's. Apart from Staite's arc-lamp, these early regulators were satisfactory only over rather brief periods of time and, therefore, other types of regulators were devised. Joseph Lacassagne and Rodolphe Thiers invented a differential arclight regulator in which the proper speed of carbon-feed was controlled by the difference of the current in 2 circuits. They illuminated a square in the city of Lyons with their arc-lamps in 1855 by using the electric power of a battery consisting of 60 Bunsen cells. They also lit the *Avenue des Champs Elysées* for 4 hours in 1856, but were not able to obtain the financial backing from the Emperor Napoleon III. The main reason for the lack of interest of the Emperor was the great expense of the source of electric power, the chemical cell. It became quite clear that arclighting required a reliable and relatively inexpensive source of power.

In 1853 the Belgian physicist Florise Nollet (1794-1853) suggested a system of electric lighting for the lighthouses in which the source of illumination was the 'limelight' consisting of a block of lime heated to incandescence by an oxyhydrogen flame. He proposed to obtain the necessary oxygen and hydrogen from the electrolysis of water, and designed a magnetoelectric generator with multiple disk armature which rotated past the poles of horseshoe magnets to supply the electric power required for such a system, but he died before it was built. The 'limelight' as a source of light for different purposes was first prosed by the English physician and steam carriage designer Goldsworthy Gurney (1793-1875) in an article published in 1823. The invention of 'limelight' is usually wrongly ascribed to Lieutenant Thomas Drummond (1797-1840) who successfully produced the 'limelight' as a light of first order in 1826. In 1824 George Birckbeck for the first time used 'limelight' in projection apparatus.

In 1853, the English engineer Frederick Hale Holmes showed how a similar magnetoelectric generator might be used better in supplying electric power to the newly developed carbon-arc lamps. In 1857 Professor Holmes demonstrated the arclighting before the Trinity House Lighthouse Board using his magnetoelectric generator in which 36 permanent magnets rotated past stationary coils. His generator weighed 4,000 pounds (1619 kgs) and produced less than 1,500 watts of power. His arclights had to be connected in series, which had the disadvantage that a failure of any individual carbon arc-lamp broke down the entire circuit. The illumination provided by the 'limelight' next to that of the arclight was like comparing a candlelight to the Sunlight.

In 1873, a Russian physician, Alexandr Nikolaevich Lodygin (1847-1923), invented an incandescent lamp consisting of incandescent retort-carbon rods encased in a glass bulb or in a glass cylinder. In 1874 Lodygin evacuated the glass bulb for improved performance of the lamp, and manufactured 200 incandescent lamps to illuminate the harbour of St. Petersburg with these lights. However, Lodygin's lamps had a very short life and his power supply was inadequate for the requirements of such a lighting system.

The first really reliable arclighting system was invented by a retired Russian military engineer Paul Nikolaevich Jablochkoff (1847-1896) who was employed in the workshop of Alfred N. Bréquet where Gramme's electric generators were being manufactured. Jablochkoff invented a new type of carbon arc-lamp, known as the 'Jablochkoff's candle', which consisted of two carbon rods mounted in Kaolin-clay insulation to separate the two electrodes and bridged by a graphite strip, all enclosed in an onyx bulb. Jablochkoff used alternating-current which burnt both carbon rods evenly without any need for a mechanical regulator. Zénobe Théophile Gramme (1826-1901), who had taken out the first patent on arclight in 1861, designed an alternator to provide the electric power for the Jablochkoff arclights, and installed in 1873 an arclighting system in his own factory. The success of Jablochkoff's arclighting system was almost immediate because it made possible the first electric illumination on a broad commercial scale. Jablochkoff was successful in solving the central problem of subdividing the electric light: to reduce the intensity of the 2,000 candle power of the arclight, he put several arclights in the same circuit for which purpose he inserted transformers in the circuit of the subdivision. The primaries of the transformers were connected in series in the circuit of the alternator, and the arclights were connected in series in the secondaries of the individual transformers. The Jablochkoff arclighting system clearly demonstrated that a central electric power station could distribute electric power to different locations. His 'candle' flickered a little, but was whiter and brighter than gaslight, and not so blinding as ordinary arclights. However, his arclight had a relatively short life: a 9-inch (0.23 m) carbon candle was consumed in an hour an a half. In 1875, American electrical inventor Moses Gerrish Farmer (1820-1883) of Newport, Rhode Island, suggested connecting arclights in parallel, and he was successful in operating 42 arclights in 42 separate parallel circuits.

The copper-plated carbons, patented in 1859 by de Fontaine-Moreau, were used by François Edouard Carré in electric-arc lamps which reduced the rate of consumption of carbon candles, and required from 8 to 9 ampere current instead of from 17 to 20 ampere current required by Jablochkoff electric-arc lamps. The tension in the electric-arc lamp circuit was about 2,000 volts.

In 1877 the Siemens & Co. brought out the wick-carbon candle in the form of a carbon rod pierced lengthwise with a hole into which was compressed the mass of wick consisting of the finely powdered coal impregnated with suitable solutions. The wick-carbon candle satisfied all the requirements of the carbon electrode in the arclight. The new arclight with wick-carbon candles was first displayed in 1879 at the Exhibition of Arts and Crafts in Berlin. In 1882 the Potsdamer Platz and the Leipziger Street to Friedrich Street were illuminated with 36 differential arclights, after which the arclighting of streets became the common system of streetlighting.

In 1878, the American engineer Charles Francis Brush (1849-1929) of Cleveland, Ohio, made the first really effective arclighting system by designing a centrally generated and transmitted electric power from high-voltage dynamos of his own invention, which were controlled by automatic regulators to maintain constant current in the arclighting circuit regardless of the load, and served a number of arclights connected in series and controlled by automatic short circuits to circumvent burnt-out arclights. Brush invented a coil-regulator for his arc-lamps to feed his less expensive carbon candles made from petroleum coke, and bound with coal tar-pitch rather than tar, and a double-carbon arc-lamp with an automatic switch to lessen the need for maintenance, and to provide an automatic replacement for a failed arc-lamp. Moreover, Brush rounded and tapered the tips of his carbon bars, and copper-plated their bases to diminish the resistance between the carbon bar and its holder, which further reduced the rate of carbon consumption in the electric arcing.

In the summer of 1879 Brush constructed the first central electric power station in United States to supply electricity to 22 outdoor electric arclights in San Francisco. Also, Niagara Falls was illuminated for the first time in 1879 by 16 arclights supplied with electric power by the Brush's high-voltage dynamo particularly manufactured for his arclighting system. The Brush high-voltage dynamo was run by a waterwheel representing the first hydroelectric power source.

Incandescent Lighting

Despite the success of arclighting in public places, it was unsuitable in domestic applications because the arclight produced eye strain owing to its intense brightness. It was obvious that for domestic use a much safer light was required, and this criterion was first completely satisfied by the incandescent light.

In 1838, a Belgian physicist Jobart returned to the idea of the incandescent light of LaRue, when he heated a thin carbon rod in vacuum to maintain the incandescence of the carbon. This method of incandescence was resumed by the American inventor J.W. Starr (c.1822-1847) of

Cincinnati, Ohio, who patented in 1845 an incandescent lamp which incorporated continuous carbon conductors enclosed in vacuum.

In 1854, an American watchmaker and optician of German stock, Heinrich Göbel (1818-1893) invented an incandescent lightbulb which incorporated an incandescent carbonised bamboo fibre. His first lightbulb consisted of an evacuated bottle into which he had fused a carbonised bamboo fibre as the incandescent element. Göbel illuminated the show-window of his modest watchmaker's shop in the Monroe Street in New York with such bamboo-fibre lightbulbs for the express purpose of attracting customers. The duration of Göbel's incandescent lightbulbs was nearly 400 hours. The electric current for his incandescent lightbulbs was produced by a large battery consisting of zinc and coal elements. He also illuminated with such 'magic lighting' a small wagon containing a 60 element battery in 2 wood boxes which he pulled along the streets of New York every evening. His lighting experiments at first attracted great public attention but as usual it was soon forgotten, because Göbel never thought of commercially exploiting his lighting invention which was not the reason why he did it in the first place. Only in 1893, when the Beacon Vacuum Pump and Electric Company in Boston in a patent litigation against the Edison Electric Light Company raised the issue of Göbel's priority in the invention of the incandescent lightbulb incorporating a carbonised bamboo-fibre filament almost 40 years before Edison, the 75-year old Göbel finally received full public recognition for his invention shortly before his death.

The English pharmaceutical chemist, Joseph Wilson Swan (1828-1914) assumed work on the incandescent light after attending one of Staite's demonstrations and having learnt of Starr's patent. By 1860 he had made a carbon filament lamp, but promptly realised that the vacuum-pumps available at the time could not satisfactorily evacuate the glass bulb of the filament lamp and, moreover, that the available electric power source, the battery, was quite inadequate for electric illumination. After the German physicist Johann Philipp Sprengel (1834-1906) had invented the mercury vacuum-pump which had been successfully used by William Crookes (1832-1919) in high-vacuum research, Swan returned to his incandescent light experiments in 1877. Swan made experiments with carbonised paper-strip filaments and with other materials without success, but finally succeeded with the carbonised cotton-fibre filaments treated with dilute sulfur acid in 1878, and in 1879 Swan perfected his lightbulb by evacuating the glass bulb with the help of Sprengel's mercury vacuum-pump. In 1880 Swan was able to exhibit to the Chemical Society at Newcastle-on-Tyne an incandescent lightbulb with a spanned carbonised fibre filament in the form of a hairpin which he later replaced by a sling type of filament. This lightbulb with a conical base was fitted into a socket with a bayonet type of locking device. He founded in 1881 the Swan United Electric Light Company Ltd. in London, and in 1881 the production of Swan's lightbulbs began in the Mawson & Swan Company in Newcastle. At first Swan had not patented his lightbulb, because he thought that all important ideas of such an incandescent electric bulb had been already anticipated by others. In 1879 Swan was able to produce a squirted cellulose filament for his lightbulb, and by 1905 Swan's electric bulb incorporating a squirted cellulose filament enclosed in vacuum had increased the lifespan of the electric lightbulb 138 times.

In United States, Thomas Alva Edison (1847-1931) experimented with the incandescent light at the same time as Swan did in England. Edison's experimental lamps with carbonised bamboo-fibre filaments arrived in London in 1880. Edison had patented all his incandescent lights, and to obviate costly and protracted litigation, Swan and Edison amalgamated in 1883 their companies in England into a joint company.

William Siemens in London kept his brother Werner Siemens in Germany informed about the experiments of Swan on incandescent lighting, which led Werner Siemens to open in Berlin the Siemens & Halske Incandescent Lightbulb Factory, the first such factory in Germany.

An English scholar, St. George Lane Fox Pitt (1856-1932), who was a researcher in psychic phenomenon and moral philosophy, was also a versatile inventor. He invented a simple mercury vacuum-pump, an automatic voltage regulator, and a few integrating energy meters. In 1878 he patented the Lane-Fox system of electric lighting and its distribution which incorporated incandescent filament lamps connected in parallel, and subsequently he took out a series of patents on electric filament lamps. He later sold his patents to the Anglo-American Brush Electric Light Corporation.

In 1896, an Italian scientist and inventor, Arturo Molignani (1865-1939), devised a procedure that used red phosphorus for the removal of the last traces of air in glass bulbs, which reduced the time to obtain vacuum in light-bulbs from half an hour to 2 minutes, a matter of great practical importance to the economical manufacture of electric light-bulbs.

Significant improvement of electric lamps began in 1902 with the invention of the osmium filament lamp in 1898 by the Austrian chemist Carl Auer von Welsbach (1858-1929), the inventor of the incandescent gas-mantle for gaslights. Tantalum filament lamps were introduced by Werner von Bolten and Otto Feuerlein in 1903. By 1905 von Bolten and von Welsbach had made tantalum so pliable that it could be made into filaments by **Siemens & Halske A.G.** in Berlin thereby opening the era of incandescent lamps with metallic filaments. Tantalum made incandescent lamps more efficient, but it was already obvious that tungsten filament lamps offered even greater advantages. The tungsten filament lamp was invented in 1902 by Alexander Just and Franz Hanamann of Vienna, Austria, but, unfortunately, their tungsten filament was fragile which made the tungsten lamps unreliable and uneconomical, although their tungsten filament lamps were introduced for street illumination already in 1907. The significant improvement in incandescent lighting was that the illuminating capacity of incandescent lights incorporating new metallic filaments had quadrupled from 1881 to 1900.

Remarkable improvements in electric lighting were achieved by William David Coolidge (1873-1975) and Irving Langmuir (1881-1957), two physicists working in General Electric Laboratories. Coolidge, after 4 years of intensive research, was able to produce a drawn-wire tungsten filament lamp in 1911. These commercial tungsten lamps were still vacuum bulbs, but soon Langmuir discovered that gas-filled lamps were superior to vacuum lamps. Although lamps filled with nitrogen were first produced in 1913, the gas-filled lamps became commercially available only after 1918. Langmuir invented the gas-filled lamp which used 85% inert gas called argon and 15% nitrogen at almost atmospheric pressure.

The gas-filled wolfram filament lamps had 10 times the illuminating power of Edison's carbon-filament lamp. In 1958 the Sun Street (*Sonnenstrasse*) in Munich, Germany, was illuminated by 3 xenon lamps which had the lighting capacity of 3,500 normal 40-watt lamps.

Fluorescent Lighting

The electric discharge tubes of glass filled with certain gas or vapour can produce a visible glow when an electric current is discharged through the gas contained in the tube. This kind of lighting goes back to a patent of MacFarlan Moore in 1902 which became better known in 1910 when the 'Moore tube' of 100-foot length filled with carbon dioxide, or nitrogen, appeared on the commercial market. The electric discharge in the 'Moore tube' was created by the application of high voltage between the electrodes infused into both ends of the tube. Soon afterwards another American inventor, Peter Cooper Hewitt (1861-1921), devised a mercury vapour arc-lamp by placing a small quantity of mercury in a sealed glass tube fitted with electric terminals at each end. When the tube was connected to an electric circuit, the mercury vapourised and an electric arc formed from one terminal to the other terminal giving off greenish blue light. This light was efficient but its colour was unsuitable for commercial illumination. In 1910, George Claude lit the ***Grand Palais*** in Paris with his electric discharge tube filled with diluted neon gas. He sold the patent rights for his 'neon discharge tube' to the General Electric Company in 1919, and after intensive experimentation which began in 1934, the practical fluorescent light was perfected by 1936. In 1938, the electric discharge tube was placed on the commercial market as a fluorescent tube, the inside of the tube is coated with a phosphoric substance that glows when struck by ultraviolet rays. The electric discharge tubes are usually filled with neon, argon or helium gas.

Transmission of Electric Power

After the initial period of the birth of industrial electricity between 1860 and 1870, the second period of development of electric power characterised by the transmission of electric power, and the rise and triumph of alternating-current took place between 1880 and 1900.

The first experimental attempt to transmit electric power over a distance was successfully accomplished by Hippolyte Fontaine (1833-1910), an associate of Gramme, at the International Electrical Exhibition in Vienna in 1873. Fontaine supplied electric current to a Gramme generator used as a motor, from a Gramme generator driven by a Lenoir gas-engine. The ½ horsepower delivered by the generator was transmitted to the motor by two 1,640-foot (500 m) long wires. The motor drove a small centrifugal pump that supplied water to an artificial 6-foot (1.83 m) high waterfall on the Exhibition grounds. It created a small sensation with the public despite the fact that the efficiency of his power transmission was only 50%. In 1878, French engineer Cadiat of the *Société du Val d'Osne*, repeated Fontaine's experiment by connecting two Gramme's machines over a distance of 492 feet (150 m) with two 0.12-inch (3 mm) wires to transmit 0.136 kW (50 kgm) of power. Another French engineer, Félix, transmitted electric power over 1,640 feet (500 m) in 1879 by following Fontaine's example. These were all empirical efforts. In 1879, Scottish physicist William Thomson [later Lord Kelvin] (1824-1907) and Charles William Siemens (1823-1883) explained in the British Parliament how electric power in the amount of 21 horsepower produced by the Niagara Falls could be transmitted 311 miles (500 km) in a 0.5 inch (12.7 mm) copper wire under a tension of 80,000 volts, a theoretical proposition which was unrealisable at the time.

The theoretical problem of transmission of electric power was taken up in earnest by the French engineer François Daniels Marcel Deprez (1843-1918), professor at the ***Conservatoire des Arts et Métiers*** (*Conservatory of Arts and Crafts*), between 1881 and 1885. He first demonstrated that the 50% efficiency in power transmission obtained by Fontaine, was a lower limit which could be improved, and he worked out the theoretical conditions of power transmission for an electric lighting system. In the late 1870's Jablochkoff had demonstrated by his system of arclight illumination that the distribution of electric light is possible. American engineers had calculated that a cable of ½ inch (13 mm) diameter is sufficient to distribute the total electric power generated by the Niagara Falls. Deprez calculated this energy to be more than 2 million horsepower, and he came to the conclusion that this electric power can be transmitted over 46.6-miles (75 km) distance under a 1 ½ million volt tension with a 50% efficiency. Deprez, moreover, derived theoretically the result that two Gramme machines of type C in the Fontaine experiment can transmit a useable 10 horsepower over 31 miles (50 km) through an ordinary telegraph line. A little later, Deprez claimed that with a second line he can recover 25% of the energy produced at the power source of the transmission line of a total length of 435 miles (700 km). In 1882, Deprez experimented at the

Conservatoire des Arts et Métiers with a line of 800-ohm resistance which was equivalent to a distance of 80 km. After this attempt, Deprez was approached by a young German electrical engineer, Oskar von Miller (1855-1934) of Bavaria, with the proposition to construct for the Electrical Exposition held in Munich, Germany in 1882, the same experiment used by Fontaine at the Vienna Exposition, in 1873.

The Bavarian government made its facilities at Miesbach available to Deprez to install a 2-horsepower Gramme dynamo driven by a hydraulic turbine and transmit its 1.1 kilowatt electric power at a tension from 1,500 to 2,000 volts over 35 miles (57 km) through existing telegraph lines to the Palace of Exposition in Munich, where it fed an electric motor that drove a hydraulic centrifugal pump which circulated water over a 6-foot (1.83 m) high artificial waterfall. The power delivered was ½ horsepower, but the efficiency of power transmission was only 22%. In 1883, Deprez installed an electric generator and an electric motor side by side at a railroad station of the Railroad du Nord, but he connected them by a circuit consisting of 2 lines each of 5-mile (8 km) length passing through Le Bourget. The motor delivered 5 horsepower, and the power delivered varied between ¼ and 4½ horsepower, with an efficiency between 30% and 48%. In 1883, Deprez gave another demonstration like the one at the Exposition in Munich in 1882, except that in this case, the distance over which electric power was transmitted through telegraph lines was 8.7 miles (14 km), from Vizille to Grenoble, where it supplied electric current to the electric motor which operated a circular saw, a printing press, and a lathe, as well as a circuit of 108 incandescent lamps of Edison. In 1884, an attempt was made in Italy to transmit electric power by direct-current 25-miles (40 km) from Lanzo to the Exhibition in Torino for the purpose of illumination. In 1885, Deprez was able to transmit 45 kilowatts of electric power by direct-current through telegraph lines from Creil to Paris-LaChapelle at 6,000-volt tension. The experimentation lasted more than a year into 1886, and Deprez was able to achieve a 50% efficiency.

Gramme company made an experiment in power transmission by using 4 Gramme dynamos connected in series which sent direct-current 35 miles (57 km) at 5,900-volt tension to feed electric current to 3 electric motors also connected in series. They achieved 50% transmission efficiency like Deprez, which appeared to be the limit of efficiency for the transmission of direct-current.

John Hopkinson (1849-1898) declared in 1894 that the efficiency of generating installations of electricity had improved from 50% in 1879 to 94%, and their power from 6 horsepower to 1,000-horsepower in the same period.

The alternating-current began to make inroads into the industrial electricity by 1883. Since 1878 Jablochkoff arclighting system using alternating-current at high voltage provided by Gramme's dynamos had demonstrated that it was possible to distribute electric power by means of alternating-current from a central power source. Low-voltage direct-current could not be transmitted much farther than 3 miles (4.8 km) for incandescent illumination, since it required heavy current and, therefore, large conductors made of expensive copper. This restriction seriously limited Edison's electric power system working with direct-current. Thomas Alva Edison invented the 3-wire distribution system for his Pearl Street Power Station in the heart of New York City's financial district, the first American commercial electric power station, which became operational in September of 1882, to reduce the size of the copper conductors in the direct-current distribution to 7,200 incandescent lamps operating at 110 volts that were connected in parallel with 3 conductors. This shortcoming of an electric power system working on direct-current placed serious limitations on the electric power system, and as a result of this limitation, Edison's electric generating station could not supply economically an area larger than about 16 square miles (41 square km). The Pearl Street Power Station was provided with 6 large 'Jumbo' direct-current generators of low internal resistance which were driven by Porter&Allen reciprocating steam-engines supplied with steam by Babcock and Wilcox boilers as well as with large storage-batteries for emergency service. Edison designed the 'Jumbo' direct-current generators with the assistance of Francis Robbins Upton (1852-1921), his scientific collaborator. The conductors consisted of half-round copper bars of about 20-foot length encased in iron tubes from which they were isolated by cardboard washers and insulated by an asphaltum compound packed into the tubes. By 1884, the Pearl Street power station supplied electric power to 11,272 incandescent lights in 500 homes.

A direct-current electric power station had been built on Holborn Viaduct in London even earlier than the Pearl Street power station in New York. This station began operation in early 1882 supplying 3,000 incandescent lights with electric power from Edison's 'Jumbo' direct-current generators driven by Armington&Sims steam-engines. Unfortunately, the gaslighting interests seriously retarded the development of electric lighting in Britain.

Arclighting system had an advantage over the incandescent lighting system in the distribution of electric power because it used high-voltage. In fact, in late 1879, a little experimental power station operating 3 Brush generators supplied alternating-current for illumination by arclighting to various customers in San Francisco. Brush used a 40 arclight dynamo to operate 30 arclights at full brilliance 10 miles away, a capability that Edison with his direct-current could not economically match. The direct-current power station lacked the most important attribute of an economical electric power system, that of transmitting electric power economically over long distances. In the electric power transmission the loss of power in the transmission lines is inversely proportional to the voltage and directly proportional to the square of the current. In contrast to the direct-current system, it is relatively easy to increase the voltage and decrease the current in the alternating-current system by means of transformers. Therefore, the realisation to use alternating-current for electric power transmission gradually occurred to a number of engineers, and in 1879 one of the first to suggest the use of a system of transformers for such a transmission was an English engineer Fuller but he never followed it up with a practical design. The long-distance transmission of electric power became feasible only when transformers became practical.

The first practical use of transformers in an electrical power system was designed by the French engineer Lucien Gaulard (1850-1888) and his associate, the English engineer John Dixon Gibbs, in 1882. In 1881, Gaulard and Gibbs established a National Society for the Distribution of Electricity by Secondary Generators in London. They proposed a 16-mile (25 km) long electric power transmission system with 4 substations in London, which supplied 30-horsepower distributed to operate Jablochkoff candles and 140 incandescent lamps of Swan. They proposed to boost the electric potential to 3,000 volts by primary transformers for the transmission of power, and then reduce it locally by means of secondary transformers to 100 volts for the feeding of electric lights. Gaulard and Gibbs designed an adjustable core-type single-phase transformer for their system which they patented in 1882. In the core-type of transformer the 2 coils, the primary coil and the secondary coil, are wrapped alternately around the same open annular core enveloped by an ebonite tube, and if an alternating-current is sent through the primary coil, an alternating-current potential is created in the other coil by induction. The ratio of the primary voltage to secondary voltage is directly proportional to the ratio of the number of turns in the coils, and the currents to the inverse ratio of the number of turns. Gaulard-Gibbs transformer consisted of 4 vertical cores supported on a wooden platform and connected in series. The power in this transformer could be controlled by a screw and the voltage by a rheostat connected to the primary coil of the transformer. Gaulard claimed a high efficiency for his transformer, between 90% and 95%, but its veracity was vehemently contested by the proponents of direct-current. In this judgement, Gaulard had overlooked most of the parasitic phenomena present in transformers, such as losses in magnetism, losses due to heating, the negative influences of the core material, and the losses due to the manner of the winding of the coils. Gaulard gave the first public demonstration of his system at the Exposition in Turin, Italy, in 1884 where he transmitted alternating-current over 50 miles (80 km) of wire, a distance which equalled the length of the transmission line for direct-current between Paris and Creil designed by Deprez.

George Westinghouse (1846-1914), famous for his invention of the pneumatic brake for trains and a mechanism for the return of derailed railroad cars to tracks as well as several automatic switching and signalling devices, bought the American rights for the Gaulard-Gibbs transformer in 1885. Westinghouse was also the foremost promoter of alternating-current electric power in preference over the direct-current electric power advocated by Edison, whom he rivalled in reputation. His chief engineer, William Stanley (1858-1916), realised that the principal deficiency of the Gaulard-Gibbs transformer was its connexion in series, which made it operationally interdependent and, therefore, prevented these transformers from working independently. Stanley subsequently made the first truly practical transformer when he improved the efficiency of the Gaulard-Gibbs transformer by replacing the iron wire in the core with iron plate, and the copper plates in the coils with copper wire. He then connected the improved transformers in parallel which rendered the transformers operationally independent, as well as making the transformers self-regulating, thereby lending each transformer independent power control over the devices it was feeding with electric current.

Stanley devised an efficient alternating-current incandescent lighting system for Westinghouse Lighting Company, and installed one at his own expense in Great Barrington, Massachusetts, which served 150 incandescent lamps for lighting the streets and the stores. In the fall of 1886, Westinghouse Lighting Company opened its first commercial alternating-current central power station in Buffalo. By 1892, more than 1,000 alternating-current central power stations were in operation.

Already in 1879, Elihu Thomson had accomplished similar works. He had designed induction coils in parallel, and made them self-regulating. Thomson invented the shell-type of transformer in which the induction coils are inside and the iron core outside the transformer. However, Thomson was deeply concerned about the safety of the transformers in case the secondary coil accidentally came into contact with the primary coil, thus allowing high voltage to enter the secondary circuit. He invented 3 methods of protection, and adopted the simplest and most reliable one which grounded the secondary wires of the transformer with a metallic conductor, a method still in current use. Only after developing his protective equipment did Thomson market his alternating-current dynamo in 1887, a dynamo which he had designed many years earlier.

Hungarian engineers Zipernowsky, Beri and Blathy invented the parallel-connexion of transformers independently, and they took out a German patent on it in 1885.

The Swiss engineer, René Thury (1860-1938) was successful in prolonging the era of direct-current transmission. In 1890, Thury installed at Geneva a 30-mile (0.8 km) long direct-current distribution circuit which worked at 12,000-volt tension and delivered 100 kilowatts of power for the Company of Industrial Electricity of Geneva. The same company installed another similar electric power distribution circuit using alternating-current. In the same year a 35-miles (56 km) long circuit for the distribution of direct-current at 22,000-volt tension was built at LaChaux-de-Fonds in Switzerland.

In 1906 a 112-mile (180 km) transmission line for the direct-current was built to supply electricity for the city of Lyon which remained in service for 30 years. The source of the power was the 213-foot (65 m) waterfall on the Isère river, which produced 6,300-horsepower in the generating plant at Moutiers. This system distributed a constant, regulated current at 75 amperes at 57,600 volt tension. The Moutiers to Lyon electric power transmission system underwent several renovations over the years before it finally gave way to alternating-current, one of the last major electric power systems using direct-current to do so. Between 1911 and 1917, the constant current was doubled to 150 amperes at 100,000 volt tension. By 1936 the tension had been raised to 125,000 volts, and the power to 18,000 kilowatts whilst maintaining the current constant at 150 amperes.

In 1891, at the Electrical Exposition in Frankfurt at Main, where German and Swiss products of electrical technology were exhibited, a triple-wire transmission line from Lauffen at the Necker river to Frankfurt, a distance of 109 miles (177 km), was constructed in which two dynamos in Lauffen driven by a hydraulic turbine produced 235-horsepower and a three-phase current which fed 1,000 incandescent lamps and powered a three-phase alternating-current electric motor driving a centrifugal hydraulic pump in Frankfurt. The transmission line was strung on 3,000 poles and delivered 60-horsepower to the centrifugal hydraulic pump in Frankfurt. The two alternating-current dynamos in Lauffen and the three-phase alternating-current transformer, which raised the tension from 55 volts to 8,500 volts, were designed by Charles Eugene Lancelot Brown (1863-1924) and constructed by the Swiss *Maschinenfabrik Oerlikon* (*Oerlikon Machine Factory*). The three-phase electric motor was designed by Michael von Doliwo-Dobrowolsky (1861-1919), who also invented a three-phase alternating-current transformer, and constructed by the German firm *Allgemeine Elektrizitätsgesellschaft* (*General Electricity Company*) known by its acronym **A.E.G.**. The efficiency of the power transmission from Lauffen to Frankfurt varied between 68.5% and 75.2%. An additional transformer at each end of the line was able to step up the tension to 28,000 volts at a frequency of 24 cycles a second, whilst the efficiency of the transmission line carrying 180-horsepower was 75%. Although the transmission line of Lauffen-Frankfurt did not resolve all the problems associated with the transmission of alternating-current, it definitely proved that the transmission of electric power can be done economically and efficiently by means of the three-phase alternating-current. This entire project was organised by Oskar von Miller, the founder of the well-known *Deutsches Museum* (*German Museum*) of technology in Munich, who had secured the collaboration of C.E.L Brown of the *Maschinenfabrik Oerlikon* and M.von Doliwo-Dobrowolsky of **A.E.G.** for this project. Since 1890, von Miller had been in charge of building an electric power plant at Lauffen which supplied the electric power to illuminate the city around Heilbronn with the three-phase alternating-current.

The success of the Lauffen-Frankfurt alternating-current transmission line influenced the Niagara Falls Electric Power Commission to decide in favour of the alternating-current generators rather than the direct-current generators. Westinghouse, following the example of the Lauffen electric power plant, designed a polyphase alternating-current system using two-phase alternating-current according to Tesla's patent that could be treated as 2 independent alternating-current circuits for lighting at 30 cycles a second frequency to lower costs of the long-distance electric power transmission. In the early 1890's, Westinghouse introduced alternating-current at 60 cycles a second frequency as a compromise between the requirements for electric lighting and electric power which by the 1920's served as the standard industrial alternating-current frequency. By 1892, it was generally recognised that the transmission of alternating-current was considerably less expensive than the transmission of direct-current, although Edison was still reluctant to admit the superiority of the alternating-current power system. However, Edison's colleagues and financial backers were eager to enter the alternating-current business because of the declining demand for direct-current power and equipment in which Edison held most of the patents. Finally in 1892, Edison electric companies merged with the Thomson-Houston Company in the formation of the General Electric Company, since Thomson-Houston and Westinghouse controlled virtually all the significant alternating-current patents. In 1894, General Electric was able to market their three-phase alternating-current power system in which one phase had a higher voltage than the other 2 phases to evade Tesla's patents owned by Westinghouse. In the late 19th century various conversion-devices began to appear on the market such as rotary-converters which could change alternating-current into direct-current and *vice versa*, phase-converters which could change alternating-current phases, and motor-generator couplers which could operate by one type of electric current whilst supplying the other type. Owing to the availability of such conversion devices on the market, customers interested in buying electric power and electric equipment no longer were forced to choose either the direct-current or the alternating-current system. Under such market conditions in 1896, Westinghouse and General Electric considered it advantageous to pool their patents which covered the entire range of direct-current and alternating-current power. For instance, this patent pooling gave General Electric access to Tesla's patents which allowed them to make all voltages the same in their three-phase power system for such balanced three-phase power system afforded a remarkable improvement in the performance of induction motors as had been clearly demonstrated at Lauffen.

Westinghouse's Niagara Falls Hydroelectric Power project, completed in 1896, was the first spectacular as well as practical large-scale hydroelectric plant in existence in which the hydraulic turbines were capable of generating 10,000-horsepower. In many electric power plants equipped with various turbogenerators the transmitted electric potential reached 40,000 volts by 1897, 60,000 volts by 1900, and about 150,000 volts by 1910. By 1903 the Canadian Niagara Power Company had installed a set of double Francis hydraulic turbines operating under 133-foot (40.54 m) head of water were able to generate 10,000-horsepower at 250 rpm. By 1923 the installed Francis hydraulic turbines in the Niagara Falls Hydroelectric Power Plant were capable of supplying 70,000-horsepower. The hydroelectric power plants were able to supply the least expensive electric power despite the fact that it was transmitted over hundreds of miles.

The hydroelectric power principle was first advocated in France by a paper-manufacturer Aristide Bergès (1833-1904) who for many years had called waterfalls the 'white coal', a natural source of power which goes to waste unless it is exploited. In 1869 he tapped a 656-foot (200 m) high waterfall in the Alps in cascades by means of penstocks which drove an electric generator that produced 1½ kilowatts of electric power. In 1882 he tapped a 1,640-foot (500 m) high waterfall with this method to produce more hydroelectric power. From 1880 to 1887 he built the first giant hydraulic accumulator in a cavern excavated 82 feet (25 m) beneath the bed of the Lake Crozet, which served as a reservoir for the overflow of the Lake Crozet driving electric generators. Bergès was responsible for the introduction of the term 'hydroelectric power' into electrical power terminology, a popular term still in current use.

The Niagara Falls Hydroelectric Power Plant which produced a total of 15,000-horsepower in 1896 was the first large hydroelectric plant in the American continent. The alternating-current generators at the Niagara Falls Hydroelectric Power Plant rotated the magnetic field outside the stationary armature, whereas the earlier Westinghouse generators had a revolving armature and a stationary magnetic field. Later the revolving-field generators in which the magnetic field revolved within the stationary armature became the standard for alternators. Since the alternators had to run at constant speed the hydraulic turbines were provided with governors which kept their speed constant regardless of the head of water. The electric power produced at the Niagara Falls Power Station was first transmitted by two-phase alternating-current to Buffalo. However, the Niagara Falls hydroelectric power transmission system was not the first in the United States, since the Westinghouse Company had installed in 1891 the first alternating-current electric power transmission system from Willamette Falls, Oregon to Portland, Oregon, a distance of 13 miles (21 km). The transformers of this alternating-current electric power system stepped-up the potential to 3,300 volts in the transmission lines and then stepped-down the potential to 1,100 volts in the primary circuits of the city, where local transformers stepped-down the electric potential to 50 or 100 volts in the neighbourhood circuits providing electric power for illumination.

The rise of the newly arrived electric power supply industry made the location of any industry quite independent of the whereabouts of its electric power sources.

CHAPTER XXIV

ELECTRONICS IN SPACE: From Wireless to Radio and Radar

Wireless Telegraphy

The idea that telegraphic signals can be sent 'without a conductor' first occurred to Samuel Finley Breese Morse (1791-1872) in 1842, when he transmitted such signals across a canal by means of 4 plates submerged in water, 2 in the transmitting circuit on one bank of the canal and the other 2 on the opposite bank in the receiving circuit. However, a great deal of his transmitting current was diffused in the water of the canal. In 1867 William Henry Preece (1834-1913), chief engineer of the British Telegraph Office, tried this type of system again in England but discovered that the Earth and water as the conducting media are of little practical value for long-distance signalling. However, 'air' as the conducting medium was much more suitable for wireless telegraphy and this method was employed by an American inventor, Dr. Mahlon Loomis (1826-1886), in 1866. Loomis built his instrument of 'aerial telegraphy', by means of which he sent a message over a distance of 14 miles (22.5 km) without the use of wires. He later successfully signalled 2 miles (3.2 km) between 2 ships in the Chesapeake Bay. Loomis patented his 'aerial telegraphy' in which he employed an 'aerial used to radiate or to receive the pulsations caused by producing a disturbance in the electric equilibrium of the atmosphere'. Therefore, Loomis is the first inventor of the 'aerial' for sending and receiving radiated telegraphic signals.

In 1879, David Edward Hughes (1831-1900) noticed that an electric spark produced a noise in his microphone located anywhere in the house. Hughes erroneously thought that the noise was produced by electric conduction through the atmosphere. From 1879 to 1886, before Hertz's experiment, Hughes worked on his electric signal generator and receiving equipment. The signal generator of Hughes was a small induction coil which transmitted signals over a distance of 1,312 feet (400 m) to his receiver consisting of an aerial connected in series to a coherer, a battery and a sensitive carbon pencil microphone, an instrument which he had invented in 1879. The electric signals transmitted by means of electric waves were received by Hughes as noise in his microphone. Hughes was discouraged from pursuing his investigations further by the wrong opinion of George Gabriel Stokes (1819-1903), a renowned Irish physicist who had attended one of the many successful demonstrations Hughes gave of his wireless experiment in 1880, and had resolutely denied Hughes' claim for the existence of conduction of electric signals through the air. Stokes had convinced Hughes that the latter's experimental findings could be explained as a phenomenon of electric induction, another wrong opinion. Hughes, therefore, committed the mistake of not publishing his results, which robbed him of the honour of being the first to develop a simple elementary operational system of wireless. English scientist William Crookes (1832-1919) wrote an article in early February, 1892, in which he made a forecast for the coming of 'electric wave telegraphy without wires', a forecast that he based on the experiments of David Hughes in 1879. This article of Crookes had an important influence on the development of the wireless telegraphy.

The 'cohering phenomenon' was discovered in 1838 by a Swedish physicist, chemist, musicologist and an accomplished pianist, Peter Samuel Munck af Rosenschöld (1804-1860), a man of refined culture who was an adjunct professor of physics at the University of Lund. This phenomenon manifests itself in the making of the loose, nonconducting iron filings enclosed in a glass vial to compact and electric current conducting after being subjected to the action of electric waves, and nonconducting again by mechanical shaking. In 1850 the French physicist Guitard noticed that when dusty air was electrified, the dust particles coalesced into strings and flakes. British telegraph engineer Samuel Alfred Varley (1830-1921) made use of this 'cohering phenomenon' in his carbon-dust lightning protector for telegraph lines. This 'cohering phenomenon' was observed again by Karl Ferdinand Braun (1850-1918) in 1876, by Hughes and Lord Rayleigh (1842-1919) in 1879, by Calzecchi-Onesti in 1884, and by the French physicist Edouard Eugene Branly (1844-1940) in 1890. However, Branly recognised that the resistance changes in such cohering phenomenon represent an extraordinarily sensitive reaction to rapid electric oscillations. Branly found that metal filings loosely packed into a glass tube, which was popularly called a 'coherer', which normally conducted little current but which began to conduct a much larger current when the tube was subjected to high-frequency electromagnetic waves. In this manner electromagnetic waves could be instantly converted into easily detectable electric current. Hence the 'coherer' was found to be a much more sensitive detector of electromagnetic waves than Hertz's spark-gap. Sometime after 1890, Branly devised an apparatus incorporating his 'coherer', which was connected to a relay device that in turn closed various local circuits through a clockwork distributor. Branly was able to light an electric lamp, start a small motor and control other instruments by means of this apparatus, which he operated from a distance by a spark transmitter.

Amos Emerson Dolbear (1837-1910), American experimental physicist who was professor of natural science at the University of Kentucky and later at the Tufts College, patented his electrostatic telephone in 1881. He discovered by accident that when a wire became disconnected his telephone still continued functioning. After 5 years of work, Dolbear patented in 1886 a system of electric communication in which he had replaced the wire by 2 conductors in the air in the form of a wire suspended from a kite with the electric circuit of his instrument grounded at both ends. In 1882, Dolbear gave a demonstration of transfer of speech by such electric communication without wires at a meeting of the Society of Telegraph Engineers and Electricians in London, England. Dolbear was first able to signal over ¼ mile (0.4 km) at Tufts College by using a 300-foot (91.44 m) long wire areal suspended from a kite. Although he preceded Marconi in his use of the wire aerial, he appears not to have recognised the phenomenon of electric radiation from the areal and, thus, he made no attempts to detect the radio signals in the sense of Hertzian radio-waves as Marconi did. Dolbear was later involved in costly litigations with Bell and Marconi over some telephone and wireless patents but he lost both of these lawsuits.

In 1885, Thomas Alva Edison (1847-1931), who was engaged by Western Union Telegraph Company to break Bell's monopoly, succeeded in sending a message from a moving train by 'wireless' transmitter. A wire was stretched on top of the train as a sort of aerial so that it was parallel to a telegraph wire strung on poles along the railroad track that made possible the transmission of telegraph signals over the short distance between the aerial and the telegraph wire by means of the static induction. Edison obtained a patent for this system in 1891, which included the aerial wire, but this system worked only over quite short distances and, therefore, was of limited practical usefulness.

In 1892, William Henry Preece employed the electric induction phenomenon and attenuating current to send wireless signals from a primary circuit to a secondary circuit located at 3.1-mile (5 km) distance from the primary circuit at Bristol canal. The primary circuit with its power supply was a giant loop of 719.6-mile (1,158 km) length raised 3,280 feet (1,000 m) above the sea level in which the Earth served as the return conductor. The secondary insulated loop incorporating a telephone receiver had a length of 1,798 feet (548 m). An interception of the current in the primary circuit changed its magnetic field, which in turn induced electric current in the secondary circuit and thereby reproduced in it the original Morse signal. Later, Preece was able to transmit wireless signals over 5 miles (8 km) by the induction method. In 1894 these wireless experiments were repeated in Germany by Emil Rathenau (1838-1915) and Heinrich Rubens.

From 1886 to 1888, Heinrich Rudolph Hertz (1857-1894) carried out his famous and decisive experiments to search for waves longer than light-waves or heat-waves. He produced the electromagnetic waves by spark discharges from the Ruhmkorff induction coil, and recaptured them from space at a short distance in the form of a small spark that jumped the gap in an open ring of wire serving as his receiver. Hertz found the wavelength of his electromagnetic waves to be 2.2 feet (66 cm), which belong to the ultrashort wavelength spectrum. In England, Oliver Joseph Lodge (1851-1940) attempted to measure similar short electric waves in wire-circuits almost at the same time, but after learning of Hertz's pioneering investigation, he duplicated Hertz's experiments.

The ingenious Nikola Tesla (1856-1943) from Croatia, an expert of high-frequency alternating-current, began his researches on wireless transmission of intelligence in 1889. By 1892, Tesla had worked out his wireless system that he demonstrated to the public in 1892 and 1893. Tesla is the true original inventor of radio telegraphy and most of its essential components. He devised for his system an aerial wire, or antenna, a ground connexion, an aerial-ground circuit incorporating an 'adjustable induction and capacitance for tuning', the transmission and the receiving set tuned to resonance with each other, and an electronic tube detector, which is the ancestor of all the later detecting and amplifying electronic tubes. Tesla studied Hertzian waves and came to the conclusion that the useful wavelengths for wireless range from 100 meters to many thousands of meters. Tesla demonstrated his wireless transmission by producing long electromagnetic waves by his high-frequency generator, the so-called 'Tesla transformer', which he captured in his laboratory with his resonant coil surmounted by one of his 'electronic brush tubes', or one of his low-pressure gas discharge tubes, and causing it to respond to signals sent out by his energising coil of similar wavelength and located at a considerable distance from the building. Tesla was convinced that these electromagnetic waves were useful both in wireless transmission of electric energy and in telephone technology. Unfortunately, Tesla could not go any further with this research because the electrotechnology at the time was not adequately sophisticated for Tesla's research. In 1898, Tesla operated his robotic boat by radio-telegraphy as a remote control device which attracted wide public attention. His electronic remote control device was a forerunner of the modern missile guidance system. In 1899, Tesla built an alternator for the transmission of electric oscillations with a frequency of 33,000 cps (cycles per second). His main purpose for the study of electromagnetic waves was the transmission of electric power. He intended to use the upper atmosphere as a gigantic vacuum tube for the distribution of cathode rays. However, he did mention that his invention could also be used for the long-distance transmission of intelligence. This scheme of Tesla was quite utopian because it would have required huge technological investments which was far beyond the capacity of the industry.

In 1890, Tesla had invented the so-called 'Tesla coil', by means of which he produced discharges of 300,000 volts. In 1899 he built a coil of 52-foot (15.85 m)diameter with which he produced high-frequency discharges of 10 million volts, resulting in long sparks in the form of 'ball-lightning' of very brief duration. In 1945 discharges of 25 million volts were reached with a 'Tesla coil' in an effort to produce long-lasting 'ball-lightning' which could have been important in fusion phenomenon, but unfortunately it failed to produce long-lasting lightning effect.

In 1896, Tesla patented the use of a rotary spark-gap attached to an alternator shaft for the generation of high-frequency electromagnetic oscillations.

In 1893, the New Zealander Ernest Rutherford (1871-1937), who became a famous nuclear physicist, demonstrated that quite feeble electromagnetic waves can demagnetise a bundle of very fine steel wire if these waves are sent through a coil of wire wrapped around the bundle. Based on this observation, Rutherford developed the first magnetic detector of Hertzian radio-waves, and by using it, he was able to send his first wireless message from one building to another at Canterbury College in New Zealand. Rutherford did not pursue the wireless transmission of electric signals any further, since he was only interested in improving the receiver of Hertzian waves. His magnetic detector was considerably more reliable in its action than the 'coherer', but its disadvantage was that its signals could only be heard as sounds which could not be recorded. Nevertheless, the magnetic detector gradually began to replace the coherer.

From 1889 to 1894, Oliver Joseph Lodge repeated Hertz's experiments. Early in 1894, Lodge put Hertz's spark-gap in a copper cylinder open at one end, and thus produced a beam of ultrashort-wave oscillations which could be transmitted in any direction. He did the same for the receiving set, which contained a Branly coherer, and was able to determine by means of it the direction of the transmitted wave. He also discovered that if 2 circuits were so constructed that if the product of the capacitance and the inductance of one circuit equals that of the other, then the effect of electric resonance is achieved which is analogous to the acoustic resonance of 2 tuning forks having equal frequencies of vibration. The equality of the product of capacitance and inductance represented the 'tuning principle' of electromagnetic circuits. He demonstrated his improved equipment at Oxford in 1894 by sending Morse signals between 2 buildings separated by several hundred feet. Lodge developed the closed oscillatory circuit to reduce energy losses, and to increase the reach of his transmitter. For some reason, Lodge failed to grasp all the technological possibilities of his research and, therefore, he did not develop his equipment into a successful practical telegraph.

In Italy, Augusto Righi (1850-1920), a professor of physics at the University of Bologna, repeated Hertz's experiments in 1889, and by improving Hertz's experimental equipment, he was able to reduce the electromagnetic wavelengths to 1.02 inches (2.6 cm), and demonstrate the relationship of Hertzian waves to light-waves. Righi was a neighbour of the wealthy Marconi family, and the elder son of the Marconis, Guglielmo Marconi (1874-1937), attended Righi's experimental lectures in 1894 after reading some articles about Hertz's experiments. Marconi, a self-taught amateur, immediately set himself the technological task of exploiting Hertzian waves in wireless telegraphy, which became his singleminded pursuit, and in this entirely practical aim, he was the first person to do so. Marconi had assembled his first wireless telegraph equipment by 1894. In his transmitting circuit, Marconi connected the secondary terminal of his induction coil across the ball-type spark-gap in the aerial circuit with the lower end of the aerial being grounded, that is, connected to a plate in the ground to complete the circuit, and by means of a transmitting key in the primary battery circuit, he was able to send rather feeble signals to his receiving circuit which had a 'coherer' in the aerial circuit with a battery and telegraph sounder connected across it. Marconi had improved the sensitivity of Branly's 'coherer' as a detecting device of high-frequency electromagnetic waves by using nickel filings with an admixture of 4% of silver filings, and by exhausting the air from the tube. The first wireless system of Marconi was defective since the plain aerial with the spark-gap at its base could not store up an adequate amount of energy to produce long trains of high-frequency electromagnetic waves and, furthermore, the receiving aerial was not tuned as a circuit, and, therefore, lacked selectivity which resulted in having the receiving antenna pick up radio signals from any other source of electromagnetic waves. Hence, this wireless telegraphic system was rather elementary, but effective enough for Marconi to achieve a range of 1½ miles (2.4 km) and to apply for a British patent. Since the vertical wire aerial of Marconi looked like the antenna of an insect, the aerial was soon called the 'antenna'.

In 1896, Marconi established the British Marconi Wireless Telegraphy Company with the help of William Preece. Marconi, a self-taught engineer as well as a very capable entrepreneur and a brilliant businessman, was virtually a one-man engineering enterprise up to 1900, after which date he hired the most talented engineers and scientific advisors available to work for his company. Altogether, 17 professional engineers, among whom were physicists John Ambrose Fleming (1849-1945) and Lodge as his advisers, worked for Marconi. Marconi discovered, as had Tesla and Lodge before him, that it is possible to 'tune' a receiver so that it is sensitive only to electromagnetic waves of a specified wavelength. This knowledge was necessary for wireless since modern radio could not be developed without it. From 1897 to 1898, Lodge perfected the coil-and-condenser tuning system that is basic to all radios, for Marconi.

In 1896, a Russian physicist, Alexandr Stepanovich Popov (1859-1905), used Lodge's arrangement in his 'lightning reporter', a receiver which consisted of a battery circuit incorporating a relay, a coherer and an electric bell. The electric disturbance created by the lightning was caught in the antenna suspended by balloons and transmitted to the receiving apparatus. In 1896, Popov carried out spark-telegraphic transmission with equipment that had both a transmitter and a receiver antenna. In 1896, Popov is reported to have established a wireless connexion between 2 ships separated by 3 miles (5 km), but it is quite improbable that Popov properly understood the function of his antenna.

Later Popov attempted to claim the credit for the invention of wireless but his attempt failed. The radio transmitter suggested by Popov could not produce the radio 'groundwaves' of Marconi's transmitter, and his idea to ground the receiver without grounding the transmitter indicated his lack of understanding of the basic principles of wireless discovered by Marconi.

From 1896 on, Marconi carried out a series of successful experiments with the help of William Preece, on Salisbury Plain and elsewhere, of the 'tuned' telegraphic wireless circuits. In 1897, a wireless telegraph circuit of 8 miles (12.9 km) was set up across the Bristol Channel and an 18-mile (29 km) circuit from Poole to the Isle of Wight. Marconi used tuning commercially in 2 simultaneous telegraphic transmissions from Poole to St. Catherine's Point, a distance of 30 miles (48.3 km). At the time, Marconi used a rough-and-ready method of tuning by means of metal strips or plates on either side of his coherer. In 1899, Marconi was able to send his wireless messages about 31 miles (50 km) across the British Channel.

In 1900, Marconi devised his tuned wireless telegraph by introducing a fundamental change in his transmitting circuit to produce long trains of electromagnetic waves. He eliminated the spark-gap from the transmitting aerial and placed it in an oscillatory circuit consisting of a condenser of several Leyden Jars in parallel and the single-turn primary of a transformer with its secondary winding of many turns being in the aerial-to-ground circuit. The electric energy was provided by the induction coil across the spark gap and the long trains of electromagnetic waves were produced by the adjustable tuning inductance in the aerial. Similar tuning facilities in the receiving aerial made the transmission so effective that Marconi was able to send wireless messages over 200 miles and, with the assistance of the English physicist John Ambrose Fleming, he began his efforts to extend the range of his wireless telegraph and, if possible, to bridge the Atlantic Ocean.

In 1901, Marconi was able to transmit a Morse telegraph signal, the letter S, from the first high-powered radio station, at Poldhu in Cornwall, England, to St. John, in Newfoundland, Canada. The power at Poldhu was supplied by an alternator designed by Fleming and driven by a 25-horsepower oil engine which excited the transformers to produce an electric potential of 15,000 volts. The aerial consisted of 50 wires suspended from 20 poles, 200-feet (61 m) high, in a 20-foot (6.1 m) circle. In this transmission, Marconi used electromagnetic waves between 2,000 and 3,000 feet (610 and 914 m) wavelengths as had been suggested by Tesla. Unfortunately, no instrumental recordings were taken by Marconi to corroborate his claim of having bridged the Atlantic Ocean with electric signals to remove the doubts raised concerning the veracity of Marconi's claim, since at the time, nobody was aware of the existence of the reflecting ionised layer in the Earth's upper atmosphere. However, nobody contested Marconi's range of 756 miles (1217 km) between Poldhu and the ship <u>Philadelphia</u> in 1902, because the telegraph messages were recorded with much better equipment and recording instruments.

In 1902, Marconi invented a rather ingenious form of magnetic detector as a replacement for the coherer which had certain practical defects. This magnetic detector was based on the principle of operation of the Rutherford magnetic detector of 1895, and it made possible for the operators with headphones to hear the signals as musical notes. Marconi's magnetic detector consisted of a loop of soft-iron wire band driven continuously by a clockwork through a permanent magnetic field and a coil carrying the aerial current to be detected. The electromagnetic wave received by the antenna demagnetised the wire band so abruptly that it made a click in the earpiece of a telephone connected to a secondary winding on a coil. Unfortunately, the magnetic detector did not record the signal. The speed of wireless telegraph messages increased to 150 words a minute with the use of Marconi's magnetic detector. The magnetic detector remained the standard equipment in Europe and Great

Britain for 10 years, whereas the Americans preferred the use of electrolytic or chemical-type detectors. The American physicist of Serbian stock, Michael Pupin (1858-1935) invented the electrolytic detector in 1898, and developed a method for the analysis of complex currents in resonant circuits.

By these achievements, Marconi had essentially overcome the fundamental technical problems of long-distance radio telegraphy. Wireless stations, after Marconi's success, were rapidly appearing around the world. Large numbers of ships were provided with Marconi's wireless equipment which became a vital component of maritime operations.

Marconi made three major discoveries in his wireless experiments: in 1901, he discovered the so-called 'groundwave', which he produced by radiating electric waves to the surface of the Earth by connecting his transmitter to an elevated and to a grounded conductor to guide the waves around the Earth's surface; in 1922, he discovered the so-called 'daylight wave', a directed shortwave beam of 328 feet (100 m) wavelength reflected from the ionosphere by using the shortwave beam transmission and superheterodyne reception over a distance of 12,428 miles (20,000 km); and in 1932, he discovered the so-called 'bending of the microwave beam'. All three of his discoveries ran counter to the accepted scientific opinions of his day. Marconi was assisted in his wireless experiments with shortwaves of wavelength from 6½ to 49 feet (2-15 m) by two very capable British electrical engineers: Charles Samuel Franklin (1879-1964) and Henry Joseph Round (1881-1966).

Contrary to current scientific opinion of the time, Marconi discovered that shortwaves were the most suitable electromagnetic waves for long-range wireless communication. Since continuous waves carry signals better, Marconi introduced in 1912, a so-called 'timed spark system' for the generation of continuous electromagnetic waves.

Marconi is one of the very few inventors who never experienced a failure or even a setback: his career as an inventor was virtually a monotonous series of successes for every venture he undertook became a success. Marconi's earnings from his successful invention were sizeable, but not enormous. Although he became famous all around the world, Marconi remained a modest and reserved person to the end of his days. In 1909, when he was 35 years of age, Marconi and Karl Ferdinand Braun, the inventor of the Braun cathode ray tube and another competing system of wireless telegraphy, shared the Nobel Prize in physics for the invention of wireless.

However, it must be recognised that Tesla by 1893 had originated much that was present in Marconi's wireless system, which Marconi, who admired Tesla, later considered his own intellectual property. After Marconi's death in 1943, his basic American patent granted in 1904 was declared void, mostly because of Tesla's prior work on wireless patented in 1897.

Marconi, a man of astounding persistence, made his technological success out of ideas and principles discovered by other men of science who apparently did not know how to utilise their discoveries in practice. Marconi, a rank amateur in electronics, was able to take the mass of scientific abstractions created by scientists and to forge them into a new form of communication, in a remarkably singleminded and unrelenting pursuit of his objective. He had only one serious competitor in wireless telegraphy, and that was the German physicist, Karl Ferdinand Braun, who was technologically Marconi's equal, but scientifically much superior.

Braun, professor of physics at the German universities in Tübingen and in Strassburg, was as remarkable a scientist as he was an engineer. In 1874, Braun made an extensive study of the electric conductivity of a number of metal sulphides, and demonstrated that such conductivity varied according to the direction in which the current passed through the solid, and that their resistance depended upon the direction of the current as well as its intensity. Besides sulphides, he also investigated copper pyrites and lead ores such as galena among other substances, all of which are now known as 'semiconductors'. The rectifying property of semiconductors was first discovered by the Swedish physicist, chemist and musicologist, Peter Samuel Munck af Rosenschöld, adjunct professor of physics at the University of Lund, in 1835. He observed that under certain conditions solids now known as semiconductors rectified the alternating current passing through them. The great importance of Munck's discovery went unnoticed for 40 years until Braun rediscovered in 1874 its practical importance to wireless. The semiconducting property of certain solids discovered by Munck has proven to be of fundamental importance to modern electronics. Munck, a man of refined culture, was also an accomplished pianist who in 1847 proposed a new theory of music.

Braun established the rectifying property of the 'cat-whisker' contact on lead sulfide known as galena, the 'first point-contact metal-semiconductor junction', and built his crimped-wire semiconductor diode which he used as a detector of radio-frequency signals generated by spark-gap transmitters. Further practical application of Braun's discoveries on conductivity to electronics took place 30 years later, when in 1904 General Henry Harrison Chace Dunwoody (1842-1933) of U.S. Army discovered that 'carborundum', a product of the electric furnace, can also be used together with a potentiometer as a detector of electric current and, therefore, replace electrolytic detectors which were in use in wireless reception at the time. Almost at the same time, in 1906, Greenlief Whittier Pickard (1877-1956) discovered in his investigations on radiophony that crystals of silicon, lead ore, and iron pyrites can be used as crystal detectors of electromagnetic waves, since they were almost as sensitive as electrolytic detectors. The function of the crystal rectifier was to change the incoming high-frequency electric signal to low-frequency signal and to transmit only signal current flowing in one direction to which the headphones were able to respond. Pickard introduced the so-called 'perikon detector' consisting of 2 metallic materials, red oxide zinc and chalcopyrite, pressed together. These crystal detectors were sold as 'cat-whisker' detectors, using a fine wire that lightly touched the surface of the crystal, which became very popular in amateur radio because they were inexpensive. The only shortcoming of the 'cat-whisker' detectors was that the crystal receivers could not amplify weak electric signals. About the same time as pyrites and galena, also selenium came under study for its rectifying property. Already by 1883, an American physicist C.E. Fritts (alias Fritz) had concocted a selenium rectifier which incorporated several modern features.

The commercial success of wireless telegraphy led to an extensive study of natural minerals and metallic sulphides as detectors of radio-waves. It was Pickard who in 1906 showed that silicon can be used as such a detector and, in 1911, the American physicist Ernest George Merritt (1865-1948) was able to receive, for its time, the remarkably short radio-waves of 15 centimeter wavelength with his silicon detectors. In the1920's lead ores, or carborundum, or iron pyrites were used together with 'cat-whiskers' for the finding of sensitive spots on the crystals as detectors of radio-waves. In 1926, Merritt introduced the metallic germanium as another new rectifying semiconductor material.

The 'cat-whisker' semiconductor detectors were gradually replaced by vacuum tube detectors, such as the diode invented in 1904 and the triode invented in 1906, which were more reliable. Unfortunately, the replacement of the semiconductor detectors by vacuum tube detectors brought with it a decline of interest in the physical operation of semiconductors, a phenomenon poorly understood at the time, until after the First World War.

Another remarkable scientific invention of Braun was the Braun cathode-ray oscilloscope, popularly known as the 'Braun tube', which Braun built in 1897 for experimental purposes to study the time-variation of the cathode rays as a stream of electrified particles.

In the same year Joseph John Thomson (1856-1940), professor of physics at the Cambridge University, measured the charge of the electron by the deflexion of the cathode ray in the electric and the magnetic field, and proved that the electron was a universal constituent of matter, the first known subatomic particle. The original Braun cathode ray tube used gas discharge phenomena for the emission and formation of the long and filamentary electron beam by the effect of ion focussing. Since Braun knew that the cathode ray has negligible inertia and, therefore, could be easily deflected by a magnetic field, he placed 2 sets of electromagnets round the neck of the tube between the cathode and the anode. One set of electromagnets deflecting the ray up and down, and the other deflecting it from side to side, were supplied by the time-base circuit [usually a condenser charged from a steady voltage through a resistance] which provided a measure of time. The deflected cathode ray impinged on a phosphor-coated fluorescent screen inside the tube which left a trace of the moving cathode ray as a glowing line on the screen.

Braun's invention led to intensive and rapid development of the cathode ray tube which made it into one of the most revolutionary devices of this century. In 1904 the German physicist,

Arthur Rudolph Berthold Wehnelt (1871-1944), invented the thermionic oxide-coated cathode which he used in his vacuum tube rectifier to transform single phase and polyphase alternating currents into direct current. In 1905, Wehnelt proposed a lime-coated platinum filament as the hot cathode to produce a satisfactory emission of electrons at a reasonable voltage for any practical application of the Braun cathode ray tube. In 1906, Wehnelt introduced his three-phase rectifying tube incorporating the hot cathode coated with calcium or barium oxide. Another significant improvement of the Braun cathode ray tube was made by the German physicist Johann Emil Wiechert (1861-1928) by incorporating in the tube the so-called 'Wiechert magnetic lens' that produced electrostatic deflexion of the cathode ray by means of condenser-like deflexion plates which gave sharper focus and greater brightness to the image of the oscillograph on the phosphorescent screen. In 1906, Max Dieckmann (1882-1960) and Gustav Glage (1882-?), a former student of Braun at the University of Strassburg, patented another type of cathode ray relay, which used the magnetic deflexion of the cathode ray imposed by a transverse magnetic field to the cathode ray. This cathode ray relay had all the components of the modern electron gun. In a few years the typical components of a cathode ray tube were established: a tubular anode located a few inches in front of the cathode, which was maintained approximately at 1,000 volts by direct current, accelerated the flow of electrons constituting the cathode ray and passed it through an electron lens consisting of a small negatively charged metal tube that crowded the stream of electrons together like a convex lens when it acts on a ray of light. The brightness of the spot on the fluorescent screen was controlled by passing the electron stream from the cathode through a grid with adjustable positive applied voltage that was opposite relative to that of the cathode thereby providing the means to control at will the number of passing electrons. The original Braun tube emitted a cathode ray of constant intensity in contrast to modern television tubes which emit cathode rays of variable intensity. The Braun tube became indispensable when rapidly varying physical processes had to be visually examined, and it has remained so till this very day, whenever wider applications are found for the Braun cathode ray oscilloscope. In modern cathode ray tubes it is possible to observe and record an event lasting only one billionth of a second. For many years, the ***Technische Hochschule*** (*Technological Institute*) in Aachen, Germany, was the main research center of the Braun tube.

Early cathode ray tubes used 'willemite' (a zinc silicate containing a manganese impurity) for a yellow-green emission, or calcium tungstate for a blue emission, both of which can be more easily photographed. The needs of radar systems led to improved screen materials among which the sulphides, silicates and fluorides predominated.

Braun became also interested in the wireless telegraphy in 1897, and began a scientific investigation of it which led him to believe that when the alternating current is employed, then the skin-effect of electricity becomes important, and therefore, greater efficiency is obtained with still higher frequencies of oscillation. He established the following facts in wireless : firstly, that a combination of capacitance such as a battery, and inductance such as a coil, constitutes an oscillating circuit; secondly, that to make the oscillations persist, the circuit has to be freed from external 'damping' by the use of inductive coupling; thirdly, that the transmission of oscillations is considerably improved if the number of windings are properly adjusted, that is, if the primary and secondary circuits are in resonance. Therefore, Braun's system of wireless telegraphy was characterised by an increase in transmitting power, employment of oscillating circuit, minimisation of damping by means of induction coupling, and resonant tuning – precisely the principles used in today's radio transmitter. Braun proved theoretically that his wireless telegraph carried signals 3 times as far as that of Marconi.

The original circuit of Marconi had an inherent limitation because the same circuit that generated the electric oscillations also radiated them, 2 functions which adversely affect each other. Braun, being scientifically aware of this adverse effect, separated the 2 functions by generating electric oscillations in a closed circuit coupled to a separate antenna circuit, which constituted a new principle in radio telegraphy. Braun's wireless circuit had the additional advantage that the decoupled antenna no longer required a high potential, which made it a great deal more safe and less susceptible to accidental short circuits.

In 1898, Braun with his coupled wireless circuits was able to signal 62 miles (100 km) whilst Marconi at that time could signal only about 30 miles (48 km).

Braun's wireless circuit was quite independent of, and superior to that of Marconi, and, therefore, his system of wireless telegraphy broke Marconi's monopoly in wireless telegraphy. It is for his superior system of radiotelegraphy that Braun shared, in 1909, the Nobel Prize in physics with Marconi. Braun's work on solid state rectifiers and the cathode ray tube, which have become very important tools of modern electronics, were far ahead of his time and their vital importance can only be properly understood today.

Radio

Nikola Tesla, in his lectures, demonstrations and laboratory experiments, had actually laid the conceptual foundation for radio broadcasting but, unfortunately, he did not pursue this line of development because he considered it scientifically unimportant as 'small-time stuff', and also because the technology at the time was quite inadequate for the task.

Tesla showed how continuous electromagnetic waves can be generated by rotary dynamos with up to 384 magnetic poles, and produce currents having up to 10,000 cps (cycles per second) frequency. Tesla realised the importance of resonance and distributed capacity in the electronic equipment, and he invented his oscillatory transformer, the so-called 'Tesla coil', a high-frequency induction coil in which the primary coil was excited by currents from the discharges of a condenser across the spark-gap. 'Tesla coil' oscillators, which are still used for long distance radio and television transmission, developed frequencies up to 2 million cycles per second, and by means of it he was able to demonstrate that continuous electromagnetic oscillations were superior to the sudden bursts of damped electromagnetic oscillations released by the spark-gap discharger of Hertz. Tesla thought that for wireless telegraphy and radio broadcasting, the useful wavelengths range from hundred meters to many thousands of meters. Mistakenly, he thought ultrashort waves to be useless for radio communication. Tesla was a pioneer in high-frequency electric phenomena and the designer of the first high-frequency alternators. Tesla also developed a system of arclighting which dispensed with the wiring, and a remote-controlled robotic boat. His most ambitious project was a communication system able to transmit sound and vision worldwide on which he began to work in 1900, but soon had to abandon this project when financial backers withdrew their support.

In 1892, American electrical engineer of British extraction, Elihu Thomson (1853-1937) showed that an electric arc does produce continuous electric oscillations if it is connected to a circuit containing a shunted inductance and capacitance, and subjected to a transverse magnetic field which makes the arc intermittent with high frequency, but he did not develop this idea any further. In 1894, English physicist William Du Bois Duddell (1871-1917) independently of Thomson arrived at the same result with the so-called 'singing arc' by making the electric arc produce a continuous sound without external source, that is, without a microphone. In 1899, a German electrical engineer, Hermann Theodor Simon (1870-1918), utilised the 'singing arc-lamp' phenomenon of Duddell to show that the electric arc-lamp can be employed both as a telephonic sender and receiver. In 1900, Duddell succeeded in increasing the frequencies of continuous undamped electric oscillations produced by an electric arc by coupling it to a resonant circuit, but the electric oscillations generated by the electric arc still had frequencies far too low to be effective in radio communication, which had to transmit speech and music without wire. Duddell pointed out that an electric arc is actually a continuous spark which could be used in wireless transmission, but he did not pursue this idea any further.

At the same time, German physicist Ernst Walter Ruhmer (1878-1913) developed telephony by means of light-waves, a method of speech transmission originally investigated by Alexander Graham Bell (1847-1922). In 1880, soon after inventing his telephone, Bell invented his 'photophone', a device transmitting speech by a beam of light. The photophone consisted of a mirrored diaphragm which converted acoustic waves directly into a modulated beam of light. Bell's receiving apparatus consisted of a selenium photocell connected to a telephone receiver which served to reconstruct the audio-wave. However, since Bell was unable to transmit speech by a beam of light beyond a few hundred feet, the telephoning by light-waves did not attain any practical importance until this idea was investigated by Ruhmer. In 1902, Ruhmer was able to transmit telephone conversation without wire over 11 miles (17 km) by light-waves, and soon after that over 22 miles (35 km), but his method of long-distance telephony could not be developed any further at the time, owing to the technological limitations of the period. The light-wave method of communication originated by Bell and developed by Ruhmer really came into its own only in the 1970's, when light-wave communication was developed which transmits voice, data and video signals as pulses of light through hair-thin glass fibres, or laser beams through space instead of electric charges through copper wire. In 1906 Ruhmer introduced the 'feedback principle' in the production of electromagnetic oscillations in electric-arc transmitters.

In 1900, an ingenious Canadian engineer of Scottish stock, Reginald Aubrey Fessenden (1866-1932), produced the first true radiotelephone consisting of a wireless apparatus which he demonstrated at Cob Point in Maryland by sending voice and music with this device over a short distance by modulating a series of electromagnetic waves produced by a Hertzian spark transmitter of his own design. In this transmission, Fessenden was able to hear faint vocal sounds but, unfortunately, it was drowned out by the noise of the static because the Hertzian waves generated by his electric spark transmitter contained frequencies in the same range as the audible sound. Since his wireless apparatus was sensitive to the electromagnetic waves with frequencies from 1,500 to 100 million cycles per second, Fessenden realised that in wireless telephony, it is impossible to change soundwaves directly into electric vibrations and then back into soundwaves, as was done in the telephony by wire. Fessenden came to the idea of sending out constant high-frequency electromagnetic waves [radio-waves] 'modulated in amplitude' by the much slower sound-waves from the telephony signal to transmit actual sound by electromagnetic oscillations. In Fessenden's system of transmission, the constant high-frequency electromagnetic waves behave as carrier-waves of the relatively low-frequency sound-waves. Fessenden's receiver consisted of a detector, his so-called 'liquid baretter', which responded to the groups of radio-waves corresponding to the individual sound-wave, and not to the individual radio-wave. To succeed in this venture, Fessenden had to find a way to generate continuous, undamped electromagnetic oscillations of such high frequencies that they do not produce audible sounds which interfere with radio listening.

In 1903, the Danish engineer, Valdemar Poulsen (1869-1942), by following the ideas of Tesla and Duddell, built an electric-arc generator which produced undamped electromagnetic oscillations that were useful in radio technology. It was only after Poulsen had enclosed the electric arc, incorporating a water-cooled copper anode enveloped in hydrogen gas, and placed it in a powerful magnetic field which extinguished the arc for part of each cycle that the electric arc was able to supply satisfactory undamped electromagnetic oscillations at frequencies from half to one million cycles per second. Unfortunately, Poulsen's electric-arc generator was only able to supply limited power, from 10 to several hundred watts. Poulsen used it in his system of wireless telephony which he had developed together with the Danish physicist, Peder Oluf Pedersen (1874-1941), professor at the Royal Technical University in Copenhagen. The main disadvantages of Poulsen's electric-arc generator were that it was unstable in its frequency range, and that it generated static noise which interfered with the sound transmission. Moreover, like the spark-gap transmitter, it was in want of a really satisfactory receiver which was devised by Pedersen who invented his 'tikker', a most efficient and sensitive receiver which reduced continuous radio-waves into Morse signals. The first telegraph station using Poulsen's electric-arc transmitter and Pedersen's receiver as a continuous-wave radio system adaptable to telegraphy and telephony was built in Denmark in 1907.

Since the electric-arc generator produced too much static noise, Fessenden decided to employ an alternate mechanical method of generating undamped, high-frequency electric oscillations: an alternator. Elihu Thomson had already built a high-frequency alternator in 1889, and a decade later, Nikola Tesla had built a more powerful, high-frequency alternator which produced electromagnetic oscillations with a frequency of 33,000 cycles per second. After his own efforts had not been quite successful, Fessenden appealed in 1903 to the General Electric Company to design and build such a generator for him. This task was first given to Charles Proteus Steinmetz (1865-1923), the German-born electrical engineer and physicist, who had discovered experimentally the laws of magnetic hysteresis [also known as the 'magnetic lag'], and produced other mathematical theories of electromagnetism. Steinmetz designed an alternator which produced alternating current at 10,000 cycles per second, much less than the frequency of a similar Tesla alternator. Since Steinmetz's alternator did not produce a sufficiently high frequency in the electric oscillations, Fessenden placed in 1904, a new order with the General Electric Company for an alternator which could produce continuous oscillating electromagnetic waves with a frequency of 100,000 cycles per second. This time, Fessenden submitted his own design specifications for the alternator, which included a stationary wooden armature in the form of a thin disk. This project was assigned by General Electric to their brilliant young Swedish engineer, Ernst Frederik Werner Alexanderson (1878-1975), whose own design for this alternator was based on a stationary laminated iron armature between 2 rotating disks providing the magnetic field. After several arguments with Fessenden and experimental failures, Alexanderson succeeded in constructing an alternator which incorporated a fixed wooden armature, as specified by Fessenden, in 1906. This alternator had 360 polar projections in the field magnet and it produced a potential of 56 volts, but its frequency had to be limited to 50,000 cycles per second on account of the gross mechanical distorsions created in the alternator by frequencies higher than this limit.

In 1907, the American Telegraph and Telephone Company (known by its acronym, **AT&T**) requested that the General Electric design an alternator as a telephone relay to step-up the current in long-distance telephone lines. This task was turned over to Alexanderson, which gave the young Swedish engineer an opportunity to build a high-frequency alternator according to his own design. After a successful demonstration of the Alexanderson alternator, which produced 2 kilowatts of power and 100,000 cps high-frequency electromagnetic waves, even Fessenden had to recognise the inferiority of his own alternator design, and to submit an order for an Alexanderson's alternator.

Alexanderson's alternator, which he patented in 1909, was the first in his series of high-frequency alternators which, by 1918 could deliver 80 kilowatts of power, and give an aerial current of 400 amperes. Alexanderson's largest alternator could deliver 200 kilowatts of power with a frequency of 100,000 cps.

Fessenden had established a 2-way communication in trans-Atlantic radio-telephony using his own high-frequency electric-arc transmitter, between the coast of Massachusetts and the west coast of Scotland in January 1906, as a publicity stunt. This was the first 2-way trans-Atlantic wireless telephony ever attempted and it remained in operation for a year until the antenna tower in Scotland was blown down in a storm.

In the meantime, Fessenden had been working on his radio broadcasting system in collaboration with Alexanderson, who worked on an alternator as the generator of high-frequency, undamped electromagnetic waves. It was Fessenden's idea to use high-frequency alternators as generators of high-frequency undamped electric oscillations. Such alternators remained in prominent use in wireless until it was discovered in 1913 that the triode vacuum tubes could be used as electronic generators of high-frequency, undamped electromagnetic waves.

In wireless communications, the detectors of electromagnetic waves are very important. In 1900, a French electrical engineer, Gustave Auguste Ferrié (1868-1932), had used an electrolytic detector of electromagnetic waves, instead of a 'coherer'. A German engineer Schlömilch and Fessenden developed their own electrolytic detectors in 1903. Fessenden's electrolytic detector, the so-called 'liquid barretter', consisted of a platinum wire placed on the surface of an acid. Direct current passing through the platinum wire formed a gas bubble which interrupted any further flow of the direct current, and when the alternating current of a wireless signal passed through the platinum wire, the bubble burst and an instantaneous flow of current could be detected in a telephone receiver. The electrolytic detector remained the standard of sensitivity for a decade until the triode replaced it both as a vacuum tube detector and an electromagnetic wave generator.

In 1901 Fessenden invented the heterodyne (in Greek, *heteros*-other, different + **dynameis**-power) principle which has remained fundamental in radio technology. The heterodyne principle consisted of superposing an inaudible high-frequency carrier-wave on an audible low-frequency electric oscillation of the sound wave that modulates the high-frequency wave into a 'beating wave' at the transmitter. This signal was mixed at the receiver with another slightly different high-frequency electromagnetic wave generated by an arcing source within the receiving set itself, which resulted in an audio beat-note and left the difference of frequency between the 2 waves as the audible sound wave of low-frequency.

Fessenden's heterodyne receiver based on the frequency 'beat' or interference reception of continuous, undamped electromagnetic waves was years ahead of the industry of the time. Continuous waves carry the signals better and tune more sharply. The basic concept of modulating a high-frequency current by speech signal did not originate with Fessenden but had been proposed by a French engineer, Maurice Leblanc (1851-1929) as long ago as in 1886, but no one before Fessenden had been able to execute this idea in practice. In 1902, Fessenden demonstrated the transmission of speech by the modulation of radio-waves. He used continuous alternating radio-waves and devised several methods of modulation among which the most successful was the direct modulation of the transmitting aerial current by a microphone. Fessenden's heterodyne wireless equipment required a local source for electromagnetic oscillations with adjustable frequencies, but at the time, only the Poulsen electric-arc as a source of undamped electromagnetic oscillations was available since high-frequency sparks were electronically too noisy for heterodyning. In fact, Fessenden's own continuous electromagnetic wave apparatus was the only wireless system able to use his method of heterodyne wireless communication. Since the electric arc generated continuous but distorted radio-waves the frequency of which was difficult to control, only after the oscillation-producing properties of the vacuum tube had been discovered by Goddard and, particularly, by Armstrong in 1912, did the heterodyne receiver of Fessenden assume its rightful importance in wireless technology. High-frequency alternators as sources of radio-waves did not exist at that time, but in 1901 Fessenden had patented one of his own design which was to be manufactured by the General Electric Company.

On Christmas Eve in 1906, Fessenden gave the first radio broadcast of speech and music in history, which was received within a radius of several hundred miles (300 km) by his standard wireless sets on ships accustomed to receiving Morse signals, to the great astonishment of the wireless operators. Fessenden used the 50,000-cycle alternator of his own design, which had been manufactured by General Electric company under the direction of Alexanderson, as the generator of high-frequency electromagnetic waves in his wireless station at Brant Rock, Massachusetts. Fessenden achieved the modulation of the alternator's kilowatt output by means of a water-cooled microphone fitted in series with 10 antenna circuits, and clear reception was established not only at many ground stations but also at ships at sea. In 1907, Fessenden extended the 12-mile (19 km) range of his experimental radio broadcasts to 500 miles (805 km). Therefore, the true 'creator of radio broadcasting and voice transmission by electromagnetic waves' is Fessenden. However, the heterodyne circuit of reception in the immediate development of radio broadcasting remained in the background mainly because it required a good, stable and compact electromagnetic wave oscillator that did not exist at the time.

Fessenden, the most versatile inventor in radio technology of his time, was one of the most prolific North-American inventors by having more than 500 patents in different fields to his credit. He was also a highly contentious individualist with an explosive temper who, during most of his professional life, was involved in litigations and feuds concerning his inventions. Fessenden vehemently opposed the National Research Council because he considered it to be a sort of 'octopus' which tends to destroy the individualism of the inventor by strangling his freedom of action. Fessenden occupied professorial chairs for a time at the University of Purdue and at the University of Pittsburgh.

In 1904, the English physicist, John Ambrose Fleming, a former student of Maxwell who was a consultant to both the Edison Company and the Marconi Company in Great Britain, devised a vacuum tube of two electrodes, known now as a 'diode', but Fleming called it 'thermionic valve', that rectified an alternating wireless signal for which it served as a detector. It consisted of a heated cathode and a cold anode, both encased in an evacuated glass bulb, that allowed the electronic radiation to pass only from the cathode to the anode which was maintained at a low positive potential. Fleming based it on the so-called 'Edison effect' of which he had been informed in 1884 by Edison himself. The theory of the electron and ion emission in such thermionic tubes with 2 electrodes were worked out by the English physicist, Owen Willans Richardson (1879-1959), a former pupil of Joseph J. Thomson, in 1901-1903 at Cambridge University which made the rapid improvement and development of vacuum tubes possible. In 1928 Richardson received the Nobel Prize in physics for this work. Fleming's thermionic valve was not very practical because it had poor sensitivity and lacked any means for amplification of the signal, but it could be used to rectify the oscillations of the high-frequency alternating current signals sent out by the transmitter of the wireless into low-frequency direct current in the receiving antenna circuit of the wireless which detected the wireless signals by operating a telephone receiver.

An Austrian physicist, Robert von Lieben (1879-1914), a self-educated man of means, also worked on the problem of amplification of telephone signals. Von Lieben based his solution of the amplification of such signals on Wehnelt's discovery that certain cathode materials emit cathode rays under small voltages in the cathode ray oscilloscope tube of Braun. In 1904, von Lieben improved the Wehnelt cathode ray tube by introducing a third electrode, a perforated grid between the cathode and the anode. This grid was fed with electric impulses coming from the telephone transmitter, and it had the purpose of modulating the stream of electrons of the cathode ray flowing from the cathode to the anode, thereby exercising a fine control on a strong current. Von Lieben probably obtained the idea of the 'principle of grid control' from the German physicist Philipp Eduard Anton Lenard (1862-1947) who used it in 1902 to control the flow of electrons from a zinc cathode exposed to ultraviolet light by making the grid positive to draw electrons from the cathode and pass them through the grid towards the anode, and negative to repel the electrons back towards the cathode. The weak electric impulses could be amplified with as great an accuracy as required by von Lieben's device. In 1906 Robert von Lieben patented his tube that in his opinion was a 'cathode-ray relay'. He improved his tube between 1907 and 1910 with the assistance of the Austrian physicists Eugen Reisz and Siegmund Strauss. By 1912, his cathode ray relay was ready for commercial use by the German *Telefunken* Company. Unfortunately, von Lieben was a sick man and he died at 35 years of age.

American engineer and inventor, Lee de Forest (1873-1961), who had been one of the last doctoral students at the Yale University still taught by the famous American mathematical physicist, Josiah Willard Gibbs (1839-1903), stumbled on a vacuum tube construction in 1906, the same year as von Lieben. De Forest, who was a scientific tinkerer more than a scientist, received his inspiration for constructing the 3-electrode vacuum tube, later called the 'triode', from reading the report of a lecture on the 'thermionic valve', the so-called 'diode', delivered by Fleming in 1905 to The Royal Society in London. He probably also obtained the idea of the control grid from Lenard's photoelectric researches in 1902. De Forest intended to use his vacuum tube, which his assistant C.D. Babcock called the 'audion', as a relay device on long-distance telephone lines. Unfortunately, de Forest triode was rather unstable owing to the ionisation of the residual gas still present in the evacuated tube.

An immigrant physicist from Germany, Fritz Löwenstein (1873-1922), who had been an assistant to Tesla, discovered in 1912 that a 'negative bias' of moderate voltage applied by a battery in series with the grid circuit produced the optimal operating conditions of the triode. Soon after, it was found that if certain values were given to the grid-leak and the capacitor in a modified triode invented by Henry J. Round, a certain steady grid bias voltage could be achieved for the continuous carrier-wave type signal used in radio-telephony.

Since de Forest himself lacked an adequate scientific understanding of the physical action of the triode, he had no appreciation of its versatility, and lacking an analytical mind, his explanation of the physical action of the triode was completely wrong and in no relation to the contemporary knowledge of physics. His erroneous explanation of the functioning of the triode actually hindered the proper understanding of how the tube's output might be increased. Curiously, the inventor of the triode neither understood its operation nor its proper use, both of which were explained by an ingenious American undergraduate student.

In 1912, Edwin Howard Armstrong (1890-1954), a brilliant undergraduate student in engineering at Columbia University, discovered that by overloading the triode tube in his radio circuit, the tube became an electromagnetic wave oscillator which could generate 'undamped' electromagnetic waves of any desired high frequency. It was the first nonmechanical generator of continuous undamped electromagnetic waves. By directing some of the current in the filament-plate circuit back to the antenna-grid circuit to increase the rectifying and amplifying effect of the grid, the effectiveness of the triode could be increased many times. This is called the 'regenerative' principle, or the principle of 'positive feedback circuit', and Armstrong discovered that with sufficient feedback, the regenerative amplifier becomes an oscillator which was a 'revolutionary' major discovery in radio technology.

The priority in the discovery of the vacuum tube as the generator of undamped high-frequency radio-waves became the most bitterly contested issue in the history of radio technology, mainly between de Forest and Armstrong. The irony in this controversy was that the priority in the discovery of the triode vacuum tube as a producer of high-frequency radio-waves belonged to the ingenious rocket pioneer, Robert Hutchings Goddard (1882-1945), who already on August 1, 1912 had applied for a patent of such a vacuum tube oscillator based on his prior research on electromagentic waves long before de Forest had done any laboratory testing which involved the wave generating capacity of his 'audion'.

The feedback principle itself was not entirely new. It had been first discovered by the ancient Greek engineer and mathematician, Archimedes of Syracuse (c.287-212 B.C.), who used it in the design of his closed-loop astronomical clock, and it was first used in electrotechnology in 1906 by the noted inventor, Frederick King Vreeland (1874-1964), an outstanding pupil of Professor Pupin at the Columbia University and a former associate of the ingenious Fessenden, and also by the German physicist Ernst Walter Ruhmer (1878-1913) in the electric-arc transmitter in the same year. However, Vreeland's radio circuit used the sluggish mercury-vapour tubes which had outside magnets as controlling devices, and he made only passing references to vacuum tubes in his patent. Feedback principle became valuable only when used by Armstrong in the vacuum tube regenerative radio circuits. This circuit not only revealed the unexpected and remarkable powers of the triode tube as a revolutionary amplifier of signals, but it also markedly sharpened the tuning of radio sets.

The 'feedback principle' for vacuum tube circuits was invented almost at the same time as Armstrong by a number of other scientists: American scientist Irving Langmuir (1881-1957) who filed for his patent on the same day as Armstrong; Austrian physicist Alexander Meissner (1883-1958) and Austrian physicist Sigmund Strauss; American physicists Harold Deforest Arnold (1883-1933) and Fritz Löwenstein. The English electrical engineer Charles Samuel Franklin also invented some kind of feedback vacuum tube circuits in 1914 but only after his visit to Meissner in Germany. Meissner had used the feedback principle in 1913 to make the von Lieben vacuum tube into a silent high frequency radio-wave generator for the heterodyne radio receiver, a problem von Lieben himself had not considered.

Armstrong published 3 masterful papers on the vacuum tube circuits in 1914 and 1915, in which, for the first time, the physical functioning of the 3-electrode vacuum tube and the amplifying and oscillating action of the feedback circuit were correctly explained. These 3 landmark papers inaugurated the real beginning of the 'age of electronics'. Armstrong's researches on the triode circuits rapidly led to the 'equivalent-circuit' representation of the triode.

Armstrong, the most outstanding student of Professor Pupin at the Columbia University, was the first to invent the regenerative circuit in his positive feedback radio receiver. In the positive feedback some of the output signal is fed back to the input to reinforce the input signal. In the positive feedback receiver, the function of the triode as an amplifier and as a detector are unified by the reaction-coupling. In this feedback receiver, feeble radio signals are increased in intensity and, thereby, a high sensitivity and selectivity is achieved. The triode tube in its application as a vacuum tube generator in the reaction-coupling, called the 'feedback', has the anode and the grid circuit of the vacuum tube inductively coupled. At a correct phase relation of the voltage at the grid and at the anode, and by a sufficiently strong coupling of the grid circuit with the oscillating circuit, constant, undamped, sinusoidal oscillations are produced. The frequency of this oscillation is determined by the capacitance and inductance of the elements in the oscillating circuit. In the feedback concept, the energy fed back into the circuit equals the resistance losses. Therefore, by means of such a feedback circuit, the triode vacuum tube could be used to produce continuous, steady and controllable high-frequency oscillations of inaudible radio-waves so that they could transmit by modulation the wave patterns of the sound waves of far slower frequencies audible to the human ear (from 20 to 20,000 cps), the impulses of which come from the microphone and are superimposed on the continuous carrier-wave emanating from the oscillating circuit by means of the grid circuit in the triode tube. In the vacuum tube receiver, the carrier-wave is filtered out, by means of a small vacuum tube generator, called the 'heterodyne', the frequency of oscillation of which is so chosen that the difference between the received signal-frequency and the frequency of the heterodyne after their superposition is equal to the low frequency oscillation that produces the audible sound. Therefore, Fessenden's heterodyne reception of radio-waves, was made really practical for the first time by the triode tube as a generator of continuous, high-frequency and stable electromagnetic oscillations.

The American Telephone and Telegraph Company bought de Forest's audion for the purpose of amplifying telephone signals in their coast-to-coast system by land-wired telephony. Their research physicist, Harold Deforest Arnold, devoted his efforts to making a better vacuum tube out of Lee de Forest's defective 'audion' tube, which owing to its low gain was only marginally better than Fleming's diode, by eliminating the detrimental ionisation of the gas still present in the imperfect vacuum of the de Forest tube. He used the molecular vacuum pump invented in 1910 by the German physicist Wolfgang Gaede (1878-1945) to produce high vacuum in the 'audion' tube. Arnold's 'hard vacuum audion' tubes used oxide-coated cathodes in place of tantalum cathodes in de Forest audion tubes, since Arnold's filaments lasted 1,000 hours instead of 50 hours for the tantalum filaments of de Forest. At the same time, unbeknownst to Arnold, similar work was carried out in the General Electric Laboratories by the physicist Irving Langmuir. Langmuir invented a high vacuum furnace, and a tungsten filament which had a good electron emission and a long life as a cathode in collaboration with his colleague William David Coolidge (1873-1975) at the General Electric Company. Langmuir discovered that a tungsten cathode containing a small percentage of thorium increased enormously its emission of electrons at a given voltage. Such a tungsten-thorium filament could be operated in a high vacuum tube at 'dull' heat instead of the usual 3632°F (2000°C) required for the pure tungsten filament which increased dramatically the life of the vacuum tube. It was superseded by oxide cathode consisting of a thin layer of mixed oxides of the alkaline earth, such as strontium and barium, coating the cathode which was first used in 1904 by Arthur Wehnelt. Langmuir proved that high vacuum was necessary for a pure electron discharge from the cathode. The vacuum tube amplifiers soon replaced the electromagnetic type amplifiers since they gave much better reception to telephone signals, permitted the use of much lighter wire and underground cable, and much better quality of transmission. It was Langmuir who was mostly responsible for clarifying the physical principles behind the physical processes taking place within the vacuum tube.

Langmuir designed a triode that could function at 250 volts and produce kilowatts of power. Previous triode tubes were limited to 30 volts and a fraction of a watt in power. With Langmuir's vacuum tube oscillators, the first high-vacuum, high-voltage tubes with drawn tungsten filaments known as 'pure electron discharge tubes' but which he himself called 'radiotrons' (*tron* in Greek means instrument), the electric impulse from the microphone could be amplified to any strength for radiation from the antenna as electromagnetic waves. The method of producing 'hard vacuum' tubes of high-voltage amplification by Arnold and Langmuir led to a lengthy litigation which the court finally declared void because their methods of improving the vacuum of the tube was deemed by the court not to constitute an original invention.

Alexanderson in the United States, Rudolph Goldschmidt (1876-1970) in Germany and Marius Latour (1875-?) in Belgium, developed high-frequency alternators capable of producing powerful undamped radio-waves having over 10,000-foot (3,048 m) long wavelengths and frequencies up to 100,000 cycles per second which were used in long-distance telephony and wireless until the advent of high-powered thermionic vacuum tubes that were superior generators of radio-waves. But it was primarily the Alexanderson's alternator that made the trans-Atlantic wireless- telephony possible. In the long-distance radio-telephony, there existed a basic problem of how to harness the weak-powered circuit of the microphone to the high-powered circuit of the antenna without having the antenna power burn out the microphone. Alexanderson solved this problem by the well-known magnetic saturation effect in soft iron. Prior to 1912, he produced a one-way, magnetic amplifier by employing an iron core as the harnessing link. In the beginning, long-distance telephony by wire was dependent on the incorporation of Pupin's loading coils into the transmission lines to increase the inductance, but by the 1950's the loading coils had virtually disappeared and had been replaced by vacuum tube amplifiers.

Alexanderson also invented the multiple tuning of the antenna, an idea ignored and scoffed at as worthless by the contemporary wireless experts. The common wireless practice at the time was to have but one connexion between the antenna and the transmitter which for long-distance radio-waves required the application of considerable power in the antenna circuit. The transoceanic radio telephony not only required an efficient transmitter but also a much more efficient antenna than the existing ones. Alexanderson introduced a series of down-wires, each equipped with a tuning coil and each leading to a ground connexion along the antenna, which greatly increased the efficiency of transmission. When in 1916, Alexanderson installed 6 tuned down-wires at regular intervals along the 900-foot (274 m) long antenna at the experimental wireless station in Schenectady, the efficiency of the antenna doubled. Marconi immediately recognised the importance of multiple-tuning of the antenna and requested Alexanderson to try out his multiple-tuning method on the mile-long (1.6 km) antenna of Marconi's New Brunswick wireless station. Alexanderson incorporated 6 tuned-down wires at regular intervals along the New Brunswick antenna which increased the power of the antenna tenfold and made Alexanderson's multi-tuned antenna system internationally famous.

Alexanderson observed that the audion vacuum tube, or the triode, could be an efficient amplifier of telephone currents were it not unduly weak in gain, and far too sluggish as a relay for the radio-frequency currents. He was convinced that this sluggishness was due to the presence of some residual ionised vapour in the audion, and that the amplifying power of the audion tube could be increased dramatically by the method of evacuation developed by Langmuir. He collaborated with Langmuir in making technical improvements in the 'audion' tube, and thereby perfecting it into an electronic amplifier which proved to be very important in the development of wireless telephony. The Langmuir vacuum tube, which Langmuir called 'radiotron', was able to step-up power into kilowatts instead of just watts. Alexanderson showed how these vacuum tubes, which soon reached a rating of 100 kilowatts, could be used in radiotelephonic transmission.

Alexanderson's system of multiple tuning of the antennae, his 200-kilowatt alternator of 100,000 frequency, and his tuned amplifier radio-frequency were tested at the New Brunswick wireless station, and it proved so successful that this station became the most powerful wireless station in the world. Moreover, Alexanderson invented his 'barrage receiver', a device to neutralise signals, or to destroy interference such as static.

Alexanderson had developed his magnetic modulator for radiotelephony but the electrical impulses coming from the microphone were so weak that his modulator did not operate properly. The Langmuir 'radiotron' provided the amplification Alexanderson's modulator required for a successful radio-telephony, as well as radio-telegraphy. Transcontinental telephone service was established early in 1915 between New York and San Francisco the operation of which depended upon regenerative repeaters. The vacuum tubes used in this transcontinental telephone link were special 'hard vacuum' devices of Arnold made by Western Electric. In 1915, an experimental 500-radiotron tube transmitter constructed in the naval radio station in Arlington, Virginia, which used regenerative circuits in its transmitter and receiver was able to transmit speech by radio-waves to Paris, France. The high-vacuum, high-voltage vacuum tubes soon replaced the alternators as generators of high-frequency undamped electric oscillations. Alexanderson remained productive in electronics engineering throughout his long life, and he took out patents for his inventions even in his very old age. He died a nonagenarian.

Armstrong invented the 'superheterodyne circuit' in 1917. In the superheterodyne receiver, the incoming signal wave is mixed, or 'heterodyned', with a wave of slightly different frequency from a local oscillator tube, producing a signal wave of intermediate frequency equal to the difference in the frequencies of the 2 mixed waves; then this wave is amplified 3 to 4 thousand times, which, in turn, is detected and rectified, and then amplified into audio frequencies and converted into sound. The superheterodyne circuit of Armstrong is a highly stable, 4-stage receiver of very high-frequency radio signals, and it stands as one of the most durable concepts of radio technology. The superheterodyne circuit provides excellent selectivity and low noise and is the basis of almost every radio, television, and microwave radar set in existence.

Claims were made by the French electrical engineer, Lucien Lévy, German physicist Walther Hans Schottky (1886-1976), and American physicist Lloyd Espenschied (1889-1986) for their priority in the invention of the superheterodyne circuit, since they also had invented circuits that vaguely resembled Armstrong's circuit, but all these circuits lacked some element or a series of elements of Armstrong's superheterodyne circuit. The superheterodyne circuit made the growth of radio phenomenal as the sale of radio sets only in United States increased 15 times from 1922 to 1929.

The use of multistage amplifiers led to some trouble in the triode vacuum tubes which stemmed from the capacitance between the anode and the grid. This capacitance tended to limit the range of frequency over which the triode could operate, and created undesirable feedback which led to self-oscillation in the signal amplifier. To obviate this trouble, American physicist Louis Alan Hazeltine (1886-1964) invented his 'neutrodyne circuit' in 1919 in which some of the signal obtained from the plate circuit was fed back by a separate route into the grid circuit in proper magnitude and phase to cancel or 'neutralise' the undesirable feedback due to the anode-to-grid capacitance inside the tube, and thereby achieve stability and prevent oscillation. Hazeltine's 'neutrodyne circuit' was basically a tuned radio-frequency amplifier using a special type of neutralisation.

A better solution to this capacitance problem was the introduction of a new tube, the 'tetrode', which had 4 electrodes to reduce the coupling between the electrodes and improve the amplification. It used an idea first advanced by the German physicist Walther H. Schottky in 1916, consisting of the insertion of an extra grid, the so-called 'screen-grid', between the grid and the anode. The English electrical engineer Henry Round was able to produce a very effective tetrode in 1926 in which the screen-grid that had a constant high voltage accelerated electrons from the cathode but being an open mesh, the screen-grid allowed electrons to pass through it and reach the anode that made the anode current almost independent of the anode voltage. Although the tetrode was quite an effective vacuum tube for both the voltage and the power amplification, its only shortcoming was the secondary emission of electrons from the anode particularly at high voltages, a defect which was eliminated by the 'pentode' vacuum tube developed from 1926 to 1927 by the Dutch physicist Bernardus Dominicus Hubertus Tellegen (1900-1990). In the pentode tube a 5th electrode, the so-called 'suppression-grid', was inserted between the screen-grid and the anode and connected directly to the cathode and maintained at the filament voltage so that the 'suppression-grid' repelled the secondary electrons emitted by the anode and drove them back to the anode. The suppression-grid being an open mesh did not interfere with the travel of the primary electrons accelerated by the screen-grid, and prevent them from reaching the anode. It proved to be an extremely versatile and successful vacuum tube, which was almost immediately incorporated into most electronic equipments until it was replaced by the transistor as late as in the 1950's.

Frequency Modulation and High Fidelity

In the transmission of intelligence by radio-waves all the information was carried by the variations of the amplitude of the electromagnetic carrier-wave. The first attempt to transmit information by modulating the frequency rather than the amplitude of the carrier-wave were made with the Poulsen arc-transmitter as the generator of the high radio frequency waves about 1902. Earlier experience with radio-waves had demonstrated that the amplitude-modulated carrier-waves travel through space with a minimum of interference and distortion, and a maximum of economy when such waves were restricted to as narrow a band of fixed frequencies as possible. In 1915, John Renshaw Carson (1886-1940) of **AT&T** carried out a mathematical analysis of the nature of modulated carrier-waves and on the basis of it he invented the 'single-sideband modulation' which saved both power and bandwidth in the wire and the wireless transmission, and patented it in 1923. In 1922, Carson proved mathematically that 'narrow-band frequency modulation' was useless, a correct mathematical deduction when the narrow-band is conserved in frequency modulation. But Carson also drew an unwarranted conclusion that the frequency modulation inherently distorts without any compensational advantages whatsoever. In 1928, Carson felt confident enough to state that 'noise, like the poor will always be with us'. The notion that radio-waves to be efficient have to be limited to a very narrow-band of frequencies came to be accepted among electronics experts as the fundamental principle of the transmission of radio-waves according to the analyses of Harry Nyquist (1889-1976) in 1924 and Ralph V.L. Hartley (1888-1970) in 1928. Hartley, who in 1915 had been in charge of the radio-receiver development for Bell System's trans-Atlantic radiotelephone for which he invented his renowned 'Hartley oscillating circuit' as well as the 'neutralising circuit' for the elimination of 'triode singing' due to internal coupling and who, during World War I, had worked out the principles on which the development of sound-type directional finders were based, gave convincing theoretical support to the idea that narrow-band amplitude modulation of radio-waves was the only practical method in radio communication. When this principle, which is valid in the amplitude-modulation of the radio-waves, was applied to frequency-modulation of radio-waves by theoreticians, such as John R. Carson in 1922, they were able to demonstrate that frequency-modulation of radio-waves was quite useless for radio communication. The early theoretical and experimental work on the frequency-modulation had been done in the research laboratories of American Telegraph and Telephone Corporation (**AT&T**) and Radio Corporation of American (**RCA**). The early experimental radio equipment was rather crude and unreliable, and when such electronic equipment was used to produce narrow-band frequency-modulated radio-waves only incredibly distorted tones could be received by this method. The theoretical conclusions reached and experimental results obtained by the frequency-modulation of radio-waves were both erroneous because they were based on an invalid hypothesis. Although one of the greatest shortcomings of the amplitude-modulated radio communication was the static interference in radio reception, everybody had given up the frequency-modulation of radio-waves as a hopelessly impractical method of communication.

Such was the state of radio technology about 1930 when Armstrong undertook a fundamental experimental investigation of the frequency-modulation of radio-waves. By 1933, Armstrong had made more than 100,000 measurements in thousands of experiments, all the work he had done himself. His persistent efforts were crowned with huge success because he had produced with these experiments what had been considered impossible by the profession, an entirely new system of radio, which overcame the static interference in radio reception. Armstrong had invented the 'wide-band frequency modulation system' in radio, known by the acronym *FM*. In 1932, Armstrong had readily discovered that the frequency-modulated radio-waves treated like the amplitude-modulated radio-waves indeed do not work just as the theoreticians such as Carson had asserted. However, Armstrong being a boldly original, imaginative and independent thinker left the trodden path and struck out in a revolutionary new direction which flew in the face of conventional wisdom: he transmitted his radio-waves over a wide-band of frequencies instead of over a narrow-band of frequencies as was the current method in radio technology. The wide-band, frequency-modulated radio system represented a new concept in radio technology which greatly extended and clarified the basic principles of communication and contradicted the accepted theories of Carson. The basic concept in *FM* was that the frequency bandwidth could be traded for a higher 'signal-to-noise ratio', which serves as the measure of quietness in communication circuits: the wider the band of frequencies used, the greater the reduction of noise or static in radio reception. The *FM* broadcasting performed with surprisingly high fidelity and freedom from noise. In 1935 Armstrong demonstrated in his frequency-modulated radio system that the power to gain in signal-to-noise ratio increases as the square of the frequency bandwidth used. This new principle revealed by Armstrong led to the revision of the accepted scientific law in communication formulated by Ralph V. L. Hartley from **AT&T** which was believed to govern the quantity of information that could be transmitted over a system of communication. The Hartley Law stated that the total amount of information that can be transmitted is proportional to the frequency range transmitted and the time of the transmission. The mathematical consequences of the new principle of Armstrong were developed in 1948 by the mathematicians Norbert Wiener (1894-1964) in his theory of cybernetics and, particularly, in 1937 and in 1949 by the ingenious Claude Elwood Shannon (born 1916) in his modern information theory, which led to a fundamental change in the analysis and design of communication systems. Information theory which involves both analysis and probability theory had its origin in the works of the Hungarian physicist Leo Szilard (1898-1964) and Harry Nyquist in the early 1920's. In 1937, Shannon made his initial landmark contribution to the information theory in one of the most important and influential Master of Science thesis ever written, by showing that the action of complex switching circuits could be analysed by means of Boolean logic and algebra which made the complex digital circuit design into an exact science. He showed that the 'fundamental unit of information' is a 'yes-and-no' situation which can be represented by binary digits '1– and – 0' called a 'bit'(*binary digit*). More complicated information is then built up of combinations of 'bits'. In 1949 he demonstrated how this quantification of information can be analysed by strict mathematical methods in which garbled information can be measured by the loss of 'bits', the distortion of 'bits', the addition of extraneous 'bits' and so on. In Shannon's information theory it was possible to express with quantitative precision such things as redundancy and noise, since information in a physical communication system is always accompanied by 'noise' or error, which materially reduces the information. Therefore, information and noise together constitute the message. He introduced a numerical measure for the randomness or uncertainty associated with a class of messages and demonstrated that this quantity measured in a real sense the extent of communication facility required to transmit messages from a given class with accuracy. He showed that this measure agreed in a certain respect with the idea of the information content of a message. Shannon's information theory could be applied to communications technology, circuit technology and computer design. The science of radio communication system was furthermore founded on the understanding of the fundamental nature of the 'background noise' that John Bertrand Johnson (1887-1970) explained in 1928 to be 'thermal noise' which is present in all electrical systems and which becomes particularly disturbing in weak-signal communication systems, and on the limits of the frequency bandwidths studied by Harry Nyquist in 1924 and Ralph V. L. Hartley in 1928.

Armstrong built his *FM* radio receiver on the basis of his superheterodyne circuit. The new *FM* radio system of Armstrong was a much more complicated invention than his feedback circuit, since his feedback circuit had only one vacuum tube whereas his early *FM* circuit often had as many as a hundred tubes in an experimental set up which took about a month to construct and test.

Armstrong's *FM* radio system improved the signal-to-noise ratio to 100-to-1, which at the time represented an incredible advance in radio technology. After much further research the signal – to– noise ratio was increased to $1000 - \text{to} -1$.

In his frequency-modulated radio system, Armstrong sent a carrier-wave of constant amplitude the frequency of which he varied between certain wide limits, that is, over a wide-band of radio frequencies. The number of times the frequency of the carrier-wave was varied from its mean-value corresponded to the frequency of the sound waves, the extent to which each variation deviated from the mean-value of the frequency of the carrier-wave corresponded to the amplitude of the sound wave. Frequency modulation eliminated most of the static interference present in radio reception because no similar electromagnetic waves exist in nature.

Armstrong's *FM* radio system had many advantages over the *AM* radio system:
(a) On account that *FM* operates on a much wider frequency band then *AM*, *FM* could reproduce almost the entire range of sound audible to the human ear, a feature called 'high fidelity';
(b) In multiplexing the wider frequency band it is possible to send more than one signal at a time;
(c) *FM* audio station can serve a greater area then an *AM* radio station with the same power or the same area with less power which made *FM* radio stations less expensive to operate;
(d) Moreover, *FM* radio stations on the same frequency can be placed closer together geographically than *AM* radio stations because unlike *AM* signals, *FM* signals do not interfere with one another, since in *FM* only the stronger station is audible.

A few years after Armstrong's revolutionary contribution to radio technology which was based on very profound physical experiments of Armstrong, the mathematical physicists were able to demonstrate mathematically that wide-band frequency modulation in radio communication does work after all. In such papers the mathematicians tried to imply that they could have been able to prove the feasibility of frequency modulation by their mathematical theory had they been really interested in this topic. This was an empty and disingenuous claim sprouting from wounded pride because mathematics proves nothing in the absence of fundamental physical experiments and reasoning such as were provided by Armstrong.

Moreover, the theoreticians asserted that very short electromagnetic waves were limited to a line-of-sight range, that is, to the horizon, from the transmitter, despite the fact that Marconi had shown in 1933 that very short electromagnetic waves can be received over a distance of 10 horizons. Armstrong demonstrated that his very high frequency *FM* radio signals were clearly received at least over three horizons, or 80 miles (129 km), but the myth that the reception for *FM* is limited to the horizon persisted perhaps because Armstrong, with his frequency-modulated system of transmission of radio signals, ran into the opposition of enormous vested interests of various radio chains, such as **NBC**, **CBS**, **ABC**, Cowles Broadcasting, Crosly, Philco, Motorola, and Du Mont, which had huge investments in amplitude-modulation radio equipment as well as in radio stations and, therefore, were dead set against the introduction of the superior frequency-modulated radio system for fear of losing their investment.

Armstrong spent over one million dollars to develop microphones, amplifiers and various other components for this *FM* system and his first *FM* radio station began broadcasting on the very high frequency spectrum early in 1939. By November, 1939, 5 *FM* radio station's were operating under the Federal Communication Commission, known by its acronym *FCC*, and 15 *FM* radio stations were operating under *FCC* construction permit. In 1941, the Yankee Network, the first *FM* radio chain, was operating in New England. After the Second World War, when *FM* began to spread owing to its technical superiority and financial economy, various radio chains, particularly **RCA &NBC** under the direction of its wily and unprincipled chairman David Sarnoff, resorted to conspiracy and collusion in an attempt to cripple Armstrong's *FM* system of broadcasting. Although in 1940 the *FCC* had assigned the frequency band between 42 and 50 megahertz to *FM* radio broadcasts which was enough for 40 *FM* channels, in 1945 the *FCC* gave this airwave space to television and moved *FM* up into the 100 megahertz range, a decision which made every existing *FM* receiver practically obsolete. To add insult to injury, the *FCC* limited the power and reach of *FM* radio stations to a single city or market and, moreover, prohibited *FM* radio stations to multiplex their signals, rulings which proved to be the *coup de grâce* to *FM* radio stations since they forced about 200 *FM* stations into bankruptcy. *FCC*, an organisation founded on the State control of the airwaves, made this preposterous and unfair decision under the pressure exerted by the unprincipled anti-*FM* lobby, and ostensibly based on the testimony of only one engineer who later had to admit after severe criticism of his opinions that his calculations in his testimony were all wrong. It is a shameful example for the evils of State control of airways, and the venality and unfairness of State commissions. **RCA** under the conniving leadership of Sarnoff had pressured the *FCC* to force *FM* broadcasting out of the airways, and to hand over the *FM*'s band of frequency to **NBC** television the audio circuit of which required *FM*. Sarnoff, a former friend of Armstrong, had offered 1,000,000 dollars as a flat fee for Anderson's basic *FM* patents, but Armstrong had rejected this offer and insisted on royalties. **RCA** had a firm policy not to pay royalties, but rather to collect them. The only exception to this rule had been the sizeable royalties **RCA** paid to Philo Farnsworth for the latter's very basic television patents. When **RCA** patented a rival *FM* system and began to manufacture *FM* receivers as well as to license it to other manufacturers, Armstrong published a technical paper in which he proved that *FM* system of **RCA** was essentially a copy of his *FM* system and, therefore, he sued the industrial giants **RCA** and **NBC** in 1948 for patent infringement. But after 6 years of intense and bitter pre-trial litigation and enormous legal costs, the exhausted Armstrong finally surrendered late in January 1954, and committed suicide by jumping to his death from his 13th floor terrace apartment in New York. A few months later **RCA** paid the estate of Armstrong $1,040,000 for his **FM** patents. This remarkable and ingenious inventor being a contentious individual who was stubborn to a fault, ultimately could not prevail over the faceless industrial colossus such as **RCA,** manipulated by its scheming chairman David Sarnoff. In all his business dealing concerning his inventions Armstrong, a confirmed individualist who knew his personal worth, was as uncompromising an inventor as Oliver Evans, Fessenden and Tesla had been before him.

In all telephonic systems signal-to-noise ratio was limited by amplifier distortion. As long as telephone and radio links covered only relatively short distances the amplifiers had to supply relatively small amount of amplification, but for longer lines of communications much higher amplifications were necessary which required the amplifiers to operate in the nonlinear part of their input-output characteristics. This nonlinear distortion became quite intolerable when 2 or more channels were fed through the same amplifier as was the case in carrier systems and resulted in 'cross modulation' which produced 'cross-talk' between the channels and imposed a serious limitation to the extension of the system. This shortcoming could be eliminated at its very source by 'negative feedback', a basic concept discovered by the American electrical engineer Harold S. Black (1898-1983) in 1927, which was a landmark and a breakthrough in communications technology. In the negative feedback amplifier of Black, some of the output signal was fed back to the input 'in opposition' to the input signal. If the original amplification was nonlinear, and tended to fluctuate, after the proper use of negative feedback which prevents unwanted gain or loss of output, the new amplification was made practically linear and constant. The negative feedback amplifier of Black had an additional advantage of having a wider bandwidth than other amplifiers. Black was also the first engineer to use 'pulse-code modulation' which later became prominent in communication technology. Harry Nyquist, who had formulated many fundamental concepts of communications and the sampling theory in its application to digital systems, established the theoretical criterion for the stability of 'negative feedback' amplifiers which promoted its acceptance in communication technology. The use of 'negative feedback' amplifiers in electronic circuitry resulted in a number of important improvements in electrical communications as well as in electronic technology in general: precise amplification, low distorsion, and good frequency response, all of which could now be theoretically evaluated and designed. In this way Black's principle of 'negative feedback' created a virtual revolution in the functioning of telephone trunk circuits.

Transistors and Solid State Revolution

The transistor is a semiconductor device which is used to rectify and amplify current in electronic systems. In general, a transistor can do what a vacuum tube was able to do.

The first solid-state rectifiers which were used in practice were the 'cat's whiskers', but in the 1920's newly developed solid-state rectifiers built of stacks of copper plates, each oxidised on one side, could operate with far more power than the sensitive 'cat's whiskers'.

Several inventors, almost at the same time, became aware of the importance of solid-state amplifiers owing to the influence of the 'field-effect principle', because its concept was related to the 'grid-control principle' of the vacuum tubes.

In 1925, 1927 and 1928, a German physicist, Julius Edgar Lilienfeld (1882-1963), a former professor of physics at the University of Leipzig in Germany who immigrated to United States, took out several patents on the design of a semiconductor amplifier using copper-sulfide, but it is doubtful that his devices could have worked in practice.

The German physicist, Oskar Heil (1908-1994), was granted a patent in 1935 for a 'field-effect' semiconductor triode. Heil used a control electrode to regulate the current flow through a thin semiconductor layer made of materials such as copper oxide, vanadium, pentoxide, tellurium, and iodine. Heil's design was a precursor of the 'insulated-gate field-effect transistor', similar to the third patent of Lilienfeld, since the control electrode was isolated from the substrate of the semiconductor. It is doubtful that Heil's 'semiconductor amplifier' ever worked in practice.

The conductivity and resistivity of semiconductors is intermediate between the conductivity and resistivity of good conductors and insulators. Metals are good electrical as well as heat conductors. Resistivity of a semiconductor becomes extremely high at low temperatures in comparison with a good conductor the resistivity of which at low temperatures remains small. The resistivity of a semiconductor varies much more with temperature than a good conductor. Careful examination of semiconductors revealed that their resistivity varied greatly from one specimen to another depending upon the extent of impurities and defects present in their crystal structures. It was of paramount importance to devise a theory of solids which could explain why semiconductors have higher resistivity than metals, and why the resistivity of semiconductors is so dependent on temperature as well as on impurities and defects incorporated in their crystal structures. The dependence of resistivity of semiconductors upon their content of impurities and structural defects offered the technological opportunity of deliberate incorporation of impurities, called 'doping', or defects in the crystal structure of the semiconductor to produce desired values of resistivity in a particular semiconductor specimen.

The first consistent solid state theory of semiconductors was developed by a Swiss-born American physicist, Felix Bloch (1905-1983), who shared the Nobel Prize in physics for new methods of nuclear magnetic measurements and nuclear magnetic resonance with Edward M. Purcell in 1952, and a few other physicists from 1925 to 1935. In 1929, Bloch, professor of physics at the Stanford University, proposed his 'band theory of solids' for semiconductors. In a solid body consisting of a large group of atoms, the *allowed* 'energy levels' for electrons in atoms are spread out to form *allowed* 'bands of energy levels'. In perfect semiconductor crystals the *allowed* 'bands of conduction' and *allowed* 'bands of valence' are separated from one another by the *forbidden* 'bands'. In semiconductors the energy gap between the 'energy bands of conduction [empty]' and the 'energy bands of valence [filled with electrons]' is small, and excitation by heat or light suffices to cause the electrons in the 'valence band' to jump into the 'forbidden band' thereby causing electric current to flow. The spaces left by the electrons in the 'valence band' behave like particles similar to electrons, except that they are positively charged and are called 'electric holes'. In a particular semiconductor, the electric conductivity may be produced by electrons, by 'electric holes', or by both. If in any particular semiconductor specimen the electric conduction is primarily by electrons, *i.e.* negative charges, then the specimen is called *n-type* (negative-type), if primarily by 'electric holes', which act like positive charges, the specimen is called *p-type* (positive-type). In metals with the lowest resistivity, electric conduction is by electrons. Technologically the importance of semiconductors derives from their property that large temporary changes in the numbers of electrons and 'electric holes' can be produced by sending electric current through the specimen, or by irradiation with light or nuclear radiation, or by some other means. Such control over the concentration of electrons and 'electric holes' is impossible for metals. In concert with the band theory of solids, the number of free electrons in semiconductors is very small as the energy bands are entirely full or empty except for a few electrons and 'electric holes' created by thermal excitation, or by the presence of impurities. The development of transistors was based on understanding the physical function of the two types of semiconductors: *n-type* semiconductors in which negative electrons move, and *p-type* semiconductors in which 'electric holes' having positive charge change position. The first transistor was developed at the Bell Telephone Laboratories in 1948, where William Bradford Shockley (1910-1989), Walter Houser Brattain (1902-1987) and John Bardeen (1908-1992) constituted the semiconductor research group. Brattain had been experimenting on semiconductors since the early 1930's, and in 1947 Brattain and Bardeen, who was the theoretician of the team, finally were able to produce the 'point-contact transistor' which harkened back to the 'cat's whiskers', and the triode. They had one thin wire contacting the surface of the semiconductor slab (germanium) and the other wire contacting the same surface at a distance of a fraction of a millimeter from the first wire. The first wire became the 'emitter', the other wire the 'collector', and the germanium slab the 'base'. Their first 'point-contact' semiconductor amplifier, which they named a *transistor* (**transit resistor**), achieved a gain of about 100 volts. John Bardeen, the theoretician of the research group, had to formulate a theory on the nature of the surface of the semiconductor, which accounted for the inability of an electric field to penetrate into the interior of the semiconductor, before their success could be achieved.

At the same time, Shockley who directed the semiconductor research at the Bell Telephone Laboratories, abandoned his work on the 'field-effect transistor' because he was unable to obtain a patent for it on account of the previous patents of Lilienfeld and Heil. He believed that a much better transistor than the 'point-contact transistor' was possible. He developed the theory of the 'junction transistor' in which the transistor action is obtained by sandwiching an *n-type* semiconductor between *p-type* semiconductors. The semiconductor theory of Shockley was rejected as unsound when he submitted it to Physical Reviews for publication. Undaunted and having a complete confidence in his theory, Shockley continued his research and finally succeeded in constructing the first junction-transistor in 1951 which functioned perfectly, and proved that his theory of junction-transistor was entirely correct. Shockley's 'junction-transistor' pointed the way to modern solid state electronics, since in the creation of his transistor Shockley had created new knowledge in solid state physics.

The usual transistor materials are germanium and silicon into which impurities are introduced by a method pioneered by Shockley. Since 1960 transistors were made of silicon which is a much more stable and abundant material that is not so sensitive to temperature variations as germanium. In 1956 Brattain, Bardeen and Shockley shared the Nobel Prize in physics for the development of the transistor. In 1972, Bardeen received his second Nobel Prize in physics for his fundamental theoretical research on superconductivity.

Although the transistors have very small size and require very small power to function, they can amplify up to 40,000 times. Transistors in contrast to vacuum tubes have excellent durability and high resistance to shock. The semiconductor technology soon led to microelectronics and integrated circuits which began to replace transistors in complex electronic equipment in 1959 since an integrated circuit which is comparable in its small size to the transistor, can perform at least the function of 15 to 20 transistors. In 1959, the first 'integrated circuit' was patented by Jack St. Clair Kilby (born 1923) of **Texas Instruments**, and 6 months later Robert Norton Noyce (born 1927) of **Fairchild Electronics**, a specialist in photolithography, patented a batch production method in which metal connexions could be evaporated onto the oxidised semiconductor surface in a planar process. Kilby had realised in early 1958 that a semiconductor crystal could function potentially like a complete electronic circuit by recognising that an undoped silicon crystal could function as a resistor, whereas particular *n-p* junctions could be used as capacitors. Inclusion of these 3 types of electronic components on a single silicon crystal constituted a complete, fast, and cool low-voltage circuit. Kilby made such a circuit in September of 1958. Kilby and Noyce patented different methods of incorporating many resistors and capacitors on a single silicon crystal chip representing their idea of a fully integrated circuit, which is called a 'microchip'. In 1968, Noyce and his fellow-employee at **Fairchild Electronics**, Gordon Moore, founded a new company, **Intel**, for the manufacture of microchips. Miniaturisation of electronic equipment using microchips soon assumed paramount importance in electronics, particularly in military and space technology, although the first application of it was to hearing aids in 1963. In 1964, **RCA** manufactured its first computers incorporating integrated circuits as microchips.

In 1950, 1,000 vacuum tubes fitted into a cubic foot, but in 1956, 10,000 transistors could be fitted into the same volume. In the late 1970's the volume was no longer relevant, as 100,000 electronic components could be squeezed onto a 5 mm square chip of silicon.

Radar

The basic idea of radar had first occurred to the German physicist Heinrich Hertz in 1887, when he observed that it was possible to bounce high-energy electromagnetic waves off a metal plate.

In 1904, Reginald Fessenden used electromagnetic pulses from a wireless oscillator for depth sounding in a contract work he carried out for the United States Navy, but he did not extend this idea any further.

In 1904, the German engineer, Christian Hülsmeyer (1881-1957), obtained a German patent for his ***Telemobiloscope***, an instrument for detecting distant metallic objects by means of electromagnetic waves which represents the basic idea of 'radio-ranging', or the so-called 'radar' (***Radio Detection and Ranging***). Unfortunately, the electrotechnology of the period was not sufficiently developed for the practical success of the ***Telemobiloscope***, for neither the electric pulse technique nor the ultrashort waves were known at the time.

In 1924, the radar principle was used by the English physicist Edward Victor Appleton (1892-1965) to bounce radio-waves off the ionised layer in the upper atmosphere, called the 'Kennelly-Heavyside Layer', thereby determined its height above the surface of the Earth to be 60 miles (97 km). In 1926 he discovered by the same ranging method another higher ionised layer, now called the 'Appleton Layer', 150 miles (241 km) above the surface of the Earth.

Effective radar requires intense short-wave emission with the wavelength that is short relative to the target being ranged. It was expedient to send short-waves in short bursts at the target which is called the 'pulsed-radar'. The successful pioneer of modern pulsed-radar was the ingenious Scottish electrical engineer, John Logie Baird (1888-1946). By 1923, Baird already had a working radar equipment at Hastings. In 1926 and 1927, Baird took out the first modern pulsed-radar patents, and his equipment was used in 1928 and 1929 to obtain image echoes for ranging. From 1936 on, Baird took out a number of secret patents on his pulsed-radar which have not been released to the public. In 1936, Baird's 6.7-meter television transmitter tubes were acquired by the British government and used as the first successful continuously operating British radar transmitters, since Baird's television transmitter tubes gave the strongest power on the wavelengths required by the long-range British pulse-radar. Baird's pulsed-radar compares favourably with the most sophisticated radar of World War II which was used to guide long-range bombers.

In a search for even shorter wavelength, stronger power, and increased speed of operation of vacuum tubes it soon became evident in the early 1930's that the vacuum tubes were inadequate for the task. The inertia of the electron with a mass as tiny as less than two-thousands of the mass of the hydrogen atom, and the ultra-high frequency such as a billionth of a second required for the centimeter wavelength seem to place it beyond the practical limits of operation of vacuum tubes which use the space charge of electrons from the cathode that is moderated in quantity during its passage to the anode. To meet the new requirements of the ultra-high frequency short-waves, a pulse technology was developed which was suitable for the radar.

Already in 1921, the American electrical engineer and a proficient inventor of various kinds of vacuum tubes, Albert Warren Hull (1880-1966), invented a low-power diode detector called the '*magnetron*' in which a powerful permanent magnet supplied a magnetic field at right angles to the electric field between the cathode and the anode. This '*magnetron*' generated sufficient power at a wavelength of less than half a meter. The design and manufacture of magnetrons led to a general advance of electronics since magnetrons required new materials and new high vacuum techniques.

In 1924 a Czech physicist A. Začek at the University of Prague designed a magnetron with a single anode to produce short-wave oscillations. In the same year a German doctoral student at the University of Jena, Erich Habann (1892-1968), obtained high-frequency electronic oscillations with a 'cavity magnetron' that had a split concentric anode and a grid as a third electrode. He observed that the mutually perpendicular electromagnetic and electrostatic fields produce a spiral path for electrons. His magnetron generated about 11 watts of power. Habann promptly patented his new high-frequency generator, his '*single-cavity magnetron*', which he had designed for use in the carrier-wave frequency technology. Many years later, during the Second World War, his '*single-cavity magnetron*' found important application in the radar technology. Habann also did important pioneering work related to transistor technology which he patented in 1942.

In 1927, the first high-frequency magnetron with multisplit anode was designed, and its action theoretically proven by the Japanese physicist Kinjiro Okabe (1896-1984). In 1930, his '*magnetron oscillator*' operating with a frequency of 400 megacycles at microwave wavelength produced an output of several watts. His oscillating magnetron was used in the Japanese radar system.

In 1934, American electrical engineer and an expert in electron tube design, Arthur Lee Samuel (1901-1990) of Bell Telephone Laboratories, who developed many electron discharge devices using space charge between parallel electrodes for operation in the ultra-high frequency range, applied the resonator principle to magnetrons, and invented and patented the **multicavity magnetron** which was capable of transmitting strong bursts of radiation of microwaves lasting about one microsecond every millisecond. It consisted of a hollow cylindrical copper block anode containing radially distributed resonating cavities and an electron-emitting cathode located concentrically in the block. The electron stream under the action of the electric field between the cathode and the anode, and a strong transverse magnetic field was subjected to the complex interaction of fields created in the cavities designed to be resonant at the frequency of the microwave oscillation extracted from the *cavity magnetron*. Samuel supplied the British radar with his *cavity magnetrons* in the beginning of the Second World War until the British scientists produced one of their own. The *cavity magnetrons* developed in England since 1940, reached a peak pulse output of 10 kilowatts at a high frequency of about 3,000 megacycles per second, a power 5 times as great and a frequency 4 times as high as the best available high-frequency triodes could deliver. Although the *cavity magnetron* was heavy and for its high power had to be artificially cooled, it proved very important for the efficiency of the long-range ground radar in the Second World War.

Another generator of microwaves called the '*klystron*' was invented in 1937 by the American physicists William Webster Hansen (1909-1949) and Russell Harrison Varian (1898-1959) at Stanford University. Its action was based upon the principle of velocity modulation of electrons, a principle which unbeknownst to Hansen and Varian had been proposed by the German physicist Oskar Heil already in 1933. The *klystron* was based on Hansen's invention, the '*rhumbatron*', consisting of a donut shaped cavity resonator in which electromagnetic waves were made to bounce back and forth at ultra-high frequency. In *klystron* a broad beam of electrons accelerated by a triode passed through two donut-shaped cavities resonating at the same frequency. In the first input cavity the magnetic field retarded the electrons entering earlier and accelerated the electrons entering later thereby bunching the stream of electrons which then entered the second output cavity where it excited an electromagnetic field and thereby extracted more power from the electron beam. A suitable coupling together of the two fields led to an oscillation which produced power at resonant frequency. The velocity modulation of a stream of electrons in the *klystron* had certain advantages over the crossed-field operation in the 'cavity magnetron', particularly in the size and the weight of the device. The *klystron* was only about 7 inches (17.8 cm) long, 3 inches (7.6 cm) in diameter, and weighed less than 6 pounds (2.7 kg). In 1940, the British 'cavity magnetrons' were more powerful than the *klystrons,* but they had a lower frequency than the *klystron* and were very heavy. In 1940 during the Battle of Britain the British aircrafts had to be fitted with airborne radar and only *klystron* tubes were suitable for it. After the Second World War, Hansen developed a 'high-powered *klystron*' that was 1,000 times as powerful as the original *klystron* for the use in the first 'linear atom smasher', or the 'linear accelerator' at Stanford University which he designed. The new linear accelerator at Stanford University which was completed in 1966 is 2 miles (3.22 km) long and powered at every 40 feet (12.20 m) with new '*klystron* tubes' which are much more powerful than the 'cavity magnetron', and about 10,000 times more powerful than the original *klystron*.

In Germany experimental tests of Hülsmeyer's method of ranging on behalf of the German Navy were begun in the 1930's. In the first tests, a wavelength of 14 centimeters was used, which for some reason was changed into a less suitable wavelength between 18 centimeters and 2 meters. Hülsmeyer's projection apparatus for a beam of electronic wave, which incorporated two adjustable lenses, was developed so far that successful tests were performed with it on a 2.4 meter wavelength before 1939. When the decimeter-long, or meter-long, radar waves were focussed by special aerials, or reflectors, and beamed to a target, it was required to send out very short electric pulses of the order of a ten-millionth of a second in a spherical wave-front 30 meters deep. This wave reflected back from the target with part of its initial energy and reached the receiver which was located next to the transmitter. The aerial then pointed in the direction of the target and the distance was calculated from the time of travel of the electronic pulse which is only a few millionth of a second since its speed is 186,000 miles (300,000 km) a second. The time of travel was measured electronically using Braun's cathode ray oscilloscope. The German radar system was developed further by the *Telefunken* company.

In England, the Scottish physicist, Robert Watson-Watt (1892-1973), designed a radar chain in 1939 which operated at first on a 12-meter wavelength, a wavelength which was far too long for its purpose. Later, he changed the radar to operate on a 9-centimeter, and then on a 3-centimeter wavelength, which was the best radar system in existence at the time.

In United States, Armstrong had not exhausted his inventive powers with his *FM* 'high-fidelity' system, but made his last significant contribution to electronics by developing the 'continuous wave *FM* radar' and filing patent applications for this system of radar in 1947 and 1948. The searching range and power of Armstrong's 'continuous wave *FM* radar' far exceeds that of the pulse-type radar. This fundamental invention of Armstrong made the pulse-type of radar system virtually obsolete. With this invention Armstrong created the basis for a system of long-range radar which is essential to any warning system in the age of supersonic missiles. Thus Edwin Armstrong proved to be a practical genius in electronics who remained inventive throughout his entire working life.

CHAPTER XXV

ELECTRIC AND ELECTRONIC TRANSMISSION OF VISION: From Phototelegraphy to Television

Photoelectric Effect

In 1817, the remarkable Swedish chemist Jöns Jakob Berzelius (1779-1848) discovered a new element called 'selenium' which offers high resistance to electric current.

In 1873, a young telegraph engineer in Ireland, Joseph May, noticed that when the Sun shone on his telegraph set it ceased to work properly. May observed that the Sunlight appreciably increased the electrical conductivity of the selenium in his instruments. Later in the same year Willoughby Smith, who worked on the laying of the Atlantic submarine telegraph cable, carried out some experiments with his chief assistant May on the variable electric resistance of selenium in which they determined that the increase of electrical conductivity in the selenium was proportional to the intensity of light which fell on it. This discovery of the photoconductivity was of fundamental importance to the invention of the photoelectric cell. Selenium as a semiconductor is also widely used to rectify alternating current in electronic equipment. Selenium occurs in many sulfide ores, mainly in iron, copper, lead, and nickel ores, and as metal selenides.

The electrochemical effect of light had already been discovered by a 19-year old French physicist, Edmond Alexandre Becquerel (1820-1891), who discovered in 1839 that when two electrodes are immersed in a particular electrolyte and exposed to a beam of light, an electromotive force was generated between the electrodes. This discovery of the photovoltaic effect by Edmond Becquerel was the first time when light radiation was connected with electricity. Actually, his experimental equipment was the first photoelectric cell, or more precisely an electrolytic photoelectric cell functioning by the photovoltaic effect. Edmond's father, Antoine César Becquerel (1788-1878), who fought in the Battle of Waterloo as an officer in the Napoleonic army and afterwards was professor of physics at the *Muséum d'Histoire Naturelle* (*Museum of Natural History*), had done fundamental research in electrochemistry, and helped to found this new science which he applied to the dressing of ores. His son, Edmond Becquerel, succeeded his father as professor of physics at the *Muséum d'Histoire Naturelle* and at the *Conservatoire des Arts et Métiers* (*Conservatory of Arts and Crafts*) where he studied fluorescence and phosphorescence [in both phenomena matter absorbs light at one wavelength and emits it at another], and also photography. Edmond's son, Henri Antoine Becquerel (1852-1908), who was appointed professor of physics at the *Muséum d'Histoire Naturelle* in 1892 and at the *École Polytechnique* (*Polytechnical School*) in 1895, discovered radioactivity of uranium in 1896.

Telegraphic Transmission of Pictures

The first system for the electrical transmission of pictures was proposed in 1843 by a Scot, Alexander Bain (1818-1903), professor of psychology at the University of Glasgow. It was based on a master-slave pendulum system consisting of electrically controlled, interconnected pendula at the sending and the receiving end of the telegraph which made the 2 pendula to be in the same relative position at any time. It was intended to transmit typeface and consisted of 2 wire combs, one at the transmission-end, the other at the receiving-end. The comb at the transmission-end was translated across a plate by the pendulum fitted with short, parallel wire sections in an insulated matrix whilst the typeface, which was connected to the battery, was brought into contact with the back-face of the plate. A similar wire comb was translated by the synchronised pendulum across a chemically treated paper which was grounded. Current passed through the corresponding tooth of the comb when a tooth of the comb at the transmitter came into contact with a part of the typeface, and as a result discoloured the chemically treated paper. Bain's automatic electrochemical recording telegraph in its initial version was never commercially constructed because it required too many circuits. In 1840, Bain also perfected an electric clock.

In about 1847, the English inventor Frederick Collier Bakewell constructed a phototelegraphic system which was able to transmit written matter by means of a single telegraph line. It was quite similar to modern apparatus of electrical transmission of photographs and printed material, and incorporated the modern concept of 'scanning'. In Bakewell's electric facsimile transmission apparatus the original was given in insulating ink on a metal foil attached to the surface of a drum turned by a clockwork. The foil was then examined by a probe translated parallel to the axis of the drum, which had the result that the original in insulated ink was scanned along a set of closely spaced, parallel lines.

The first time mechanical scanning principle was used commercially in the transmission of electric pictures was in 1862. A French Abbé of Italian stock, Giovanni Caselli (1815-1891), was the first to transmit drawings and written messages electrically over a distance, in this case from Amiens to Paris. Caselli's mechanical scanning apparatus consisted of a tiny light bulb mounted on a movable carriage which translated back-and-forth across the picture, and after each crossing the light bulb was moved up to the next horizontal line to be scanned. It took about 10 minutes to scan a single picture. The reflected light from the picture was caught by a photosensitive element which produced electricity. At the receiving-end a synchronised motion of the chemically treated paper reproduced the image. The synchronisation in this picture telegraph was based on tuning forks acting as speed governors for spring-driven mechanisms. With the financial support of Napoleon III, Caselli set up a number of commercial stations for his picture telegraph, but other telegraph messages interfered with his picture transmission which made his picture-telegraph a commercial failure.

In 1878, de Paiva, a Portuguese physicist in Oporto, suggested that the pictures to be transmitted electrically be first projected onto a selenium-coated metal plate, and then the metal plate scanned by a metal point. The metal plate and the metal point were to be connected in series in a circuit with a battery and relay, in which the relay was controlled by the current from an electric light bulb to be translated synchronously with the metal point. The bulb would light up in positions which correspond to the bright areas of the picture, and thus replicate the original picture to be transmitted.

Another system of distant vision by means of electricity which made use of May's discovery of the photoconductivity of the selenium was proposed by a French inventor Senlacq of Ardres in 1878. He proposed to reproduce telegraphically at a distance the images obtained in a *camera obscura*. His scheme also contained the idea of scanning. In the focus of his *camera obscura* he proposed to place a ground glass screen. A small piece of selenium held by two springs as pincers, insulated and connected, one with an electric pile and the other with the line of any system of autograph-telegraphic transmission apparatus, traversed the surface of the ground glass in scanning the image. His receiver consisted of a pencil attached to a soft iron which was operated by electromagnets like in the Bell telephone. In 1880, after realising the impracticality of his first scheme, Senlacq proposed a new system in which the transmitter screen was made of a large number of selenium cells in which each cell was connected to a distributor which could bring the cell into contact with the single telegraph line by means of a falling slider, an arrangement which scanned each cell only once. The receiver also consisted of a mosaic of cells, each cell having a fine platinum wire and connected to contacts on a distributor plate which was driven by a clockwork mechanism that maintained synchronism with the falling slider in the transmitter. This scheme of Senlacq for his *telectroscope* attracted wide attention in England and America, and many inventors such as Ayrton and Perry in England, Sawyer in New York, Carey in Boston, Sargent in Philadelphia, Tighe in Pittsburgh, and Alexander Graham Bell proposed their own versions of distant vision.

An American inventor, George R. Carey, was the first to propose a system in 1875, which constituted a kind of television system. In his first scheme, Carey suggested that the image of the picture to be transmitted be projected on a photosensitive insulating surface such as a layer of silver halide. Closely spaced pairs of terminal platinum wires were to be imbedded in the photosensitive layer, so that when light impinges on the insulating surface the pair of wires would be short-circuited with the partial reduction of silver halide to silver. Each circuit was controlled through a relay connecting with an electric light source in a mosaic of light sources arranged in the same geometric order as the terminal pairs of platinum wires embedded in the photosensitive layer. Therefore, a projected light pattern on the photosensitive plate would be recreated in a similar light pattern on the receiving mosaic panel. In this system every element of the image had a separate channel and each circuit controlled by means of a relay the light source in the receiver panel. Although his system was never built, and as proposed could not have functioned, it did correspond in its fundamentals to the technique used in the animation of modern advertising signs. It was the only system that could have functioned with the slow-response of the early selenium cell to transmit moving pictures.

In 1880, Carey proposed 2 other schemes for the electric transmission of pictures. In the first scheme he proposed to project an image on the multicellular mosaic screen made by selenium cells, each cell separately connected by wire to corresponding points in the receiver panel where the replica of the image was to be produced by means of electrochemical decomposition.

Carey's second scheme was based on an elementary system of scanning, whereby a selenium cell traversed a spiral scanning path, and transmitted the electrical signals to the receiving panel by using only a single-line wire. Carey used clockwork mechanisms at the transmitter and at the receiver, but he failed to provide for their synchronisation.

In 1877, Ayrton and Perry in England proposed the use of a mosaic panel made of selenium cells in place of Carey's photosensitive plate.

However, the American W.E. Sawyer pointed out in 1880, that to provide sufficient details in the picture a mosaic panel consisting of 10,000 photosensitive elements is necessary, but this would require huge numbers of parallel transmission lines and shutters, or relay devices which makes such a system completely impractical. Apart from Caselli, it is improbable that any inventor of a distant electric vision system actually constructed a transmission equipment which led to successful practical results. However, an English inventor, Shelford Bidwell, did construct in 1881 a practical apparatus to send silhouettes with his apparatus which consisted of a transmitter and a receiver. For the transmitter, a shadowgraph image was projected onto the front of the box with a pinhole aperture containing a selenium cell. This image was scanned by the vertical motion of the box controlled by a cam, and the horizontal motion controlled by a horizontal shaft with fine screen threads. In the receiver, the receiving copper drum was covered with platinum on which a platinum point on a flexible brass arm pressed on the surface of the cylinder on which was wrapped a paper soaked in a solution of potassium iodide. The cylinder rotated in synchronism with the shaft of the transmitter, and the arm traced out brown lines upon the paper, the intensity of which varied with the current passing through the selenium cell. Bidwell gave public demonstrations of his apparatus, which is still extant.

In 1880, the French engineer, Maurice Leblanc (1851-1923) proposed an ingenious method for the scanning of pictures with a small mirror. He proposed to scan the image with a small mirror which vibrated simultaneously about two axes in which one frequency was much larger than the other. Such a vibrating mirror is able to scan each small neighbourhood of an image, and project it onto a selenium cell. It was a very efficient optical scanning method, but the mechanical difficulties and synchronisation problems encountered in using this particular scanning method with a mirror made it unsuccessful in practical applications.

However, another mirror-scanner, the 'mirror-drum' which was invented and used by the English inventor L.B. Atkinson in 1882, and successfully applied to scanning in 1889 by Professor Lazare Weiller in France. The polygonal drum had as many mirrors as there were scanning lines, and as the mirror-drum rotated the image was scanned along a number of parallel lines and projected onto a selenium cell.

Television

In 1884, a German engineering student, Paul Gottlieb Nipkow (1860-1940), invented his mechanical scanning disk which gave television the first really practical means of transmitting pictures at an adequate speed for continuous observation of motion over a single circuit. In 1885 he patented his scanning disk in the form of a round disk pierced with 24 holes equally spaced along a spiral near the periphery of the disk, each hole containing a lens for the increase of the amplitude of the image signal. The transmitted image was focussed on a small region at the periphery of the disk which spun at 600 revolutions a minute. As the disk rotated, the spiral sequence of holes scanned the image in a succession of lines. The lenses collected the sequential lights transmitted by the series of holes and focussed them on a single selenium photocell which would then produce a succession of currents, each proportional to the intensity of the light on a different element of the image. Nipkow designed a receiver which used the magneto-optical light-modulator based upon the Faraday effect to vary the intensity of light in the reconstructed image. It depended on the property of flint glass to rotate its plane of polarisation of light in a magnetic field. A second disk at the receiver end was used to give the transmitted image. The receiving disk was identical to the transmitting disk and rotated in synchronisation with it. Unfortunately, Nipkow was unable to build his television, since the technology at the time was not up to this task.. His light-modulator alone would have required about 10 watts of control power. Also the selenium photocells were too sluggish in their response to the changes in light intensity. For lack of amplification, Nipkow's system was also too insensitive for practical results, so that Nipkow himself could do no more than to give lectures on it, yet Nipkow's disk proved to be the most durable mechanical scanning device invented. Nipkow also suggested the use of phonic wheels for the synchronisation of scanning in his transmitter and receiver. Although Nipkow's idea for mechanically scanned television was about

40 years before its time, he did live long enough to see it realised in the 1920's, in the hybrid electromechanical era of television.

The selenium cells were rather slow in responding to changes of light intensity. The Polish inventor, Szczepanik, attempted to improve this shortcoming of selenium with his patent of 1897, in which he proposed to construct the selenium cell in the form of a ring which rotated whilst a pair of oscillating mirrors scanned the image and projected it through an aperture onto the rotating selenium cell ring which exposed continually fresh surface of selenium to the aperture and, thereby, was capable of responding to higher frequencies. His receiver, however, was based on incorrect ideas, and it is doubtful that his apparatus was ever built.

A new photoelectric cell was invented and the first completely successful technological application of photo-electricity was made by the German physicist and mathematician, Arthur Korn (1870-1945), who was the first physicist to develop a technologically effective phototelegraphy in 1902. Korn devised a fundamental type of photoelectric cell for phototelegraphy based on the discovery of photoemission by Heinrich Rudolph Hertz (1857-1894) in 1887: the photoemissive cell, which consisted of a highly evacuated glass cell coated inside with an alkali metal except for a narrow 'window' through which the light could enter the cell. The cell contained a thin metal ring which was kept at a positive potential whilst the coating was kept at a negative potential. The light reaching the coating gave off electrons in proportion to its intensity, the resulting electron flow reached the metal ring, where the current created by the flow of electrons was amplified. In Korn's phototelegraph the still picture was scanned point-by-point by a photoelectric cell and the modulated current issued by the cell was transmitted to the receiver by wire where the light intensity of the picture was recovered by modulating a lamp which exposed a photosensitive paper or film rotating on a drum point-by-point and line–by–line. Korn's phototelegraphy is still in current use, which is a measure of its effectiveness. Beside photoemissive cells, photoconductive selenium cells are still used for switching services, and photovoltaic cells which are composed of 2 materials separated by a specially treated boundary, are used producing voltaic current when exposed to light. Photoelectric cells turn light into electricity, and sound into optical soundtracks and back into a number of other vital technical services in modern technology.

Another vital component of modern television, the Braun cathode-ray tube, was invented in 1897 by a German physicist, Karl Ferdinand Braun (1850-1918), professor of physics at the universities in Marburg, Karlsruhe, Tübingen and Strassburg, for research purposes. Braun used the gas discharge phenomena for the emission and formation of an electron beam just like in the early X-ray tubes which produced a relatively feeble charge of electrons. From 1903 to 1905 Braun's tube was improved by another German physicist, Arthur Rudolph Wehnelt (1871-1944), who in 1898 had built the electrolytic current breaker, by introducing a thermionic cathode consisting of the platinum filament coated with lime, the so-called 'Wehnelt oxide cathode', which markedly increased its sensitivity to electron emission at reasonable voltages. Within a few years the electron flow was accelerated by a tubular anode situated a few inches (5 cm) in front of the cathode which was maintained at 1,000 volts or more by positive direct-current. The charge produced by the output from a secondary high voltage transformer passed through a vacuum tube rectifier. The Braun-Wehnelt cathode-ray tube was first used as an electron beam oscilloscope to demonstrate the oscillations of electronic waves.

Since the use of Nipkow's disk as a mechanical scanner was too clumsy for television, a German physicist and former pupil of Braun, Max Dieckmann (1882-1960), was the first scientist to suggest the use of Braun-Wehnelt cathode-ray tube for the reception of pictures in television. In 1906 Dieckmann, together with his associate Gustav Glage (1882-?), built a system of facsimili-transmission of 20-line shadow pictures of metal patterns, in which the receiver was a genuine electronic cathode-ray tube receiver. Since at the time there were no means available for amplifying photoelectric currents, Dieckmann used Nipkow disk for the mechanical scanner with 20 contact brushes replacing 20 apertures on the disk. The Nipkow disk was mechanically coupled to a generator which supplied sawtooth current to the horizontal deflexion coils in the cathode-ray tube receiver and to the contact brush on a slide-wire potentiometer supplying the vertical deflexion current to the vertical coils. When the brushes on the Nipkow disk contacted a point on the metal pattern, they allowed the current to flow through the coiled electromagnets which, in turn, deflected the electron beam in the 'Braun-Wehnelt cathode-ray oscilloscope' to make it miss an aperture in its path. In this way, the conducting portions of the metal pattern were reproduced dark on a luminous background of the oscilloscope screen, thus creating a shadowgraph. Since a complete revolution, that is, the complete scan, was made in one-tenth of a second, the movement of the metal patterns could be observed on the cathode-ray oscilloscope screen, which made it into a television receiver. In 1906 Dieckmann received a patent for his electronic television receiver, the Braun-Wehnelt cathode-ray tube, which was a landmark in television technology.

In 1907, Boris Rosing (1869-1933), a Russian scientist of German stock who was professor of physics at the St. Petersburg Technical Institute, proposed a closed circuit television system which consisted of a mechanical transmitter and an electronic receiver. The mechanical scanner in his transmitter consisted of 2 octagonal mirror-drums of Weiller, rotating about 2 mutually perpendicular axes. The vertical cylinder rotating at high speed scanned the picture horizontally, whereas the horizontal cylinder rotating at a much slower speed raised and lowered the horizontal scanning lines. The optically focussed image was reflected from the drums through a hole in a screen on an inertia-free alkali phototube, which was much more sensitive than the selenium photocell and represented Rosing's most important improvement in television transmission that removed one major obstacle in the single channel television transmission. His receiver consisted of a Braun-Wehnelt Cathode-Ray oscilloscope tube fitted with condenser plates (between which the electron beam passed) that were connected in series with the alkali photoelectric cell and a battery. The deflexion of the electron beam was produced by the current induced in the pick-up coils by permanent magnets associated with the individual mirror faces of his octagonal mirror drums of Weiller. The mirror coils were connected in series with the deflexion coils in the Braun-Wehnelt cathode-ray tube receiver. In all other details Rosing's television receiver was similar to that of Dieckmann. However, since the signals from the alkali photocell were far too weak because of the lack of means for amplification, Rosing could not achieve practical transmission in his experimental apparatus which he built in 1910 with the assistance of his student Vladimir K. Zworykin and exhibited in St. Petersburg. However, Rosing's system was still, like the Dieckmann apparatus, a hybrid electromechanical television system, and it did not really function properly, since in 1911, he was only able to transmit primitive shadows of 3 or 4 lines.

After reading Shelford Bidwell's discussion of a high-definition, point-by-point, remote electric vision system using a multiwire cable of 90,000 conductors which was obviously far too complicated and expensive to be practical, the Scottish physicist Alan Archibald Campbell-Swinton (1862-1930), professor of physics at the London University, analysed the problem of electric vision at distance and proposed in his letter to the magazine Nature in 1908 the elimination of all mechanical devices in such a system by employing an entirely electric television system which used Braun-Wehnelt cathode-ray tubes both in the transmitter and in the receiver. He insisted that the only workable solution for television must be based on 'electronic scanning', and that the deflexion of the electron beam in the transmitting and the receiving cathode-ray tube, which is achieved by electromagnets worked by alternating currents of 2 greatly differing frequencies, was the only way to obtain high definition in the transmitted picture. In 1911, Campbell-Swinton gave further details of his all-electric television system. Compared to Dieckmann, the novel part of Campbell-Swinton television system was his transmitter, now called the 'camera tube'. It incorporated a mosaic screen consisting of mutually insulated photosensitive elements, such as rubidium cubes, on which the picture was optically projected, and scanned on the back side by a beam of cathode-rays controlled by the currents from the alternators. The cathode-ray beam in the receiver was synchronised with the cathode-ray beam in the transmitter by deflexion coils connected to the same alternators. The other side of the mosaic screen formed a wall of a receptacle filled with a gas such as sodium vapour, which conducts negative electricity more readily under the influence of light. This receptacle contained a storage screen for the negative charge. A conductor connecting this storage screen to the capacitor plates modulated the electron flow in the cathode-ray tube receiver, so that the quantity of the current collected by the storage screen determined the brightness of the scanning spot on the screen of the cathode-ray receiver tube. He regarded the great number of photoelectric cells in parallel, each one of which had to deliver one electric pulse once in a picture period of one-tenth of a second, to be the great advantage of his system. The parallel arrangement of the photosensitive elements in the photosensitive mosaic proposed by Campbell-Swinton indeed has proven to be an essential feature of television tubes with high-sensitivity. The introduction of high-vacuum cathode-ray tube into the television technology by Dieckmann and Rosing removed the slow response of photoelectric devices as a technical barrier. Another important principle in Campbell-Swinton's television system was the 'storage principle': the electric charge which produces the picture signal is not just the charge liberated by photoemission from the picture element at the instant of scanning, but rather the charge accumulated on the insulated photosensitivity mosaic during the entire period between two subsequent scans. The 'storage principle' in the storage device yields an increase in sensitivity of a system over a nonstorage device by a factor equal to the number of picture elements in the transmitted picture, or, approximately the square of the number of scanning lines. For modern television this factor lies between 100,000 and 1,000,000.

Campbell-Swinton carried out some experiments with cathode-ray tubes using metal plates coated with selenium as the photosensitive screen in the transmitter with some success. He published a new elaboration of his 1911 proposal in 1924, and in 1926 he gave an account of some television experiments he had performed. Campbell-Swinton died a few years before his system of electronic television was successfully demonstrated.

Campbell-Swinton's proposal for an entirely electronic television system ended the speculative era of television in which all proposals dealt with closed-circuit transmission. The transmission of television by broadcasting was a much later development.

Ever since 1887, the photoelectric vacuum tube based on the 'inertia-free photoemissive effect' was being developed in Germany by the versatile physicists Julius Elster (1854-1920) and Hans Friedrich Geitel (1855-1923). They not only constructed a transformer for Tesla current, but also carried out detailed investigations of the electric charges on atmospheric rain, of the conduction and ionisation by electricity in gases, of cathode-rays, and of the natural radioactivity that proved valuable to the research on atomic decay. Elster and Geitel also discovered in 1893 that the photoelectric current emitted by a potassium coating within a hydrogen vacuum is proportional to the intensity of the light that falls on the coating. On this discovery, Elster and Geitel based their design of a sensitive photoelectric photometer which was the first practical photoelectric cell capable of converting a flux of light into a strong electric current. By 1913 the rapid and sensitive potassium-hydride photoelectric cell of Elster and Geitel had been perfected, and was available for general practical applications.

In 1906, the electronic amplification of the strength of the electron-beam was made possible by the vacuum tube of 3 elements of Lee de Forest as well as by the 3-element cathode-ray relay tube incorporating a grid control of Robert von Lieben. In 1917 a light source permitting rapid modulation was invented by Daniel MacFarlan Moore.

The 4 inventions made before the First World War were all important components in a practical television system.

Mechanical Scanning Era of Television

By 1920, all the means necessary for the development of a practical television system were available.

An entirely electronic television was not considered possible in 1920 mostly because Dieckmann's cathode-ray tube had to rely on gas-focussing of the scanning beam which restricted the sharpness of the light spot of the electron stream and limited the permissible speed of deflexion of the electron beam. Moreover, such tubes also had a short life. The sharp and bright images under high scanning speeds became possible only after 1926, when the German physicist Hans Walter Busch (1884-1973) had shown that rotationally symmetric electromagnetic fields act on electrons as glass lenses act on light rays, and had laid the theoretical foundation for electron optics which could be applied to the focussing of electron beams in vacuum. For this reason only mechanical scanning methods were used both in the transmitter and in the receiver of any practical television system of the period. The mechanical scanning systems used in that period were the 'Nipkow disks', rotating mirror-drums and disk prisms, mirror screens, rotary commutators and other devices, all of which were developed to a high state of refinement.

The Scottish electrical engineer, John Logie Baird (1888-1946) became the first inventor to succeed in transmitting a picture of a human face by means of television in January, 1926, which made this imaginative Scot famous overnight. Baird began his experiments on television transmission with mechanical scanning in 1923, and early in 1925 he was able to transmit by means of the Nipkow disk transmitter, a tiny, shaky, pink image of a Maltese cross over a distance of 6 feet (1.83m). The light source in his receiver was the Moore negative-glow neon lamp and a gas-filled potassium photocell, and the frequency of his television pictures was 5 frames a second. Since his Nipkow disk had 20 holes, Baird's pictures were made of 20 vertical scanning lines, but the lighting and definition of his transmitted television image, which flickered, was of poor quality. In April, 1925, Baird gave the first public demonstration of his television. In 1927 he began operating the first television station in the world. It was located in London from which he transmitted his television programme over 12 miles (19.3 km) to Harrow on a wavelength of 200 meters. By that time Baird still had rather low definition in his television pictures produced with 30 scanning lines and 12½ frames a second frequency.

Almost at the same time as Baird, 4 other men were working on mechanically scanned television. The German physicist, August Karolus (1893-1972) who developed his own system of phototelegraphy for the *Telefunken* (*Wireless*) company, used mechanical scanning with the Nipkow disk and the 'Karolus cell', a photoelectric cell which transmitted light variations by means of the 'Kerr effect' based on optical double-refraction of certain materials in the electric field, to obtain good television transmission in 1925. Karolus employed electrically driven tuning forks to ensure synchronisation between his transmitter and receiver. His experimental apparatus was exhibited at the Wireless Exhibition in Berlin in 1929, and it gave the first practical television transmission in Germany. Later Karolus used the Weiller mirror-drum in mechanical scanning.

In Austria, the Hungarian engineer, Dénes von Mihály, had already in 1919 been able to transmit shadowgraphs electrically with the Braun-Wehnelt cathode-ray oscillograph tube. In 1923, Mihály experimented with a system in which scanning was done by means of oscillating mirrors in front of powerful magnets, but he failed to produce a practical television system with it. In the late 1920's, Mihály was able to build a mechanico-optical television system of 180 scanning lines of fairly high definition, but by that time mechanical scanning was not able to meet the competition provided by the all-electronic cathode-ray tube system.

In United States, the motion picture pioneer Charles Francis Jenkins (1867-1934), physicist Herbert Eugene Ives (1882-1953) of the Bell Telephone Laboratories, and the Swedish engineer Ernst Fredrik Warner Alexanderson (1878-1975) of General Electric also worked on television systems using mechanical scanning. Jenkins was mainly interested in showing motion pictures at home by means of television. Jenkins' mechanical scanning system consisted of a pair of bevel-edged glass disks in which the bevel changed continuously along the perimeter of the disks. The bevelled edges of the disk formed prisms which deflected the beam of light when the disk rotated. By rotating one disk several times faster than the other, the entire surface of the image could be scanned. It was able to transmit film but could only produce shadowgraphs of real objects. It was Jenkins who gave the second public demonstration of practical television in June,1925 by transmitting and receiving a shadowgraph of a slowly revolving model of a windmill by means of wireless, in which he used between 30 and 60 scanning lines in his picture. In 1928 Jenkins undertook experimental broadcasting of motion pictures, but bankruptcy forced him to abandon the field. Alexanderson who used the Weiller mirror-drum for scanning gave up his efforts before reaching the broadcasting stage.

Ives, who was trying to develop 2-way video telephony for Bell Telephone Company, used 50 scanning lines. In 1927, he transmitted television pictures over 25 miles (40 km) by wireless, and over 265 miles (330 km) by telephone line. Ives used the rapidly moving spotlight scanner similar to the one used by Baird, only he transmitted at the frequency of 17.7 frames a second. He also experimented with a colour television. Ives gave a public demonstration of his television on April, 1927, in which the receiver screen used a spirally moving neon tube with electrodes constituting a 2,500-element picture grid.

Upon learning of this accomplishment, Baird who recognised the great importance of publicity, transmitted his television pictures from Glasgow to London over the service telephone line in May, 1927, and sent his television signals from London to New York, and from London to the ocean liner Berengaria in the middle of the Atlantic ocean by wireless in February, 1928. In the same year Baird developed the first video-recording method on disks. In an effort to keep ahead of the competition, Baird demonstrated a crude mechanically scanned tricolour-television and stereoscopic television in August, 1928, and a large screen television consisting of a multicellular lamp screen of 2,100 flash-lamp bulbs in July, 1930.

Baird operated the first regular daily experimental television programme at **BBC** from 1929 to 1934. Initially, Baird's television had a definition of 30 lines and a frequency of 12½ frames a second. To increase the brightness of the picture, Baird began in 1932 to use the Weiller mirror-drum scanner working on the flying spotlight principle of von Ardenne by means of which he was able to scan 60 lines. In April, 1932 Baird used ultrashort-waves of 6.1-meter wavelength to broadcast his television programme. In 1931, Baird had applied for the priviiege of building a coast-to-coast television network in the United States, but his effort was frustrated by a court injunction requested by **RCA**, his American rival.

In his hectic activity, Baird established an amazing number of television 'firsts': Transmission of television pictures by telephone lines from London to Glasgow in 1927; telecast pictures of 30-line resolution by shortwave radio across the Atlantic and to a ship, Berengaria, at sea in 1928; daylight television transmission in 1928, video recording in 1928; regular television service, including the transmission of synchronising electric pulses, in 1929; simultaneous transmission of television pictures and sound in 1930; large-screen television in 1930; televising a daytime public event, the Epsom Derby, in 1931; televising of motion-picture film in 1931; and television transmission by ultrahigh frequency in 1932. These were the foremost achievements in electromechanical television.

Beginning in 1932, Baird's mechanically scanned television had to compete with an entirely electronic television system developed by Marconi-Electrical and Music Industries Ltd., which was based on an improved Braun-Wehnelt cathode-ray tube transmitters and receivers. Baird's mechanical system had already lost this competition by 1937, despite the fact that in 1934 Baird had been able to increase the number of his scanning lines to 240 by introducing a mirror screw-scanner of Okolicsanyi, a flying spotlight scanner working at a frequency of 25 frames a second. The mirror screw-scanner was large and yet small enough for a television camera, but it made the television camera so bulky, heavy and awkward that it had to be bolted to the floor during its operation. In comparison the **Marconi-EMI** television camera incorporating the electronic **Emitron** camera tube was light and mobile on its wheeled dolly. By 1938 Baird had come to the conclusion that mechanical scanning had no future because of the awkwardness and slowness of its equipment, and the low definition of its reproduced video picture. Only by means of an intermediate movie camera, which filmed the scene and then passed the wet film in front of his electromechanical television camera for transmission, was Baird able to send televised pictures which were a little better in definition than the pictures transmitted by the electronic television cameras of the period. Of course, the cathode-ray television tubes at the time were not yet sufficiently developed. Baird finally abandoned the mechanical scanning in television in 1936, and began work on an all-electronic television system under a licence from the Farnsworth Television Laboratory in Philadelphia. There were some difficulties in developing this electron-camera which had initially a 150-line definition of the image and no storage, and he only used it in its original form for a brief period in late 1936. In 1935, Baird was able to transmit 700-line images at 25 images a second with the redesigned electronic 'image dissector' television camera of Philo Farnsworth, which Baird had obtained in the summer of 1934 and improved its resolution. The shortest wavelength used in the radio-transmitters in 1935 and 1936 was 6.7 meters, which could only transmit a 240-line image at 25 images a second. That was a problem for Baird's high definition television. Even for a definition of 405 lines at 25 images a second a 1.5 meter wavelength is necessary.

Then Baird developed his own 2-colour, high-definition electronic television camera, called ***Telechrome***, which he patented in 1940 and demonstrated to the press in 1941. The ***Telechrome*** worked with 600 scanning lines and had 2 electron-guns, one for each colour, but it still had certain shortcomings which Baird was determined to remove by its further development. In 1944, Baird, after years of ingenious experiments, demonstrated the first high-definition, multi-electron beam shadow-mask television receiver, his 3D-colour ***Telechrome***. All modern television receivers are in fact developments of this triply interlaced 3D-colour receiver using either 2 or 3 electron-guns. In 1946, at the time of his death, Baird was working on a single electron-gun television receiver which was quite similar to the more recent ***Trinitron*** television receiver of the Sony Company.

By 1944, Baird had developed an ultra-fast television signalling system which transmitted 750,000 words a minute, or a page of text in one 25th of a second, using the all-electronic flying-spot scanner invented by von Ardenne in 1931. Three years later, in 1947, **RCA**, Baird's rival company in the United States, produced their 'ultra-fax' system, which was remarkably similar to Baird's fax system, without recognising Baird's priority in this field.

Baird's most outstanding achievement in signalling was his slow-scan television along metal telephone cables transmitting 30-line images at 12 ½ images a second which he was able to reduce to one image every 10 seconds. In this system, which was already in operating condition by 1927, Baird employed the first memory tubes ever used.

During World War II, Baird was consultant to the Cable and Wireless Company in London for whom he developed a system of speeded-up cable telegram signalling which was completely safe from enemy interception.

In 1940, Baird's company founded in 1927 became known as 'Rank *Cinetel* (*Cinema Television*)' which still supplies worldwide some of the best equipment for recording and replaying television programmes.

Baird made fundamental contributions to video-recording, modern pulsed radar, military night-vision, and ultra-secure secret television signalling either 'ultra-fast' or 'ultra-slow', in all of which he was a pioneer.

It was evident that the heydays of individual inventors such as Baird were rapidly coming to a close as collective research by teams of scientists and engineers took over the development of television. Baird, among all the pioneers of television, is mostly responsible for creating a general public interest in television. Right from the beginning Baird had recognised the vital importance of public promotion of television as a public medium, and he was equally capable as a promoter and as an inventor.

In Germany, Karolus of the *Telefunken* (*Wireless*) company had developed an experimental television equipment in 1926, the transmitter and receiver of which used the Nipkow scanning disks with 30-line resolution. In 1929 Karolus used the Weiller mirror-drum scanners in his experimental television equipment to demonstrate telecinema and television at the Wireless Exhibition in Berlin. In 1934 the German Post Office purchased from the *Fernseh* (*Television*) company a television transmitter working by the 'flying light-spot method', and began the first television transmission in Germany on ultrashort waves. It was a high-definition television system using 180 lines sequentially scanned. Although an experimental regular hour-long programme of silent films, and current events had been televised in 1929 by the middle-wave radio station in Witzleben, Berlin, the first regular television shortwave transmitter began its operations in 1934 at the same location. In 1935, the first public television station was opened in the Postal Museum in Berlin which used a Nipkow disk-scanner in its transmitter. It remained in service until 1944. By 1936 Berlin had a regular 2-hour television programme every evening. The Berlin Olympic Games in 1936 were televised in the same year since Berlin had 28 television saloons by the time of the Olympic Games.

A public television-phone service was opened between Berlin and Leipzig in 1937, and between Berlin and Munich in 1938. In 1939 the German Post Office gave television demonstrations in every large city in South America. Television service in Hamburg commenced in 1941, and in the late 1930's the commercial electronic television receivers began to make their appearance. Regular television programming in France commenced in 1937, in Russia in 1938, and in United States in 1939.

The electromechanical scanning era began slowly to give way to the 'electronic scanning era' from 1930 to 1938 during which time the cathode-ray tube underwent gradual improvements through the research and development done by scientific teams in United States, in England and in Germany.

However, it was in the mechanical scanning era that the technical requirements of television such as the number of scanning lines, field frequency, frame frequency, brightness for flicker-free transmission, and reception of video signals were determined. It was discovered that higher field frequencies admit flicker-free television at high picture brightnesses. Following the motion picture practice, the frame frequency was taken to be half of that of the field, and this was made feasible in the alternate fields by the interlacing method of scanning in which odd lines are scanned first in the picture and the even lines second with the resulting refinement in vertical definition. It was also discovered that the scanning speed and the frequency band necessary to transmit the video signals are determined by the product of the number of scanning lines and the least frame frequency. Also precise theoretical studies were carried out for the video reconstruction which set down the relationship between horizontal resolution and the frequency bandwidth of the transmission channel between the picture pickup and reproducer, and the effect of the size of scanning spot on the electronically reproduced picture. In addition, suitable methods of mechanical synchronisation were worked out.

By the time electronic methods in television became dominant, in the receivers about 1930 and in the transmitters about 1936, the basic problems of television had been laid bare in the electromechanical scanning period of television. It was Nipkow himself who emphasised in 1934 that much better scanning methods than his mechanical scanning disk have to be devised for improved brightness and resolution of the reproduced image, after he had observed the first experimental broadcasts of television which employed his scanning disks both for the transmission and the reception of video signals.

Electronic Scanning Era of Television

The Braun-Wehnelt cathode-ray tube was the fundamental component in the electronic distant vision and it had been introduced into electronic image reproduction by Dieckmann already in 1906 when he took out a patent in Germany on an electronic television system incorporating cathode-ray tubes for the transmission of images. In 1907, Boris Rosing in Russia used Dieckmann's cathode-ray receiver in his experimental television system which still relied on an electromechanical television camera incorporating 2 mirror-drum scanners of Weiller.

Dieckmann's cathode-ray tube employed gas focussing of the electron scanning beam which limited its spot sharpness and permissible speed of deflexion. The scanning spot was modulated by deflecting the electron beam across an aperture, and the electrons passing through the aperture were collected into a picture signal. The pioneering development of the television receiver and transmitter is due to the work of two young men in United States. An American physicist of Russian extraction, Vladimir Kozma Zworykin (1889-1982), a former pupil of Rosing in St. Petersburg, who earned his doctorate at the University of Pittsburgh in 1926, filed a patent for a cathode-ray receiver in 1923, which used an axially symmetric grid and the 'intensity modulation of the electron beam'.

Zworykin constructed an experimental television system in 1923 which was quite imperfect, and he was only able to transmit with it a cross-mark between his cathode-ray pick-up tube and a cathode-ray receiver, the first of its kind. Unfortunately, his transmitted picture was very crude and not much better than that of Rosing in 1911. Zworykin's tube incorporated a thin aluminium signal plate oxidised on one side for insulation. The insulated side was photosensitised and provided with a metal-grill facing which served as an electron storage. The picture to be transmitted was optically projected through the metal grill onto the photosensitive layer which acquired through photoemission a charge image. The electron beam issuing from an electron-gun in an oblique tube then scanned the signal plate on the same side and penetrated the insulating oxide-layer thereby creating a conducting path between the signal plate and the photosensitive element on the other side of the plate. Although the resulting transmitted pictures were poorly defined, Zworykin had demonstrated with this experimental tube the feasibility of Campbell-Swinton's all-electronic television system: the 'practical utility of electronic scanning' and the 'importance of the electronic storage principle'.

The electronic camera without storage is the so-called 'image-dissector'. This type of an electron cathode-ray transmitter was first developed by Max Dieckmann and his assistant Rudolf Hell (1901- ?) and patented in 1925. In the image-dissector the image of a picture is optically projected onto a photocathode and then deflected by orthogonal magnetic fields across an aperture in such a manner that the aperture picks in succession electrons from different picture elements along a scanning line, which pass through a minute aperture, and are collected to produce the picture signal. This transmitter of Dieckmann could only produce rather crude images in an electronic receiver. A young ingenious American farm boy from Iowa, Philo Taylor Farnsworth (1906-1971), who in 1922 as a teenager had conceived an all-electronic television system, working at first alone and in secret, but later was assisted by a few collaborators, invented an important improvement to Dieckmann's image-dissector. Farnsworth's improvement consisted of a longitudinal magnetic focussing field which formed a sharp electronic image of the picture in the plane of the aperture, an improvement he patented in 1927 and 1928. He was able to demonstrate a 150-line image resolution scanned 30 times a second. After 1930 Farnsworth replaced the electron collector by an 'electron multiplier', an ingenious device, acting as a noise-free amplifier, which gave the maximum sensitivity to a television camera operating with photocathodes. The electron multiplier tube, which Farnsworth called the 'multipactor tube', made use of the secondary electron emission, that is, the emission of electrons from the photosensitive material produced by the impinging high-velocity electron beam, for current amplification within the image-dissector tube when placed behind the image receiving aperture. The image-dissector tube was entirely linear in operation: its output signal was strictly proportional to the light falling on the photocathode. The 'electron multiplier tube' was a valuable invention because it had many other applications in electronics industry.

In the receiving cathode-ray tube, named the 'oscillight' by Farnsworth, the electron beam of which is focussed on the fluorescent screen by a short axial magnetic solenoid around the neck of the tube at the electron-gun. The horizontal scanning is accomplished by magnetic deflexion coils at right angles to the axis of the electron-gun and were actuated by the same sawtooth wave form and frequency as the current through corresponding deflexion coils in the 'image dissector tube'. The low picture frequency scanning was produced by a magnetic field orthogonal to the frequency displacement field. The low frequency field employed pole pieces of special shape.

Farnsworth was assisted in his work by Professor Carl J. Christensen of the University of Utah and for 2 years by Russell Harrison Varian (1898-1959), an electrochemist. Varian experimented with phosphors in an effort to find the materials and the technique which would give the most light in the receiving tube, and also examined different kinds of electronic wave-generators, known as 'oscillators', for television transmission. Varian's contributions proved valuable to Farnsworth's television system. Farnsworth was granted many patents in 1930, among which was a patent for the circuits by which the signals sent by the transmitter were synchronised at the receiver which kept the corresponding electronic scanners in step with each other. The electronic scanning and synchronisation invented by Farnsworth made modern television possible. Farnsworth's television system was the 'first successful operational electronic television system in the world', and it was first publicly demonstrated in July, 1929 – the 'birthday' of the all-electronic television. Farnsworth had already proven in September, 1927 that his all-electronic television system works.

In 1930, Manfred von Ardenne (born 1907), a self-educated German physicist, built a similar picture tube. The synchronising pulses to the control oscillators of the deflexion generator provided the correct timing. In 1931 Ardenne developed a cathode-ray tube with a flying-spot scanner In this tube a scanning pattern of uniform intensity on the face of a cathode-ray tube is imaged onto the subject to be transmitted and the light reflected from the subject, or passing through it, is collected by a phototube. The photocurrent is instantaneously proportional to the transmission, or reflexion, of the point being scanned and generates the picture signal. The time for the light emission of the phosphor phototube must be shorter than the time of transit of the picture element for the flying-spot scanner to be effective, because otherwise the scanning spot becomes a line. Therefore, these phototubes became practical only when efficient short-persistence phosphorus and highly sensitive multiplier tubes were available. Von Ardenne demonstrated the 'velocity modulation' method, in which the variation of light intensity at the receiver screen is obtained by varying the velocity of the scanning spot instead of its actual intensity. It had been vaguely described by Rosing in 1911, and then revived by R. Thun in 1929. In early 1934 von Ardenne gave the first public demonstration of his television in which the picture scanning in the transmitter was still done with the Nipkow disk.

Vladimir Zworykin of the Radio Corporation of America, known as **RCA**, the radio producing division of General Electric Company and Marconi Company in U.S.A., came to examine Farnsworth's electronic television in 1930. After returning to his laboratory, Zworykin, who had been working on a television system which used an oscillating mirror scanner, first suggested by the French engineer Maurice Leblanc (1851-1923) in 1880, in the transmitter and a cathode-ray tube as the receiver, eliminated the mechanical scanner and set to work on his own ideas of an all-electronic television system using the storage principle.

In 1931, Zworykin and his scientific team at **RCA** built the first partly effective storage-type television tube, which Zworykin named the *Iconoscope*. In this camera tube the picture was projected onto a mosaic screen faced with silver globules photosensitised with calcium, a screen which had also been used by Farnsworth in his 'image dissector'. The back of the mica screen was made into a metal signal plate. The picture to be transmitted was optically focussed onto the front of the mica screen which created an emission of electrons from the photosensitive elements and gave the globules positive charge relative to the signal plate. The electron beam from the cathode, representing the electron-gun in the oblique side-tube, then scanned the insulated side of the photosensitive mosaic thereby discharging the globules of the mosaic. This discharge caused a corresponding pulse of current to flow from the signal plate to the ground through a resistance, and the resulting potential change across the resistance constituted the television signal. The charge on the photosensitive mosaic was being stored up in accordance with the light falling on it from the optical image projected on it during the entire period between scans, which made the *Iconoscope* much more sensitive than the 'image dissector tube' of Farnsworth. The *Iconoscope* was satisfactory up to 500 scanning lines, and it was the first television camera tube, which had adequate definition for practical applications and, therefore, it opened up a new era of television technology. Regrettably, the *Iconoscope* tube still had a number of practical shortcomings: only 5% to 10% of the possible increase in sensitivity due to the charge storage was actually realised; the secondary electron emission created electrons which either fell back on the mosaic thereby partially neutralising its elements and thus reducing the tubes sensitivity, or were redistributed onto the adjacent elements of the mosaic by the braking field between the mosaic and the charge collector which created light and dark spots, and reduced the sharpness of the video picture by this spurious electron shading. *Iconoscope* tube also had trouble picking up clear pictures in dim light.

In 1933, British physicists A.G. Lubszynski and S. Rodda were able to improve the *Iconoscope* with their picture tube called *Emitron*, or *Super-Iconoscope*, a tube which was otherwise identical to the *Iconoscope* tube. In this pick-up tube there were 2 separated cathodes. In the *Super-Iconoscope* tube the optical image of the picture was focussed onto a transparent photosensitive cathode and the stream of photoelectrons was accelerated by a high voltage onto the anode, a separate nonphotosensitive mosaic storage plate. An electron beam from an oblique electron-gun scanned the mosaic storage plate. Owing to the high potential between the photocathode and the mosaic that released the 'secondary emission', a much stronger electronic image was obtained which yielded a much higher definition in the television picture, and increased the sensitivity of the *Iconoscope* about 10 times.

By 1934 Farnsworth had brought his television system to a high level of perfection, and he held over 300 U.S. and foreign patents in television, radio and electronics. In the same year he was invited to give demonstrations of his television system in London and in Berlin. He sold licenses to Baird and to another British company, as well as to a German company, for the right to use his patents in producing their own television systems. **RCA** and its wily Chairman David Sarnoff's devious legal attempt to appropriate Farnsworth's fundamental television patents in 1935 failed, and **RCA** was finally compelled to obtain a nonexclusive license from Farnsworth in 1939 for which **RCA** had to pay huge royalties. It was a very painful experience for **RCA**, since Sarnoff took great pride in collecting rather than paying royalties.

In 1949, Philo Farnsworth sold his patents to the International Telephone and Telegraph Company (known by its acronym, **IT&T**), and served as the head of the research laboratories of this company where he supervised work that at times involved certain aspects of nuclear physics.

Like Baird in 'electromechanical television', the ingenious Philo Farnsworth was the last great individual inventor in 'all-electronic television'. Beginning with Zworykin, invention in television technology became a collective enterprise carried out by teams of physicists, engineers and mathematicians.

A considerable improvement was introduced in 1939 to television camera tubes by the American physicists, Albert A. Rose and A. Iams of **RCA** with their *Orthicon* (*orthiconoscope*) tube. In this tube the photosensitive layer and the mosaic plate were also separated like in the *Super-Iconoscope* tube, but the *Orthicon* tube obviated the signal loss from the redistribution and the shading created by secondary emission, and its operation was nearly linear. In this tube the electrons in the scanning beam had low velocity which, to a great extent, avoided secondary electron emission in the mosaic plate so that no light or dark spots were produced in the transmitted picture.

However, the *Orthicon* tube exhibited instability at very high levels of frequency, yet required a minimum of light for its operation, but its long storage time left the detailed definition of very rapid motions in the transmitted video picture somewhat blurred.

A great improvement and a significant advance in television camera tubes was attained by **RCA** physicists Albert Rose, Paul K. Weimer and Harold B. Law in 1943, when they invented the *Image Orthicon*, or the *Super-Orthicon* tube which is still the basic universal camera tube in contemporary broadcast television. The *Super-Orthicon* tube incorporated the advantages of both the *Super-Iconoscope* and the *Orthicon* tube. In this complicated tube the separation of the charging and discharging sides of the mosaic, the projection of the electronic image and amplification of the signal through an electron multiplier were successfully accomplished. The instability and the low output signal level of the *Orthicon* tube were eliminated by such means, whilst the inherent advantages of low-velocity scanning were retained.

In the *Super-Orthicon* tube the optical image was projected onto a continuous, transparent photocathode, and the photoelectrons emitted by the photocathode are accelerated and focussed by a longitudinal magnetic field onto a thin glass target. The secondary electrons from the target were collected by a closely spaced, fine-mesh target screen. A positive charge image to the original picture on the photocathode was stored in the target. The opposite side of the target was scanned by a low-velocity electron beam which neutralised the stored charge whilst the conductivity between the two sides of the target resulted in a return of electron beam which was modulated with this discharge current. The secondary emission from the beam electrons turned back to the disk electrode of the electron gun by the positive part of the target were amplified in 5 stages by secondary emission multiplication by one electron multiplier which produces an output current 500 times as large as the incident return beam current. The varying component of this output current corresponded to the charge stored on the target by the photoelectrons and, therefore, constituted the picture signal. At low light levels the charge stored on the target and, therefore, the picture signal was directly proportional to the brightness of the picture. The *Super-Orthicon* tube which transmits a satisfactory picture of a scene at very low light levels and over an extraordinarily wide range of illumination is most commonly used in all studio cameras because it is about 100 times as sensitive as any other camera tube. The *Super-Orthicon* tube is a large and complicated tube requiring auxiliary circuits and numerous adjustments, and it is restricted to a narrow range of temperature for its optimal operation. Moreover, this tube tends to create halos round very strong lights in the scene. It is also difficult and expensive to manufacture.

In 1950, American physicists Paul K. Weimer, Stanley V. Forgue and Robert R. Goodrich introduced a photoconductive television camera tube called the *Vidicon* tube, which differed from the *Super-Orthicon* tube by being small, simple and rugged. The construction of the *Vidicon* tube is quite similar to that of the *Orthicon* tube, but it functions by photoconduction rather than by photoemission. The target in the *Vidicon* tube is a thin homogeneous semiconducting photosensitive selenium layer deposited on a transparent signal plate which is positively biased with respect to the cathode of the electron-gun (from 15 to 50 volts). In the dark the photoconductive layer behaves as an insulator. The low-velocity electron beam sweeps over its back surface and drives it to the cathode potential, depositing just enough electrons to compensate for the current which has leaked through the photosensitive layer in the preceding frame period. Whilst the target is illuminated, the current flow from the front surface to the back surface increases, since the resistance of the layer is reduced by light and as a consequence, the number of electrons deposited by the scanning beam on the target increases in proportion, and this increase induces current pulses in the signal plate lead which constitute the picture signal. The return current, like in the *Orthicon* tube, is not used. The *Vidicon* tube, unlike the *Orthicon* tube, exhibits no stability problems, and it has high sensitivity because photoconductive films can be given quantum efficiency (ratios of the number of electrons transferred to the number of light quanta or photons incident on the target) close to unity, unlike the photoemissive targets the quantum efficiencies of which are about one-fifth. In general, the *Vidicon* tube is free of disturbance effects, and it is outstandingly useful for the scanning of motion pictures run by the ordinary movie projector. The *Vidicon* tube also uses a fraction of the current required for its electron-beam deflexion, which makes it suitable for miniaturisation.

Television technology has found wide use in television microscopes, industrial closed circuit television, medical practices, biology, radiology, remote control, industrial supervision, astronomy, aeronautics, astronautics, military uses, and meteorology.

Television Broadcasting

Wireless telephoning was carried on very long electromagnetic waves. For instance, the giant transmitter at Bordeaux, France, operated in the 1920's on a wavelength of 27,000 meters and the transmitter in Nauen, Germany, operated on a wavelength of 13,000 meters. The medium wavelengths became popular after 1918. At present, the long-wave stations transmit on the 1,500-meter wave-band. The short-waves have wavelengths below 60 meters. In 1925, the new electronic circuits invented before 1920 made it possible to transmit even under the 10-meter wavelength, such waves are classified as 'ultrashort waves'.

German physicist Abraham Esau (1884-1955) carried out experiments in 1925, and he was successful in telegraphing over 25 miles (40 km) on a wavelength of 3 meters, the first successful experiment on ultrashort waves, which have enormously large frequencies. For instance, electromagnetic waves with a wavelength of one meter oscillate 300 million times a second. This property of the 'ultrahigh frequency' waves permits a large number of broadcasting stations to transmit over a very short wavelength without interference. About 3,000 ultrashort wave transmitters could operate between the 9-meter and the 10-meter band. John Logie Baird was the first to transmit his television broadcast in 1932 by the ultrashort, 6.1-meter wavelength. The German television transmitter in Berlin-Witzleben began broadcasting by ultrashort waves in 1934.

It was soon discovered that ultrashort electromagnetic waves travel as far as a beam of light, yet sometimes the signals of ultrashort electromagnetic waves were received thousands of miles away by being reflected from the Kennelly-Heaviside ionised layer located between 60 and 70 miles (97 and 172 km) above the surface of the Earth. It was also discovered that ultrashort electromagnetic waves were split by mountain ridges and other solid objects, which led to the concept of focussing such electromagnetic waves as streams of particles by means of reflectors. The shorter the wave the sharper they could be focussed. It became possible to concentrate electromagnetic waves into 'directional beams' by focussing the ultrashort waves.

The discovery of the usefulness of the ultrashort waves led to a virtual international contest to produce shorter and shorter electromagnetic waves. In 1938, American scientists were able to transmit on an 8 millimeter wavelength, whereas Esau was able to produce a 4.4 millimeter wavelength. In 1939, Esau was able to reduce the wavelength to 2.2 millimeters. Soon the wavelengths were reduced so drastically that the gap between electromagnetic waves and heat radiation was closed. Russian physicist Maria A. Levitska (1883-1963) was able to produce an electromagnetic wavelength of 30 microns by means of a surface transmitter, whereas the wavelengths of 750 microns were achieved for infrared heat radiation. In 1922 the microwave band of wavelengths from 30 microns to 750 microns was overlapped from both sides by another Russian physicist Alexandra A. Giagoleva-Arkadyeva (1884-1945).

Microwave Communication

Microwaves which represent a form of electromagnetic radiation quite similar to light radiation but having a considerably longer wavelength possess some of the properties of both the radio and the infrared waves. Unfortunately, microwaves were difficult to generate by electronic circuits.

In 1951, the American physicist Charles Hard Townes (born 1916) had a brilliant idea: the molecules of some substance could be made into microwave generators by exciting the molecules to a higher energy state and then allowing the molecules to return to their original energy state that would result in the emission of molecular energy in the form of radiation which could be stored. This stored radiation could be used to stimulate other molecules to emit radiation of the same wavelength in a cascade reaction which produces a highly amplified beam of microwave radiation consisting of a series of superimposed microwaves with equal frequency. Townes designed a microwave generator using ammonia molecules which was a huge practical success, since it could be used in long distance communication because the amplified microwave beam could not only be accurately directed at a distant receiver, but it could also be used to amplify minute radio signals. It is called by its acronym

maser (***m****icrowave **a**mplification **b**y **s**timulated **e**mission of **r**adiation*). In 1957 Townes suggested the use of the same kind of technique to amplify light-waves. An American physicist Theodore Maiman (born 1927) took up this challenge in 1960 and built the first optical maser, called by its acronym ***laser*** (***l****ight-wave **a**mplification **b**y **s**timulated **e**mission of **r**adiation*) by exciting the molecules of a synthetic ruby crystal with a flash-tube and reflecting the molecules of the ruby crystal back-and-forth between mirrors which amplified it into an intense beam of red light, the most intense light ever produced. Since laser beams can be aimed accurately over enormous distances with negligible diffusion, lasers have become important in communication systems, as well as in precision welding, drilling of heat resistant materials, and highly accurate methods of measurement. Laser technology has led to new navigation systems and medical practices.

Lasers are also used to create 3-dimensional images by a method called 'holography'. Holography, or 'interference photography', records visual information in terms of phase and amplitude of a 'highly coherent and phased' light such as the light from the laser beam. The object-wave itself is recorded in holography rather than in the optically formed image, usually on the photographic plate, in such a way that an illumination of this record, called a 'hologram', made by exposing a photographic plate to the 'intersecting' laser lights reflected from the subject and from the reference source reconstructs the original object-wave itself. A visual observation of the reconstructed object wave-front gives a photographic view of the object which is for all practical purposes indiscernible from the original, including the 3-dimensional parallax effects.

The 2-step procedure of recording the interference pattern of 2 intersecting laser light-waves photographically, and then reconstructing the original object wave-front by illuminating the recording was invented in 1948 by Professor Dennis Gabor in England.

REFERENCES and SELECT BIBLIOGRAPHY

The best that can be done here apart of giving virtually an interminable list of works which touch on some aspects of the topics covered in this panoramic overview of the history of technology and engineering is to refer the reader to the monumental encyclopædic work of seven volumes on the history of technology which contains many further references on multifarious topics of technology:

< *A History of Technology* >,
Oxford at the University Press, 1978:
Volume I - *From Early Times to Fall of Ancient Empires*, Edited by Charles Singer, E. J. Holmyard and A. R. Hall,
Volume II - *The Mediterranean Civilizations and the Middle Ages, c.700 B C. to c.A.D.1500*,Edited by Charles Singer, E. J. Holmyard, A. R. Hall and Trevor I. Williams,
Volume III - *From the Renaissance to the Industrial Revolution, c.1750 to* 1850,Edited by Charles Singer, E. J. Holmyard, A. R. Hall and Trevor I. Williams,
Volume IV - *The Industrial Revolution, c.1750 to c.1850,* Edited by Charles Singer, E. J. Holmyard, A. R. Hall and Trevor I. Williams,
Volume V - *The Late Nineteenth Century, c.1850 to c.1900,* Edited by Charles Singer, E. J. Holmyard, A. R. Hall and Trevor I. Williams,
Volume VI - *The Twentieth Century, c.1900 to c.1950*, Part I, Edited by Trevor I. Williams,
Volume VII - *The Twentieth Century, c.1900 to c.1950,* Part II, Edited by Trevor I. Williams.

Another encyclopædic work in several volumes (French original has five volumes) on the history of technology and invention worthy of attention is:
Maurice Daumas (Editor): < *A History of Technology and Invention* >,
Crown Publishers Inc., New York,1968.

Additional Select Bibliography:

Duhem, Pierre, M. M.: < *The Origines of Statics.* The Sources of Physical Theory >,
Kluwer Academic Publishers, 1991.

Duhem, Pierre, M. M.: < *The Evolution of Mechanics* >,
Sijthoff & Noordhoff, 1980.

Kline, Morris: < *Mathematical Thought from Ancient to Modern Times* >,
Oxford University Press, New York, 1972.

Pedersen, Olaf and Pihl, Mogens: < *Early Physics and Astronomy.* A Historical Introduction >,
MacDonald and Janes, London & American Elsevier Inc., New York, 1974.

Clagett, Marshall: < *The Science of Mechanics in the Middle Ages* >,
University of Wisconsin Press, 1960.

Clagett, Marshall: < *Greek Science in Antiquity* >,
Abelard-Schuman, Inc., 1955.

Crombie, Alistair C.: < *Augustine to Galileo* >,
        Vol. I : *Science in the Middle Ages, V-XIII Centuries,*
        Vol. II: *Science in the Later Middle Ages and Early Modern Times, XIII-XVII Centuries.*
Mercury Books, 1961.

De Camp, L. Sprague: < *The Ancient Engineers* >,
Doubleday & Company, 1963.

De Camp, L. Sprague: < *Great Cities of the Ancient World* >,
Doubleday & Company, 1972.

Klemm, Friedrich: < *A History of Western Technology* >,
Charles Scribner's Sons, 1959.

Straub, Hans: < *A History of Civil Engineering*, An Outline from Ancient to Modern Times >,
Leonard Hill, Ltd., 1952.

Forbes, Robert J. : < *Studies in Ancient Technology* >, 7 volumes,
E. J. Brill, 1955-1963.

Aitchison, L.: < *A History of Metals* >, 2 volumes,
Macdonald & Evans, 1960.

Finch, James Kip: < *The Story of Engineering* >,
Anchor Book, Doubleday & Company, 1960.

Gille, Bertrand: < *Engineers of the Renaissance* >,
The MIT Press, 1966.

Forbes, Robert J.: < *Notes on the History of Ancient Roads and their Construction* >,
Adolf M. Hakkert, 1964.

Rolt, Lionel T. C.: < *A Short History of Machine Tools* >,
MIT Press, 1965.

Stodola, Aurel : < *Steam and Gas Turbines* >, 2 volumes,
McGraw-Hill, 1927.

Dunsheath, Percy: < *A History of Electrical Engineering* >,
Faber & Faber, 1962.

Strandh, Sigvard: < *A History of the Machine* >,
A. & W. Publishers Inc., 1979.

Dijksterhuis, Edward J. & Forbes Robert J. < *A History of Science and Technology* >,
Vol. I - *Ancient Times to Seventeenth Century*,
Vol. II -*The Eighteenth and Nineteenth Centuries*,
Penguin Books Ltd, 1963.

Sandström, Gösta E.: < *The History of Tunnelling* >,
Barrie and Rockliff, 1963.

Gibbs-Smith, Charles H.: < *Flight Through the Ages* >,
Thomas Y. Crowell Company, Inc., 1974.

Dennis, W. H.: < *A Hundred Years of Metallurgy* >,
Duckworth, 1959.

Ashby, Thomas: < *The Aqueducts of Ancient Rome* >,
Clarendon Press, Oxford, 1935.

Gest, Alexander P.: < *Engineering*. Our Debt to Greece and Rome >,
Cooper Square Publishers, Inc., 1963.

Reynolds, Terry S.: < *Stronger Than Hundred Men* >,
The Johns Hopkins University Press, 1983.

Kilon, I.: < *The Evolution of the Heat Engine* >,
Longman, 1973.

Rolt, Lionel T.C.: < *Great Engineers* >,
G. Bell and Sons, Ltd, 1962.

Rolt, Lionel T. C.: < *Navigable Waterways* >,
Longman, 1969.

Rolt, Lionel T. C.: < *From Sea to Sea.* The Canal du Midi >,
Allen Lane, 1973.

Rolt, Lionel T. C.: < *Victorian Engineering* >,
Penguin Books, 1970.

Galuzzi Paolo (Editor):< *Leonardo da Vinci. Engineer and Architect* >,
The Montreal Museum of Fine Arts, 1987.

von Braun,Wernher & Ordway,Frederick I.:< *History of Rocketry & Space Travel* >,Rev. Ed.,
Thomas Y. Crowell Company, 1969.

Ucceli, Arturo: < *Storia della Tecnica dal Medio Evo ai nostri Giorni* (History of the Technics
    from the Mediaeval Age to Our Time) >, 2nd Edition,
Milan, 1941.

White, Jr., Lynn: < *Medieval Religion and Technology*. Collected Essays >,
University of California Press, 1978.

Gimpel, Jean: < *The Medieval Machine.* The Industrial Revolution of the Middle Ages >,
Holt, Rinehart and Winston, 1976.

Gimpel, Jean: < *The Cathedral Builders* >,
Harper Colophon Books, Harper & Row, Publishers, 1983.

Whittaker, Edmund T.: < *A History of the Theories of Aether and Electricity* >,
    Vol. I : *The Classical Theories,*
    Vol. II: *The Modern Theories ,1900 -1926,*
Thomas Nelson and Sons Ltd, 1953.

Graf von Klinckowstroem, Carl: < *Knaurs Geschichte der Technik* (*Knaurs History of Technic*) >,
Droemersche Verlagsanstalt Th. Knaurs Nachfolger, 1959.

Dijksterhuis, Eduard J.: < *The Mechanization of the World Picture* >,
Oxford at the Clarendon Press, 1961.

Finch, James Kip: < *Engineering and Western Civilization* >,
McGraw-Hill Book Company, Inc., 1951.

Sandström, Gösta E.: < *Man the Builder* >,
McGraw-Hill Book Company, 1970.

Smiles, Samuel: < *Industrial Biography.* Iron Workers and Tool Makers >,
Introduction by L.T.C. Rolt,
Augustus M. Kelley - Publishers, 1968.

Smiles, Samuel:<Selections from *Lives of the Engineers* with an Account of Their
    Principal Works>,
The M.I.T. Press, 1966.

Burstall, Aubrey F.: < *A History of Mechanical Engineering* >,
Faber and Faber, 1963.

Kirby, R.S. ; Withington, S. ; Darling A.B. & Kilgour, F. G.: < *Engineering in History* >,
McGraw-Hill Book Company, 1956.

Kirby, Richard S. & Laurson, Philip G.: < *The Early Years of Modern Civil Engineering* >,
Yale University Press, 1932.

Calvani, Vittoria: < *Lost Cities* >,
Editions Minerva S. A., 1976.

Neuburger, Albert: < *The Technical Arts and Sciences of the Ancients* >,
Methuen & Co, Ltd, 1969.

Forbes, Robert J.: < *The Conquest of Nature*. Technology and Its Consequencies >,
A Mentor Book, The New American Library, 1968.

Rouse, Hunter & Ince, Simon: < *History of Hydraulics* >,
Iowa Institute of Hydraulic Research, State University of Iowa, 1957.

Goldstine, Herman H.: < *The Computer* from Pascal to von Neumann >,
Princeton University Press, 1972.

Drachmann, Aage G.: < *The Mechanical Technology of Greek and Roman Antiquity* >,
The University of Wisconsin Press, 1963.

Bird, Anthony: < *Roads and Vehicles* >,
Longmans, 1969.

Forbes, Robert J.: < *Man the Maker.* A History of Technology and Engineering >,
Abelard-Schuman Limited, 1958.

Whitney, Charles S.: < *Bridges.* Their Art, Science & Evolution >,
Crown Publishers, Inc., 1983.

Hodges, Henry: < *Technology in the Ancient World* >,
Alfred A. Knopf, 1970.

Smith, David E.: < *History of Mathematics* >,
        Vol. I - *General Survey of the History of Elementary Mathematics*,
Dover Publications, Inc., 1958.

Heath, Thomas L. : < *A Manual of Greek Mathematics* >,
Dover Publications, Inc., 1963.

Heath, Thomas L. : < *A History of Greek Mathematics* >,
        Vol. I - *From Thales to Euclid*,
        Vol. II - *From Aristarchus to Diophantus*,
Dover Publications, Inc., 1981.

Lampl, Paul: < Cities and Planning in *The Ancient Near East* >,
George Braziller, 1968.

De Camp, L. Sprague: < *The Heroic Age of American Invention* >,
Doubleday & Company, Inc., 1961.

Oliver, John W.: < *History of American Technology* >,
The Ronald Press Company, 1956.

Hall, A. R.: < *Ballistics in the Seventeenth Century* >,
Cambridge at the University Press, 1952.

McCloy, Shelby T.: < *French Inventions of the Eighteenth Century* >,
University of Kentucky Press, 1952.

Lanczos, Cornelius : < *Space Through the Ages.* The Evolution of Geometrical Ideas from
 Pythagoras to Hilbert and Einstein >,
Academic Press, 1970.

Kline, Morris: < *Mathematics for the Nonmathematician* >,
Dover Publications, Inc., 1967.

Hofmann, Joseph E.: < *Classical Mathematics* . A Concise History of the Classical Era in
 Mathematics >,
Philosophical Library, 1959.

Boyer, Carl B.: < *A History of Mathematics* >,
John Wiley & Sons, Inc., 1968.

Struik, Dirk J.: < *A Concise History of Mathematics* >, 4th Revised Edition,
Dover Publications, Inc., 1987.

Struik, Dirk J.: < *The Land of Stevin and Huygens* >,
D. Reidel Publishing Company, 1981.

Sarton, George: < *A History of Science.* Ancient Science Through the Golden Age of Greece >,
Harvard University Press, 1966.

Feldhaus, Franz M.: < *Die Maschine im Leben der Völker.* Ein Überblick von der Urzeit bis zur
 Renaissance (*Machines in the Life of Peoples. An Overview from Primitive
 Times to Renaissance*) >,
Birkhäuser Verlag AG., 1954.

Feldhaus, Franz M.: < ***Lexikon*** der Erfindungen und Entdeckungen auf den Gebieten der Naturwissenschaften und Technik in chronologischer Übersicht mit Personen- und Sachregister (***Lexicon*** *of Inventions and Discoveries in the Field of Natural Sciences and Technics in Chronological Overview with the Index of Persons and Subjects*) >,
Carl Winter's Universitätsbuchhandlung in Heidelberg, 1904.

Rolt, Lionel T.C.: < ***The Mechanicals*** >,
Heinemann, 1967.

Rolt, Lionel T.C.: < ***Thomas Newcomen.*** The Prehistory of the Steam Engine >,
MacDonald, 1963.

Cadogan, Gerald: < ***Palaces*** of Minoan Crete >,
Barrie & Jenkins, 1976.

Straub, Hans: < ***Die Geschichte der Bauingenieurkunst.*** Ein Überblick von der Antike bis in die Neuzeit (*History of the Art of Structural Engineering. An Overview from Antiquity to the Present Time*) >, 2nd Revised Edition,
Birkhäuser Verlag, 1964.

Landels, J. G.: < ***Engineering in Ancient World*** >,
University of California Press, 1978.

Zammatio, Carlo ; Marinoni, Augusto & Brizio, Anna M.: < ***Leonardo the Scientist*** >,
McGraw-Hill Book Company, 1980.

Heydenreich, Ludwig H. ; Dibner, Bern & Reti, Ladislao : < ***Leonardo the Inventor*** >,
McGraw-Hill Book Company, 1980.

Sarton, George: <***A History of Science.*** Hellenistic Science and Culture in the Last Three Centuries B.C. >,
John Wiley & Sons, Inc., 1959.

Sarton, George: < ***Six Wings.*** Men of Science in the Renaissance >,
The Bodley Head, 1958.

Pacey, Arnold : <***The Maze of Ingenuity.*** Ideas and Idealism in the Development of Technology>,
Allen Lane, 1974

Vitruvius: <***The Ten Books on Architecture*** >,
Dover Publications, Inc., 1960.

Greaves, W. F. & Carpenter, J. H.: < *A Short History of Mechanical Engineering* >,
Longmans, 1969.

Norrie, Charles M.: < *Bridging the Years*. A Short History of British Civil Engineering >,
Edward Arnold (Publishers) Ltd, 1956.

Hill, Donald R.: < *A History of Engineering in Classical and Medieval Times* >,
Open Court Publishing Company, 1984.

White, Jr., Lynn : < *Medieval Technology and Social Change* >,
Oxford Paperbacks, Oxford University Press, 1962.

Boyer, Carl B.: <*The Historical Development of the Calculus and its Conceptual Development*>,
Dover Publications, Inc., 1949.

Carnot, Sadi : < *Reflections on the Motive Power of Fire* >,
Dover Publications, Inc.,1960.

Everitt, C.W.F.: < *James Clerk Maxwell* >,
Charles Scribner's Sons, 1975.

Rolt, Lionel T.C.: < *Isambard Kingdom Brunel* >,
Penguin Books, 1957.

Rolt, Lionel T.C.: < *Thomas Telford* >,
Longmans, 1958.

Steinman, David B. & Watson, Sara R.: < *Bridges and Their Builders* >,
Dover Publications Inc. 1957.

Armytage, W.W.G.: < *A Social History of Engineering* >,
Faber and Faber, 1961.

Planck, Max : < *A Survey of Physical Theory* >,
Dover Publications, Inc., 1960.

Gadol, Joan : < *Leon Battista Alberti.* Universal Man of the Early Renaissance >,
The University of Chicago Press, 1969.

Prager, Frank & Scaglia, Gustina: < *Brunelleschi .* Studies of His Technology and Inventions >,
The MIT Press, 1970.

Wheeler, Lynde P.: < *Josiah Willard Gibbs.* The History of a Great Mind >, 2nd Edition,
Yale University Press, 1952.

Wheeler, L. P.; Waters, E. O. & Dudley, S. W.: < The Early Works of ***Willard Gibbs*** in Applied Mechanics >,
Henry Schuman, 1947.

Rukeyser, Muriel : < ***Willard Gibbs*** > ,
Doubleday, Doran & Company, 1942.

Wilson, Mitchell: < ***American Science and Invention***. A Pictorial History >,
Simon and Schuster, 1954.

Sarton, George: < Appreciation of ***Ancient and Medieval Science During the Renaissance*** >,
A Perpetual Book, A. S. Barnes & Company, Inc., 1955.

Sarton, George: < ***The Study of the History of Mathematics*** >,
and
< ***The Study of the History of Science*** > ,
Dover Publications, Inc., 1957.

De Bono, Edward (Editor): < ***Eureka !*** An Illustrated History of Inventions from the Wheel To the Computer >,
Holt, Rinehart and Winston, 1974.

Street, Arthur & Alexander, William: < ***Metals in the Service of Man*** >, 5th Edition,
Penguin Books, 1972.

Turner, Roland & Goulden, Steven L.(Editors): < ***Great Engineers and Pioneers of Technology*** >,
Vol. I : From Antiquity through the Industrial Revolution,
St. Martin's Press, 1981.

Brumbaugh, Robert S. < ***Ancient Greek Gadgets and Machines*** >,
Thomas Y. Crowell Company, 1966.

Drummer, G. W. < ***Electronic Inventions and Discoveries, 1745-1976*** >, 2nd Edition,
Pergamon, 1978.

Heiberg, Johan L. < ***Mathematics and Physical Science in Classical* Antiquity** >,
Oxford University Press, 1922.

Davey, Norman : < ***A History of Building Materials*** >,
Drake Publishers Ltd., 1971.

Cohen, Morris R. & Drabkin, I.E. : < ***A Source Book in Greek Science*** >,
Harvard University Press, 1948.

Alberti, Leon B. : < *On Painting* > , Revised Edition,
Yale University Press, 1966.

Oravas, Gunhard Æ. & McLean, L.: < *Historical Development of Energetical Principles in Elastomechanics* >
Part I, August, No.8, pp.647-658,
Part II, November, No.11, pp.919-932,
Applied Mechanics Reviews, vol. 19, 1966.

Oravas, Gunhard Æ. : Essay Review,< *Geschichte der Mechanischen Prinzipien* und ihrer wichtigsten Anwendungen (*History of the Mechanical Principles* and Their Most Important Applications) >
by István Szabó, Birkhäuser Verlag, Basel and Stuttgart,1977,
Historia Mathematica, vol.9, 1982, pp.81-94.

INDEX

Abel, Niels Henrik, 613
Abelard, Pierre, 318-320, 323, 343, 346, 417
Acheson, Edward Goodrich, 842
Ackeret, Jakob, 1015, 1017
Ackermann, Rudolph, 981
Adelard [alias, *Æthelard*] of Bath, 345, 346, 350
Agatharchos of Samos, 105, 106
Ageladas of Argos, 117
Agricola, Georgius [alias, Georg Bauer], 451
Agrippa, Marcus Vispanus, 231, 235-237, 239, 240, 244, 248, 256
Aiken, Howard, 574
Aischylos of Eleusis, 75, 105, 121
Alban, Ernst, 789, 794, 798, 801, 860
Albert von Bollstadt [alias, *Albertus Magnus*], 321, 322, 324, 350-352, 355, 404, 447
Albert von Rückmersdorf [alias Albert of Saxony], 379, 381, 384, 385, 392
Alberti, Leon Battista, 260, 441, 451, 454-459, 461, 463, 467, 495
Alcuin of York [alias, *Flaccus Albinus*], 298-300, 314, 318
Alexander of Hales, 322, 345
Alexanderson, Ernst Frederik Werner, 1096-1098, 1103, 1104, 1129
Alexandros III of Macedon [alias, Alexander the Great], 17, 112, 153, 155, 175, 176, 178-182, 184, 508
Alexandros of Aphrodisias, 272
Al-Haytham, 349
Alhazen, 349
Alighieri, Dante, 444, 479, 480
Alkmaion of Kroton, 95, 96, 132
Allen, John, 845, 850
Alphand, Jean Charles Adolphe, 899
Amen-Hotep, son of Hapu, 34, 35, 38
Ammanati Battiferri da Settignano, Bartolomeo [alias, *Ammanati*], 473
Amontons, Guillaume, 564, 593, 603, 689, 693
Ampère, André Marie, 670, 677, 709-711, 719, 721, 1040
Anacharsis of Scythia, 110
Anaxagoras of Klazomenai, 79, 103-107, 109, 121, 124, 126, 128, 131, 132, 134, 142, 504
Anaximandros [in English, Anaximander] of Miletos, 78, 89-91, 102, 108, 151, 484
Anaximenes of Miletos, 90, 91, 103, 106, 108, 109

INDEX

Andronikos of Kyrrhestes, 209
Anselm, 317, 319
Anthemios of Tralles, 286, 287
Antiphon of Athens, 127, 148, 149
Antisthenes of Athens, 130
Antz, Hans Martin, 1027
Apelles of Kolophon, xii, 174, 176
Apollodorus of Damascus, 245, 246, 248-250
Apollodoros, 106
Apollonios of Perge, xiii, 153, 187, 203, 204, 205, 264, 265, 275, 278, 287, 288, 433, 473, 477, 518, 538, 540
Appleton, Edward Victor, 1115
Aquinas, Thomas, 322-326, 330-332, 350-352, 355, 358, 362-365, 380, 390, 397, 415, 424, 435, 479
Arago, Jean Dominique François, 711, 715, 720, 1057, 1058
Archimedes of Syracuse, xiii, 146, 148, 188, 191-205, 204, 205, 207, 208, 210, 213-216, 253, 264, 275, 288, 359, 370, 391, 433, 455, 460, 473, 492, 494, 496, 505-507, 516, 526, 528, 537, 539, 542-544, 555, 597, 613, 761, 1044, 1101
Archytas of Taras, xiii, 134-139, 141-143, 145, 147, 151, 153, 169, 188, 189, 191, 195, 203, 213-215, 292, 359, 386, 390, 484, 535, 590, 760, 782, 845, 996, 1015, 1017, 1035
Ardenne, Manfred von, 1130, 1135, 1136
Aristaios of Kroton, 153, 204
Aristarchos of Samos, xii, xiii, 189, 190, 199, 202, 206, 208, 265, 487, 488
Aristippos of Kyrene, 130
Aristotle [real name, *Aristoteles*] of Stageira, xix, xxxii, xxxv, xxxvii, 74, 76, 82-84, 95, 118-120, 130-133, 135, 145, 147, 152, 154-171, 183, 187, 194, 208, 209, 217, 270, 273, 278, 281-283, 292, 293, 316, 317, 319, 321-327, 329, 330, 336, 339, 341, 346-348, 350-352, 355-357, 362-372, 378-384, 386, 392, 393, 424, 425, 433-436, 451, 479, 485, 489-493, 504, 528, 529, 576
Aristoxenos of Taras, xii, 155
Arkwright, Richard, 749, 750
Armati, Salvino degli, 354, 399
Armengaud, René, 1013, 1014
Armstrong, William George, 815
Armstrong, Edwin Howard, 1098, 1100, 1102, 1105, 1107-1111, 1118
Arnault of Zwolle, Henri, 407, 466
Arnold, Harold Deforest, 1101-1103, 1104
Arnold, Engelbert, 1064
Arnold, John, 555
Arnold of Westphalia, 460

INDEX

Artachaies, 71
Atkinson, L.B., 1123
Auer, Freiherr [Baron] von Welsbach, Karl, 967, 1072
Aurelius Augustinus [alias, St. Augustine, Bishop of Hippo], 315-317, 322-326, 330, 332, 348, 349, 448
Austin, Lord, 982
Autolykos of Pitane, 152, 265
Autrecourt, Nicolas de, 358
Auzout, Adrien, 622
Averlino, Antonio Francesco [alias, *Il Filarete*], 461
Avery, William, 997, 998
Ayrton, 1121, 1122

Babbage, Charles, 570-575, 818, 1045, 1057, 1058
Babbage, Benjamin Herschel, 887
Bacon, Roger, 343, 347, 352-354, 363, 378, 380, 386, 397, 531, 761
Bailey, Walter, 1058
Bain, Alexander, 946, 947, 1120
Baird, John Logie, 1115, 1128-1132, 1137, 1139
Baker, Benjamin, 809, 894, 900, 929
Bakewell, Frederick Collier, 1120
Baldwin, Frank Stephen, 569
Baliani, Giovanni Battista, 523, 524
Ballo, Giuseppe, 492
Bar-le-Ducin, Errard de, 624
Barber, John, 966, 996, 1010
Bardeen, John, 941, 1113, 1114
Barlow, William A., 702, 929
Barlow, Peter, 836, 893, 894, 1050
Barnett, William, 970, 973
Barrow, Isaac, 543, 584
Barsanti, Eugenio Niccolò, 970
Bartholomew the Horologist, 410
Bartlett, Thomas, 895
Baudot, J. M. Émile, 952
Bauersfeld, Walther, 1017
Bayfius, 451
Bayton, Henry, 692
Bazalgette, Joseph William, 899
Bazin, Henri Émile, 637, 649
Beach, Alfred Ely, 894

INDEX

Beaufoy, Mark, Col., 854
Becquerel, Antoine César, 716, 953, 954, 1119
Becquerel, Alexandre Edmond, 738, 739, 954, 1042, 1119
Becquerel, Henri Antoine, 738, 1119
Bede, The Venerable [alias, *Beda Venerabilis*], 296-299
Beeckman, Isaac, 464, 497, 517, 522, 549, 560, 561, 592, 614
Behrend, Bernhard, 1065
Belanger, Jean Baptiste, 646
Belgrand, Marie François Eugene, 898
Belidor, Bernard Forest de, 626, 627, 628, 638, 646, 653, 660, 661, 666, 668, 675, 747, 803, 807
Bell, Alexander Graham, 962-964, 1084, 1094, 1095, 1121
Bell, Henry, 858
Benedetti, Giovanni Battista, 209, 491, 492, 528, 543
Bennett, Abraham, 705
Bentele, Max, 1023, 1029, 1032
Bentham, Jeremy, 827
Bentham, Samuel, 816, 827
Bentley, Edward M., 1054, 1055
Benz, Carl, 976, 978, 980-982, 984
Bérard, Jacques Étienne, 695
Berenger of Tours, 316, 317, 319
Bergès, Aristide, 1081
Bergman, Torbern Olof, 757, 907
Berkeley, George, Bishop, 677
Berkley, J. J., 890
Berliner, Emil, 964
Berlioz, Auguste, 1042
Bernard Sylvester, 318
Bernard of Chartres, 342, 343
Bernard of Clairvaux, 309, 310, 320, 398, 446
Berneval, Alexandre de, 416
Bernoulli, Jacob [in English, James], 464, 589-592, 595, 598, 601, 608, 681
Bernoulli, Johann [in English, John], 371, 464, 561, 567, 592, 595-597, 601, 613, 642, 649, 668
Bernoulli, Daniel, 567, 592, 597-601, 604, 608, 611-613, 638, 641, 670, 846, 850
Bernoulli, Nikolaus, 597
Bernoulli, Johann Jr., 724
Berthoud, Ferdinand, 554, 555
Berthoud, Pierre Louis, 555
Bertot, Jean Henri, 687

INDEX

Berzelius, Jöns Jakob, 758, 1119
Bessel, Friedrich, 189
Bessemer, Henry, 759, 908-914, 921, 928, 930, 998
Besson, Jacques, 451, 627
Bétancourt, Augustin de, 685
Betz, Albert Johann, 1017, 1020, 1026
Beyer, Johann Matthäus, 782
Bezold, Friedrich Wilhelm von, 729
Bidone, Giorgio, 646,
Bidwell, Shelford, 1122, 1126
Binet, Jacques, 679
Bion, Nicolas, 623
Birckbeck, George, 1068
Biringuccio, Vannoccio, 451
Black, Harold S., 1111
Black, Joseph, 774
Blanc, Honoré Le, 826, 828
Blanchard, Jean François, 650
Blanchard, Thomas, 821
Blenkinsop, John, 863-865
Bloch, Felix, 1112
Blondel, François, 619
Blondel, André Eugène, 1064
Boccaccio, Giovanni, 479, 481
Bocket, Adrien, 860, 992
Bodmer, Johann Georg, 750, 818, 832
Boethius, Ancius Manlius Severinus, 292-296, 298, 319, 341, 348, 366, 409
Böhm-Bahwerk, Eugen von, 425
Boistard, 669
Bollée, Amédeé, 650, 880, 980
Bollée, Léon, 980
Bolten, Werner von, 1072
Bolzano, Bernardus Placidus Johann, 199, 507, 525, 678
Boltzmann, Ludwig, xxviii
Bombelli, Raffaello, 476, 477
Bond, William C., 821
Bonrepos, Pierre Paul Riquet de, 622, 625
Boole, George, 576, 577
Borda, Charles Jean de, 606, 612, 634, 642, 643, 645, 646, 652, 689
Born, Max, iv

INDEX

Borough, William, 702
Borro, Girolamo, 526
Bosch, Robert, 983, 994
Boscovich, Ruggiero Giuseppe, 468, 648
Bossut, Charles, 634, 637, 641, 644, 645, 652
Bouch, Thomas, 928
Bouchon, Basile, 828, 830
Bouguer, Pierre, 854
Boulton, Matthew, 775, 778, 779, 783, 807, 813, 816, 855, 864, 895, 974, 993
Bourne, William, 844
Bourseul, Charles, 961
Bouton, George Thadée, 979, 986
Boyden, Uriah Atherton, 655, 656
Boyle, Robert, 102, 547, 548, 551, 557, 561, 566, 607, 766
Bozek, Josef, 872
Brabant, Siger de, 321
Bradley, James, 488, 489
Bradley, Charles, 1058, 1060
Bradwardine, Thomas, 200, 366-372, 376, 377, 379, 380, 384-386, 525, 535
Brahe, Tyge [alias, *Tycho Brahe*], 154, 490, 491, 498, 502, 503, 508, 519
Braithwaite, John, 869
Bramah, Joseph, 814-818, 827, 883
Branca, Giovanni, 451, 763, 781, 996-998, 1010
Brand, Henning, 757
Brandt, Alfred, 898
Brandt, Georg, 757
Branly, Edouard Eugene, 1083, 1086
Brassey, Thomas, 889
Brattain, Walter Houser, 941, 1113, 1114
Braun, Wernher von, 1032, 1038
Braun, Karl Ferdinand, 1083, 1089-1093, 1099, 1118, 1124-1126, 1129
Brayton, George B., 988
Brearley, Harry, 758
Breguet, Louis, 971
Bréquet, Alfred N., 1069
Bresse, Jacques Antoine, 687
Brett brothers, Jacob and John Watkins, 958, 959, 1043
Bright, Charles Tilson, 959
Brighton, Henry, 770
Brindley, James, 746, 747, 770, 805, 810

INDEX

Brouncker, William, 551
Brotherhood, Peter, 797
Brown, Joseph Rogers, 820, 841, 843
Brown, Charles Eugene Lancelot, 860, 1050, 1061, 1062, 1079
Brown, Samuel, Capt., 674, 866, 969, 970
Brown, Charles, 798, 860, 1061
Browne, Thomas, 700
Brunel, Isambard Kingdom, 723, 817, 820, 854, 882, 884-889, 893, 950, 960
Brunel, Marc Isambard, 675, 816, 827, 828, 833, 834, 854, 859, 882, 885, 891-893
Brunelleschi, Filippo, 249, 453-455, 460, 467
Bruno, Giordano, 520
Brunton, John, 887
Brunton, William, 836, 864, 873
Brush, Charles Francis, 1055, 1070, 1076
Brustlein, Henri, 837
Bryson of Herakleia, 127
Bubares, 71
Büchi, Alfred J., 993, 1012, 1030
Buffon, George Louis LeClerc, Comte de, 630, 634, 638
Bull, Edward, 780
Bullet, Pierre, 660
Bunsen, Robert Wilhelm Eberhard, 954, 955, 971, 979, 982, 1045, 1067
Burdin, Claude, 653, 1010
Bürgi, Joost [alias, *Iustus Byrgius*], 515, 516, 530, 531
Bürgin, Emil, 1046
Buridan, Jean, 379, 381-385, 524, 581, 615, 616
Burr, Thomas, 814
Burstall, Timothy, 869, 872, 881
Burton, John, 751, 789
Busch, Hans Walter, 1128
Busemann, Adolf, 1026, 1033
Bushnell, David, 848, 854, 855, 857
Butler, Edward, 980

Cabeo, Niccolo, 701, 716
Caesar, Gaius Iulius, 210, 225-227, 229, 232, 236, 243, 297
Callaud, A., 955
Calley (alias, Cawley), 769
Calvin, John [alias, Jean Caulvin], 448
Campbell-Swinton, Alan Archibald, 1126, 1127, 1134

INDEX

Camus, Charles Étienne Louis, 563
Canton, John, 705
Cantor, Georg Ferdinand Ludwig Philip, 578
Capella, Martianus Minero Felix, 292, 295, 296
Cardano, Girolamo, 470, 471, 475-478, 699, 761
Carey, George R., 1121, 1122
Carlisle, Anthony, 708, 709
Carnot, Lazare Nicolas Marguerite, 634, 643, 653, 676, 688, 689, 968
Carnot, Sadi Nicolas Léonard, 688-695, 698, 990, 991
Carré, François Edouard, 1069
Carson, John Renshaw, 1106, 1107
Cartwright, Edmund, 751, 789
Carus, Titus Lucretius, 103, 108
Casali, Giovanni di, 386, 616
Cascoigne, William, 622, 814
Caselli, Giovanni, 1120, 1122
Cassini, Giovanni Domenico [alias, Jean Dominique], 152, 533, 625, 626
Cassiodorus Senator, Flavius Magnus Aurelius, 292, 294-296, 298
Castelli, Benedetto, 541, 542, 564
Castiglione, Baldassare, 313
Castner, Hamilton Young, 935
Cauchy, Augustin Louis, 197, 199, 374, 507, 613-615, 674-676, 678, 681-684, 685-688
Cautley, Proby Thomas, Col., 900
Cavalieri, Francesco Bonaventura, 197, 371, 388, 507, 522, 534-537, 542, 544, 583
Cavendish, Henry, 705, 706, 712
Caxton, William, 450
Cayley, George, 580, 606, 969, 985
Cecil, William, 969
Celaya, Juan de, 392
Cellini, Benvenuto, 439
Cessart, Louis Alexandre de, 667
Chang Heng, 429
Chanute, Octave, 986
Chaperon, Georges, 956
Chapman, William, 864
Chappe, Claude, 946
Chares of Lindos, 184, 185
Charlemagne [alias, *Carolus Magnus*], 299-301, 342, 394, 417, 760
Charles, Jacques Alexandre César, 650

INDEX

Chatelier, Henri Louis Le, 929, 930
Chersiphron of Knossos, 113
Chézy, Antoine de, 631, 637, 640, 641
Chiao Wei-Yo, 427
Chladni, Ernst Florens Friedrich, 561, 671
Chryses of Alexandria, 287
Chu Yu, 429
Chuquet, Nicolas, 474
Cicero, Marcus Tullius, xvii, xxii, 83, 133, 210, 225, 226, 233, 272, 284, 315, 330, 333, 480
Clairaut, Alexis Claude, 494
Clapeyron, Émile Benoît Paul, 685-687, 695
Clark, Edwin, 883
Clarke, Edward M., 1041, 1053
Clarke, J. Latimer, 957
Clarke, Uriah, 1052
Claude, George, 1073
Clausius, Rudolf, 695
Clemens, Titus Flavius, 315
Clement, Joseph, 571, 817, 818, 838
Clement, William, 552
Clément, Nicolas, 689
Clerk, Dugald, 978, 993, 1012
Cochrane, Thomas, 892-894
Cockcroft, John Douglas, 740
Cockerill, John, 970
Colbert, Jean Baptiste, 555, 556
Colechurch, Peter, 418
Collin, Alexandre, 663, 664
Colmar, Charles Thomas de, 569
Colt, Samuel, Col., 821, 833
Colton, G. Q., Dr., 1052
Columbus, Christopher, 268
Combes, Charles, 698
Condorcet, Antoine Marquis de, 643
Congreve, William, Col., 1035
Constant, Hayne, 1031
Constantinus, Flavius Valerius Aurelius [Roman Emperor 'Constantine the Great'], 246, 252, 261-263, 272, 284, 285
Cook, James, Capt., 553

INDEX

Cooke, William Fothergill, 949, 950
Coolidge, William David, 1073, 1102
Cooper, Theodore, 929
Copernicus, Nicolaus, 189, 389, 474, 475, 487-491, 499-501, 520, 790
Cordes, James Jamieson, 997
Coriolis, Gustave Gaspard de, 612, 659, 683, 687
Corliss, George Henry, 795, 796
Cornell, Ezra, 957
Cort, Henry, 754-757, 922
Coster, Laurenz Janszoon, 449, 450
Cotes, Roger, 703
Cotton, Arthur Thomas, Col., 900
Couch, Jonathan, 895
Coulomb, Charles Augustin de, 633, 661-663, 668-669, 674, 681, 686, 706, 709-711, 740
Count Rumford [alias, Sir Benjamin Thompson], 635, 684, 778
Countess of Lovelace [alias, Byron, Augusta Ada], 573, 574
Couplet de Tartereaux [alias, Tartreaux], Pierre, 625, 633, 638, 660, 661, 667-669
Cowles brothers, Eugene H. and Emil A., 935
Cowper, Edward Alfred, 920, 921
Crampton, Thomas Russell, 958, 959
Cranage brothers, Thomas and George, 754
Craponne, Adam de, 620-621
Crassus, Appius Claudius, 233, 235, 236
Crompton, Rookes Evelyn Bell, 798, 1046
Crompton, Samuel, 750
Cronstedt, Axel Fredrik, 757
Crookes, William, 735, 736, 739, 1071, 1082, 1083
Cruickshanks, William, 953
Cubitt, William, 810, 890, 891
Cugnot, Nicholas Joseph, 782, 783, 826, 846
Cunard, Samuel, 859
Curie, Irene, 740
Curie, Marie, 739
Curie, Pierre, 739
Curtis, Charles Gordon, 658, 1007, 1008, 1013
Curtiss, Glenn, 986
Cuthbertson, John, 706

d'Agnolo, Donato [alias, *Bramante*], 467
d'Alameida, José, 957

INDEX

d'Alembert, Jean Le Rond, 592, 595, 599, 607, 611, 643, 670, 679
d'Auxiron, J. B., 847
Daft, Leo, 1054, 1055
Dahlén, Nils Gustaf, 1011
Daimler, Gottlieb, 975, 976, 978-984
Daimler, Paul, 983
Dance, Charles, 875
Danforth, Charles, 751
Daniell, John Frederic, 954, 1068
Danisy, Augustine, 668
Dante Alighieri, 444, 479, 480
Darby, Abraham, 752, 753, 756, 906
Darby II, Abraham, 753
Darby III, Abraham, 753
Darcy, Henri Philibert Gaspard, 637, 664, 665
Darios, Emperor of Persia, 69-72, 98, 183
Davenport, Thomas, 838, 1051-1053
Davidson, Robert, 1052, 1053
Davies, Daniel, 1041, 1053
Davy, Humphry, 708, 714, 758, 871, 934, 935, 953, 954, 1067
Day, Joseph, 978
Dedekind, Richard, 148, 374, 578
Deinokrates of Rhodos, 112, 174, 179, 180
Deinostratos of Alopekonnesos, 153, 154
Delaroche, Francis, 695
Deleuil, Joseph, 1067
DeMeritens, 1048
Demetrios of Ephesos, 113
Demetrios of Phaleron, 156, 182
Demokritos of Abdera, xxvi, xxxii, 83, 95, 105-109, 122, 127, 131-134, 143, 148, 154, 161, 170, 171, 195, 196, 343, 358, 371, 374, 536
Demoustier, Pierre Antoine, 630, 631
De Paiva, 1121
DePoele, Charles Joseph van, 1054, 1055
Deprez, François Daniels Marcel, 1048, 1058, 1074, 1075, 1077
Derand, François, 619
Desaguliers, Jean [or, John] Théophile, 703, 768
Desargues, Gérard, 510, 522, 537, 538
Descartes de Perron, René, 213, 276, 360, 364, 388, 475, 477, 494, 497, 521, 522, 534, 536-538, 540, 541, 545, 548, 556, 558, 561, 562, 567, 575, 582, 584, 586-588, 615, 682

INDEX

Deville, Henri Etienne Sainte-Claire, 934, 935, 942
Dieckmann, Max, 1092, 1125-1128, 1133, 1134
Diels, Hermann, iii
Diesel, Rudolf Christian Carl, 694, 860, 861, 989-995, 1006, 1019, 1021
Dietrich [alias, *Theodoric*] of Freiberg, 364
Dietrich, Günther, 1033
Dietz, Jean Christian, 880
Dietz, Charles, 880
Dietz, Christian, 880
Digges, Leonard, 531
Digges, Thomas, 531
Dikaiarchos of Messina, 203
Diokles, 275, 287
Dion, Albert Comte de, 979, 986
Diophantos of Alexandria, 273-275, 278, 361, 362, 433
Dirichlet, Peter Gustav Lejeune, 614, 670
Doble, William Abner, 657
Dolbear, Amos Emerson, 963, 1083, 1084
Doliwo-Dobrowolski, Michael von, 1061-1063, 1079
Domitius Ulpianus, 329
Dondi, Giovanni di, 411, 469
Dondi, Jacopo di, 411
Donkin, Bryan, 572, 814, 835, 836
Donovan, 972
Drake, Dr., 970
Droz, Pierre Jaquet, 829
Drummond, Thomas, 1042, 1068
Du Buat [alias, Dubuat], Pierre Louis Georges, Compte, 641
Du Petit-Vandin, Robert Xavier Ansart, 638, 639
Duboscq, Jules, 1068
Duddell, William Du Bois, 1094, 1095
Dudley, Dud, 752
Dufay [alias, Du Fay], Charles François de Cisternay, 703
Duhamel du Monceau, Henri Louis, 633
Duhem, Pierre Marie Maurice, iii, iv, v, 728
Dullaert, Jean, 392
Dumbleton, John, 366, 372, 377, 379, 380, 385, 386, 616
Duncan, Louis, 1063
Dunlop, John Boyd, 984
Duns Scotus, John, 324-328, 356, 364, 377, 380
Dunwoody, Henry Harrison Chace, 1090

INDEX

Dupin, Charles, 685
Dupuit, 637
Durant, William, 1012
Dürer, Albrecht, 455, 456, 475, 482, 495
Dyar, Harrison Edward, 947, 950
Dyckhoff, Friedrich, 860, 992

Eads, James Buchanan, 929
Earl of Stanhope, Charles, 576
Earnshaw, Thomas, 555
Edgar, Graham, 987
Edison, Thomas Alva, xxvii, xxxiii, 187, 737, 740, 741, 924, 952, 956, 963, 964, 967, 1048, 1049, 1053-1055, 1057, 1071-1073, 1076, 1077, 1080, 1084, 1099
Edmundson, T., 1051
Egbertus, 416
Ehrenhaft, Felix, 737
Eiffel, Alexandre Gustave, 185
Einhard [alias, *Eginhard*], 299
Einstein, Albert, xxvi, 728, 73, 733, 738
Ekphantos of Syracuse, 96
Elder, John, 797, 860
Elhuyar, brothers Don Fausto and Don José, de, 758
Elias, P., 1043, 1052
Elling, Ægidius, 1010, 1032
Ellyot, William, 903
Elster, Julius, 732, 739, 1127
Empedokles of Akragas, 101-106, 109, 123, 143, 154, 160, 186
Encke, Walter, 1017, 1020, 1023, 1029
Englesson, Olov, 658
Epikouros of Samos, 108, 343
Erasmus, Desiderius, 433, 442, 443
Eratosthenes of Kyrene, xiii, 146, 152, 195, 200-203, 208, 211, 268, 622
Ercker, Lazarus, 451
Ericsson, John, 854, 869, 985
Esau, Abraham, 1139, 1140
Espenschied, Lloyd, 1105
Eudemos of Rhodos, 155
Eudokios of Askalon, 288
Eudoxos of Knidos, 90, 97, 128, 131, 132, 136, 141-153, 163, 186, 190, 195, 196,

INDEX

201, 202, 204, 207, 208, 264, 275, 298, 338, 367, 369, 373, 494, 506, 542, 613, 676

Eukleides [in English, Euclid] of Athens, 125, 158, 185-187, 194, 199, 204, 215, 265, 275, 277, 278, 281, 288, 293, 338, 346, 349, 361, 433, 474, 486, 509, 614

Euler, Johann Albrecht, 612, 642

Euler, Leonhard, 361, 388, 567, 577, 592, 595, 596, 598-615, 636, 659, 670, 676, 681, 682, 684, 718, 723, 997

Eupalinos of Megara, 114

Euripides of Salamis, 103, 106, 121, 699

Eustachio, Bartholommeo, 95

Evans, Arthur John, iii

Evans, Oliver, 784, 785, 787-789, 792-794, 799, 800, 803, 826, 832, 849, 852, 853, 856-997, 1111

Fahrenheit, Gabriel Daniel, 533
Fairbairn, William, 800, 810, 818, 882-884
Falcon, Henri, 828, 830
Faraday, Michael, 711, 714-719, 721, 723-727, 733, 734, 737, 758, 873, 914, 949, 954, 958, 971, 1040, 1050, 1057, 1058, 1123
Farmer, Moses Gerrish, 952, 1044, 1053, 1069
Farnsworth, Philo Taylor, 1131, 1134-1137
Faure, Camille A., 956
Favre, Louis, 898
Feddersen, Berend Wilhelm, 722
Feldhaus, Franz Maria, iii
Fellenius, Wolmar Knut Axel, 664
Fermat, Pierre Siméon de, 283, 362, 388, 477, 522, 534, 536, 538-540, 549, 556, 584
Ferranti, Sebastian Ziani de, 943, 1049, 1050
Ferrari, Lodovico, 476
Ferraris, Galileo, 1058-1061, 1063
Ferrié, Gustave Auguste, 1097
Féry, Charles, 957
Ferrieres, Etienne de, 419
Ferro, Scipione del, 475
Fessenden, Reginald Aubrey, 1095-1099, 1101, 1102, 1111, 1115
Feuerlein, Otto, 1072
Ficino, Marcilio, 433
Field, Cyrus, 888, 959, 1054
Field, Joshua, 816, 875, 888
Fieschi, Sinibald, 332

INDEX

Fieviez, Charles, 733
Fioravanti, Ridolfo [alias, *Aristotile*], 460, 461, 481
Fiore, Antonio Maria, 475
Fischer, Johann Conrad, 914
Fitch, John, 784, 800, 846, 848-850, 854, 856
FitzGerald, George Francis, 733, 736
Fizeau, Armand Hyppolyte Louis, 720
Flamsteed, John, 533
Fleming, Sandford, 890
Fleming, John Ambrose, 741, 1087, 1088, 1099, 1100, 1102
Focq, Nicolas, 815
Follenay, Count Charles M. de, 847, 848
Fontaine, Hippolyte, 1045, 1046, 1074, 1075
Fontaine-Baron, Pierre Lucien, 656
Fontana, Carlo, 598
Fontana, Domenico, 468
Fontana, Niccolo [alias, *Tartaglia*], 451, 475-477, 484-486, 489, 491, 519
Fontana, Giacomo, 408, 1035
Ford, Henry, 982
Ford, Richard, 752
Forest, Lee de, 1100-1102, 1128
Forest, Fernand, 977
Forgue, Stanley V., 1138
Forq, Nicholas, 625
Foucault, Jean Bernard Léon, 719, 720, 721, 1067, 1068
Fourier, Jean Baptiste Joseph, 635, 670, 671, 679, 683
Fourneyron, Benôit, 654-656
Fowle, Joseph, 895, 896
Fowler, John, 890, 894, 929
Fox, Charles, 890, 891
Fox, James, 818
Fracastoro, Girolamo, 614
Fra Giocondo [alias, Monsignori, Giovanni of Verona], 467, 618
Francesca, Piero della, 455
Francis, James Bichens, 656, 657, 658, 1000
Français, 663
Franklin, Charles Samuel, 1089, 1101
Franklin, Benjamin, 353, 644, 704, 705, 714, 722, 775, 850, 1050
Franz, Anselm, 1029-1031
Frege, Friedrich Ludwig Gottlob, 578, 579
Fresnel, Augustin Jean, 611, 680-682

INDEX

Frézier, Amadée François, 668
Friedrich, Rudolf, iv, 1025, 1027, 1029
Frisius, Gemma, 534, 544, 548, 550
Fritts [alias, Fritz], C.E., 1091
Froissart, Jean, 1035
Froment, Gustave, 1045, 1052, 1053
Frontinus, Iulius Sextus, 227, 234, 235
Froude, William, 644, 926, 961
Fulton, Robert, 751, 846, 849, 854-858
Fusoris, Jean, 407

Gabor, Dennis, 1141
Gabriel I, Jacques, 689, 629
Gabriel II, Jacques, 629
Gaddi, Taddeo, 251, 472, 629
Gadroy, Col., 660, 661
Gaede, Wolfgang, 1102
Gahn, Johann Gottlieb, 757
Galbraith, 895
Gale, Leonard Dunnell, 950
Galenos [in English, Galen] of Pergamon, Klaudios, 271, 339, 350
Galileo Galilei, xxiv, 213, 360, 374, 380, 387, 472, 490, 492, 494, 496-498, 502,
    508, 511, 513, 518-532, 534, 536, 541, 542, 549, 564, 565, 593, 619, 627, 634,
    701, 765
Gallmayr, Joseph, 829
Galvani, Luigi, 707
Gassend [alias, Gassendi], Pierre, 522, 541
Gassner, C., 957
Gaulard, Lucien, 1077
Gauss, Carl Friedrich, 613, 615, 671, 672, 683, 687, 948
Gauthey, Emiland Marie, 621, 631-633, 661, 663, 669, 670, 673, 675
Gautier, Hubert, 626, 627, 667
Gautier, Joseph, 846
Gay-Lussac, Louis Joseph, 650, 711
Geissler, Heinrich, 734
Geitel, Hans Friedrich, 732, 739, 1127
Geminos of Rhodos, 264, 265
Geoffrey, Claude François, 757
Gerard of Brussels, 340, 362, 370
Gerbert of Aurillac, 341, 342, 395, 396, 409, 530

INDEX

Gergonne, Joseph Diez, 684
Germain, Sophie, 671, 672
Gerstner, Franz Joseph von, 645
Gesner, Abraham, 988
Ghiberti, Lorenzo, 453
Giagoleva-Arkadyeva, Alexandra A., 1140
Gibbs, Josiah Willard, iv-vii, 609, 681, 695, 930, 1100
Gibbs, John Dixon, 1077
Giddy [alias, Gilbert], David, 692, 779
Giesenberg, 977
Giffard, Henri Jacques, 696-698, 801, 802
Gilbert, William, 397, 398, 471, 472, 502, 504, 699-702
Gilchrist, Percy Carlisle, 927
Giles of Rome, 397
Giorgio Martini, Francesco di, 461, 462, 466
Girard, Pierre Simon, 636, 640, 641, 675, 805
Girard, Louis Dominique, 656, 698
Gisborne, Frederick Newton, 959
Gislebertus, 412, 416
Glage, Gustav, 1092, 1125
Glaukos of Chios, 114
Göbel, Heinrich, 1071
Goddard, Robert Hutchings, 1036, 1037, 1039, 1098, 1101
Gödel, Kurt, 579
Goethal, George Washington, 40
Goethe, Johann Wolfgang, xxii-xxiv, xxxvii, 711
Goldschmidt, Rudolph, 1103
Goldschmidt, Hans, 942
Goldstein, Eugen, 735-737
Goodrich, Robert R., 1138
Göransson, Göran Fredrik, 913
Gordon, James Edward Henry, 1049
Gordon, Andrew, 715
Gordon, David, 873
Görges, Johannes, 1066
Gorgias of Leontini, 119, 123, 130, 316, 317
Gorrie, John, 794
Graef, Rudolf, 931
Graham, George, 552, 553
Gramme, Zénobe Théophile, 838, 839, 1042, 1045-1048, 1069, 1074, 1075
Grassi, Orazio, 519

INDEX

Gratianus, 243
Gray, Elisha, 962, 963
Gray, John McFarlane, 960
Gray, Stephen, 703
Greathead, James Henry, 893-895
Green, George, 682
Gregor, William, 758
Gregory of Tours, 296
Gresham, Thomas, 387, 487
Gribeauval, Jean Baptiste de, 826
Griffith, Alan Arnold, 1017, 1018, 1031
Griffith, Julius, 880
Grimaldi, Francesco Maria, 561
Grimaldi, P. M., 781
Groot, Jan Cornets de, 493, 496
Groot, Hugo de [alias, *Grotius*], 493
Grosseteste, Robert, 130, 321, 327, 346-352, 355, 358, 363, 364, 366, 371, 503, 518
Grove, William Robert, 954, 1053, 1067
Gudea, 12, 17
Guericke, Otto von, 545-547, 555, 556, 702, 765
Guest, James J., 843
Guidi, Tomasso, 455
Guidobaldo [alias, Guidobaldi] del Monte, 451
Guillaume de Conches, 343
Guillaume, Maxime, 1016
Guillaume, Charles Edouard, 932
Guillaume de Champeaux, 318, 319
Guldin, Paul, 277, 535
Güldner, Hugo, 993
Gundermann, Wilhelm, 1021
Gundissaunus, Dominicus, 396
Gurney, Goldsworthy, 800, 872-877, 1068
Gutenberg, Johann, 449-452

Habann, Erich, 1116
Hackworth, Timothy, 865, 869
Hadfield, Robert, 758
Hadrianus, Publius Aelius [Roman Emperor- architect], 229, 236, 239, 244-251, 259, 284
Hagen, Gotthilf Heinrich Ludwig, 647
Hagenbach, Eduard, 648

INDEX

Hahn, Max, 1020-1022
Hall, Charles Martin, 935-938
Hall, Joseph, 757, 922
Hall, John A., Capt., 828
Hall, Samuel, 800
Hall, Thomas, 1053
Halley, Edmund, 489, 511, 513, 532, 533, 559, 677
Halske, Johann Georg, 924
Hamilton, William Rowan, 283
Hanamann, Franz, 1072
Hancock, Walter, 872, 876-878
Hansen, William Webster, 1117
Hargreaves, James, 748, 750
Harpales of Tenedos, 70
Harriot, Thomas, 479, 511, 515, 532, 534
Harris, E. J., 1016
Harrison, John, 553, 554
Hartley, Ralph V.L., 1106, 1108
Hartmann, Georg, 701
Haselwander, Friedrich, 1058, 1060
Hatey, 36
Hauksbee, Francis, 702, 703
Haussmann, George Eugene, Baron, 899
Hautefeuille, Jean de, Abbé, 546, 547, 765
Hayes, James, 844, 850
Haynes, Elwood, 758
Haynes, Edward, 839
Haywood, William, 899
Hazeltine, Louis Alan, 1105
Heath, J.M., 759, 911
Hedley, William, 788, 865
Hefner-Alteneck, Friedrich von, 1047
Hegel, Georg Wilhelm Friedrich, 323
Heil, Oskar, 1112, 1114, 1117
Heinkel, Ernst, 1021-1023, 1027-1029, 1038, 1039
Hekataios of Miletos, 151, 152
Hell, Rudolf, 1134
Helmholtz, Hermann Ludwig Ferdinand von, v, xxix, 722, 726, 728, 963, 1024
Helmont, Johann Baptist van, 391
Henlein [alias, Hele], Peter, 466
Henry, Joseph, 713-715, 717, 722, 730, 949, 950, 962, 1040, 1051, 1052

INDEX

Henry of Sens, Bishop, 413
Henry the Navigator, 452
Henry, William, 846
Henry, J.C., 1054
Henschel, Carl Anton, 656, 801, 1000
Herakleides of Pontos, 144, 154, 189
Herakleitos of Ephesos, 78, 97, 98, 99, 106, 108, 109, 133, 592
Herodotos of Halikarnassos, 18, 32, 118, 152
Heron [in English, Hero] of Alexandria, 212-218, 264, 273, 278, 283, 287, 359-361, 395, 396, 408, 464, 473, 493, 523, 529, 542, 593, 761, 846, 853, 996, 997, 1015
Héroult, Paul Louis Toussaint, 935-938
Herschel, Friedrich Wilhelm, 570
Herschel, John Frederick William, 570, 1045, 1057, 1058
Hertel, Heinrich, 1038
Hertz, Heinrich Rudolph, 728-733, 735, 740, 1082-1084, 1086, 1094, 1115, 1124
Hesiod [real name, *Hesiodos*] of Kyme, 78, 91
Hewitt, Peter Cooper, 1073
Heyland, Alexander, 1065
Heytesbury, William, 366, 372-374, 377, 379, 388, 389, 616
Hilbert, David, xxiv
Hiketas of Syracuse, 96
Hildebrand, Heinrich Rupert, 979
Hill, F., 880, 881
Hindley, Henry, 803, 814
Hipparchos of Nikaia, 154, 206-209, 211, 212, 214, 264-268, 270, 271, 281, 282, 367
Hippasos of Metapontion, 96
Hippias of Elis, 123-125, 153, 154, 275
Hippodamos of Miletos, 115, 122, 175, 179
Hippokrates of Chios, 125-127, 135, 141, 153, 185
Hippokrates of Kos, xii, 138, 156, 219, 350
Hirn, Gustav Adolf, 698, 798, 860
Hittorf, Johann Wilhelm, 734
Hjelm, Peter Jakob, 757
Hjorth, Søren, 1043, 1044, 1052
Hoch, Julius, 988
Hodgkinson, Eaton, 882
Holden, Major, 982
Holley, Alexander Lymon, 913
Holmes, Frederick Hale, 1042, 1048, 1068

INDEX

Holtzer, Jacob, 758, 837, 914
Holzwarth, Hans G., 1014
Homer [real name, *Homeros*] of Smyrna, 47, 75, 76, 91, 481
Honnecourt, Villard de, 410, 416
Hooke, Robert, 466, 491, 498, 513, 514, 547, 548, 551, 552, 557-562, 564, 566, 581, 585, 589, 590, 592, 598, 601, 602, 648, 674, 680, 702, 727, 766, 769
Hooper, Stephen, 773
Hopkins, William, 728
Hopkinson, John, 925, 1047, 1049, 1064, 1075
Hornblower, Jonathan Carter, 692, 777, 779, 780, 785, 792
Horner, William George, 362
Horrock, William H., 752
Horrocks [alias, Horrox], Jeremiah, 515, 814
House, Royal Earle, 951
Howd, Samuel B., 656
Howe, Frederick W., 820
Hudde, Johannes, 556
Hudson, Henry, 701
Hugh of St.Victor, 396
Hughes, David Edward, 728-730, 951, 963, 1082, 1083
Hugon, Pierre, 972
Hull, Albert Warren, 1116
Hulls, Jonathan, 777, 845, 846
Hülsmeyer, Christian, 1115, 1118
Humboldt, Alexander von, 708
Hunnings, Henry, 964
Huntsman, Benjamin, 754, 756, 759, 903, 904, 911
Hutin, Maurice, 1063
Huxley, Thomas Henry, 729
Huygens, Christiaan, 466, 522, 530, 532, 533, 543, 546, 548-552, 555-557, 560, 563, 564, 566, 567, 581, 586, 589-591, 593, 594, 598, 607, 667, 680, 765, 766, 966, 968, 970
Hypatia of Alexandria, 277-280
Hypsikles of Alexandria, 206

Iamblikos of Chalkis, 277
Iams, A., 1137
Ibn al-Haitham [alias, *Alhazen*], 336, 349, 509, 510
Ibn Badja [alias, *Avempace*], 337, 363, 367
Ibn Hadjan [alias, *Geber*], 337
Ibn Rushd [alias, *Averroës*], 321, 323, 324, 337, 351, 363, 423, 510

INDEX

Ibn Sina [alias, *Avicenna*], 321, 323, 337, 349, 363, 367, 380
Ibn Yunus, 396, 516, 530
Iktinos, 13, 116, 118, 120
Im-Hotep, 25, 26, 34
Ineni, 33, 38
Isidoros of Miletos, 286-288
Isokrates of Athens, 79
Ives, Herbert Eugene, 1129, 1130

Jablochkoff, Paul Nikolaevich, 1048, 1069, 1070, 1075, 1077
Jacobi, Carl Gustav Jacob, 679, 1051
Jacobi, Moritz Hermann von, 1051, 1053, 1067
Jacobs, Charles B., 842
Jacobsen, Thorkild, iii
Jacomo, Mariana di [alias, *Taccola*], 460
Jacquard, Joseph Marie, 570, 572, 831, 832, 883
Jacquier, François, 468, 648
James, William Henry, 872, 879
Jean de Mirecourt, 358
Jean Roscelin of Compiègne, 318, 319
Jedlik, Anyos, 1044
Jefferson, Thomas, xl, 826, 854
Jenkins, Charles Francis, 1129
Jerome, Chauncey, 828
Jessop, William, 804, 806, 810, 862, 886
Jevons, William Stanley, 577
Johansson, Carl Edvard, 843
John of Salisbury, 310, 330, 332
Johnson, John Bertrand, 1108
Joliot, Frédéric, 740
Jonval, Nicolas J., 656
Jordan de Nemore [alias, *Jordanus Nemorarius*], 214, 340, 359-361, 367, 436, 474, 476, 486
Jouffroy d'Abbans, Claude Marquis de, 847, 848
Joule, James Prescott, 693, 698, 1052
Joy, David, 871
Junkers, Hugo, 1012
Just, Alexander, 1072

Kallikrates, 116-118, 120

INDEX

Kalling, Bo, 930
Kallinikos of Syria, 283
Kallippos of Kyzikos, 152
Kant, Immanuel, xxviii, xxx, 613, 709
Kaplan, Victor, 656-658
Kapp, Gisbert, 1064
Karavodin, 1033
Karmarsch, Karl, ii
Karolus, August, 1129, 1132
Karr, Alphonse, 98
Karsten, 907
Kay, John, 747, 748
Keefer, Thomas Coltrin, 884
Keill, John, 595, 596
Kelly, William, 907-910, 912, 913, 923
Kempelen, Wolfgang von, 776, 996
Kendall, Larcum, 553, 554
Kennedy, John, 751
Kepler, Johann, xxxv, 105, 153, 206, 391, 472, 491, 497-518, 520-522, 532, 534-539, 558, 559, 585, 586, 588, 814
Kerr, John, 715, 718, 1129
Kettering, Charles Franklin, 985
Kilby, Jack St. Clair, 1114
Kiliani, Martin, 935
Kilson, James, 1010
Kilwardby, Robert, 397
King, Walter H., 1054
Kinsey, Apollos, 784
Kircher, Athanasius, 829, 830
Kirchhoff, Gustav Robert, 672
Kirk, A.C., 797, 860
Klaproth, Martin Heinrich, 758
Klau [alias, *Clavius*], Christopher, 477
Kleist, Ewald Georg von, 703
Kleomedes of Lysimachia, 211, 269
Knight, John H., 982, 1055
Kohlrausch, Rudolf H.A., 719
Koldewey, Robert, iii
Köller, Franz, 758, 914
Konon of Samos, 197
Korn, Arthur, 1124

INDEX

Krates of Chalkis, 179, 180
Kroisos of Lydia, 69, 110, 112
Krupp, Alfred, 791, 912, 916, 924
Ktesibios of Askra, 187-190, 278, 283, 761, 762
Kyeser, Konrad, 406-408, 460, 1035

L'Orange, Prosper, 987, 994
Lacassagne, Joseph, 1068
La Hire [alias, Lahire], Philippe de, 538, 563, 564, 593, 594, 626, 627, 630, 631, 633, 667, 669, 670
La Place, Jean Baptiste Meusnier, 651,
La Rive, Auguste de, 953
La Rive, Lucien de, 731
Lacer, Gaius Iulius, 246, 247
Lacroix, Silvestre François, 678
Ladd, William, 1042, 1043, 1045, 1047
Lagerhjelm, Per, 686
Lagrange, Joseph Louis, 361, 610, 613, 615, 635, 643, 671, 674-678, 683
Lalande, Félix de, 956
Lambert, Johann Heinrich, 807
Lamblardie, Jacques Elie, 634, 636
Lamé, Gabriel, 685-687
Lanchester, Frederick W., xxxi, 982
Lanfranc, 317, 319
Langdon-Davies, Walter, 1065
Langen, Eugen, 974-976, 979
Langley, Samuel Pierpont, 985, 986
Langmuir, Irving, 1073, 1101-1104
Lankensperger, Georg, 880, 981
Lapicide, Gerradus, 416
Laplace, Pierre Simon, 610, 615, 635, 671, 673, 674, 682, 711
LaRue, Warren de, 1067, 1070
Latour, Marius, 1103
Laval, Carl Gustaf Patrik de, 658, 998, 1000, 1003, 1004, 1006, 1007, 1011
Lavoisier, Antoine Laurent de, 102, 191, 634, 651, 706
Law, Harold B., 1138
Lawrence, Richard, 833
Lax, Gaspar, 392
Le Blanc, Honoré, 826, 828
Le Roy [alias, Leroy], Pierre, 554
Le Seur, Thomas, 468, 648

INDEX

Lebedev, Pyotr V.N., 728
Leblanc, Maurice, 1063, 1098, 1122, 1136
Leblanc, Nicholas, 680
Lebon d'Humbersin, Philippe, 778, 967-969, 971, 985
Lecher, Ernst, 731
Leclanché, Georges, 956, 957
Lecreulx, François Michel, 630, 669
Lee, Edmund, 773
Lee, William, 466, 467
Leibniz [also spelled Leibnitz], Gottfried Wilhelm, 167, 328, 371, 376, 423, 494, 535, 544, 562, 567-569, 575-577, 580, 581, 584, 585, 589, 590, 592, 596-597, 601, 605, 613, 671, 678, 768, 985
Lemale, Charles, 1013, 1014
Lenard, Philipp Eduard Anton, 735, 736, 739, 740, 1099
Lenoir, Jean Joseph Étienne, 971-975, 978, 1074
Lenz, Heinrich Friedrich Emil, 717, 1045
Leodamas of Thasos, 141, 142
Leonardo Fibbonacci of Pisa [alias, Leonardo of Pisa], 359, 361, 362, 474
Leonardo da Vinci, xxviii, 360, 395, 455, 456, 458, 463-466, 470, 471, 480, 517, 542, 544, 560, 620, 761, 813, 889, 892
Lesseps, Ferdinand, 40
Lesage, Georges Louis, 945
Leukippos of Miletos, 101, 106, 107, 109, 131
Leupold, Jacob, 782, 824
Levitska, Maria A., 1140
Lévy, Lucien, 1105
Leyner, John George, 897
L'Huilier, Simon Antoine Jean, 677
Lieben, Robert von, 1099-1100, 1128
Liebig, Justus von, 907
Lilienfeld, Julius Edgar, 1112, 1114
Lindbergh, Charles Augustus, 1036
Linde, Carl von, 989
Lindley, William, 890, 899
Lintlaer, Jean, 620, 628
Lippershey [alias, Lipperhey], Hans, 532
Lippisch, Alexander Martin, 1038
Livingston, Robert R., 854-858
Ljungström brothers, Birger and Fredrik, 659, 999, 1008, 1009, 1014
Locke, Joseph, 871, 882, 889
Locke, Edward, 997

INDEX

Lodge, Oliver Joseph, 732, 1084, 1086, 1087
Lodygin, Alexandr Nikolaevich, 1069
Lohneyss [alias, Lohneiss], G. E. von, 451
Lombard, Peter, 320
Loomis, Mahlon, 1082
Lorentz, Hendrik Antoon, 726, 733, 737, 738
Lorenzen, Christian, 1030
Lorin, René, 1016, 1033
Lorini, B., 451
Lovelace, Countess [alias, Augusta Ada Byron], 573, 574
Low, George, 897
Löwenstein, Fritz, 1100, 1101
Lubszynski, A.G., 1137
Lull, Ramón [alias, **Ramundus Lullius**], 422, 423, 575
Luther, Martin, 442-449, 474, 489
Lysholm, Alf, 1014
Lysippos of Sikyon, 176, 184
Lysistratos of Sikyon, 184

MacAdam, John Louden, 809
Maceroni, Francis, Col., 872, 874, 876
MacIntosh, Charles, 984
MacNeill, John B., 810
Magnus, Heinrich Gustav, 606, 916
Maiman, Theodore, 1141
Malderen, Joseph van, 1042, 1045
Mallet, Alain Manesson, 622
Malus, Étienne Louis, 680
Manby, Aaron, 858, 859
Mandroklos of Samos, 70
Manly, Charles Matthews, 986
Mannoury Dectot [alias, d'Ectot], Jean Charles Alexandre François, Marquis de, 651
Mansart, Jules Hardouin, 618, 628
Manuzio, Aldo [alias, **Aldus Manutius**], 451
Marchia, Francesco de, 380, 382, 384
Marchi, Francesco di, 466, 623
Marconi, Guglielmo, 1083, 1084, 1086-1090, 1093, 1104, 1109
Marconnet, 1033
Marcus, Siegfried, 975, 976, 981
Marggraf, Andreas Sigismund, 757
Marinos of Tyre, 268

INDEX

Mariotte, Edmé, 543, 547, 564-566, 581, 589, 590, 593, 594, 625, 627, 636, 764
Marsiglio of Padua, 331
Marsilius van Inghen [alias, Marsile van Inghen], 381, 383-385
Martin, Piérre Émile, 922, 923
Marum, Martin van, 706
Maskelyne, Nevil, 554
Masson, Antoine, 971
Matteucci, Carlo Felice, 970
Maudith, John, 366, 491
Maudslay, Henry, 795, 814-819, 827, 828, 835, 859, 885, 888, 892, 915
Maupertuis, Pierre Louis Moreau de, 283
Maurice, Peter, 745, 746
Maurolico, Francesco, 510, 511
Maury, Matthew Fontaine, 959
Maxwell, James Clerk, v, 686, 710, 714, 716, 718, 723-731, 733, 734, 736, 741, 1063, 1064, 1099
May, Joseph, 1119, 1121
Maybach, Karl, 982, 984
Maybach, Wilhelm, 975-980, 982-984
Mayniel, K., 663
Mayr [alias, *Marius*], Simon, 532
McNaught, John, 794
Mead, Thomas, 773
Medeleev, Dimitrii I., 1036
Medhurst, 872
Meikle, Andrew, 773, 776, 807
Meissner, Alexander, 1101
Melissos of Samos, 101, 107, 109
Menabrea, Luigi Federigo, 573
Menaichmos of Alopekonnesos, 136, 142, 153, 204, 205, 337, 518, 536
Mendeleev, Dimitrii Ivanovich, 737, 740
Menelaos of Alexandria, 264, 265, 276
Merckel, Curt, ii
Meri, 37
Merritt, Ernest George, 1091
Mersenne, Marin, 521, 522, 536, 537, 561, 614
Messerschmitt, Willy, 1023, 1026, 1027, 1038
Metagenes of Knossos, 113
Metcalfe, Charles, 890
Metcalfe, John, 808, 809
Meton of Athens, 297

INDEX

Meucci, Antonio, 962
Meyer, Adolf, 1014
Meysey, Mathias, 903
Michelangelo Buonarroti [alias, Michelangelo], 241, 468
Michelin, Edouard, 984
Michell, Anthony George Maldon, 889
Midgley Jr., Thomas E., 987
Mihály, Dénes von, 1129
Mill, John Stuart, 579
Miller, Oskar von, 1061, 1075, 1079
Miller, Patrick, 851, 852
Miller, Samuel, 854
Millikan, Robert Andrews, 738
Millward, William, 1041, 1042
Mirecourt, Jean de, 358
Mitchell, John, 705
Mnesikles, 116-118
Moissan, Henri, 943
Molignani, Arturo, 1072
Monge, Gaspard, 634, 635, 651, 684
Monge, Louis, 634
Monsignori, Giovanni of Verona [alias, *Fra Giocondo*], 467, 618
Montgolfier, Jacques Étienne de, 650
Montgolfier, Joseph Michel de, 649, 650
Montricher, Jean François Mayor de, 665
Moore, Gordon, 1114
Moore, MacFarlan, 1073, 1128
Moray, Robert, 551, 552
Mordey, William M., 1049, 1050
Morgan, Augustus de, 573, 577
Morland, Samuel, 568, 764, 816
Morley, Samuel, 969, 970
Morse, Samuel Finley Breese, 950-952, 957, 1082
Morveau, Guyton de, 907
Moss, Sanford Alexander, 1011-1013
Mouillard, Louis, 986
Moyes, H., Dr., 708
Mudge, Thomas, 554, 555
Müller, Johann [alias, *Regiomontanus*], 473, 491
Müller, Max Adolf, 1023, 1025-1029
Müller, Franz Josef, 757

INDEX

Munck af Rosenschöld, Peter Samuel, 1083, 1090
Muncke, Georg Wilhelm, 947, 949
Murdock, William, 778, 783-785, 790, 838, 859, 895, 996
Murray, Matthew, 790, 791, 816, 817, 863-865, 887
Mushet, Robert Forrester, 758, 759, 836-838, 909-915, 922
Mushet, David, 910, 915
Musschenbroek, Pieter van, 597, 602, 631, 633, 703
Myddleton, Hugh, 742, 745

Nagler, Forrest, 657
Napier [or Neper], John, 515, 516
Napier, Charles, 858
Napier, David, 858, 859
Napier, Robert, 859, 900
Nasir Eddin at-Tusi, 790
Nasmyth, James, 796, 819, 820, 832, 833, 838, 841, 851, 882, 909, 910, 912
Navier, Louis Marie Henri, 645, 646, 670, 672-675, 682-684, 687
Negrelli, Alois, 39
Negro, Salvatore Dal, 1050
Neilson, James Beaumont, 756, 919, 920
Nek-Hebu, 28, 37
Neumann, Franz Ernst, 648, 719, 721
Newcomen, Thomas, 692, 768-774, 776, 777, 780, 781, 789, 793, 799, 806, 845, 846, 850
Newton, Isaac, 105, 358, 376, 382, 383, 386, 436, 491, 496, 498, 506, 513, 514, 533, 535, 543, 549, 556, 559, 562, 567, 581-590, 592, 595-597, 601, 606, 608, 613, 633, 642, 678, 679, 681, 705, 709, 710, 717, 719
Nicholas of Cusa, 331, 385, 389-391, 473, 487, 506, 507, 510, 520, 528, 539, 701
Nicholson, William, 708, 709
Nicolas de Autrecourt, 358
Niepce brothers, Joseph Nicephore and Claude, 968, 969
Nikomachos, 277, 281
Nikomedes, 275
Nipkow, Paul Gottlieb, 1123, 1125, 1128, 1132, 1133, 1136
Nobili, Leopoldo, 712
Nollet, Florise, 1041, 1042, 1068
Nollet, Jean Antoine, 634
Norman, Robert, 701, 702
Normand, Charles Benjamin, 781, 797, 860

INDEX

North, Simeon, Col., 828
Norton, Charles A., 842
Norton, W. P., 839
Noyce, Robert Norton, 1114
Nunes, Pedro [Latin name, ***Petrus Nonius***], 486, 622
Nyquist, Harry, 1106, 1108, 1111

Oberth, Hermann Julius, iv, 1037, 1038
Odhner, Willgodt Theophil, 569
Ohain, Hans Joachim Pabst von, 1016, 1020-1023, 1026, 1027, 1032
Ohm, Martin, 916
Ohm, Georg Simon, 599, 712, 713, 916, 1063
Oinopides of Chios, 125, 142, 297
Okabe, Kinjiro, 1116
Okolicsanyi, 1131
Oldenburg, Henry, 551, 552, 560
Olympiodoros, 283, 349
Omar Khayyam, 337, 338
Onions, Peter, 754
Opel, August, 983
Opel, Fritz von, 1038
Orata, Gaius Sergius, 239
Oresme, Nicole, 379, 384-389, 487, 507, 527, 528, 531, 535, 539, 616
Origen [alias, ***Origenes Adamantius***], 315
Ørsted, Hans Christian, 709, 711, 716, 718, 758, 934
Otto, Nikolaus August, 972-979, 986-988, 991, 992
Overton, Thomas, 863

Pacinotti, Antonio, 838, 1043-1045
Pacioli, Luca, 474, 475
Page, Charles Grafton, 961, 1041, 1053
Pain, Albert Charles, 1001
Paionios of Miletos, 113
Palladio of Vicenza, Andrea, 451, 467
Pamphilos of Amphipoles, 176
Papin, Denis, 556, 580, 765-768, 770, 771, 777, 778, 781, 782, 845, 848, 895, 966, 970, 985
Pappos of Alexandria, 275-277, 359, 360, 433, 477, 509, 540
Parcieux, Antoine de, 639, 640, 644, 652
Pardies, Ignace Gaston, 557, 592, 637
Parent, Antoine, 594, 595, 598, 604, 611, 628, 637-640, 642, 645, 668, 776, 805,

INDEX

807, 996
Parmenides of Elea, 97-102, 105, 108, 109, 128, 160, 166
Parry, George, 921
Parsons, Charles Algernon, 860, 993, 1000-1003, 1005-1010
Parter, Peter, 416
Pascal, Blaise, 361, 465, 494, 522, 537, 540, 543-545, 567, 568, 575
Patte, Pierre, 630, 632
Paul, Lewis, 748, 749
Paul of Venice, 388, 391
Paxton, Joseph, 890, 891
Peano, Giuseppe, 578, 579
Pearson, George, 708
Pedersen, Peder Oluf, 965, 1095, 1096
Peltier, Jean Charles Athanase, 712
Pelton, Lester, 657, 658, 1003
Penn, John, 776, 859, 888, 960, 961
Pererius, Benedictus, 526
Périer, Jacques C., 847, 855
Perikles of Athens, xix, 79, 81, 84, 103, 104, 116, 117, 120, 122, 133, 172, 173
Perkins, Jacob, 789, 793
Perrin, Jean Baptiste, 735
Perronet, Jean Rodolphe, 252, 630-634, 640, 669-670, 825
Perry, 1121, 1122
Perseus, 275
Peter of Crescenzi, 404
Peterson, Karl, 687
Petit, Alexis Thérèse, 646, 692
Petit-Vandin, Robert Xavier Ansart Du, 638, 639
Peto, Morton, 889
Petrarca, Francesco, 479-481
Petrie, William Mathew Flinders, iii, 1068
Pettit-Smith, Francis, 854
Peugot, Armand, 983
Peurbach, Georg von, 473
Pfaff, Johann Friedrich, 679
Pheidias of Athens, 116, 117, 173
Phelps, George M., 951
Phillips, Edouard, 555
Philolaos of Kroton, xiii, 96, 134, 137, 144
Philon of Athens, 175
Philon of Byzantion, 190, 191, 278, 359, 471, 524, 561, 611, 651, 683, 761, 776,

INDEX

996, 1015
Philoponos [alias, 'Grammatikos'] Ioannes of Alexandria, 209, 281-283, 363, 367, 368, 370, 380-382, 493, 541, 616
Picard, Jean, 533, 622,
Pickard, Greenlief Whittier, 1090, 1091
Pickard, James, 777
Pierre de Maricourt [alias, *Petrus Peregrinus*, that is:'Peter the Pilgrim'], 340, 397, 398, 701
Pi Sheng, 449
Pitot, Henri de, 628, 637-640, 644, 665
Pitrou, Robert, 629
Pitt, St. George Lane Fox, 1072
Pixii, Hippolyte, 1040, 1048
Plana, Giovanni, Baron, 573
Planté, Gaston, 955, 956
Plato [Greek nickname, *Platon*; real name, *Aristokles*], 73, 79, 115, 118, 119, 130, 131, 133, 134, 136-147, 153-156, 159, 160, 166, 171, 183, 186, 194, 278, 293, 315, 317, 318, 329, 342, 343, 346, 348, 366, 371, 394, 409, 433, 434, 440, 500, 518, 525, 541
Plinius Secundus [in English, Pliny], Gaius, 227
Ploutarchos of Chaironeia, xxiv, 119, 247, 264
Plotinos, 273, 315
Plücker, Julius, 734, 735
Plumier, Charles, 813
Poggendorff, Johann Christian, 948, 955
Pohl, Robert Wichard, 1020, 1021
Poincaré, Henri, 698, 731
Poinsot, Louis, 670, 683, 701
Poiseuille, Jean Lois, 648
Poisson, Denis Siméon, 191, 671-673, 678, 679, 683, 721
Pole, William, 883
Poleni, Giovanni, 468, 648
Polhammer, Kristofer [alias, Polhem, Christopher], 624, 755, 756, 805, 823-825, 833, 834
Polykleitos, xii, 117, 184
Poncelet, Jean Victor, 634, 648, 652-654, 657, 664, 674, 684, 685
Popov, Alexandr Stepanovich, 1087
Porphyrios, 273, 277, 293, 294
Porta, Giovanni Battista [alias, Gianbattista] della, 441, 469-471, 531, 533, 761-763, 767
Poseidonios of Apameia, 178, 206, 209-211, 264, 268, 269, 271, 272, 513, 622
Poulsen, Valdemar, 964, 965, 1095, 1096, 1098, 1106

INDEX

Power, Henry, 547, 566, 607
Prandtl, Ludwig, 1016, 1017, 1024
Preece, William Henry, 741, 1082, 1084, 1087
Priestly, Joseph, 705, 774, 907
Proklos of Byzantion, 273, 276, 280, 281, 509
Prony, Gaspard François Clair Marie Riche de, 630, 631, 635, 655, 662, 684
Protagoras of Abdera, 73, 79, 122, 123, 130, 167
Proust, Joseph Louis, 102
Ptolemaios (in English, Ptolemy) of Alexandria, Klaudios, xxvi, 207, 265-272, 278, 281, 339, 349, 353, 433, 457, 473, 487, 488, 490, 491, 509, 519, 540, 548
Pupin, Michael Idvorsky, xxvii, 1089, 1101, 1103
Purcell, Edward M., 1112
Pyrgoteles, 176
Pythagoras of Samos, 83, 91-97, 100, 108, 124, 131, 293, 371, 440, 500
Pytheas of Massilia, 203, 508
Pytheos, xii, xiii, 173, 175

Radcliffe, William, 752
Ramelli, Augustino, 451
Ramsbottom, John, 871, 872
Ramsden, Jessie, 814
Ramsey, David, 763, 844, 845
Rankine, William John Macquorn, 662, 664, 689, 722, 729, 797
Rastrick, John Urpeth, 853, 868, 894
Rateau, Auguste Camille Edmond, 659, 999, 1005-1008, 1012, 1013
Rathenau, Emil, 924, 1084
Rawlinson, Henry Creswicke, 72
Rawlinson, Robert, 899
Read, Nathan, 784, 800, 854
Reauleaux, Franz, 976
Réaumur, René Antoine Ferchault de, 637, 756, 834, 906, 907
Recorde, Robert, 495, 518
Redmund, 880, 981
Reech, Ferdinand, 644, 698
Regement, Jean Baptiste de, 629
Regnault, Henri Victor, 697
Reinecker, J. E. R., 841
Reinhold, Erasmus, 489
Reis, Philipp Johann, 722, 961, 962
Reisz, Eugen, 1099
Reithmann, Christian, 972, 973

INDEX

Reitsch, Hanna, 1034
Renault, Louis, 983
Rendel, A. M., 890
Rendel, James Meadows, 810, 890
Rennie, John, 776, 777, 804, 806-808, 810, 818, 863, 892
Rennie, George, 818, 910
Rennie, John Jr., 810
Ressel, Joseph, 854
Reynolds, Osborne, 647, 998-1000
Reynolds, William, 759, 910
Reynolds, Richard, 753, 806
Rhoikos of Samos, 111
Ricardo, David, 425
Ricardo, Harry Ralph, 987, 994
Richard of Wallingford, 410
Richardson, Owen Willans, 741, 1099
Richmann, Georg Wilhelm, 353
Rickett, Thomas, 880
Riemann, Georg Friedrich Bernhard, 578, 613, 679
Righi, Augusto, 732, 1086
Rinaldo, George, 620
Ritchie, William, 1040, 1041, 1051, 1053
Ritter, Johann Wilhelm, 707, 708
Rivault, David, 762, 844
Rivaz, Isaac de, 968
Robert [alias, Robertson], Etienne Gaspard, 708
Robert, Nicolas Louis, 834
Roberts, Richard, 750-752, 818, 880, 883
Roberval, Gilles Personne de, 522, 534, 536, 537, 543
Robins, Benjamin, 606, 607, 641
Robinson, John, 806
Rochas, Adolphe Eugène Beau de, 973, 974, 977
Rodda, S., 1137
Roebuck, John, 775
Roland of Cremona, 345
Romain, François, 618, 619, 629
Rømer, Øle Christensen, 103, 533
Ronalds, Francis, 945, 946, 949
Rondelet, Jean Baptiste, 630, 632, 633
Röntgen, Wilhelm Conrad, 738
Roosevelt, Nicholas J., 856, 858

INDEX

Root, Elisha K., 833
Roscelin of Compiègne, Jean, 318, 319
Rose, Albert, 1137, 1138
Rosing, Boris, 1125-1127, 1134, 1136
Ross, Alexander McKenzie, 884
Rosselino, Bernardo, 459
Round, Henry J., 1089, 1100, 1105
Rowland, Henry Augustus, 733, 1064
Rubens, Heinrich, 1084
Rugerius, Ludovico, 526
Rühlmann, Moritz Christian, ii
Ruhmer, Ernst Walter, 1094, 1095, 1101
Ruhmkorff, Heinrich Daniel, 730, 971, 972, 1045, 1084
Rumsey, James, 800, 850, 851, 856
Russell, Arthur William Bertrand, 578, 579
Russell, John Scott, 872, 875, 876, 888
Rutherford, Ernest, 735, 739, 740, 1085, 1088

Sagebien, Alphonse, 652, 655
Saint-Claire Deville, Henri Étienne, 934, 935, 942
Saint-Venant, Jean Claude Barré de, 464, 687
Saint-Vincent, Gregoire de, 128, 149, 539, 555, 688
Salomon de Caus, 471, 762, 763, 781, 844
Salvá, Francisco Don, 945, 946
Samuel, Arthur Lee, 1116
San Michele, Michele, 462
Sander, Friedrich Wilhelm, 1038
Sangallo, Giuliano da, 459, 467
Salario, Pietro Antonio, 461
Santi, Raffaelo [alias, 'Raphael'], 467
Sarasin, Edouard, 731
Sarnoff, David, 1110, 1111, 1137
Sauvage, Pierre Louis Frederic, 854
Savery, Thomas, 766-770, 773, 778, 780
Sawyer, W.E., 1122
Saxton, Joseph, 1040, 1041
Scheele, Karl Wilhelm, 907
Scheiner, Christopher, 532
Schelp, Helmut, 1025-1028, 1032
Scheutz, Edvard, 572
Scheutz, Per Georg, 571, 572, 836

INDEX

Schickard, Wilhelm, 516, 517
Schilling von Canstadt, Paul Ludwig, 947-949, 957
Schlömilch, 1097
Schmidt, Gustav Johann Leopold, 973
Schmidt, Paul, 1027, 1033
Schott, Kaspar, 546, 547, 829
Schottky, Walther Hans, 1105
Schuckert, Johann Siegmund, 924
Schweigger, Johann Salomo Christoph, 711, 946, 947
Scot, Michael, 397
Scott, Charles F., 1060, 1063
Scott-Moncreiff, Colin Campbell, Col., 900
Scotus Erigena, John, 314-319
Seebeck, Thomas Johann, 711-713, 716
Seftström, Nils Gabriel, 758
Segner, Johann Andreas [alias, János András], 608, 611, 651
Seguin, Marc, 800, 870
Sen-Mut, 33, 34, 38
Seneca, Lucius Anneus, 226, 333
Senlacq, 1121
Sennacherib, King of Assyria, 56-60
Serpollet, Léon, 881
Serrin, Victor, 1042, 1068
Shannon, Claude Elwood, 1108
Shaw, William Napier, 695
Shaw, C.A., 897
Shockley, William Bradford, 941, 1113, 1114
Shreve, Henry M., 858
Sickels, Frederick Ellsworth, 795
Siemens, Ernst Werner, 916-918, 923-926, 951, 958, 1043, 1044, 1047, 1048, 1054, 1072
Siemens, Friedrich, 918-922
Siemens, Karl Wilhelm [alias, Charles William], 724, 916-919, 921-926, 928, 935, 943, 951, 961, 1044, 1054, 1072, 1074
Siemens, Karl, 924
Sieur de Joinville, 313
Simms, Frederick, 983
Simon, Hermann Theodor, 1094
Simplikios of Cilicia, 132, 281
Sinsteden, Wilhelm, 1044
Skopas of Paros Island, 173, 174, 179
Smeaton, John, 606, 640, 644, 747, 753, 771-773, 789, 793, 803-806, 810, 811, 816,

INDEX

    836, 892
Smee, Alfred, 955
Smith, Adam, 425
Smith, Robert, 900
Smith, Willoughby, 1119
Snel van Roijen, Willebrord, 267, 534, 544
Soddy, Frederick, 737, 739
Soemmering, Samuel Thomas, 946-948, 957
Sokrates of Athens, 74, 76, 83, 84, 94, 96, 104, 108, 128-131, 133, 138, 142, 147, 159, 334
Solla Price, Derek de, iv, 428
Solon of Athens, 77, 78, 110
Solvay, Ernest, 680
Somerset, Edward [alias, Second Marquis of Worcester], 470, 763, 845
Sommellier, Germain, 896
Sophokles of Kolonos, 73, 76, 79, 121
Sorby, Henry Clifton, 907
Sorocold, George, 746
Sosigenes of Alexandria, 227, 297
Sostratos of Knidos, 183
Soto, Domingo Francisco de, 392, 393, 489, 490, 527, 528
Southern, John, 777
Soufflot, Jacques Germain, 630-632
Spencer, Christopher M., 821
Spina, Alessandro della, 354, 399
Spinoza, Baruch, 556
Spottiswoode, William, 729
Sprague, Frank Julian, 1054-1057
Sprengel, Johann Philipp, 1071
St. Francis of Assisi [alias, Giovanni Francesco Bernadone], 344, 345
St. Dominicus [alias, Domingo de Guzman], 345
St. Jerome, 294, 295, 443
St. Clement of Alexandria [alias, *Titus Flavius Clemens*], 315
St. Augustine [alias, *Aurelius Augustinus*], 294
St. Bénezèt, 417, 419
St. Benedict of Nursia, 294-296, 307, 394, 417
Stabernack, Hans, 1025, 1027, 1029
Staite, W. Edward, 1067, 1068
Stanley, William, 1077, 1078
Starr, J.W., 1070
Stearns, Joseph B., 952

INDEX

Steinbach, Erwan von, 416
Steinheil, Carl August, 948, 949
Steinmetz, Charles Proteus, 1064, 1065, 1096
Stephenson, George, 800, 811, 863-868, 870-872, 881, 883-885, 889, 915
Stephenson, Robert, 787, 800, 820, 866-872, 875, 881-885, 887, 889, 928, 950
Stern, W.J., Dr., 1015, 1017
Stevens, Robert L., 857
Stevens, John, 800, 846, 854, 856-857, 997
Stevin, Simon, 209, 213, 360, 391, 478, 492-497, 517, 531, 537, 539, 542, 543, 557, 563, 582, 587
Stewart, Allan, 929
Stifel, Michael, 474, 516
Stirling, Robert, 580, 917, 985
Stodola, Aurel, iv, 1008, 1011
Stöhrer, Emil, 1041
Stokes, George Gabriel, 729, 730, 737, 1082
Stoney, George Johnstone, 737
Strabon [in English, Strabo] of Amasia, 211, 267, 268
Strachy, Richard, 899
Straton [in English, Strato] of Lampsakos, 135, 168-171, 181, 185, 188, 189, 191, 213, 215, 359, 360, 464, 549
Strauss, Siegmund, 1101
Street, Robert, 966
Stromeyer, Friedrich, 758
Struik, Dirk, iv
Strutt, John William [alias, Lord Rayleigh], 999, 1083
Stuart, Herbert Akroyd, 694, 989, 992
Sturgeon, William, 712, 713, 953, 1040, 1051
Sualem brothers, René Rennequin and Paul, 623, 628, 764
Suger, Abbé, 412-414, 416
Sulzer, Johann Georg, 707
Su Sung, 429
Swammerdam, Jan, 707
Swan, Joseph Wilson, 1071, 1072, 1077
Swedenborg, Emanuel, 825
Swineshead, Richard, 366, 372, 375-377, 379, 380, 385, 386, 616
Symington, William, 784, 851, 852
Sylvester, Bernard, 318
Synesios of Kyrene, 278
Szczepanik, 1124
Szent-Györgi, Albert, xxvii

INDEX

Szilard, Leo, 1108

Tacquet, André, 535, 536, 544, 583
Tangye bothers, James and Joseph, 817
Tartreaux, Pierre Couplet de [alias, Tartereaux], 625, 633, 638, 660, 661, 667-669
Taylor, Brook, 595
Taylor, Frederick Winslow, 836-839, 841, 916
Telekles of Samos, 111
Telford, Thomas, 665, 675, 747, 804, 806, 808-811, 825, 836, 868, 881-883, 885, 889
Tellegen, Bernardus Dominicus Hubertus, 1106
Terry, Eli, 828
Terzaghi, Karl, 664
Tesla, Nikola, 732, 738, 1053, 1058-1063, 1079, 1080, 1084, 1085, 1087-1089, 1093-1096, 1100, 1111
Thales of Miletos, 70, 77, 78, 80, 87-91, 108, 109, 151, 152, 699
Theaitetos of Herakleia, 147, 150
Themistios, 272
Theodoros of Kyrene, 138, 147
Theodoros of Samos, 110-113, 116, 179
Theon of Alexandria, 277, 278
Theophilus, 311
Theophrastos of Eresos [alias, *Tyrtamos*], 155, 156, 168, 170, 182, 183, 270
Theudios of Magnesia, 125, 185
Thévenot, Melchisédech, 622
Thiene, Gaetano da, 388, 392
Thierry of Chartres, 343
Thiers, Rodolphe, 1068
Thiot, Antoine, 814
Thomas of Aquino [alias, *Thomas Aquinas*], 319, 322-326, 329-332, 350-352, 354, 355, 358, 362-365, 380, 390, 397, 415, 424, 435, 479
Thomas, Sidney Gilchrist, 927, 928, 935
Thomaz, Alvaro, 392
Thomson, Robert William, 983
Thomson, Elihu, 998, 1055, 1056, 1078, 1094, 1096
Thomson, Joseph John, 733, 736-738, 741, 1091, 1099
Thomson, James, 656, 999
Thomson, William [alias, Lord Kelvin], v, 656, 695, 721-723, 728, 730, 959, 960, 999, 1049, 1074
Thorp, John, 751
Thoukideides, xix, xxxii, 104

INDEX

Thury, René, 1078
Thyssen, August, 792
Timaios of Lokri, 138
Toletus, Francisco, 393, 490
Toogood, Thomas, 844, 850
Torres y Quevedo, Leonardo, 574
Torricelli, Evangelista, 191, 464, 534, 535, 537, 539, 542, 543, 550, 564, 587, 593, 594, 638, 697, 765
Toscanelli, Paolo Dal Pozzo, 453
Tosi, Franco, 993
Towneley, Richard, 547, 566, 607
Townes, Charles Hard, 932, 1140
Traianus, Marcus Ulpius [Roman Emperor], 39, 229, 234, 235, 238, 244-248, 251, 257, 258
Tredgold, Thomas, x
Trésaguet, Pierre Marie Jerôme, 665, 808
Trevithick, Richard, 697, 778, 780, 784-787, 789, 791, 792, 795, 799, 800, 853, 854, 863-865, 872, 885, 895, 997
Trudaine, Daniel Charles, 629, 665
Truesdell, Clifford Ambrose, iv, vii
Tsai Lun, 429
Tsiolkovskii, Konstantin E., 1035
Turgot, Anne Robert Jacques, 617, 618, 629
Turriano, Juanelo [alias, Giovanni Della Torre Gianello], 469, 625, 825
Twedell, R. F., 817

Uccelli, Paolo, 455
Ulloa, Antonio de, 757, 940
Upton, Francis Robbins, 1048, 1055, 1076
Urban, 290

Vail, Alfred, 950, 951
Valier, Max, 1038
Valla, Paolo, 526
Valturio, Roberto, 406, 844
Vandermonde, Alexandre Théophile, 831
Vanvitelli, Luigi, 468, 648
Varian, Russell Harrison, 1117, 1135
Varignon, Pierre, 563
Varley, Cromwell Fleetwood, 735, 955, 959, 960
Varley, Samuel Alfred, 1044, 1083

INDEX

Varro, Marcus Terentius, 103, 284
Vasari, Georgio, 413, 454
Vauban, Sebastian Le Prêstre, 466, 619, 623, 625, 659, 666
Vaucanson, Jacques de, 813, 829-831
Vauquelin, Louis Nicolas, 758
Venerable Bede [alias, *Beda Venerabilis*], 296-299
Venn, John, 577
Venturoli, Giuseppe, 646,
Veranzio, Fausto, 451
Verbiest, Ferdinand, 781
Verdun, Mathieu de, 419
Vergilius [in English, Virgil] Maro, Publius, 320, 479, 480
Verne, Jules, 1035
Vernier, Pierre, 487, 622
Viète, François, 361, 477, 478
Vigevano, Guido da, 399, 406
Vignoles, Charles Blacker, 810, 890
Villard, Paul Ulrich, 739
Ville, Arnold de, 623, 764
Vincent de Beauvais, 352
Vitelleschi, Muzio, 526
Vitruvius Pollio, Marcus, x, xiii, xiv, xvi, 227, 228, 234, 253, 299, 408, 454, 458, 760, 761
Voigt, Woldemar, 682
Volta, Alessandro Giuseppe Antonio Anastasio, 707-709, 953, 967
Vreeland, Frederick King, 1101

Wagner, Herbert, 1016, 1023-1029, 1032
Walker, James, 868
Wallis, John, 582-584
Walter, Hellmuth, 1038, 1039
Walter of Henley, 404
Walton, Ernest Thomas Sinton, 740
Warsitz, Erich, 1022, 1038
Wasbrough, Matthew, 777
Washington, George, xvii, 850
Watkins, Francis, 1052
Watson, William, 704, 705, 948
Watson, Thomas August, 963
Watson-Watt, Robert, 1118
Watt, James, 692, 695, 747, 750, 753, 755, 771-784, 786-788, 790, 792-794,

INDEX

    799, 800, 807, 813, 816, 821, 847, 849, 851, 852, 855, 864, 895, 974, 990, 993, 996, 998
Weber, Wilhelm Eduard, 647, 718, 719, 721, 948
Weber, Ernst Heinrich, 647
Wedgewood, Josiah, 747
Wehnelt, Arthur Rudolph Berthold, 1092, 1099, 1103, 1124-1126, 1129
Weierstrass, Karl Theodor Wilhelm, 137, 148
Weiller, Lazare, 1123, 1125, 1129, 1130, 1132, 1134
Weimer, Paul K., 1138
Weisbach, Julius Albin, 648
Welch, Charles Kingston, 984
Wenström, Jonas, 1046, 1047, 1058
Westinghouse, George, 1057, 1060, 1077, 1079, 1080
Weston, Edward, 957, 1046, 1054
Wheatstone, Charles, 949-951, 958, 1042, 1044, 1045
White, James, 880
White, John, 790
White, Maunsel, 837, 838
Whitehead, Alfred North, 579
Whitehurst, John, 649
Whitney, Eli, 819, 820, 828, 830, 840
Whitney, Eli Jr., 833
Whittle, Frank, 1016, 1017, 1018-1022, 1026, 1031, 1032, 1034
Whitworth, Joseph, 571, 818-820, 832, 833, 915
Wiechert, Johann Emil, 736, 737, 1092
Wieleitner, Heinrich, ii
Wien, Wilhelm, 736
Wiener, Norbert, 1108
Wilcox, Stephen, 801
Wilde, Henry, 1042, 1043, 1048
Wilkinson, David, 813
Wilkinson, John, 753, 754, 772, 775, 776, 780, 814, 816-818, 858
Wilkinson, William, 754
Willans, Peter William, 797, 798, 1001
William of Sens, 413, 416
William of Ockham, 324-328, 331, 340, 351, 355-358, 364-366, 370, 372, 379-383, 385, 443
Williamson, Joseph, 880
Wilm, Alfred, 937
Wilson, Edwin Bidwell, iv, v
Wilson, Ernest, 1065

INDEX

Wilson, Robert, 997, 1000
Witelo [alias, *Vitellio*], 509
Wöhler, Friedrich, 758, 934
Wohlwill, Emil, ii
Wolfmüller, Alois, 979
Wollaston, William Hyde, 722, 938, 953
Woltmann, Reinhard, 641, 662, 663
Wood, Nicholas, 866
Woolf, Arthur, 779, 786, 792, 793, 795
Woolley, Charles Leonard, iii
Woolrich, John Stephen, 1041, 1045, 1048
Worthington, Henry Rossiter, 801
Wren, Christopher, 557-559
Wright, Edward, 745
Wright brothers, Wilbur and Orville, 696, 986
Wright, Lemuel Wellman, 970
Wyatt, John, 749
Wycliff, John, 446

Xenophanes of Kolophon, xiv, 76, 77, 90, 91, 99
Xenophon of Athens, 119

Yen Su, 428
Young, Thomas, 679, 680

Začek, A., 1116
Zarathushtra Spitama [alias, Zoroaster the White], 66-68, 76, 85, 291
Zeeman, Pieter, 733
Zelander, Willem [alias, *Guillelmus Zelanderus*], 411
Zenodoros, 276
Zenon of Kition, 130, 226
Zenon of Elea [in English, Zeno], 99-101, 106, 128, 146, 166, 198, 292, 430, 539
Zeppelin, Ferdinand von, 982
Zeuner, Gustav, 698
Zhuravskii, Dimitrii I., 674, 883
Zoelly, Hans, 659, 1008
Zonca, Vittorio, 451
Zworykin, Vladimir Kozma, 1126, 1134, 1136, 1137